D1265827

Dr. Z. Ahmed
Chem. Eng.
Cooper Union

THEORETICAL ASPECTS OF HETEROGENEOUS CATALYSIS

VAN NOSTRAND REINHOLD

CATALYSIS SERIES

Burtron Davis, Series Editor

Metal-Support Interactions in Catalysis, Sintering, and Redispersion, edited by Scott A. Stevenson, R.T.K. Baker, J.A. Dumesic, and Eli Ruckenstein

Molecular Sieves: Principles of Synthesis and Identification, R. Szostak

Raman Spectroscopy for Catalysis, John M. Stencel

Theoretical Aspects of Heterogeneous Catalysis, John B. Moffat

Biocatalysis, edited by Daniel A. Abramowicz

THEORETICAL ASPECTS OF HETEROGENEOUS CATALYSIS

John B. Moffat

Van Nostrand Reinhold Catalysis Series

Library of Congress Catalog Card Number 89-75819
ISBN 0-442-20528-7

Printed in the United States of America

Van Nostrand Reinhold
115 Fifth Avenue
New York, New York 10003

Van Nostrand Reinhold International Company Limited
11 New Fetter Lane
London EC4P 4EE, England

Van Nostrand Reinhold
480 La Trobe Street
Melbourne, Victoria 3000, Australia

Nelson Canada
1120 Birchmount Road
Scarborough, Ontario M1K 5G4, Canada

16 15 14 13 12 11 10 9 8 7 6 5 4 3 2 1

Library of Congress Cataloging-in-Publication Data

Theoretical aspects of heterogeneous catalysis / John B. Moffat [editor].
p. cm. — (Van Nostrand Reinhold catalysis series)
ISBN 0-442-20528-7
1. Heterogeneous catalysis. I. Moffat, John B. II. Series.
QD505.T47 1990 89-75819
541.3′95—dc20 CIP

Contents

Series Introduction

Catalysis involves just about every field of scientific study. This means that a multidisciplinary approach is needed in catalytic studies. Catalysis involves breaking and forming new bonds and this requires an understanding of either adsorption by bonding to an extended structures or bonding in a coordination sphere. Any understanding of catalytic action must necessarily involve an understanding of this bonding.

Even 200 years ago scientists were aware that a properly treated material, such as charcoal, could adsorb an enormous quantity of gas. In 1812, de Sassasure (English translation, *Annal Philosphy*, *6*, 241 (1815)) proposed that the ability of a material to increase the rate of chemical reaction was due to adsorption of the material in the fine structure of the solid so that the concentrations of the reactants were significantly increased, and this increase in concentration led to an increase in reaction rate. During the 1800s, little advance was made in the understanding of adsorption.

Sabatier (*Bull. Soc. Chim. France*, *6*, 1261 (1939)) must be among the first to utilize chemisorption concepts in catalysis research. He recalls, "Like my illustrious master, Marcellin Berthelot, I always assumed that the fundamental cause of all types of catalysis is the formation of a temporary and very rapid [chemical] combination of one of the reactants with a body called the catalyst . . . . I have tenaciously held to my theory of a temporary combination. It has guided my work both in hydrogenations and in dehydrations." And a good guide it was since he was awarded the Nobel Prize in 1912 for his work in catalysis. This award was followed three years later when Ostwald was awarded the Nobel Prize for his work in catalysis and provided an unprecedented period for catalysis.

Langmuir (*J. Am. Chem. Soc.*, *38*, 2221 (1916)) advanced his views of adsorption in the 1910s. His model was simple: an atom, such as formed by the dissociation of hydrogen, was confined to the space occupied by a surface atom of the adsorbent. Thus, Langmuir advanced his famous checkerboard model; the squares of the board represented the space assigned to each surface atom and the checker piece represented the individual adsorbed atom held to its space by a chemical bond. Langmuir (*Trans. Faraday Soc.*, *17*, 621 (1921)) developed very successful kinetic equations based upon his adsorption model. In 1932 he was awarded the

Nobel Prize for his work on adsorption so that his views had great popularity. However, some held the view that adsorption involved multilayers rather than the monolayer as required by Langmuir's theory. Sir Hugh Taylor complicated the situation even more by advancing his views of activated adsorption. Gradually, the scientific community realized that there was not a simple model that was adequate to describe the totality of adsorption just as the advances of quantum mechanics led the community to realize that there is not just one simple description of chemical bonds.

The application of quantum mechanics to provide a description of adsorption on catalytic surfaces was slow to materialize. Thus, it was not until 1950 that a very influential and, by today's standards, rather elementary paper appeared that attempted to bring together the theoretical ideas of the solid state, adsorption, and catalytic reactions. While several papers appeared at about the same time, Dowden's paper (*J. Chem. Soc.*, *1950*, 242) appears to have had the major impact. And just as it was the ability of Langmuir to present complex theoretical ideas in terms of simple models that were understandable to working catalytic scientists, so it was with Dowden and his electronic views of catalysis. But simple models that are able to bridge the gap between the world of complex mathematical equations and the art of the practicing catalytic scientist have been slow to emerge. In the nearly 40 years since Dowden first advanced his ideas, tremendous advances have occurred in the theories of chemical bonding, the physics of the solid state, and the dynamics of chemical reactions. There is a need to reduce this mass of sophisticated mathematical material to models that can guide catalytic scientists. It is intended that this book will provide a step in that direction.

Burtron H. Davis

Preface

Catalysis may well be one of the most practical of the many subdisciplines of physical science. More than 90% of the chemical processes worldwide involve heterogeneous catalysts and more than one billion dollars were spent in 1988 in the United States alone on catalysts for the industrial, automotive, chemical, and petroleum sectors. The gasoline burned in internal combustion engines, the precursors for a variety of plastics and polymers, fertilizers, edible oils, pharmaceutical and medicinal products, dyestuffs, food additives, cosmetics, pollution control both in automobiles and in stationary plants, all involve the use of catalysts.

As with much of science, experiment has preceded theory in catalysis. Many of the catalysts referred to in the present volume have been in industrial use for a considerable length of time, but nevertheless, understanding of the theory of their catalytic activity is not complete.

It is evident that the surface of a heterogeneous catalyst is one of its more important features. While theoretical techniques such as those of quantum mechanics are most conveniently applied to isolated molecules. The application to surfaces is considerably more complex. The present volume provides examples of methods for the theoretical treatment of surfaces both by the more exact *ab initio* approaches as well as semiempirical methods, illustrates the information that may be obtained from these techniques, and provides a new insight into the nature of catalysis.

Theoretical understanding of a complex system such as is encountered in heterogeneous catalysis can frequently be gained by systematic examination of experimental data obtained with a number of related catalysts and/or related reactant systems. Correlations may be found which provide valuable insights concerning the important parameters in the operation of the catalyst and the catalytic process. The advantages of such an approach are obvious. Detailed, time-consuming (and frequently expensive) calculations are avoided in favor of direct reliance on experimental information. The knowledge gained is frequently no less valuable than that resulting from the application of more theoretical approaches.

The present volume should provide the reader with a deeper understanding of the way in which catalysts operate. In so doing it is hoped that the reader may not only discern why one catalyst may be superior to another, but may be provided with sufficient theoretical background to aid in both the improvement of current catalysts but also the design of entirely new catalysts.

J.B. MOFFAT

Contributors

Alfred B. Anderson, Department of Chemistry, Case Western Reserve University, Cleveland, Ohio

Jean-Marie Andre, Dr.Sc., Institute for Studies in Interface Sciences, Laboratoire de Chimie Théorique Appliquée, Facultés Universitaires Notre-Dame de la Paix, Namur, Belgium

R.C. Baetzold, Corporate Research Laboratories, Eastman Kodak Company, Rochester, New York

Stanislav Beran (deceased), J. Heyrovsky Institute of Physical Chemistry and Electrochemistry, Czechoslovak Academy of Sciences, Prague, Czechoslovakia

R.R. Chianelli, Corporate Research and Development Laboratories, Exxon Research and Engineering Company, Annandale, New Jersey

Eric G. Derouane, Dr.Sc., Institute for Studies in Interface Sciences, Laboratoire de Catalyse, Facultés Universitaires Notre-Dame de la Paix, Namur, Belgium

Joseph G. Fripiat, Dr.Sc., Institute for Studies in Interface Sciences, Laboratoire de Chimie Théorique Appliquée, Facultés Universitaires Notre-Dame de la Paix, Namur, Belgium

P. Galet, Institute for Studies in Interface Sciences, Laboratoire de Chimie Théorique Appliquée, Facultés Universitaires Notre-Dame de la Paix, Namur, Belgium

Patrick Geneste, Ph.D., Ecole Nationale Supérieure de Chimie, C.N.R.S., Université de Montpellier, Montpellier, France

Helmut Haberlandt, Ph.D., Central Institute of Physical Chemistry, Academy of Sciences of the GDR, Berlin, German Democratic Republic

Suzanne Harris, Corporate Research Laboratories, Exxon Research and Engineering Company, Annandale, New Jersey

P.A. Jacobs, Ph.D., Laboratorium voor Oppervlaktechemie, K.U. Leuven, Leuven, Belgium

Laurence Leherte, Institute for Studies in Interface Sciences, Laboratoire de Chimie Théorique Appliquée, Facultés Universitaires Notre-Dame de la Paix, Namur, Belgium

J.A. Martens, Ph.D., Laboratorium voor Oppervlaktechemie, K.U. Leuven, Leuven, Belgium

John B. Moffat, Ph.D., Department of Chemistry, University of Waterloo, Waterloo, Ontario, Canada

Claude Moreau, Ph.D., Ecole Nationale Supérieure de Chimie, C.N.R.S., Université de Montpellier, Montpellier, France

Wilfried J. Mortier, Ph.D., Basic Chemicals Technology, Exxon Chemical Holland, Rotterdam, The Netherlands, and K.U. Leuven, Laboratorium voor Oppervlaktechemie, Heverlee, Belgium

Evgeny Shustorovich, Ph.D., Corporate Research Laboratories, Eastman Kodak Company, Rochester, New York

Didier Vandervecken, Institute for Studies in Interface Sciences, Laboratoire de Chimie Moleculaire Structurale, Facultés Universitaires Notre-Dame de la Paix, Namur, Belgium

Daniel P. Vercauteren, Dr.Sc., Institute for Studies in Interface Sciences, Laboratoire de Chimie Moleculaire Structurale, Facultés Universitaires Notre-Dame de la Paix, Namur, Belgium

William A. Wachter, Exxon Research and Development Laboratory, Baton Rouge, Louisiana

Satohiro Yoshida, Ph.D., Department of Hydrocarbon Chemistry and Division of Molecular Engineering, Kyoto University, Kyoto, Japan

THEORETICAL ASPECTS OF HETEROGENEOUS CATALYSIS

1
A Theoretical View and Approach to the Physics and Chemistry of Zeolites and Molecular Sieves

E.G. Derouane, J.-M. André, L. Leherte, P. Galet, D. Vanderveken, D.P. Vercauteren, and J.G. Fripiat

INTRODUCTION

The understanding and modeling of catalysts and catalytic behavior at the molecular level can benefit from accurate and significant quantum chemical calculations or molecular statistics and dynamic simulations or all three procedures. Although theory has so far had a limited impact on the understanding and prediction of catalysis, the advent of more powerful computing facilities, which permit more complex calculations and simulations representing more realistically the systems of interest, makes this area of research one of potential and rapid growth.

Zeolites appear as particularly interesting systems for such studies, since their structures are defined and known in most instances, in contrast to the amorphous catalysts. The zeolites' fundamental chemistry and physical properties are rather well assessed by the full arsenal of spectroscopic and other physicochemical techniques.

The quantum chemical evaluation of potential surfaces for such three-dimensional systems is highly complex. Therefore, only rather small representative entities (pseudomolecules, clusters, etc.) can be treated adequately. The geometry of such entities can be ideal and chosen *a priori*, for example, to derive and support general concepts, or the geometry can be selected on the basis of crystallographic data to represent given domains of the zeolite framework.

In contrast, recourse to molecular statistics and molecular dynamics enables the description of much larger systems for which even border effects can be taken into account rather satisfactorily. Some of the parameters necessary to perform the latter calculations are available from experimental data, but, interestingly enough, can also be evaluated from quantum mechanical calculations on small species of zeolites.

This chapter examines practical examples of the preceding situations:

1. The capabilities of quantum mechanical calculations for the description of Al-site-related properties in zeolite frameworks. Such properties are the well-known aluminum avoidance principle (Loewenstein's rule), the possible preferential siting of Al at given crystallographic tetrahedral sites, the possibility of long-range interactions between Al sites leading to the formation of Al–O–Si–O–Al pairs (as it is observed in mordenite and could occur in ZSM-5), and the acidity evaluation of the charge-compensating proton associated with the presence of structural Al. Applications to a variety of zeolites including ZSM-5, mordenite, ferrierite, gismondine, laumontite, and bikitaite are presented and discussed.

2. The utilization of advanced molecular graphics, statistical mechanics, and molecular dynamics to approach and describe the mobility of sorbed species within the free intracrystalline pore volume of zeolites. Such mobilities in self-diffusion conditions are readily measured by pulsed-field-gradient NMR measurements, which can serve as quality and reliability tests for the development of new theoretical methods and approaches that are still in their infancy with respect to their application to zeolitic systems. As an example, water (structure, mobility, etc.) in ferrierite will be discussed in detail.

For both strategies, there are still some limitations. These limitations are discussed in each section, and the practical problems limiting the present ability of the theoretical chemist in this field are set forth.

It is clear, however, that the successful application of theory to other areas of molecules and materials design will result in continuous development and improvements. The latter should keep such theoretical work most fertile and render it an essential tool for the refinement of our fundamental understanding and modeling of zeolite chemistry and catalytic behavior.

QUANTUM MECHANICAL DESCRIPTION OF Al-SITE-RELATED PROPERTIES IN ZEOLITES

The catalytic activity of zeolites is known to depend largely on their acidic properties, which arise from the presence of trivalent Al atoms, replacing tetravalent Si in their framework. The environment and symmetry of the lattice sites that accommodate these Al atoms are not all equivalent and may vary from one zeolite to another. Recent experimental results, mainly based on high-resolution solid-state ^{29}Si NMR studies,[18,61,69,78,104] have revived the interest for the type of siting that aluminum atoms will adopt in the zeolite aluminosilicate framework. Indeed, the magnitude of the ^{29}Si isotropic chemical shift reflects the structural surroundings of the silicon atom in the tetrahedral sites (T-sites) of the silicate and aluminosilicate framework. However, the correlation between the measured chemical shifts and the occupancy of T-sites is often ambiguous.

Various models have been proposed, describing either partially ordered or random distribution of Al in different zeolites. These models generally obey the Loewenstein aluminum avoidance principle.[71] It is possible to define two groups of zeolites.[46] In the first group, where the ratio of Si/Al is close to 1, the ordering is dominated by the Loewenstein rule, so that a simple alternation of Si and Al in the T-sites is preferred to disorder. Gismondine and zeolites A and X with suitable composition belong to this group. In the second group, where the ratio of Si/Al is higher than 1, these authors have defined four classes, which are characterized by the degree of order: a complete order as in laumontite and bikitaite, a variable degree of order from complete to partial as in natrolite, a partial order as in mordenite, and a complete disorder as in offretite.

Quantum chemical calculations may be a help in investigating the siting and the pairing of Al in zeolite frameworks, since such properties are not always easily quantified by experimental means.[94] Quantum chemical *ab initio* calculations become reasonable when the cluster model is adopted. The restricted Hartree–Fock–Roothaan (LCAO–SCF–MO) approach has proved to be successful in zeolite and silica chemistry as shown, for example, by the works of Sauer et al.[92,93,95] on the stabilizing effect played by the counterions in aluminosilicate framework clusters, Hass et al.[48,49] about the validity of the aluminum avoidance concept, and Geisinger et al.,[43] Mortier et al.,[81] and Derouane et al.[31–33,39,40] on other aspects of this subject. Generally, in these studies, geometries based on standard bond lengths and angles were used for model clusters.

Our knowledge that all the T-sites in a zeolite framework are not equivalent persuaded us to investigate clusters reflecting more closely the structures of ZSM-5 (MFI), mordenite (MOR), laumontite (LAU), offretite (OFF), gismondine (GIS), bikitaite (BIK), and ferrierite (FER) zeolites, in which the Si, Al, and O atoms would occupy their framework positions. The characteristics of these zeolites are described in Tables 1.1 and 1.2.[3,17,41,55,75–77,86] Table 1.1 gives the space groups, the number of topologically inequivalent T-atoms, the nature of the secondary building units (SBU) of which their three-dimensional arrangement leads to the different zeolite structures, the Si/Al ratios, and the framework density. Table 1.2 describes the channel system of these zeolites using the notation introduced by Meier and Olson.[75]

The present review will demonstrate that LCAO–SCF–MO calculations are able to predict the preferential siting and pairing of aluminum in these zeolites. It will be shown in addition that the presence of charge-compensating protons is critical to the stabilization of some of the Al-containing clusters.

Table 1.1. Topological description of typical zeolite structures.

Type Code	Name	Space Group	Q[a]	Number of T-Atoms in Unit Cell	SBU[b]	Si/Al Ratio	Framework Density[c]
BIK	Bikitaite	*P*1	6	6	5-1	2	19
FER	Ferrierite	*Immm*	4	36	5-1	5	17.7
GIS	Gismondine	$P2_1/c$	4	16	4	1	15.4
LAU	Laumontite	*Am*	6	24	6	2	17.7
MFI	ZSM-5	*Pnma*	12	96	5-1	±30	17.9
MOR	Mordenite	*Cmcm*	4	48	5-1	5	17.2
OFF	Offretite	*P*6*m*2	2	18	6	3.5	15.5

[a] Number of topologically inequivalent atoms (per unit cell).
[b] Secondary building units.
[c] Framework density, number of T-atoms per 1000 Å^3.

Method and Models

Restricted Hartree–Fock–Roothaan (LCAO–SCF–MO) calculations were performed using the GAUSSIAN program package.[15,16] All two-electron integrals larger than 10^{-6} a.u. were explicitly taken into account and the convergence threshold on the density matrices was fixed to 5×10^{-5}. The program calculates the electronic structure and total energy of a molecule (cluster), determines its equilibrium geometry by allowing a relaxation from the set atomic positions, evaluates the forces acting on the various atoms, evaluates the charges and overlap populations, and so forth.

Two atomic orbital basis sets were used. The first set is the so-called minimal STO-3G basis set,[50,51] which uses three gaussian functions to

Table 1.2. Pore structure of the typical zeolites.

Type Code	Name	Channel System (Å)[a]
BIK	Bikitaite	8 3.2 × 4.9*
FER	Ferrierite	10 4.3 × 5.5* ↔ 8 3.4 × 4.8*
GIS	Gismondine	(8 3.1 × 4.4 ↔ 8 2.8 × 4.9)*
LAU	Laumontite	10 4.0 × 5.6*
MFI	ZSM-5	(10 5.4 × 5.6 ↔ 10 5.1 × 5.5)***
MOR	Mordenite	12 6.7 × 7.0* ↔ 8 2.9 × 5.7*
OFF	Offretite	12 6.4* ↔ 8 3.6 × 5.2**

[a] The underlined number indicates the number of oxygen atoms constituting the smallest ring determining the pore size. The other two numbers separated by "×" give the free diameter of the pore. The number of dimensions in which the channel runs are given by the number of asterisks. The ↔ sign indicates that the channels are interconnected.

represent each atomic orbital. In a minimal basis set, the number of atomic orbitals corresponds to the number of atomic orbitals belonging to the occupied shells in the free atom. This basis set has met considerable success in the comparison of energies between reactions in which the number of each kind of formal chemical bond is conserved,[52] as it occurs for example in the present study when silicon is replaced by aluminum in the various clusters. The second set, the 6-21G basis set,[14,45] is a "split-valence" basis set in which two basis functions describe each valence atomic orbital. This kind of basis set is known to give a better evaluation of the relative energies and of the geometrical features of molecules. However, its use is generally limited to relatively small entities, since the number of functions is almost twice the number of basis functions per atom as compared to a minimal STO-3G basis set.

For clusters containing a large number of atomic orbitals, the calculated total energy values are large compared to the energy differences on which the description of the behavior of aluminum is based. Two types of errors can affect the calculation of the total energies: first, errors introduced by the Hartree–Fock model and the LCAO approximation, such as the correlation error or the basis set error; second, the numerical errors due to the finite representation of numbers in the computer. Our experience has shown that the computed total energy has at least seven significant digits, since all the computations are performed in the double precision mode. It corresponds in our calculations to a numerical error of about 1 kcal mol^{-1} for a cluster with 137 orbitals. Since all comparisons are made between systems having the same number of electrons, we feel that 6 kcal mol^{-1} is a realistic lower limit for significant differences between computed energies.

The selected models were topological open monomer, open pentamer (linear chains) (ZSM-5), cyclic tetramer (mordenite), or cyclic hexamer (ferrierite) clusters of which the geometries were derived from the structure of selected zeolites (Figs. 1.1–1.3). The T-atoms (Si or Al) and the oxygen atoms, positioned at their location in the framework, are identified by the usual symbols TX and OY. Terminal hydrogen atoms, needed to maintain the (near) neutrality of the clusters and to ensure a higher reliability of the calculations, are also positioned at T-sites in order to simulate the outer silicon atoms.

The smaller clusters were monomers of the type H_4SiO_4 (mono-X-Si) or $H_4AlO_4^-$ (mono-X-Al) centered on each T-site, X, belonging to the different symmetry types.

The second type of model (pentamers) consists of one T-site surrounded by four tetrahedra (T-sites) as schematized in Figure 1.1c and is of formula $H_{12}Si_5O_{16}$, $H_{12+m}Si_{5-m}Al_mO_{16}$, or $H_{13}Si_4AlO_{16}$. For mordenite, the pentamers centered on sites 3 and 4 present a geometry that differs

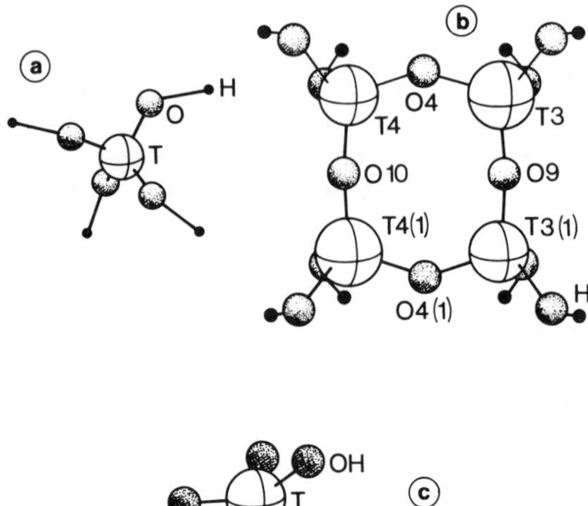

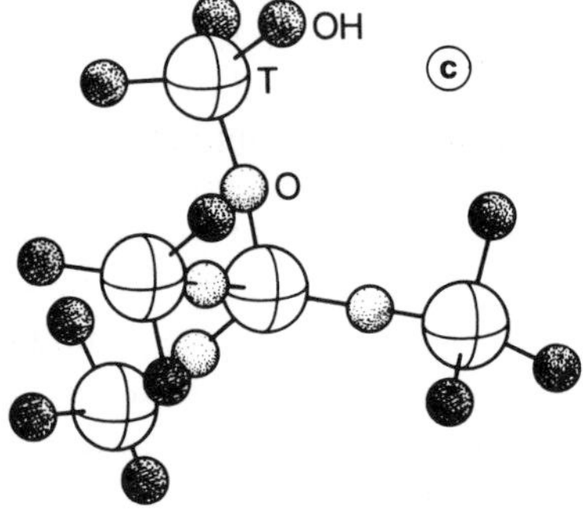

Figure 1.1 Model clusters: (a) monomer, (b) tetramer (mordenite), (c) pentamer.

slightly from the one derived from the structure: one of the outer T-sites, simulated by hydrogen atoms, was split into two virtual T-positions to achieve the same configuration (number of atoms and orbitals) for all the pentameric clusters. The same procedure was applied to the two T-sites of offretite. Both monomeric and pentameric models were used to evaluate the relative energy variations occurring upon substitution of Si by Al at various T-sites in the different zeolites.

Another type of pentamer is derived from the ZSM-5 framework. It is characterized by a chain of T-sites as shown in Figure 1.2. Two different chains were selected, each consisting of the same T-sites (T12, T12, T3, T2, and T8) but organized in different ways. One has the sequence T12-T12-T3-T2-T8, which is designated as *cis*-nAl(x,y)-mH(p,q), whereas the other cluster has the sequence T12-T12-T8-T2-T3 and is named *trans*-nAl(x,y)-mH(p,q), n indicating the number of Al atoms (sited at positions x and y) and m indicating the number of charge-compensating protons

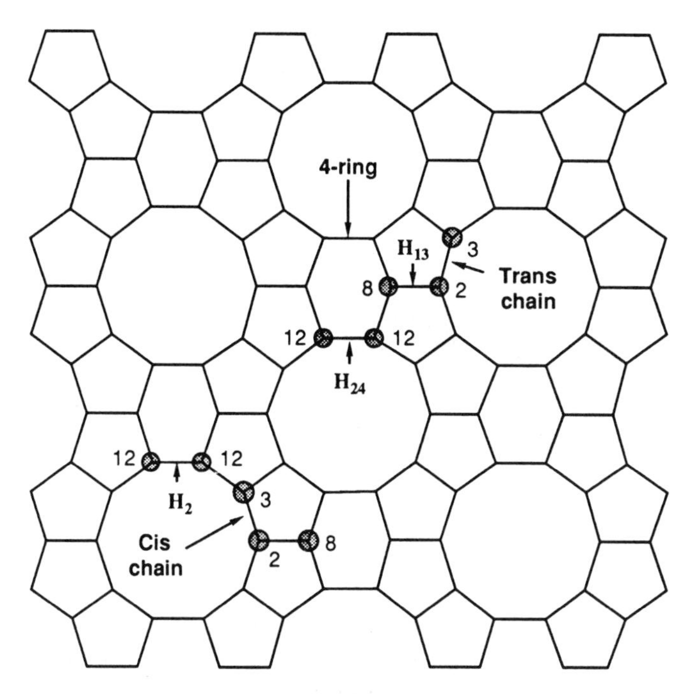

Figure 1.2 Schematic representation of the linear pentameric chain clusters in the ZSM-5 framework.

bound to oxygen atoms located at given positions p or q or both. Numbers in parentheses differentiate the symmetry-related sites. The first chain is called "*cis*" because all its T-sites are located along the boundary of the channel, which runs parallel to [010]. The "*trans*" chain crosses the (alumino)silicate layer separating two channels.

Mordenite tetrameric four-membered ring clusters are designated by *tetra-n*Al(x,y)-mH(p,q). Their general formula is $H_{8+m}Si_{4-n}Al_nO_{12}^{(m-n)+}$. All tetrameric clusters were of the type schematized in Figure 1.1b in which the numbering of the atomic positions is that used for mordenite.

Ferrierite hexameric six-membered ring clusters are designated by *hexa-n*Al(x,y,z), n being the number of Al atoms sited at position x, y, and z. All hexameric clusters were of the type schematized in Figure 1.3.

These last three models were used to investigate the problem of Al pairing in the different zeolites. Substitution of Si by Al generates framework negative charges that may be partially or totally compensated by linking protons to bridging oxygen atoms. These O–H bond lengths were assumed equal to 0.0975 nm and the protons were located in the corresponding T–O–T plane.

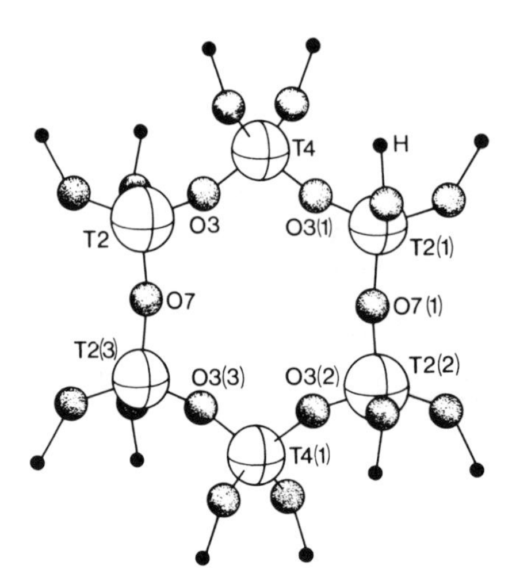

Figure 1.3 Hexameric model cluster (ferrierite).

Results and Discussions

The determination of preferred sites for substitution of Si by Al and the investigation of pairing of Al in the various frameworks require the knowledge of the total energy of the different clusters. The position of the Al atom(s) was not optimized; separate calculations[34] on model dimers indicate that this leads to an estimated error of about 2–3 kcal mol^{-1} in the evaluation of the cluster total energies. Therefore, an important and novel element in our approach is the use of differences between the total energies of equivalent clusters to predict the siting and behavior of aluminum at particular T-sites. For all cluster calculations described in this chapter, no occupied molecular orbital has a positive energy. Consequently, the calculated wave functions are stable and correspond to a minimum in the total electronic energy of the system.[1]

Siting of Aluminum in the Zeolite Frameworks

Monomer Clusters. The total energies, computed using the STO-3G and 6-21G atomic basis sets, for the monomer clusters centered on the T-sites of the different zeolites are given in Table 1.3, which shows that energies obtained from the "split-valence" basis set calculations are significantly lower than those computed with the minimal basis set. This trend is expected since the 6-21G basis set involves more flexibility in representing the valence electrons and provides a better representation of the core atomic orbitals.

Table 1.3. Calculated total energies of monomeric clusters (in atomic units).

Cluster Type	STO-3G		6-21G	
	Si	Al	Si	Al
GIS-1	−582.35610	−535.76832	−589.62838	−542.67788
GIS-2	−582.34934	−535.76105	−589.62492	−542.67220
GIS-3	−582.48576	−535.93535	−589.73500	−542.84034
GIS-4	−582.48868	−535.93753	−589.73169	−542.83971
BIK-1	−582.43643	−535.83781	−589.68503	−542.72401
BIK-2	−582.48114	−535.88333	−589.72372	−542.75954
BIK-3	−582.42275	−535.82638	−589.67630	−542.71731
BIK-4	−582.48173	−535.88434	−589.72446	−542.76119
BIK-5	−582.55013	−535.96996	−589.77557	−542.87495
BIK-6	−582.55671	−535.99546	−589.77744	−542.87792
OFF-1	−582.49237	−535.91342	−589.73522	−542.79884
OFF-2	−582.42884	−535.84121	−589.68980	−542.73449
FER-1	−582.35695	−535.75909	−589.61206	−542.63865
FER-2	−582.56605	−535.98787	−589.80325	−542.86094
FER-3	−582.43885	−535.84924	−589.68806	−542.72143
FER-4	−582.54591	−535.95757	−589.79037	−542.82718
MOR-1	−582.49182	−535.90257	−589.73826	−542.78261
MOR-2	−582.48545	−535.89492	−589.73424	−542.77336
MOR-3	−582.47934	−535.89917	−589.73146	−542.78725
MOR-4	−582.49059	−535.90428	−589.73774	−542.78531
LAU-1	−582.44116	−535.83956	−589.68864	−542.72216
LAU-2	−582.53013	−535.96102	−589.75818	−542.84361
LAU-3	−582.46275	−535.86286	−589.70306	−542.74107
LAU-4	−582.49943	−535.90592	−589.73526	−542.78177
LAU-5	−582.57102	−536.00932	−589.78902	−542.88753
LAU-6	−582.47883	−535.88447	−589.71888	−542.76493
MFI-1	−582.47325	−535.87906	−589.71446	−542.75259
MFI-2	−582.55059	−535.96237	−589.78594	−542.82887
MFI-3	−582.50802	−535.91286	−589.75099	−542.78184
MFI-4	−582.46862	−535.87021	−589.71183	−542.73913
MFI-5	−582.49037	−535.89184	−589.73311	−542.76107
MFI-6	−582.48946	−535.88536	−589.72223	−542.74592
MFI-7	−582.52369	−535.92790	−589.75944	−542.79246
MFI-8	−582.44159	−535.83750	−589.68478	−542.70472
MFI-9	−582.48483	−535.88747	−589.72769	−542.76051
MFI-10	−582.49331	−535.88973	−589.72835	−542.75164
MFI-11	−582.48507	−535.88520	−589.72517	−542.75087
MFI-12	−582.54637	−535.96078	−589.77648	−542.82621

The relative energies obtained in the two basis sets are presented in Table 1.4. This table shows the Al content; the mean value of the four T–O tetrahedral bond lengths around a given T-site; and the relative substitution energy, E_R, of Si by Al, which is the relative energy variation of a cluster upon replacement of Si by Al at its central T-site, referred to the most favorable substitution site:

$$E_R = (E_{\text{mono-X-Al}} - E_{\text{mono-X-Si}}) - (E_{\text{mono-X*-Al}} - E_{\text{mono-X*-Si}}) \quad (1.1)$$

where $E_{\text{mono-X-Al}}$ and $E_{\text{mono-X-Si}}$ represent the total energy of the cluster with Al or Si at its central TX-site, respectively (X* meaning the reference site).

The analysis of these data indicates the following:

1. The four different T-sites of gismondine fall into two classes. The first class is constituted by sites T3 and T4, which present the lowest substitution energies and consequently correspond to the preferred sites for Al in agreement with the X-ray or neutron diffraction measurements. The siting of Al is less probable in sites T1 and T2.
2. The calculations show that the sites T5 and T6 of bikitaite have a substitution energy slightly higher than that of sites T3 and T4 of gismondine and are the preferred substitution sites, while the other sites present a substitution energy still higher by an amount of ~24 kcal mol^{-1} or 40 kcal mol^{-1} according to the basis set.
3. The substitution energy order of the different T-sites of the laumontite is T5 > T2 > T4 > T6 > T3 > T1. Al-siting is strongly preferred at site T5 and less at site T2. It is much less probable at sites T4 and T6, and unlikely at sites T3 and T1.
4. In ZSM-5 (MFI), the siting of Al is more probable at sites T2 and T12, with a slight preference for the latter. The substitution of Al by Si at sites T6, T8, and T10 has the lowest probability. Compared with the data obtained for the T4 site of gismondine, the lowest relative substitution energy in ZSM-5 is equal to 37 kcal mol^{-1} (6-21G) or 22 kcal mol^{-1} (STO-3G), which indicates a low probability for the siting of Al in this zeolite. These observations are consistent with the observed Si/Al ratio (see Table 1.1).
5. The calculations on mordenite show no significant differences (6 kcal mol^{-1} for STO-3G or 10 kcal mol^{-1} for 6-21G) in the substitution energies of the four T-sites and indicate only that Al atoms would be preferentially located at site T3, characterized by a value of E_R equal to 33 kcal mol^{-1} (6-21G) or 18 kcal mol^{-1} (STO-3G). These data are

Table 1.4. Relative substitution energies,[a] E_R (kcal mol^{-1}), of monomeric clusters, averaged T–O bond distances (nm), and Al occupancy probabilities.[b]

	Substitution Energy			
Cluster	STO-3G	6-21G	$R_{T\text{–}O}$	Al Content[c]
GIS-1	23.5	36.7	0.162	0.15
GIS-2	23.9	38.1	0.162	0.11
GIS-3	0.0	1.7	0.174	0.94
GIS-4	0.5	0.0	0.175	0.99
BIK-1	30.3	43.3	0.162	0.11
BIK-2	29.6	45.3	0.161	0.05
BIK-3	28.9	42.1	0.162	0.12
BIK-4	29.5	44.8	0.161	0.07
BIK-5	18.7	5.4	0.174	0.95
BIK-6	6.8	4.7	0.175	0.98
OFF-1	17.9	27.9	0.166	0.39
OFF-2	23.4	39.8	0.161	0.09
FER-1	29.8	51.1	0.160	0.0
FER-2	17.4	31.6	0.162	0.15
FER-3	24.6	46.9	0.160	0.0
FER-4	23.8	44.7	0.161	0.07
MOR-1	24.4	40.0	0.162	0.13
MOR-2	25.2	43.3	0.161	0.07
MOR-3	18.7	32.8	0.164	0.27
MOR-4	22.5	37.9	0.163	0.20
LAU-1	32.1	46.8	0.160	0.0
LAU-2	11.7	14.2	0.171	0.73
LAU-3	31.0	44.0	0.161	0.07
LAU-4	27.1	38.6	0.163	0.20
LAU-5	7.1	6.0	0.174	0.93
LAU-6	27.6	38.9	0.163	0.20
MFI-1	27.5	43.9	0.160	0.0
MFI-2	23.7	40.9	0.162	0.13
MFI-3	28.1	48.4	0.159	0.0
MFI-4	30.1	50.7	0.158	0.0
MFI-5	30.2	50.3	0.158	0.0
MFI-6	33.7	52.9	0.158	0.0
MFI-7	28.5	47.1	0.160	0.0
MFI-8	33.7	55.3	0.158	0.0
MFI-9	29.5	47.2	0.160	0.0
MFI-10	33.4	53.2	0.158	0.0
MFI-11	31.1	51.7	0.158	0.0
MFI-12	22.1	36.6	0.163	0.20

[a] Referred to sites 3 or 4 in gismondine (see explanation in the text).
[b] Estimated from the Smith and Bailey[98] diagram.
[c] The Al occupancy probability of sites with an averaged T–O bond length smaller than 0.16 nm is set to zero.

consistent with the observation that mordenite is characterized by a partial order.

6. The substitution energy order of the different T-sites of ferrierite is T2 > T4 > T3 > T1. It spreads over 12 (STO-3G) to 20 (6-21G) kcal mol^{-1}, suggesting a low order in the siting of Al in the framework.
7. The energy difference between the two sites in offretite is equal to 5 kcal mol^{-1} (STO-3G) or 12 kcal mol^{-1} (6-21G) indicating a low order, the site T1 being the preferred site.

The substitution energy data confirm the classification of Gottardi and Alberti.[46] Indeed, Table 1.4 shows that gismondine, bikitaite, and laumontite are characterized by a high order in the distribution of aluminum in the framework, the other zeolites presenting a variable degree of disorder.

The relative energies obtained from the two basis sets differ and, since no experimental data are available to assess the validity of the individual basis sets, no definite choice can be made in favor of one particular set of data. Since it is known from the literature that the 6-21G basis set usually gives a better description of the relative energies and of various geometrical features of molecules, we better trust the energies computed with this basis set. However, let us recall that, for practical reasons, the 6-21G basis set can be used only for small clusters and that the recourse to a minimal STO-3G basis set is often imposed by the large size of the systems that are investigated. Since it is apparent that a linear correlation exists between the 6-21G and the STO-3G energies (the correlation coefficient is equal to 0.9908 for Si clusters and to 0.9757 for Al clusters) and since we are mainly interested in the relative stability of the different clusters, we feel that the results obtained using the minimal STO-3G basis set can be utilized for the classification of the various clusters and the identification of preferential Al substitution sites in the framework.

Figures 1.4 and 1.5 show the correlation between the computed substitution energy of Si by Al and the average T–O bond length for the different T-sites of the various zeolites, obtained with both types of orbital basis sets. The relative substitution energies are referred to the energy for sites T3 or T4 of gismondine, which are the most favored substitution sites given by the STO-3G and the 6-21G calculations, respectively. Figure 1.4 illustrates the absence of correlation between the STO-3G results and the average T–O bond lengths, the correlation coefficient being equal to 0.9115. In contrast, the correlation coefficient for the data plotted in Figure 1.5 is equal to 0.9895, indicating a correlation between the substitution energies obtained with the 6-21G basis set and the averaged T–D bond lengths. From the slope of the plot, it is seen that the substitution energy decreases by about 30 kcal mol^{-1} when the averaged T–O bond is elongated by 0.01 nm. As expected, Al substitution is favored for longer T–O bonds and vice versa.

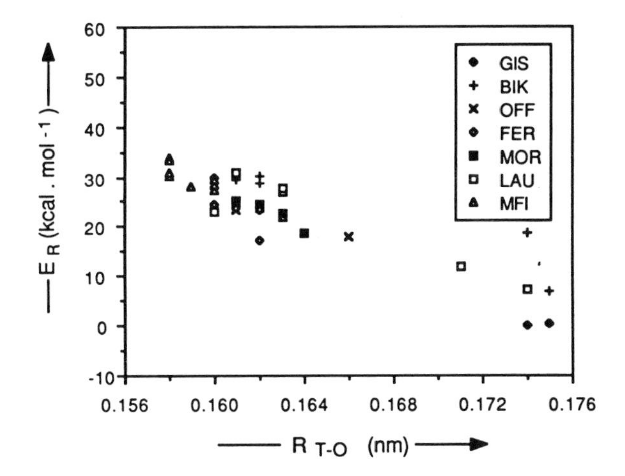

Figure 1.4 STO-3G substitution energies for the studied zeolites as a function of the mean T–O distance of their various T-sites.

Smith and Bailey[98] have proposed that the mean value of the four T–O tetrahedral bond lengths around a given T-site can be correlated linearly with its aluminum occupancy probability. On the basis of this approach, we may estimate the approximate Al content probabilities for the T-sites of the various zeolites, which are given in Table 1.4. From a comparison of these numbers with the energies derived from the molecular orbital calculations, we conclude that the elongation by 10^{-3} nm of the mean T–O

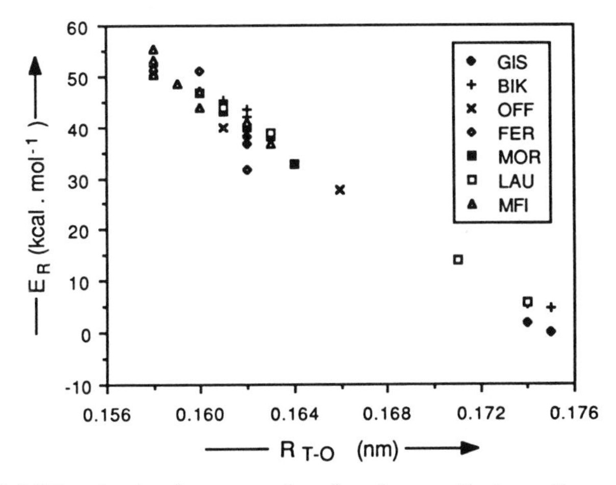

Figure 1.5 6-21G substitution energies for the studied zeolites as a function of the mean T–O distance of their various T-sites.

bond around a given T-site, as measured from X-ray diffraction data, decreases by 3 kcal mol^{-1} the replacement energy of Si by Al, which in turn increases its Al occupancy probability by about 7%. However, it must be stressed that this kind of correlation based only on averaged T–O bond distances does not take into account other short-range (like bond angles) or long-range effects (due to the finite size of cluster models). Also, the validity of such conclusions depends directly on the quality of the X-ray data used to derive the geometries of the model clusters.

Pentamer Clusters. In order to identify the effect of longer-range interactions, topologically open pentameric clusters were also considered. They consist of one T-site surrounded by four tetrahedra (T-sites) as schematized in Figure 1.1. Penta-X-Si or penta-X-Al designate these pentameric clusters centered on the atomic positions TX (Si or Al).

The relative stability of Al in the T-sites can be evaluated in two ways. The first method is identical to the procedure used for the study of the monomeric clusters, i.e., the evaluation of relative stabilities by calculating the relative energy variation E_R of a pentameric cluster upon replacement of Si by Al at its central T-site using as reference the most stable configuration. Relative stabilities evaluated in this manner (Table 1.5) confirm, for all the studied zeolites, the same preferential substitution sites as the one derived from the monomeric clusters calculations.

Another more formal approach is to consider as a model for solid-state reactions the following Si-Al hypothetical exchange processes occurring at site TX:

$$\text{penta-X-Si} + \text{mono-X-Al} = \text{penta-X-Al} + \text{mono-X-Si}$$

with the reaction energy E_S:

$$E_S = (E_{\text{penta-X-Al}} + E_{\text{mono-X-Si}}) - (E_{\text{penta-X-Si}} + E_{\text{mono-X-Al}}) \qquad (1.2)$$

Substitution energies E_S calculated in this manner (Table 1.5) confirm the former results. However, it is important to note that the values of E_S do not correspond to the real energy changes occurring upon substitution of Si by Al in the zeolite structures. The E_S values apply strictly to this formal process that considers, among others, monomeric and charged species.

As we have mentioned, these data should give a reliable trend for the siting of Al, since all occupied molecular orbitals have negative energies. Furthermore, in all our comparisons of corresponding Si- and Al-containing clusters, the total number of electrons is kept constant, whereas the total nuclear charge is reduced by one unit each time Al replaces Si.

Table 1.5. Relative substitution energies, E_R (kcal mol^{-1}), and pentamer–monomer substitution energies, E_S (kcal mol^{-1}), for the pentameric clusters.

Cluster	E_R	E_S
GIS-1	30.5	−1.6
GIS-2	32.0	−0.4
GIS-3	0.0	−8.6
GIS-4	1.7	−7.4
BIK-1	47.6	8.7
BIK-2	52.6	14.3
BIK-3	43.4	5.9
BIK-4	52.6	14.5
BIK-5	7.3	−20.0
BIK-6	11.8	−3.6
OFF-1	20.1	−6.4
OFF-2	37.2	5.2
FER-1	82.5	44.1
FER-2	12.6	−13.4
FER-3	78.8	45.6
FER-4	40.8	8.4
MOR-1	49.2	16.2
MOR-2	51.3	17.5
MOR-3	34.1	6.8
MOR-4	38.1	6.9
MFI-1	59.4	23.3
MFI-2	45.2	12.9
MFI-12	41.5	10.8

Consequently, the correlation energy and errors introduced by the utilization of the Hartree–Fock model and small basis sets should be comparable for all clusters.

Charge Distributions. Figure 1.6 illustrates the distribution of the negative charge generated by the replacement of Si by Al in the pentamers centered on site T12 of the ZSM-5. The atomic charges of the Si, Al, O, and H atoms are small compared to their formal oxidation numbers, indicating the strongly covalent character of the zeolite framework. The first histogram (Fig. 1.6a) shows the repartition of the excess negative charge

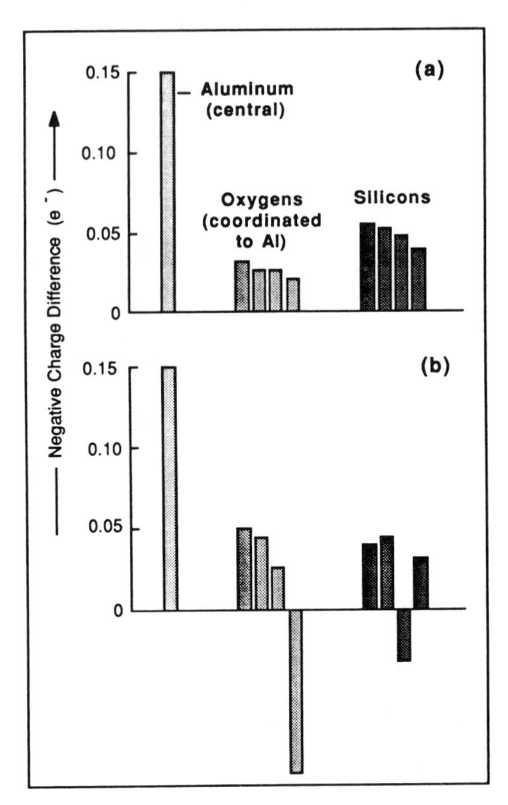

Figure 1.6 STO-3G charge distributions in the T12-centered pentameric cluster for ZSM-5. (a) Distribution of the framework negative charge in Al-containing cluster with no compensating proton (net charge = −1). (b) Negative charge difference between the cluster with Si at the central site and the (Al,H) corresponding cluster.

present in the pentamer when Si is replaced by Al. Most interestingly, this charge is spread over the whole cluster, with 0.15 e^- being retained by the central Al atom and 0.44 e^- being delocalized over the central AlO_4Si_4 unit. This conclusion is consistent with the fact that the base associated to a zeolitic Brønsted site is very weak. Such a high degree of delocalization of the negative charge, which cannot be explained simply on the basis of electronegativity differences, is not surprising, however. Indeed, since the total number of electrons is kept fixed when Si is substituted by Al, only the positive charge at the Al site is decreased, and it is thus expected that the negative charge will spill away from this site, since it has a smaller nuclear attraction potential.

The effect of cluster neutralization by a proton has been examined by adding a proton onto an oxygen that bridges two T-sites. Addition of the proton decreases by 0.120 e^- the negative charge of the oxygen to which it is bound and of all Si atoms (in particular by 0.078 e^- if bound to the oxygen bearing the proton). It also relocalizes the negative charge on the other oxygens. The net proton charge has a value of +0.294 e^-, only.

Finally, one can also compare the charge densities of the neutral Si- and (Al, H)-containing pentamers. Replacement of Si by (Al, H) increases the negative charge of the central T-site, of the central TO_4 unit, and of the TO_4T_4 unit by 0.138, 0.161, and 0.243 e^-, respectively (Fig. 1.6b). We therefore conclude that the framework negative charge (due to Al) remains largely delocalized in the presence of the proton, the charge of the central TO_4 unit being only one-half of the proton charge. Hence, delocalization of the proton is substantiated by the theoretical calculations, which agrees with experiment.

The strong acidic character of zeolites is then explained not only by the polar character of the bridging hydroxyl group but also by the ability of the aluminosilicate framework to accept and delocalize easily the negative charge(s) generated by framework aluminums. In other words, the (negative) zeolite framework acts as a weak but soft base. This conclusion can be invoked to justify the cation selectivities of zeolites and, in particular, their affinity for large, soft, and complex cations.

Pairing of Aluminum in the Zeolite Framework. Some authors[21,70] have felt the need to question the validity of the "aluminum avoidance rule," also known as "Loewenstein's rule."[71] This rule states that in aluminosilicate tetrahedral frameworks no oxygen should be linked to two Al atoms when Si is available.

Mordenite. Four-membered ring clusters (Fig. 1.1b) were used to investigate the Al pairing in the mordenite structure. These clusters are designated by tetra-nAl(x,y)-mH(p,q), n being the number of Al atoms sited at positions x and y and m being the number of charge-compensating protons bound to the oxygen atoms located at given positions p or q or both.

As proposed previously, one can calculate the energy variations for hypothetical Si–Al exchange processes involving one or more T-positions [Eq. (1.2)] and the possible addition of the corresponding charge-compensating protons. As in the pentamer cluster calculations, the reacting mono-X-Al entities replace Si by Al at sites TX leading to mono-X-Si clusters. Exchange energies E_S are compiled in Table 1.6, in which the reactions are identified by the number of the cluster generated in the exchange processes.

The first and most striking observation is the profound effect of the protons that compensate the anionic framework charge. The reaction is endothermic for clusters containing one or two Al atoms when no proton is present; it is exothermic in all other cases. Indeed, when the cluster contains one Al only, the presence of the proton stabilizes the system by about 380 kcal mol^{-1} if it is located on an oxygen bound to the Al atom (R-3 to R-3-9, R-4 to R-4-10), but by 332 kcal mol^{-1} only when linked to

Table 1.6. Energy variation, E_S (kcal mol^{-1}), upon substitution of Si by Al in different clusters modeling four-membered rings of mordenite.

Number	Cluster	Charge	E_S
R-3	Tetra-1Al(3)-OH	−1	1.5
R-4	Tetra-1Al(4)-OH	−1	1.9
R-34	Tetra-2Al(3,4)-OH	−2	97.3
R-34-10	Tetra-2Al(3,4(1))-OH	−2	69.0
R-3-9	Tetra-1Al(3)-H(9)	0	−379.3
R-3-10	Tetra-1Al(3)-H(10)	0	−330.4
R-4-10	Tetra-1Al(4)-H(10)	0	−377.1
R-34-910	Tetra-2Al(3,4)-H(9,10)	0	−733.8
R-341-910	Tetra-2Al(3,4(1))-H(9,10)	0	−771.6
R-331-910	Tetra-2Al(3,3(1))-H(9,10)	0	−756.3
R-441-910	Tetra-2Al(4,4(1))-H(9,10)	0	−757.2
R-341-9	Tetra-2Al(3,4(1))-H(9)	−1	−384.2
R-341-10	Tetra-2Al(3,4(1))-H(10)	−1	−382.2

the opposite oxygen atom (R-3 to R-3-10). This indicates that the "delocalization" of the proton will be restricted essentially to the immediate environment of the Al site, if the latter is strictly isolated. When the cluster contains two diagonally opposed Al-sites, the addition of the first proton stabilizes the cluster by about 452 kcal mol^{-1} (R-341 to R-341-9 or R-341-10), while the second proton addition lowers the energy further by about 388 kcal mol^{-1} (R-341-910). Nearly identical E_S values for the clusters R-341-9 and R-341-10, which are mononegatively charged and differ by the proton position only, suggest that the remaining proton is probably able to circulate rather freely over the four-membered ring oxygens.

Second, as already concluded from the pentamer cluster calculations, it is confirmed that equivalent configurations that have Al in site 3 or site 4 (R-3 and R-4, R-3-9 and R-4-10, R-341-9 and R-341-10, R-331-910 and R-441-910) have comparable substitution energies, a slight preference always being given to the siting of Al at T3 or, when two Al are present, to the binding of the charge-compensating proton to O9. This observation suggests that the stability of Al at a given T-site is mainly determined by its immediate environment.

Finally, our results are in agreement with the "aluminum avoidance rule." Neutral ring clusters that have a direct Al–O–Al linkage are less stable by 14 kcal mol^{-1} (R-331-910) or 38 kcal mol^{-1} (R-34-910) than the cluster having the aluminum atoms diagonally opposed in the four-membered ring (R-341-910).

ZSM-5. Assuming that Loewenstein's rule holds, siting of Al at sites T12 and T2 allows a maximum of 12 Al per unit cell, corresponding to a lower Si/Al limit of 7, because of the T2 and T12 sites connection pattern. Occupation by Al of sites T12 and T2 can lead to the formation of Al–O–Si–O–Al pairs somewhat analogous to those formed by diagonal pairing of Al in the mordenite four-membered rings. The size of the preceding arrangements (eight T-sites) does not allow calculations on topologically closed clusters as performed in the case of mordenite. Therefore, we have considered two linear pentameric chains, the *cis* chain (T12-T12*-T3-T2*-T8) connecting T-sites located along the boundary of a given channel and the *trans* chain (T12-T12*-T8-T2*-T3) connecting sites that bridge two adjacent channels (see Fig. 1.2). It is worth noting that these pentamers contain the same sites but with different arrangements; they can therefore be considered as isomers, which enables a more reliable comparison of their calculated properties. Al atoms, when replacing Si, were located at the T-sites indicated by an asterisk in order to minimize the impact of boundary effects. The substitution energies E_S are listed in Table 1.7.

Table 1.7. Energy variation, E_S (kcal mol^{-1}), upon substitution of Si by Al in "linear" clusters of ZSM-5.

Number	Cluster	Charge	E_S	E_E
(a) *Cis*				
C-12	*cis*-1Al(12(2))-OH	−1	7.6	0.0
C-2	*cis*-1Al(2)-OH	−1	8.2	0.6
C-122	*cis*-1Al(12(2),2)-OH	−2	68.7	61.1
C-12-24	*cis*-1Al(12(2))-H(24)	0	−369.1	−376.7
C-2-2	*cis*-1Al(2)-H(2)	0	−377.3	−384.9
C-122-24	*cis*-2Al(12(2),2)-H(24)	−1	−354.1	−361.7
C-122-2	*cis*-2Al(12(2),2)-H(2)	−1	−389.6	−397.2
C-122-242	*cis*-2Al(12(2),2)-H(24,2)	0	−752.0	−759.6
(b) *Trans*				
T-12	*trans*-1Al(12(2))-OH	−1	−2.4	0.0
T-2	*trans*-1Al(2)-OH	−1	5.0	7.4
T-122	*trans*-1Al(12(2),2)-OH	−2	65.3	67.7
T-12-24	*trans*-1Al(12(2))-H(24)	0	−373.6	−371.2
T-2-13	*trans*-1Al(2)-H(13)	0	−355.4	−353.0
T-2-2	*trans*-1Al(2)-H(2)	0	−375.3	−372.9
T-122-24	*trans*-2Al(12(2),2)-H(24)	−1	−355.1	−352.7
T-122-13	*trans*-2Al(12(2),2)-H(13)	−1	−368.7	−366.3
T-122-2	*trans*-2Al(12(2),2)-H(2)	−1	−364.4	−362.0
T-122-2413	*trans*-2Al(12(2),2)-H(24,13)	0	−731.1	−733.5
T-122-242	*trans*-2Al(12(2),2)-H(24,2)	0	−744.6	−742.2

As noted from the ZSM-5 tetrahedral pentamer and the mordenite tetramer cluster calculations, the reaction is endothermic for linear clusters containing one or two Al atoms without any proton (except for T12); it is exothermic in all other cases. Indeed, when the cluster contains one Al only, the presence of the proton stabilizes the system by about 375 kcal mol^{-1}. When the clusters contain two Al atoms, addition of the first proton stabilizes the system by about 440 kcal mol^{-1}, whereas addition of the second proton lowers the energy further by about 380 kcal mol^{-1}. When comparing the E_S values, energy differences between the *cis* and *trans* clusters may not be physically meaningful if no account is made of their finite size. For a better comparison between the two models, we have listed in the second column of Table 1.7 the corrected exchange energies E_E, which are the differences between E_S for the various clusters and the substitution energy of the C12 and T12 clusters. These data show that the substitution of a second Si by Al in the site T2 when the T12 is already occupied by an Al atom is favored in the *cis* chain, i.e., when the pairing of Al takes place in the same channel.

Ferrierite. The T2 and T4 sites, favored for the substitution of Si by Al, are located along the boundary of the six-membered channels. Keeping in mind that the ferrierite unit cell contains 36 T-sites distributed as 16 T1, 8 T2, 8 T3, and 4 T4 sites, the siting of Al at sites T2 and T4 in accordance with Loewenstein's rule allows at maximum 6 Al per unit cell. It corresponds to a lower Si/Al ratio limit of 5, because of the T2-T4 connection pattern. Such a value agrees with the experimental compositional range of ferrierite. In order to investigate the problem of Al pairing in the ferrierite framework, calculations were performed on six-membered ring (six-ring) clusters (Fig. 1.3) with different Al siting and pairing situations. The six-ring consists of four T2 and two T4 sites.

The values of E_S reported in Table 1.8 show that equivalent configura-

Table 1.8. Energy variation, E_S (kcal mol^{-1}), upon substitution of Si by Al in six-membered ring clusters of ferrierite.

Number	Cluster	Total Charge	E_S
R-2	Hexa-1Al(2)	−1	8.6
R-4	Hexa-1Al(4)	−1	5.8
R-222	Hexa-2Al(2,2(2))	−2	64.8
R-441	Hexa-2Al(4,4(1))	−2	60.7
R-221	Hexa-2Al(2,2(1))	−2	73.7
R-241	Hexa-2Al(2,4(1))	−2	70.9
R-240	Hexa-2Al(2,4)	−2	111.2
R-22141	Hexa-2Al(2,2(1),4(1))	−3	191.3

Table 1.9. Relative stabilities, E_R (kcal mol^{-1}), and Al–Al distance (nm) for the different di-substituted six-membered ring clusters of ferrierite.

Number	Al–Al Distance	E_R
R-222	0.66	0.0
R-441	0.62	8.7
R-221	0.57	9.0
R-241	0.55	12.6
R-240	0.32	52.9

tions with Al in site T2 or site T4 have comparable substitution energies, a slight preference being given to the siting of Al at T4. It is also worth noting that the di-substitution is favored in the *para* position relative to the *meta* position by about 10 kcal mol^{-1}. The di-substitution in *ortho* position is much less stable (by about 50 kcal mol^{-1}). This result is in agreement with the "aluminum avoidance" principle. Tri-substitution (R-22141) is destabilized by about 130 kcal mol^{-1} compared to the preferential di-substitution (R-441). The general trend is readily rationalized in terms of electrostatic repulsions. Table 1.9 lists the relative stability energies, E_R, of the di-substituted hexamers and the corresponding Al–Al distances. The relative stability energy of a di-substituted hexameric cluster is calculated as the difference between its total energy and the total energy of the most stable di-substituted hexamer (R-222). The data show that the relative stability energies are negatively correlated to the Al–Al distances, the correlation coefficient being equal to −0.988.

Conclusions

From the present restricted Hartree–Fock calculations on model clusters, selected to represent the real geometry and environment of tetrahedral framework sites of selected zeolites, we conclude the following:

1. The calculations predict the siting preference of aluminum for the various zeolites in agreement with the available experimental facts. In particular, the predicted siting is consistent with the observed Si/Al ratio for these zeolites. The present calculations further indicate that diagonal pairing is favored relative to the *meta* or *ortho* pairings in accordance with electrostatic principles and Loewenstein's aluminum avoidance rule.

2. The analysis of the atomic charge distribution in the clusters indicates the highly covalent nature of aluminosilicate frameworks. The framework negative charge, associated with the presence of structural Al, is largely delocalized in the presence of the charge-compensating proton and even more so upon proton abstraction. Hence, the anionic framework of a zeolite must be considered as a weak but soft base. The high acid strength of zeolites therefore stems, at least partially, from the ability of their framework to accommodate and to delocalize easily such negative charges.

STATISTICAL MECHANICS AND MOLECULAR DYNAMICS SIMULATIONS APPLIED TO A WATER–FERRIERITE SYSTEM

Water–Zeolite Interaction Studies

The role of water molecules in zeolites is twofold[2]: to complete the coordination of the cations present in the pores and to minimize the electrostatic repulsion between the framework oxygens. The amount of adsorbed water depends essentially on the Si/Al ratio of the zeolite. Indeed, aluminum atoms substituting for silicon imply the presence of a framework negative charge counterbalanced by cations or protons.

In heterogeneous catalysis, as in many other fields such as electrochemistry or biochemistry, a precise understanding at the microscopic level has become necessary. In that sense, computer experiments such as statistical mechanics and molecular dynamics simulations have increasingly shortened the gap between the theoretical descriptions and the experimental characterizations of liquid–liquid, liquid–solid, or gas–solid interactions. Some pertinent examples are the studies of water structuring and water–ion–biomolecule interactions by Clementi and coworkers,[23,24,57,108] Jorgensen and Gao,[54] Mezei and coworkers,[80,89] Wong and McCammon,[109] Kollman and coworkers,[9,97] and Karplus and coworkers.[36,87] Similarly, modeling zeolites may provide information on the structure and properties of their bulk and surface sites, some of which are not easily accessible by experimental determinations, and thereby explain the physicochemical phenomena related to adsorption.

An early attempt to evaluate thermodynamic properties of molecules adsorbed in a zeolite structure using a Monte Carlo (MC) experiment was made by Stroud et al.[101] In their work, a MC simulation of the canonical ensemble was performed to compute the heat capacity, the isosteric heat of adsorption, and adsorption isotherms of methane, between 194 and 300 K, in a Linde Sieve 5A. Electrostatic forces were neglected, the methane–methane interactions were described by Lennard-Jones parameters, while the adsorbate–adsorbent interactions were derived from a combined Lennard-Jones and Devonshire model. Similar computations were extended to alkanes–NaCaA systems by Kretschmer and Fiedler.[63] A lattice sum over nearest neighbors was applied to calculate both dispersion and electrostatic contributions to the potential energy of the adsorbed molecules. However, the simulations were restricted to one molecule per cavity. MC simulations of the grand canonical ensemble were also performed by Soto and Myers[100] to generate adsorption isotherms, adsorption heats, radial distributions, and average number of krypton atoms in an MS-13X cavity. Their model based on hard spheres and Lennard-Jones potentials included both dispersion and electrostatic contributions for the gas–solid interaction energy. Heterogeneity of the zeolite cavity and interatomic forces between the adsorbate atoms were found to be equally important in determining the adsorption properties. A recent paper by Kono and Takasaka[62] describes isotherms and isosteric heats of Ar and N_2 in dehydrated zeolite 4A where the interaction potentials between the adsorbate and the adsorbent are a combination of the London dispersion energy and a point charge model for polarization energy. The effect of three-body interactions is examined as the work of Takaishi[103] concerning statistical thermodynamics calculations of a one-dimensional gas adsorbed in zeolitic pores.

To our knowledge, the first molecular dynamics (MD) study of a fluid characterized by a continuous interaction potential function was realized by Rahman[88]; he showed that the motion equations for several hundred particles (Ar atoms) can be integrated directly. The application of such simulation to silicates is recent. Mulla et al.[82,83] analyzed the behavior of rigid water molecules interacting with silicate surfaces, through a Lennard-Jones potential, as a function of their distance from the surface. Coulombic interactions between the surface oxygens and the water molecules were omitted in order to study the influence of the short-range hard-core repulsive forces separately from the long-range coulombic ones. Demontis et al. investigated water structure interacting with a phyllosilicate structure[26] and more recently with natrolite[27–30] using an interaction potential based on *ab initio* calculations. They also simulated IR spectra by Fourier transformation of dipole autocorrelation functions. The dynamics of cations inside a zeolitic framework has also been described by MD simulations as proposed by Shin et al.[96] in their work on Na atoms within a zeolite A.

Meaningful statistical mechanics or molecular dynamics simulations or both necessitate, in our opinion, the following strategy: (i) choose or determine the potentials between all the interacting species, (ii) test the quality of the interaction potentials by energy profiles or isoenergy maps calculations, and (iii) perform MC and MD simulations based on the previously cited potentials in order to evaluate thermodynamic properties, significant adsorbate repartitions, or diffusional behavior in the adsorbent or all three.

In the following sections we apply this three-point strategy to simulations of the water distribution and behavior within a ferrierite-type zeolite model in order to understand the interaction between the water molecules and the zeolite channel at the molecular level. Particularly, in relation with previous studies on the characterization of structural and thermodynamic properties of zeolites,[34] we detail and complete some statistical mechanics simulations described previously.[65,66] We discuss (i) structural and thermodynamic results; (ii) ensemble average positions and probability densities computed through our MC simulations; and (iii) time autocorrelation functions, self-diffusion coefficient, linear velocity time autocorrelation functions, libration angle, and dipole–dipole autocorrelation functions computed through our MD simulations.

Water–Ferrierite Model and Interaction Potentials

The starting point for a computer experiment is the development of a mathematical model based on physical laws that idealizes the chosen system.

Water–Ferrierite Interaction Potential. Atomic contributions are considered in order to take into account the heterogeneity of the zeolitic channel. The total interaction energy U is limited to a two-body term:

$$U = \sum_{ij} U_{ij} \tag{1.3}$$

where i and j denote water atoms and zeolite atoms, respectively. Each contribution U_{ij} is composed of a dispersion, a repulsion, and an electrostatic interaction energy term:

$$U_{ij} = -U_{ij\,D} + U_{ij\,R} + U_{ij\,E} \tag{1.4}$$

Dispersion Energy U_D. Only the first-order terms corresponding to the induced dipole–dipole interactions are considered. We have adopted the Kirkwood and Müller formula[58,84] as proposed by Salem,[91] which involves experimentally measured or computable parameters:

$$U_D = \sum_{ij} (6mc^2\alpha_i\alpha_j)/[(\alpha_i/\chi_i) + (\alpha_j/\chi_j)]r_{ij}^6 = \sum_{ij} A_{ij}/r_{ij}^6 \tag{1.5}$$

where m is the electron mass, c is the velocity of light in vacuum, r_{ij} is the interatomic distance between atoms i and j, α_i and α_j are their polarizabilities, and χ_i and χ_j are their diamagnetic susceptibilities. This formula has been used widely for the particular calculation of potential interaction energies of various molecules in several zeolites.[11,12,19,25,59,60,73]

Electrostatic Energy U_E. As for the dispersive contribution, this term is also computed within the pairwise approximation between all water molecule atoms and all zeolite atoms:

$$U_E = \sum_{i=j} C_{ij} q_i q_j / r_{ij} \tag{1.6}$$

where q_i and q_j are the Mulliken *ab initio* STO-3G atomic charges (Table 1.10) determined for significant zeolite pentamers.

Repulsion Energy U_R. The repulsion term is evaluated as

$$U_R = \sum_{ij} (A_{ij}/2)(r_i + r_j)^6 / r_{ij}^{12} = \sum_{ij} B_{ij} / r_{ij}^{12} \tag{1.7}$$

where r_i and r_j are the van der Waals radii of water and lattice atoms, respectively. The B_{ij} constants are determined by considering the potential energy of a water atom i facing an isolated lattice atom j and placed in its stable equilibrium between attractive and repulsive forces at the distance $r_e = r_i + r_j$, where the first derivative of the potential vanishes automatically. The electrostatic contribution is neglected, since its consideration in the calculation of the repulsion constants leads to unrealistic values for B_{ij}.[12]

The values of the polarizabilities α, of the diamagnetic susceptibilities χ, and of the van der Waals radii r of the water and zeolite atoms are listed in Table 1.11. The A_{ij} and B_{ij} constants are listed in Table 1.12; $C_{ij} = C$ is equal to 1 in the cgs unit system. Equation (1.4) for the potential energy

Table 1.10. STO-3G net charges of the lattice atoms and water atoms (in units of the proton charge); the positions of the corresponding zeolite atoms are shown in Figure 1.7.

Si_1	(T1)	1.5634	O_1	−0.7267	O_7	−0.7464
Si_2	(T2)	1.4247	O_2	−0.6786	O_8	−0.7936
Si_{3c}[a]	(T3)	1.5812	O_3	−0.7416	O_9	−0.7060
Si_{3nc}[a]	(T3)	1.5257	O_4	−0.6996	O_{10}	−0.6195
H		0.2755	O_5	−0.7110	O_{water}	−0.6802
Al	(T4)	1.3638	O_6	−0.7201	H_{water}	0.3401

[a] The index c stands for the site T3 directly connected to an oxygen atom linked to a proton H^+. In the other case, the site is indexed T_{3nc}.

Table 1.11. Polarizabilities (α in $cm^3\ mol^{-1}$), diamagnetic susceptibilities (χ in $10^{-6}\ cm^3\ mol^{-1}$), and van der Waals radii (r in Å) for the water and zeolite atoms.

	Water		Zeolite			
	O	H	Si	Al	O	H
α	0.3612[a]	0.16106[a]	0.01204[b]	0.00993[c]	0.994[b]	0.259[d]
$-\chi$	7.09566[e]	0.9379[e]	1.0[b]	1.0	12.58[b]	2.0[d]
r	1.40	1.10	0.42	0.50	1.40	1.10

[a] Rhee et al.[90] [b] Mayorga and Peterson.[73] [c] Bosacek and Dubsky[19] [d] Kiselev et al.[59]
[e] The diamagnetic susceptibilities are computed using the atomic polarizabilities.[59,60]

describing an adsorbed molecule interacting with the zeolitic framework thus becomes

$$U = -\sum_{ij} A_{ij}/r_{ij}^{6} + \sum_{ij} B_{ij}/r_{ij}^{12} + \sum_{ij} Cq_i q_j/r_{ij} \qquad (1.8)$$

The use of the proposed Kirkwood–Müller parameters in order to estimate the interaction energies of various species with zeolites calls for the following remarks. As the values for the polarizabilities, the susceptibilities, and the van der Waals radii for the Si and Al atoms are relatively low (Table 1.11), their influence was usually neglected by most authors.[6,7,11,12,20,25,59] The potentials based on the Kirkwood–Müller parameters are also found to overemphasize the attractive term as compared, for example, to the London dispersion term.[6,7,62] The Henry constant is also overestimated as calculated by Bezus et al.[13] for the system methane–zeolite A.

Water–Water Interaction Potential. The chosen water–water interaction potential is described by the well-tested Matsuoka–Clementi–Yoshimine (MCY) potential[72] adapted from nonempirical CI calculations

Table 1.12. Calculated constants A_{ij} (kcal mol^{-1} Å^6) and B_{ij} (kcal mol^{-1} Å^{12}).

i	j	A_{ij}	B_{ij}	i	j	A_{ij}	B_{ij}
O	H	101.2108	12354.8392	H	H	27.0283	1532.2324
O	Si	13.4848	245.0430	H	Si	2.0596	12.7002
O	Al	11.5073	270.6863	H	Al	1.7184	14.4148
O	O	539.3729	129959.2894	H	O	124.6165	15211.9763

on water dimers. This formalism has been considered previously in the study of aqueous solutions of many important biological molecules, such as proflavin,[56] gramicidin A,[38,57] and DNA.[24]

The MCY water–water potential is known to give relatively good agreement with experimental radial distribution functions in simulations carried out at experimental density.[67] Errors arise in the neglect of many-body effects, estimated to be a 13% error in computed internal energies for liquid water, and lead to a too high calculated pressure indicating deficiencies in the representation of the potential curvature. This latter problem is controlled by working at experimental densities in the so-called (N, V, T) ensemble simulations.[53]

Ferrierite Model. The crystallographic structure of a ferrierite-type zeolite[106] consists of a two-dimensional system of intersecting channels: ten-membered ring channels (4.3–5.5 Å) along the c axis (Fig. 1.7a) and eight-

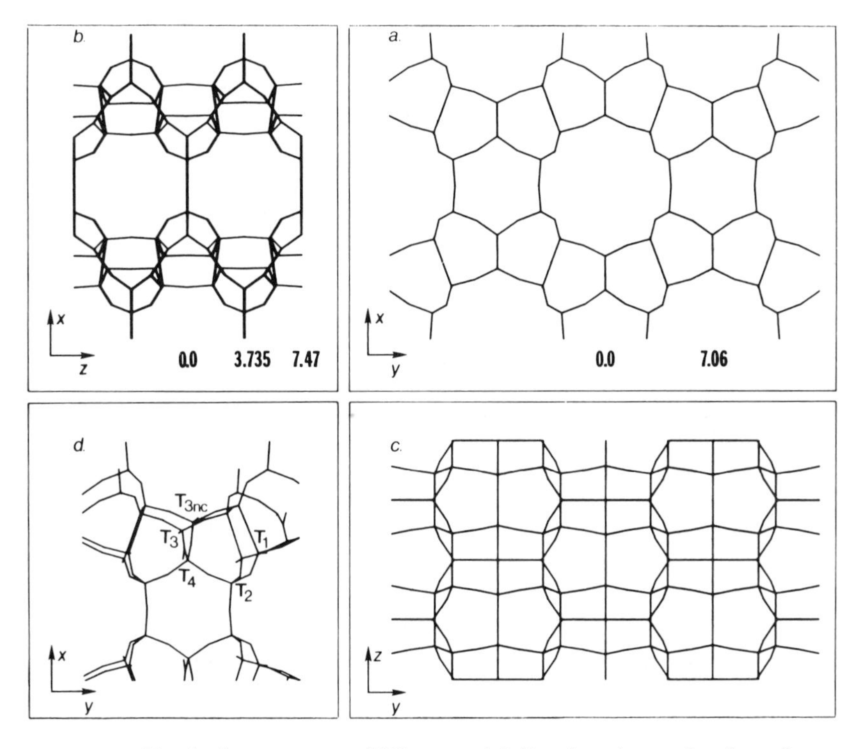

Figure 1.7. Ferrierite structure (522 atoms) following the projection planes (a) xy showing the 10-T and 6-T membered ring channels, (b) xz showing the 8-T ring channels, and (c) yz; (d) description of the labels for each ferrierite tetrahedral site.

membered ring channels (3.4–4.8 Å) parallel to the b axis (Fig. 1.7b). Ferrierite crystallizes in the orthorhombic system with the *Immm* space group (a = 19.15 Å, b = 14.12 Å, and c = 7.47 Å). One unit cell contains 36 tetrahedral sites whose symmetries are T_1 (1), T_2 and T_3 (m), and T_4 (mm) (Fig. 1.7d). This porous system has a high hydrophilic behavior[47] relative to other zeolites with a higher Si/Al ratio. In our mode simulated by 1250 atoms, the set Si/Al ratio is equal to 8. The aluminum atom distribution has been fixed as previously determined with the help of *ab initio* calculations[40] showing that silicon atoms are substituted preferentially by aluminum at sites $T4$. To cancel the negative charges introduced by these substitutions, protons are linked to the oxygens adjacent to the aluminum atoms located along the eight-membered ring channels (Fig. 1.7b). The dimensions of our model are: $-9.57 \leq X \leq 9.57$ Å, $-11.39 \leq Y \leq 11.39$ Å, and $-21.10 \leq Z \leq 21.10$ Å. The X, Y, and Z axes are associated with the a, b, and c unit-cell parameters.

Isoenergy Maps

Isoenergy maps are used to check the interaction potential and to locate the stabilization sites in a first noncooperative approach. Some authors also used the information these maps provide to study translational and rotational diffusion of the molecules interacting with the zeolite framework.[85]

The position of the interaction sites within ferrierite can be approximated by using one water molecule as a probe in the channel. The interaction energy is computed for an optimally oriented water molecule with its oxygen atom placed at the grid points on a predetermined plane (rotation displacement is 30°). This data set is used to obtain the isoenergy maps analyzed below.

The positive energies (repulsive energies) are not shown (the contour corresponding to 0 kcal mol^{-1} excepted), since we look for the most stable positions of water within the framework. These positions correspond to adsorption sites if the absolute value of the interaction energy is greater than the water–water interaction energy, 6 kcal mol^{-1}.[38] Isoenergy maps for XY cross sections of the channel (grid meshes are 0.33 Å) for Z = 0.0, 1.745, and 3.735 Å are represented in Figure 1.8. The calculations are performed with the 1250 zeolite framework atoms. First, we observed that the center of the ten-membered ring channel does not correspond to an equilibrium and stable position as shown by first-principles van der Waals interactions.[35] Isoenergy maps for XZ cross section in the main ten-membered ring channel (Y = 0.0 Å) and in the six-membered ring channel (Y = 7.06 Å, corresponding to the Y coordinate of the protons) are shown in Figure 1.9. As in Figure 1.8 particular stable positions are localized in

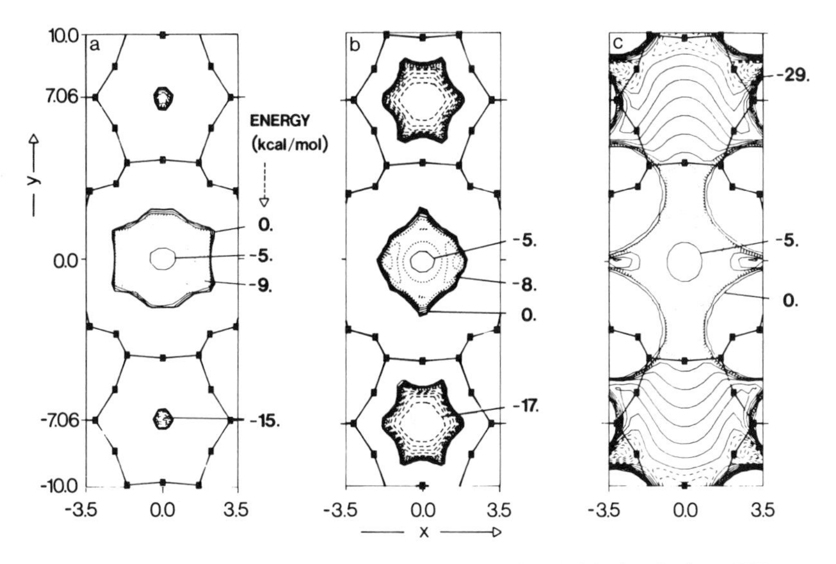

Figure 1.8. Isoenergy maps for water interacting with ferrierite; XY cross sections (Å); plane defined by (a) $Z = 0.0$ Å, dashed lines from -15 to -11 ($\Delta = 2$), dotted from -9 to -6 ($\Delta = 1$), full from -5 to -1 ($\Delta = 2$) kcal mol^{-1}; (b) $Z = 1.745$ Å, dashed lines from -17 to -9 ($\Delta = 2$), dotted from -8 to -6 ($\Delta = 2$), full from -5 to 0 ($\Delta = 1$) kcal mol^{-1}; and (c) $Z = 3.735$ Å, dashed lines from -29 to -23 ($\Delta = 1$), full from -22 to -10 ($\Delta = 2$), dotted from -9 to -6 ($\Delta = 1$), full from -5 to 0 ($\Delta = 1$) kcal mol^{-1}.

the main channel near the negative AlO_4^- region and on either side of the H^+ in the eight-membered ring channel.

To take into account the surroundings and to cancel the border effects resulting from a finite size box, minimal image conditions[105] are often used to introduce the periodicity of the system. When periodic boundary conditions are employed, the pattern of the isopotential energy maps is changed rather substantially as shown by the maps corresponding to $Z = 3.735$ Å (compare Figs. 1.10 and 1.8c). The positions corresponding to the more attractive sites are nearly the same but the energetic values are considerably lowered by 3–10 kcal mol^{-1} indicating important effects of the infinite lattice on the energetic values. This stresses the importance of taking into account long-range effects.

Monte Carlo Simulations of Water Interacting with Ferrierite

In order to obtain a more realistic representation of the structure and interactions of many water molecules with ferrierite, we must introduce

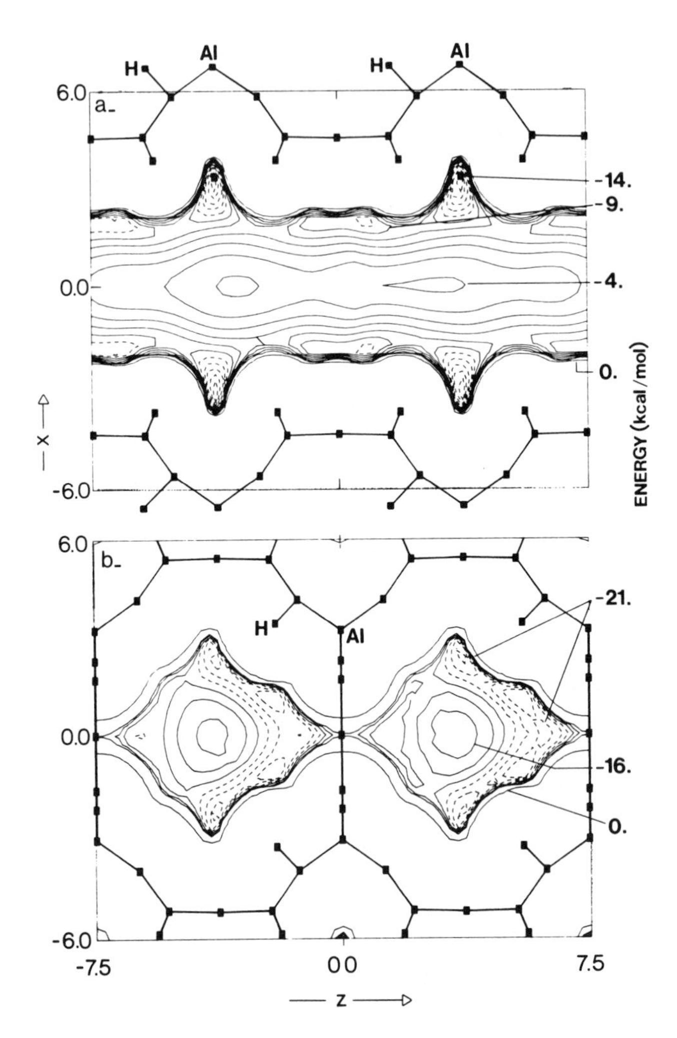

Figure 1.9 Isoenergy maps for water interacting with ferrierite; *XZ* cross sections (Å); plane defined by (a) $Y = 0.0$ Å, dashed lines from -21 to -19 ($\Delta = 1$), full from -18 to -16 ($\Delta = 1$) kcal mol^{-1} and (b) $Y = 7.06$ Å (*Y* coordinate of the protons), dashed lines from -14 to -10 ($\Delta = 1$), full from -9 to -4 ($\Delta = 1$) kcal mol^{-1}.

temperature effects by considering explicitly distributions of Boltzmann's weighted configurations and interactions of the water molecules not only with the zeolite framework but also among themselves. This goal is reached by performing MC simulations.

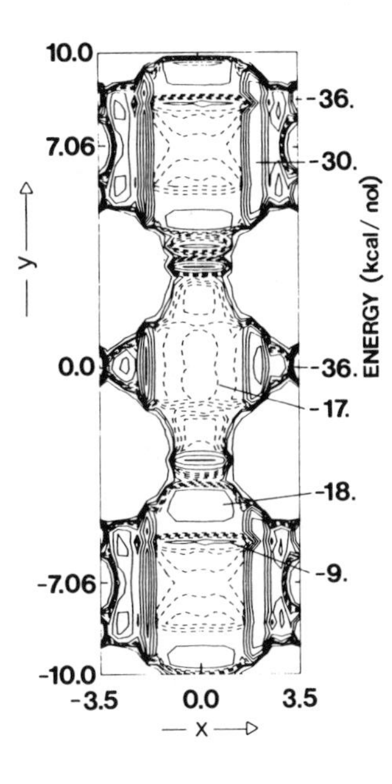

Figure 1.10 Isoenergy map for water interacting with ferrierite using periodic boundary conditions; XY cross section (Å); plane defined by $Z = 3.735$ Å, full lines from -28 to -16 ($\Delta = 4$), dashed from -14 to -11 ($\Delta = 1$), full from -9 to -7 ($\Delta = 2$) kcal mol^{-1}.

Methodology. Reference to a MC method generally implies the use of random sampling techniques to estimate averages by evaluating integrals. The MC method has been introduced by Metropolis et al.[79] It is a very efficient importance sampling technique particularly designed for statistical mechanics involving generations of a chain of configurations of a many-body system and for which the probability of a particular configuration of energy U is proportional to the Boltzmann factor $\exp(-U/kT)$. The average of any function X over the chain of configurations gives an estimation of the canonical average of that function, as formally stated in the following equation:

$$\langle X \rangle = \int X(q^N) \exp[-U(q^N)/kT] dq^N \Big/ \int \exp[-U(q^N)/kT] dq^N \quad (1.9)$$

where q^N is the number of coordinates describing the system.

A crude approach would be to estimate the canonical average by selecting a large number of configurations randomly and averaging the function X over these configurations, weighting each configuration by the Boltzmann factor. However, this method requires the generation of too large a number of configurations to sample the system, while importance sampling increases the convergence rate of the function X by only selecting the configurations with a sufficient weight in the average calculation. This method consists in evaluating averages of Eq. (1.9) as

$$\langle X \rangle = \frac{1}{M} \sum_{M} X(q^N) \tag{1.10}$$

where M is the number of configurations used to compute the average. Each configuration is obtained from a previous one by random selection of one water molecule and by its random displacement; the maximum allowed translation distances, 0.3 Å, and rotation angles, 18°, are multiplied by random numbers between −0.5 and 0.5. The generated configuration is accepted only if the energy difference ΔU between two consecutive states is negative or if the new state has a sufficiently high occurrence probability [$\exp(\Delta U/kT)$ must be greater than a random number chosen between 0 and 1]. The process is repeated several thousand times to form a sequence of states, and the properties are averaged over the successive states after their convergence.

Simulation Conditions. In the following simulations, the temperature has been set to 298 K. Within a cylindrical volume coaxial to the zeolite main channel, we placed a number (N) of water molecules, varying from 40 to 80, in order to reproduce a weight percentage of water adsorbed in ferrierite varying between 7 and 13%. The radius and length of the cylinder were chosen as 9.28 and 37.35 Å (five times the c parameter of the unit cell). The water molecules, supposed to be rigid (HOH angle = 109.5°, OH distance = 0.9572 Å), were constrained to move randomly within the cylindrical volume of the fixed zeolite framework. For each density, we reported the number of MC steps generated to reach equilibration, the number of configurations used to analyze the energetic and structural results, the percentage of accepted movements after equilibrium, and the corresponding CPU time (Table 1.13). Usually, in MC experiments, the percentage of accepted configurations is set around 50%.[5]

Table 1.13. Simulations conditions for the various MC runs for each water density.

N	Number of MC Steps for Equilibration	Number of MC Steps for Analysis	Percentage of Accepted Movements	FPS or IBM CPU Time[a] (hr)
40	2.0×10^5	1.5×10^5	50.5	4.90 (FPS)
50	6.0×10^5	1.5×10^5	49.6	14.77 (IBM)
65	4.5×10^5	2.5×10^5	48.0	7.30 (FPS)
80	2.0×10^5	2.5×10^5	46.6	7.45 (FPS)

[a] Computation time on two coupled IBM 4341 computers connected to a FPS 164 attached processor.

Table 1.14. Water–water, W–W, zeolite–water, Z–W, total *U* internal energies (in kJ mol^{-1}), and average number of water molecules per unit cell computed throughout the MC runs for each density; all values are reported per water molecule.

N	W–W	Z–W	*U*	Number Molecules/Cell
40	−11.70 ± 1.62	−63.14 ± 2.65	−74.84 ± 2.39	5.88
50	−22.02 ± 1.77	−50.79 ± 1.28	−72.80 ± 1.70	7.42
65	−18.50 ± 0.92	−59.71 ± 1.37	−78.21 ± 1.33	10.17
80	−19.79 ± 1.92	−55.98 ± 1.46	−75.93 ± 1.30	11.95

Thermodynamic Results. To avoid the influence of border effects, the various interaction energies were computed considering the statistical configurations generated by taking into account the total number of water molecules introduced in the cylinder but considering the two central cells only ($-9.57 \leq X \leq 9.57$ Å, $-7.06 \leq Y \leq 7.06$ Å, and $-7.47 \leq Z \leq 7.47$ Å). The configurational internal energies and heat capacities computed throughout the analysis part of the MC run are reported in Table 1.14.

The experimental water–ferrierite adsorption heats for a temperature equal to 40°C show a plateau at 59.9 kJ mol^{-1} (from four to nine molecules sorbed per unit cell).[33] Translational, rotational, and vibrational contributions, not included in our model, would decrease the internal energies. As noted earlier, the dispersion energies estimated with the Kirkwood–Müller parameters are overestimated. The variation in configurational energy as a function of the water density will be discussed together with the structural results.

Structure Analysis. We recall that the MC method permits the study of an equilibrium state without any dynamical considerations. Plots of one significant configuration per density (Figs. 1.11 and 1.12) show that the water molecules tend to form cage structures in the cavities and remain far from the center and the walls of the ten-membered ring channel. However, in the eight-membered ring channels containing the protons, the molecules are closer to the walls.

The variations in the energy values as function of the number of water molecules versus each density (Table 1.14) are related to the differences between the various configurations. The increase of the total internal energy *U* from $N = 40$ to $N = 50$ and from $N = 65$ to $N = 80$ is associated with an increase in the water–zeolite interaction energy, since the water molecules are closer to the channel walls, but the water–water energy decreases, since hydrogen bonds are created at higher coverages. How-

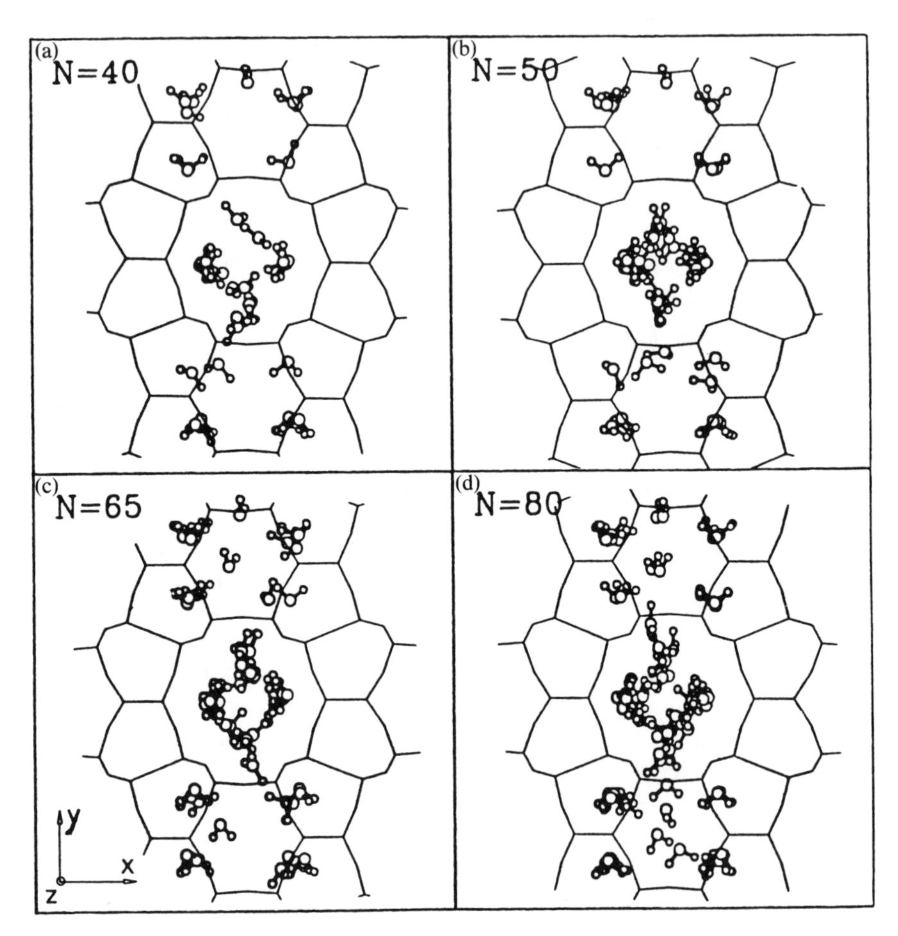

Figure 1.11. Plots following the XY plane of one significant configuration for all water molecules corresponding to each density: (a) $N = 40$, (b) $N = 50$, (c) $N = 65$, and (d) $N = 80$. MC simulation.

ever, a change in the distribution of water molecules between $N = 50$ and $N = 65$ is observed (Fig. 1.13). New adsorption sites, shown by the appearance of peaks between $Y = -5$ and 5 Å in the occurrence probability histograms, are occupied by water molecules related to a more stable interaction energy. The repartition of water is greatly governed by the negative framework oxygens: the water-hydrogen–framework-oxygen interactions are clearly observed (Fig. 1.14) by plotting the hydrogen distributions along the X and Y axes. The main peaks of the hydrogen atom distributions are positioned around the oxygen framework coordinates.

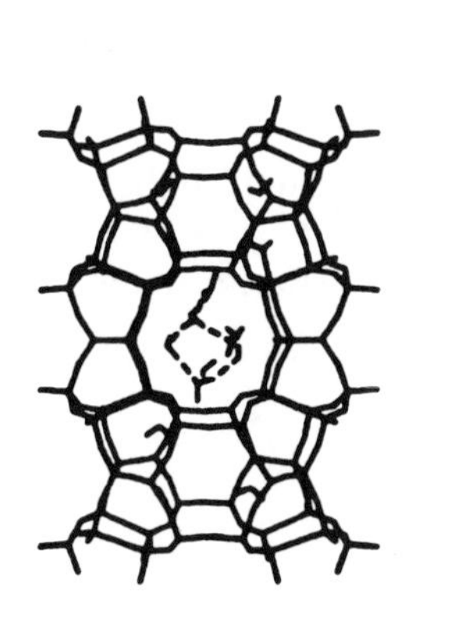
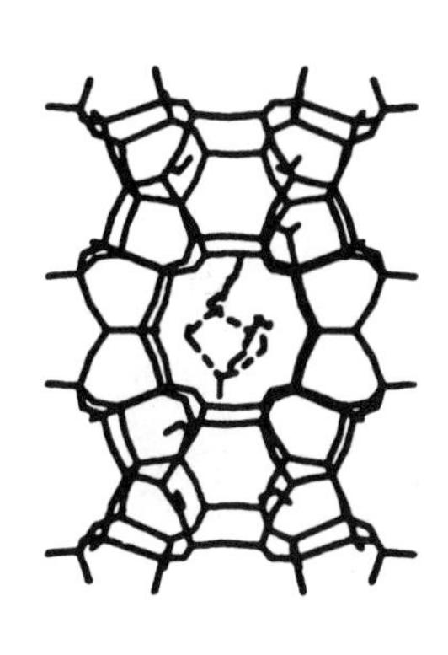

Figure 1.12 Stereoview showing the hydrogen bonds (dotted lines) of one significant water configuration for $N = 80$ in an 8-membered ring channel ($-7.47 \leq Z \leq 0.0$ Å). MC simulation.

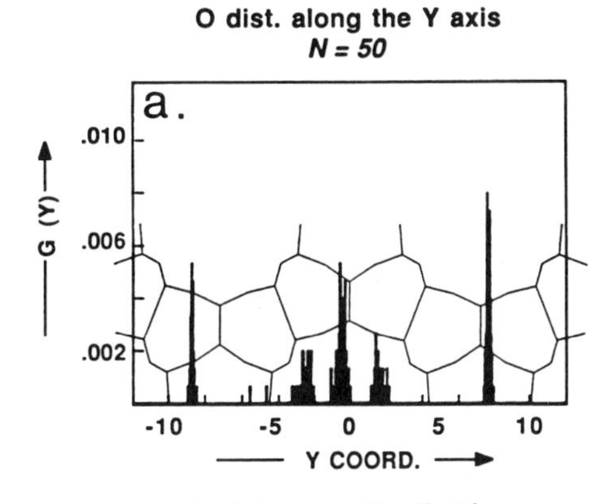

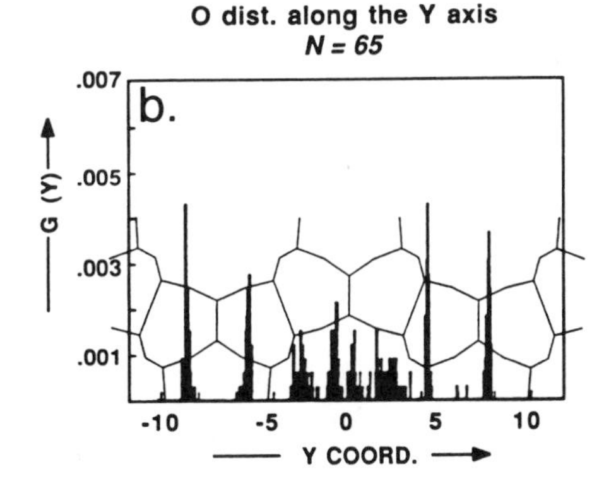

Figure 1.13 Water oxygen occurrence probability histograms superposed to the ferrierite framework for (a) $N = 50$ and (b) $N = 65$. MC simulation.

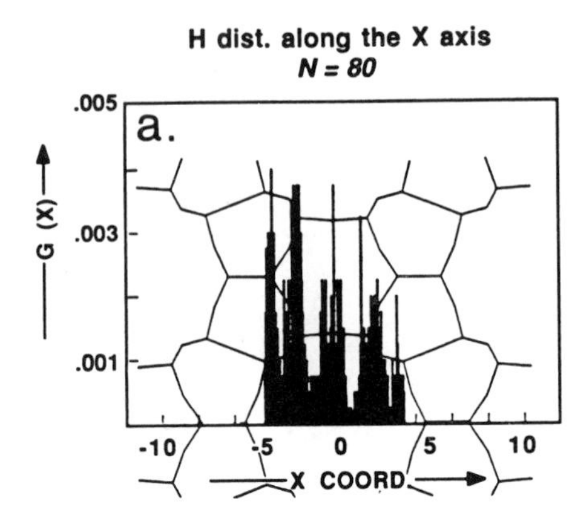

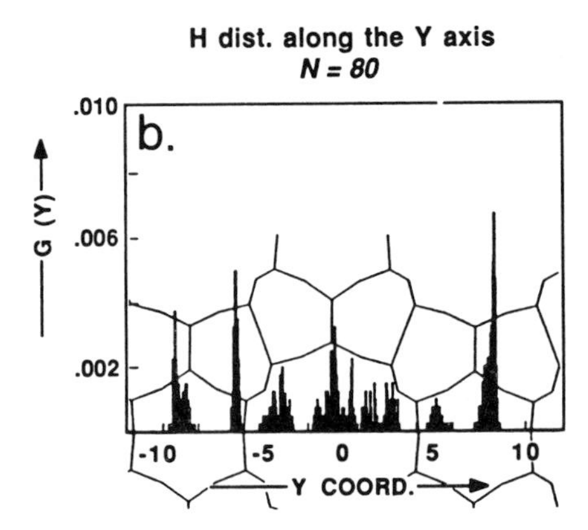

Figure 1.14 Water hydrogen occurrence probability histograms along the (a) X and (b) Y axes, for all water molecules corresponding to $N = 80$, superposed to the ferrierite framework. MC simulation.

Molecular Dynamics Simulations of Water in Ferrierite

Methodology. From interaction potentials, molecular dynamics methods create points in phase space by solving numerically the classical motion equations. Positions and velocities for each molecule can thus be known for discrete points in phase space separated by a time interval Δt such as

$$\Delta t = t_{n+1} - t_n \tag{1.11}$$

The evolution of molecules in phase space is expressed by the mean of Newton's classical equation to describe the translational motion:

$$\frac{d^2\mathbf{r}_i}{dt^2} = \frac{\mathbf{F}_i}{m_i} \tag{1.12}$$

where $\mathbf{F}_i$ stands for the force acting on the molecule i, m_i is its mass, and r_i is its center-of-mass position. The Euler equations are used to describe the rotational behavior of the water molecules:

$$I_{xx}\omega_{xi}^{(p)} = \Gamma_{xi}^{(p)} + \omega_{yi}^{(p)}\omega_{zi}^{(p)}(I_{yy} - I_{zz}) \tag{1.13}$$

$$I_{yy}\omega_{yi}^{(p)} = \Gamma_{yi}^{(p)} + \omega_{xi}^{(p)}\omega_{zi}^{(p)}(I_{zz} - I_{xx}) \tag{1.14}$$

$$I_{zz}\omega_{zi}^{(p)} = \Gamma_{zi}^{(p)} + \omega_{xi}^{(p)}\omega_{yi}^{(p)}(I_{xx} - I_{yy}) \tag{1.15}$$

where Γ_i corresponds to the torque acting on the molecule i. I_{xx}, I_{yy}, and I_{zz} are the diagonalized inertia tensor elements and ω_i are the angular velocities. The superscript (p) indicates that the quantities are given in the principal framework of the molecule (Fig. 1.15). The orientation of this framework with respect to the box framework is described by the so-called quaternions (ξ_i, η_i, ζ_i, χ_i) defined by the mean of the Euler angles.[44] The quaternions allow the treatment of the rotational motion.[99]

A typical MD experiment generally consists in three consecutive parts: an initiation stage followed by an equilibration part, and, finally, by a production stage.

Initiation Stage. N water molecules are placed in the zeolitic framework following an equilibrated, i.e., low energy, MC configuration. Random linear and angular velocities are attributed to the molecules with respect to the simulated temperature. The kinetic energies must verify the relations

$$E_{\text{translation}} = \tfrac{3}{2}NkT = \tfrac{1}{2}\sum_{i=1}^{N} m_i\mathbf{v}_i \cdot \mathbf{v}_i \tag{1.16}$$

$$E_{\text{rotation}} = \tfrac{3}{2}NkT = \tfrac{1}{2}\sum_{\alpha}\sum_{i=1}^{N} I_{\alpha\alpha i}^{(p)}\omega_{\alpha i}^{(p)}\omega_{\alpha i}^{(p)} \tag{1.17}$$

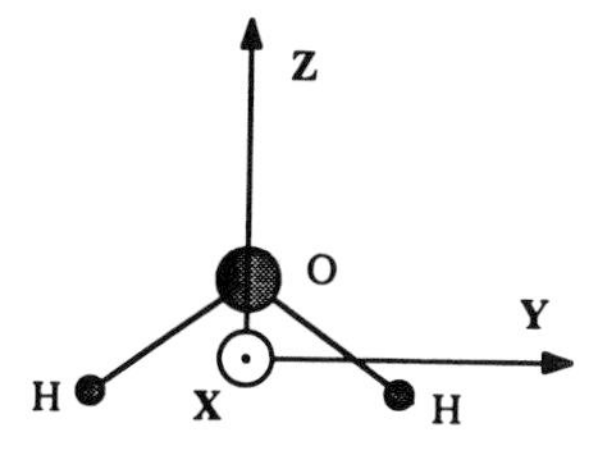

Figure 1.15 Orientation of the principal framework fixed to the water molecule.

where $I_{\alpha\alpha}$ are the diagonalized inertia tensor elements ($\alpha = x$, y, or z) and $\mathbf{v}_i$ and ω_i are the linear and angular velocities associated with the molecule i. During the simulation, the average temperatures are computed as functions of the kinetic energy.

Equilibration Stage. The equilibration stage involves movement of the system for a sufficiently long time to suppress any dependence on the starting point of phase space. At each step, the new positions and velocities are corrected to satisfy the kinetic energy relations. When the system is equilibrated, the temperature scaling algorithm is removed and the system behaves in phase space without any change in the total energy. This corresponds to a simulation in the microcanonical ensemble.

Production Stage. The production stage is achieved without the temperature scaling; the velocities, the positions, etc., at each step are then stored to permit the analysis of the system in terms of time averages.

Simulation Conditions. The ferrierite model described previously (in the section titled Ferrierite Model) has been reduced to 604 fixed framework atoms; 35 water molecules, representing a density of 6.6 molecules per unit cell, were placed in the zeolite channels. The simulations were performed by an adapted version of the BIOMOL program.[102] First, during the equilibration stage, the Newton and Euler equations of motion are integrated by means of a leap-frog algorithm based on the Verlet[107] and Fincham methods[37] with a time step Δt of 5 fs, during 3 ps, to randomize the system. Thereafter, both Δt and the integration algorithm were modified to pursue the equilibration stage during 3.5 ps ($\Delta t = 2.5$ fs) with a Nordsiek predictor–corrector algorithm.[42] The translational equations of motion are solved using a fifth-order predictor–corrector method. The rotational motion of water is described using quaternions[44] and the rotation equations are solved using a sixth-order predictor–corrector method. The production stage was achieved with the same integration algorithms and time step for 5 ps.

Thermodynamic Results. Based on time averages over 5 ps in phase space, the mean thermodynamic properties are presented in Table 1.15. Values are reported per water molecule. It is observed that due to the low water density, the translation and rotation temperatures fluctuate by 50 K around their average values.

A static representation of the system is obtained by plotting one configuration corresponding to one point in phase space (Fig. 1.16). Occurrence probability histograms may be established by averaging the water configurations (Fig. 1.17). Following these results, it is observed that the water

Table 1.15. Mean thermodynamic properties of water inside a ferrierite channel computed throughout an MD run over 5 ps; all values are reported per water molecule.

Water–water potential energy	−11.57 ± 0.77 kJ mol^{-1}
Zeolite–water potential energy	−65.61 ± 0.90 kJ mol^{-1}
Total potential energy	−77.18 ± 0.90 kJ mol^{-1}
Translational kinetic energy	4.49 ± 0.62 kJ mol^{-1}
Rotational kinetic energy	4.37 ± 0.62 kJ mol^{-1}
Total energy	−68.32 ± 0.57 kJ mol^{-1}
Translation temperature	359.77 ± 49.88 K
Rotation temperature	350.12 ± 49.88 K

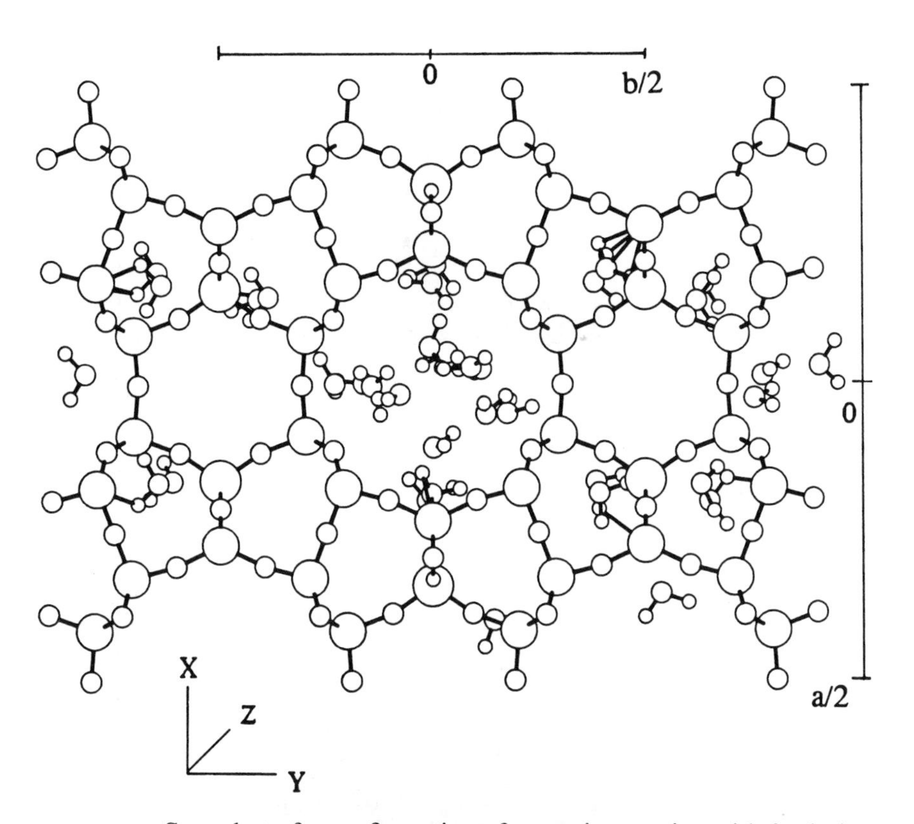

Figure 1.16 Snapshot of a configuration of water interacting with ferrierite after a 5-ps trajectory in the phase space at 350 K. MD simulation.

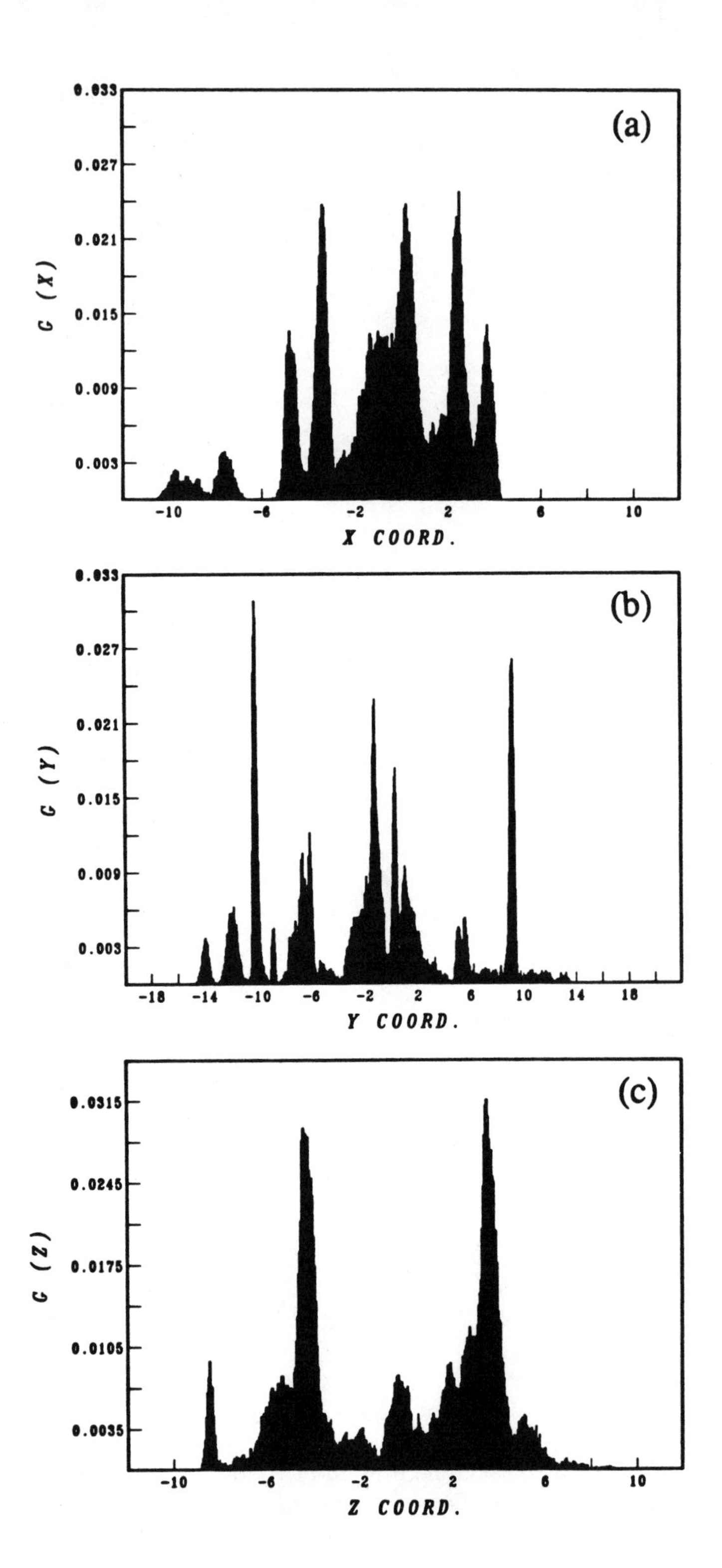

(a)
0.033
0.027
0.021
0.015
0.009
0.003
C (X)
-10
-6
-2
2
6
10
X COORD.
(b)
0.033
0.027
0.021
0.015
0.009
0.003
C (Y)
-18
-14
-10
-6
-2
2
6
10
14
18
Y COORD.
(c)
0.0315
0.0245
0.0175
0.0105
0.0035
C (Z)
-10
-6
-2
2
6
10
Z COORD.

molecules are closer to the walls of the eight-membered ring channels containing the additional protons, a situation consistent with hydrogen bonding. These features, potential energies and probability histograms, are quasiexactly comparable with those observed by the MC experiments discussed previously in the sections titled Thermodynamic Results and Structure Analysis.

Dynamical Properties. In this section, we will focus successively on time autocorrelation functions, self-diffusion coefficient, linear velocity time autocorrelation functions, libration angle, and dipole–dipole time autocorrelation functions estimated throughout the MD experiment.

Time Autocorrelation Functions. Theories to evaluate time autocorrelation functions have already been largely illustrated.[10,74] The time autocorrelation function $C_{bb}(t)$ associated with a property b

$$C_{bb}(t) = \langle b(0) \cdot b(t) \rangle = \frac{1}{T} \int_0^T b(t)b(t + \tau)dt \qquad (1.18)$$

provides information on the magnitude of the fluctuations of b and their time dependence over a time interval T. More precisely, it measures the "rapidity" or conversely the "persistence" of the fluctuations. Following the ergodic hypothesis,[74] $C_{bb}(t)$ defined as an ensemble average is calculated over a large number of measurements of the time average $\langle b(\tau) \cdot b(t + \tau) \rangle$, each measurement extending over the time interval T. When the system evolves in phase space, each point or system replica becomes the origin of a trajectory. Averaging $b(0) \cdot b(t)$ over all replicas of the system corresponds to an average over a distribution of initial states. These time-averages are calculated by the formula[64]

$$C(t) = \frac{1}{NN_t} \sum_{i=1}^{N} \sum_{j=1}^{N_t} [b_i(\tau) \cdot b_i(t + \tau)]_j \qquad (1.19)$$

where N stands for the molecule number and N_t stands for the number of projection values corresponding to a time interval t.

Self-Diffusion Coefficient and Linear Velocity Autocorrelation Function. Two methods are usually used to evaluate the self-diffusion coeffi-

---

Figure 1.17 Water's hydrogen occurrence probability histograms along the (a) X, (b) Y, and the (c) Z axes, superposed to the ferrierite framework. MD simulation.

cient D. One is based on the estimation of the linear velocity time autocorrelation function, LVCF(t):

$$\text{LVCF}(t) = \frac{\langle \mathbf{v}(0) \cdot \mathbf{v}(t) \rangle}{\langle \mathbf{v}(0) \cdot \mathbf{v}(0) \rangle} \tag{1.20}$$

where $\mathbf{v}$ is the velocity of the center of mass (COM) of each molecule. By integration of the LVCF(t) curve, i.e., by Fourier transformation at a frequency $\omega = 0$, D can be obtained as

$$D = \frac{\langle \mathbf{v}(0) \cdot \mathbf{v}(0) \rangle}{3} \int_0^\infty \text{LVCF}(t)dt \tag{1.21}$$

Unfortunately, the too small number of water molecules considered and the too short simulation time prevent the use of this first method, the statistics being not good enough.

Another method is based on the fact that the self-diffusion can be considered as the net result of a random walk process due to thermal movements. The probability P of finding a molecule at position r from an initial position r_0 after a time t is given by

$$P(r_0, r, t) = \frac{1}{(4\pi Dt)^{1/2}} \exp\left(\frac{-(r - r_0)^2}{4Dt}\right) \tag{1.22}$$

representing a Gaussian function, where D, the self-diffusion coefficient, characterizes completely that radial distribution. This leads to the Einstein relation[4]

$$D = \lim_{t \to \infty} \frac{\langle (\Delta r)^2 \rangle}{6t} \tag{1.23}$$

Computing the mean square deviation (MSD) of the trajectory of a molecule will permit the estimation of D and the comparison of its value with results obtained from NMR experiments. This is easily done by averaging over all water molecules and taking each discrete point in phase space as the origin of the trajectory (Fig. 1.18). Values beyond 3.5 ps are not shown as they lack statistics.

The quasilinear behavior of the MSD shows that the molecular movements begin to be dominated by a random process after about 1 ps. This is almost twice as long as the time observed for pure water, about 0.6 ps.[68] By linear regression between 1 and 3.5 ps, the following result is obtained:

$$\begin{aligned} \langle (\Delta r)^2 \rangle &= 6Dt + \text{constant} \\ &= 0.3016t + \text{constant} \end{aligned} \tag{1.24}$$

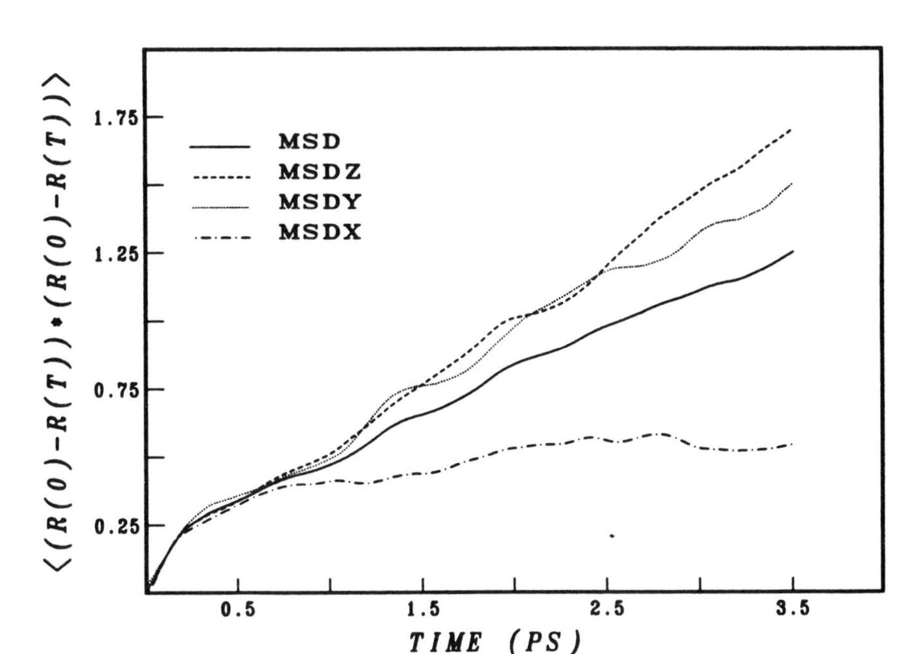

Figure 1.18 Mean square deviation (in Å^2) for water inside ferrierite as a function of time (in ps) computed throughout a 5-ps MD simulation.

with a regression coefficient equal to 0.997. The resulting self-diffusion coefficient D for water in the ferrierite channel is 0.5×10^{-5} cm^2 s^{-1}, about a factor four times smaller than that for liquid water.[68] This is to be expected, since the motion of the water molecules is hindered inside the channel. NMR experiments have led to a D value of 0.5×10^{-5} cm^2 s^{-1} for water in ZSM-5 at room temperature.[22] It was also observed that D for water was not significantly dependent on the pore size, since water is a small molecule. All the experimental values obtained by NMR spectroscopy for various pore size zeolites lie between 10^{-5} and 10^{-6} cm^2 s^{-1}.[8] The differences in behavior between the straight line for the MSD along the X axis and the two other MSDs along the Y and Z axes are because there are no connections between the ferrierite channels along the X axis. There is thus no way for the molecules to move from one channel to another following that direction and, consequently, their mobility is lowered.

Normalized by the factor $\langle \mathbf{v}(0) \cdot \mathbf{v}(0) \rangle$, time autocorrelation functions are presented in Figure 1.19. To explain the behavior of these functions, the movement of the fluid particle is described as follows. In the case of a Brownian particle, the movement is governed by the Langevin equation:

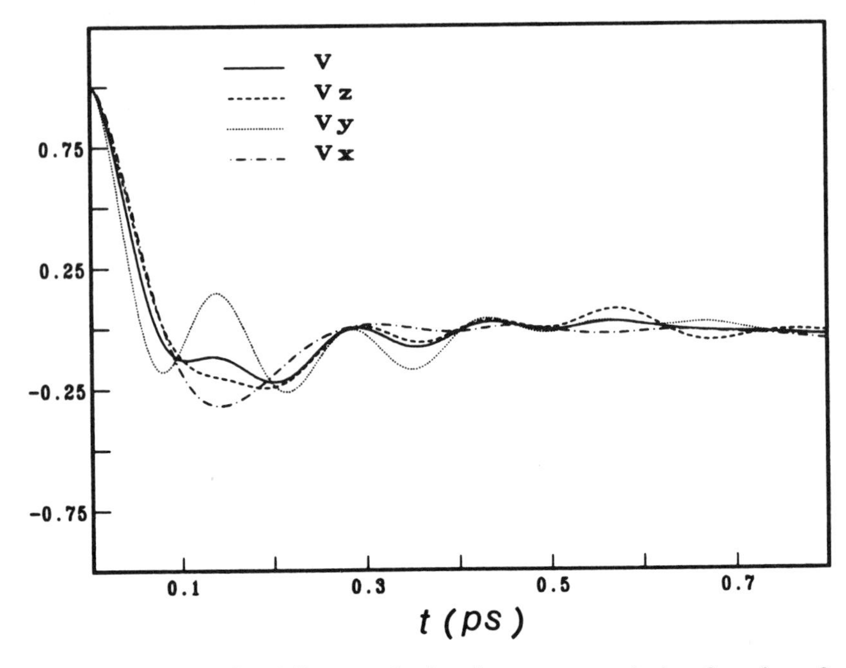

Figure 1.19. Normalized linear velocity time autocorrelation functions for water inside ferrierite computed throughout a 5-ps MD simulation.

$$M \frac{d\mathbf{v}}{dt} + \zeta \mathbf{v} = \mathbf{F}(t) \tag{1.25}$$

where $\mathbf{F}(t)$ is the stochastic force whose time average is zero and ζ is the friction coefficient. The linear velocity autocorrelation function then adopts a decreasing exponential form. In contrast, if we consider the fluid as an ideal Einstein solid constituted by a collection of independent oscillators vibrating at the same frequency ω_1, the velocity autocorrelation function has the behavior of a cosine function. From the damped oscillatory shape of the LVCF(t) presented in Figure 1.19, it is observed that the motion of a fluid particle has both a brownian character and a simple solid nature. The functions are also characterized by negative values showing that a molecular displacement toward the nearest neighbors is followed by a back movement to the initial position.

The Fourier transform of LVCF(t)

$$J(\omega) = \int_0^\infty \exp(-i\omega t) \frac{\langle \mathbf{v}(0) \cdot \mathbf{v}(t) \rangle}{\langle \mathbf{v}(0) \cdot \mathbf{v}(0) \rangle} \, dt \tag{1.26}$$

provides velocity fluctuations frequency spectrum (Fig. 1.20) which shows a high COM vibrational band around 70 cm^{-1}, a higher frequency with respect to liquid water, i.e., 53 cm^{-1}.[68]

In conclusion, the mean jump length of the water molecules in the zeolitic system with respect to liquid water is lowered, but the frequency of these movements is higher.

Libration Angle and Dipole–Dipole Autocorrelation Functions. Dipole–dipole autocorrelation functions give information about the molecular orientational motion. The first- and second-order Legendre polynomial functions P1DDCF(t) and P2DDCF(t)

$$\text{P1DDCF}(t) = \frac{\langle \mu(0) \cdot \mu(t) \rangle}{\langle \mu(0) \cdot \mu(0) \rangle} \tag{1.27}$$

$$\text{P2DDCF}(t) = \frac{\langle \frac{1}{2}[3(\mu(0) \cdot \mu(t))^2 - 1] \rangle}{\langle \frac{1}{2}[3(\mu(0) \cdot \mu(0))^2 - 1] \rangle} \tag{1.28}$$

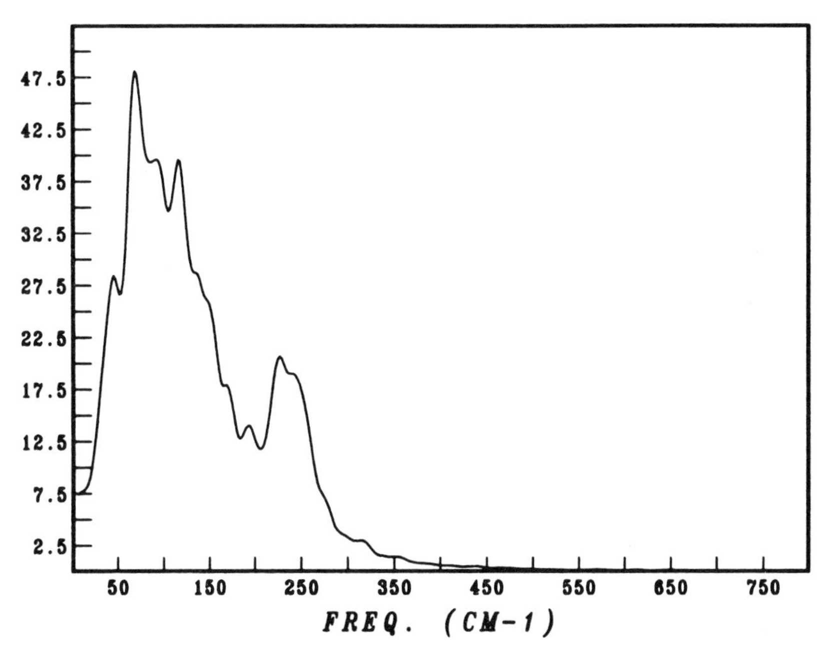

Figure 1.20 COM vibration frequency spectrum computed as Fourier transform of the linear time autocorrelation function represented in Figure 1.20.

where $\mu(t)$ is the dipole moment vector along the molecular axis, measure the molecular reorientation velocity in the fluid (Fig. 1.21). The minima of these curves are associated to the negative part of the angular velocity autocorrelation functions AVCF(t) (Fig. 1.22) defined as:

$$\mathrm{AVCF}(t) = \frac{\langle \omega(0) \cdot \omega(t) \rangle}{\langle \omega(0) \cdot \omega(0) \rangle} \tag{1.29}$$

The minimal value of the P2DDCF(t) curve provides a way to estimate the mean libration angle given by

$$\tfrac{1}{2}(3 \cos^2 \theta - 1) = 0.393 \tag{1.30}$$

θ is thus equal to 39.5°.

The behavior of the dipole–dipole autocorrelation function curves is due to two types of motion: (a) the libration movement represented by the minimal values of the curves, which is at longer times dominated by (b) structural changes detected by the monotonously decreasing part of the two curves. The negative part of the AVCF(t) is due to the fact that molecules are propelled in the original direction of rotation because of the energetic barrier.

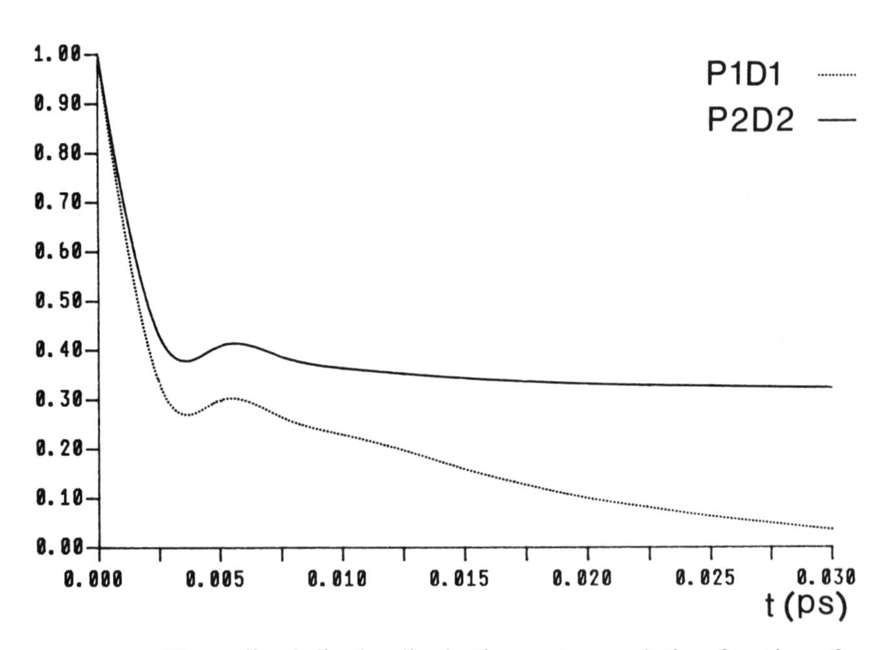

Figure 1.21 Normalized dipole–dipole time-autocorrelation functions for water inside ferrierite computed throughout a 5-ps MD simulation.

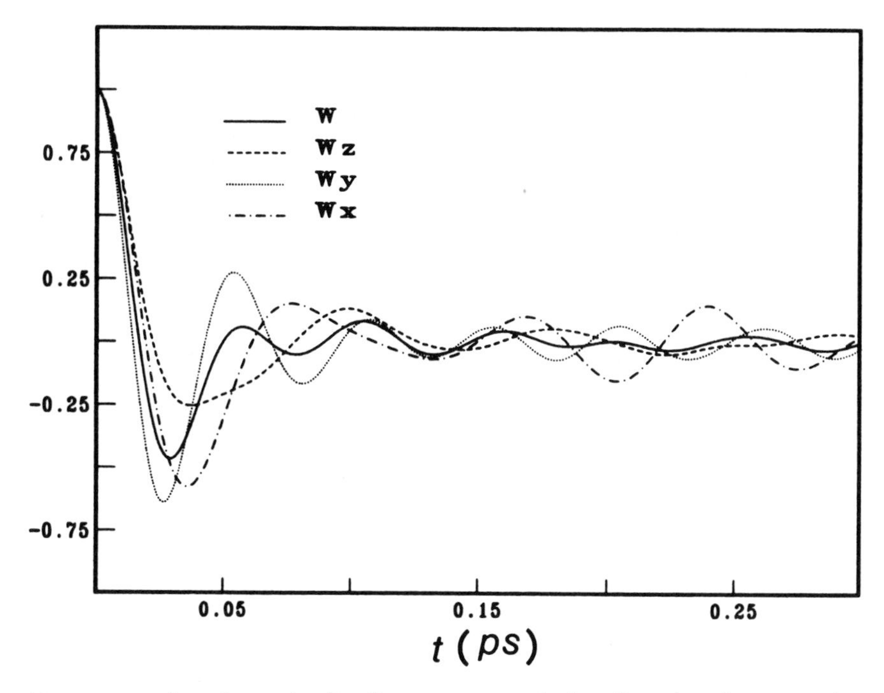

Figure 1.22 Angular velocity time–autocorrelation function for water inside ferrierite computed throughout a 5-ps MD simulation.

Perspectives and Conclusions

The chosen water–ferrierite potential, although partially empirical, reflects compositional variations of the framework, since the compensation of the negative charges introduced by the substitution of silicon by aluminum implies a change in the water molecule orientation related to the channel structure. Comparison of the computed theoretical energy values with experimental values shows a too attractive potential. This phenomenon is especially explained by the fact that the model contains approximately six unit cells along the *Z* axis without consideration of border effects. By limiting the statistical analysis to the two central unit cells along the *Y* and *Z* axes, the results are noticeably improved.

In spite of the limitations of the model, the Monte Carlo simulations, which give only a static representation of adsorption, permit the association of a particular distribution at a microscopic level to macroscopic quantities. The molecular dynamics results provide a first sketch of the microscopic behavior of water moving within the zeolite channels and allow the reproduction and thus the interpretation of the measured data such as self-diffusion coefficient from NMR experiments or libration motions from IR spectroscopy.

An important part of our on-going work consists in improving all parts of the three-step strategy developed in the Introduction. Mainly, we investigate the improvement of the interaction potential and how to consider all important macroscopic quantities in order to compare them to the available experimental data. The use of analytical potentials obtained by the fitting of a large number of *ab initio* interaction energies is currently under way. Structural results will probably be relatively insensitive to the choice of potential options, but interaction energies could show marked differences. Comparisons with other experimental structure determinations, such as neutron scattering or Raman spectroscopy, could give additional information on the quality of the potential and on its ability to simulate such systems. The application of the Monte Carlo approach to determine entropy and free energy changes and to relate these to the observed molecular distributions as a function of the nature and density of adsorption sites is also of interest.

ACKNOWLEDGMENTS

J.G.F. thanks the Scientific Affairs Division of NATO for a fellowship in the area of their International Intersectorial Exchanges in Oriented Research. P.G. and L.L. acknowledge the Institut pour l'Encouragement de la Recherche Scientifique dans l'Industrie et l'Agriculture (IRSIA) for a doctoral fellowship. L.L. thanks also the Belgian National Foundation for Scientific Research (FNRS) for the financial support of her stay at IBM-Kingston. Members of "Facultés Universitaires de Namur (FUNDP)" acknowledge Profs. E. Clementi and G.C. Lie for continuous collaboration and fruitful discussions. This work was rendered possible thanks to the generous access to the extended computing facilities of Mobil Research and Development Corporation and to the Namur Scientific Computing Facility, which is a common project of the FNRS, IBM Belgium, and the FUNDP. Part of this work concerning Statistical Mechanics Simulations was funded by the Belgian Program on Interuniversity Attraction Poles initiated by the Belgian State Prime Minister Office (Science Policy Programming).

REFERENCES

1. Ahlrichs, R., *Chem. Phys. Lett.* **34:** 570 (1975).
2. Artioli, G., Smith J.V., Kvick, A., Pluth, J.J., Ståhl, K., *Zeolites*. Amsterdam: Elsevier, 1985, p. 249.
3. Artioli, G., Rinaldi, R., Kvick, A., Smith, J.V., *Zeolites* **6:** 361 (1986).
4. Atkins, P.W., *Physical Chemistry*. Oxford: Oxford University Press, 1978.
5. Barker, J.A., Watts, R.O., *Chem. Phys. Lett.* **3:** 144 (1969).
6. Barrer, R.M., Gibbons, R.M., *J. Chem. Soc., Trans. Faraday Soc.* **59:** 2569 (1963).

7. Barrer, R.M., Gibbons, R.M., *J. Chem. Soc., Trans. Faraday Soc.* **61:** 948 (1965).
8. Barrer, R.M., *Zeolites: Science and Technology,* NATO ASI Series. The Hague: Martinus Nijhoff, 1984.
9. Bash, P.A., Singh, U.C., Langridge, R., Kollman, P.A., *Science* **236:** 564 (1987).
10. Berne, B.J., Harp, G.D., *Adv. Chem. Phys.* **17:** 63 (1970).
11. Bezus, A.G., Kiselev, A.V., Lopatkine, A.A., Du, P.Q., *J. Chem. Soc., Faraday Trans. II* **74:** 367 (1978).
12. Bezus, A.G., Kocirik, M., Kiselev, A.V., Lopatkine, A.A., Vasilyeva, E.A., *Zeolites* **6:** 101 (1986).
13. Bezus, A.G., Kocirik, M., Vasilyeva, E.A., *Zeolites* **7:** 327 (1987).
14. Binkley, J.S., Pople, J.A., Hehre, W.J., *J. Amer. Chem. Soc.* **102:** 939 (1980).
15. Binkley, J.S., Whiteside, R.A., Krishnan, R., Seeger, R., Defrees, D.J., Schlegel, H.B., Topiol, S., Kahn, L.R., Pople, J.A., Gaussian 80 *QCPE* **13:** 406 (1981).
16. Binkley, J.S., Frisch, M.J., Defrees, D.J., Raghavachari, K., Whiteside, R.A., Schlegel, H.B., Fluder, E.M., Pople, J.A., Gaussian 82 Chemistry Department, Carnegie-Mellon University, 1983.
17. Bissert, G., Liebau, F., *N. Jb. Miner. Mh.* **6:** 241 (1986).
18. B'Nagy, J., Gilson, J.P., Derouane, E.G., *J. Chem. Soc., Chem. Comm.* 1129 (1981).
19. Bosacek, V., Dubsky, J., *Coll. Czech. Chem. Commun.* **40:** 3281 (1975).
20. Broier, P., Kiselev, A.V., Lesnik, E.A., Lopatkine, A.A., *Russian J. Phys. Chem.* **42:** 1350 (1968).
21. Bursill, L.A., Lodge E.A., Thomas, J.-M., Cheetham, A.K., *J. Phys. Chem.* **85:** 2409 (1981).
22. Caro, J., Hocevar, S., Kärger, J., Riekert, L., *Zeolites* **6:** 213 (1986).
23. Clementi, E., *Computational Aspects for Large Chemical Systems, Lecture Notes in Chemistry* (G. Berthier et al., eds.). Berlin: Springer-Verlag, 1980, Vol. 19.
24. Clementi, E., *Structure and Dynamics: Nucleic Acids and Proteins* (E. Clementi and R.H. Sarma, eds.). New York: Academic Press, 1983, p. 321.
25. Cohen de Lara, E., Vincent-Geisse, J., *J. Phys. Chem.* **80:** 1922 (1976).
26. Demontis, P., Fois, E.S., Gamba, A., Manunza, B., Suffritti, G.B., *J. Mol. Struct. Theochem* **93:** 245 (1983).
27. Demontis, P., Suffritti, G.B., Alberti, A., Quartieri, S., Fois, E.S., Gamba, A., *Gazz. Chim. Ital.* **116:** 459 (1986).
28. Demontis, P., Suffritti, G.B., Quartieri, S., Fois, E.S., Gamba, A., *Dynamics of Molecular Crystals* (J. Lascombe, ed.). Amsterdam: Elsevier, 1987, p. 699.
29. Demontis, P., Suffritti, G.B., Quartieri, S., Fois, E.S., Gamba, A., *Zeolites* **7:** 522 (1987).
30. Demontis, P., Suffritti, G.B., Quartieri, S., Fois, E.S., Gamba, A., *J. Phys. Chem.* **92:** 867 (1988).
31. Derouane, E.G., Fripiat, J.G., *Proceedings of the 6th International Zeolite Conference* (D.H. Olson and A. Bisio, eds.). Guilford: Butterworths, 1984, p. 717.
32. Derouane, E.G., Fripiat, J.G., *Zeolites* **5:** 165 (1985).
33. Derouane, E.G., Fernandez, C., (unpublished results, 1987).
34. Derouane, E.G., Fripiat, J.G., *J. Phys. Chem.* **91:** 145 (1987), and references therein.
35. Derouane, E.G., André, J.-M., Lucas, A.A., *J. Catal.* **110:** 58 (1988).
36. Elber, R., Karplus, M., *Science* **235:** 318 (1987).
37. Fincham, D., *CCP5 News Letter* **2:** 6 (1981).
38. Fornili, S.L., Vercauteren, D.P., Clementi, E., *J. Mol. Catal.* **23,** 341 (1984).
39. Fripiat, J.G., Berger-André, F., André, J.-M., Derouane, E.G., *Zeolites* **3:** 306 (1983).
40. Fripiat, J.G., Galet, P., Delhalle, J., André, J.-M., B'Nagy, J., Derouane, E.G., *J. Phys. Chem.* **89:** 1932 (1985).

41. Gard, J.A., Tait, J.-M., *Adv. Chem. Ser.* **101:** 230 (1971).
42. Gear, C.W., *Numerical Value Problems in Ordinary Differential Equations*. Princeton, NJ: Prentice-Hall, 1971.
43. Geisinger, K.L., Gibbs, G.V., Natrotsky, A., *Phys. Chem. Minerals* **II:** 266 (1985).
44. Goldstein, H., *Classical Mechanics*. Reading, MA: Addison-Wesley, 1957.
45. Gordon, M.S., Binkley, J.S., Pople, J.A., Pietro, W.J., Hehre, W.J., *J. Amer. Chem. Soc.* **104:** 2797 (1982).
46. Gottardi, G., Alberti, A., *Bull. Geol. Soc. Finl.*, **57,** 197 (1985)
47. Harrison, I.D., Leach, H.F., Whan, D.A., *Zeolites* **7:** 421 (1987).
48. Hass, E.C., Mezey, P.G., Plath, P.J., *J. Molec. Struct., Theochem.* **76:** 389 (1981).
49. Hass, E.C., Mezey, P.G., Plath, P.J., *J. Molec. Struct., Theochem.* **87:** 2261 (1982).
50. Hehre, W.J., Stewart, R.F., Pople, J.A., *J. Chem. Phys.* **51:** 2657 (1969).
51. Hehre, W.J., Ditchfield, R., Stewart, R.F., Pople, J.A., *J. Chem. Phys.* **52:** 2769 (1970).
52. Hehre, W.J., *Accounts Chem. Res.* **9:** 399 (1976).
53. Jayaram, B., Mezei, M., Beveridge, D.L., *J. Comput. Chem.* **8:** 917 (1987).
54. Jorgensen, W.L., Gao, J., *J. Phys. Chem.* **90:** 2174 (1986), and references therein.
55. Kerr, I.S., *Nature* **210:** 294 (1966).
56. Kim, K.S., Clementi, E., *J. Phys. Chem.* **89:** 3655 (1985).
57. Kim, K.S., Vercauteren, D.P., Welti, M., Chin, S., Clementi, E., *Biophys. J.* **47:** 327 (1985).
58. Kirkwood, J.G., *Phys. Z.* **33:** 57 (1932).
59. Kiselev, A.V., Du, P.Q., *J. Chem. Soc., Faraday Trans. II* **77:** 1 (1981); **77:** 17 (1981).
60. Kiselev, A.V., Lopatkine, A.A., Shulga, A.A., *Zeolites* **5:** 261 (1985).
61. Klinowski, J., Ramdas, S., Thomas, J.M., Fyfe, C.A., Hartman, J.S., *J. Chem. Soc., Faraday Trans. II* **78:** 1025 (1982), and references cited therein.
62. Kono, H., Takasaka, A., *J. Phys. Chem.* **91:** 4044 (1987).
63. Kretschmer, R.G., Fiedler, K., *Z. Phys. Chem.* **258:** 1045 (1977).
64. Kushick, J., Berne, B.J., *Modern Theoretical Chemistry, Statistical Mechanics, Part B:Time-Dependent Processes* (B.J. Berne, ed.). New York: Plenum, 1977.
65. Leherte, L., Lie, G.C., Swamy, K.N., Clementi, E., Derouane, E.G., André, J.M., *Chem. Phys. Lett.* **95:** 237 (1988).
66. Leherte, L., Vercauteren, D.P., Derouane, E.G., André, J.-M., *Innovation in Zeolite Materials Science, Studies in Surface Science and Catalysis* (P.J. Grobet et al., eds.). Amsterdam: Elsevier, 1988, p. 293.
67. Lie, G.C., Clementi, E., Yoshimine, M., *J. Chem. Phys.* **64:** 2314 (1976).
68. Lie, G.C., Clementi, E., *Phys. Rev. A* **33:** 2679 (1986).
69. Lippmaa, E., Magi, M., Samoson, A., Engelhardt, G., Grimer, A.R., 1980, *J. Am. Chem. Soc.* **102:** 4889 (1980).
70. Lippmaa, E., Magi, M., Samoson, A., Tarmak, M., Engelhardt, G., *J. Am. Chem. Soc.* **103:** 4992 (1981).
71. Loewenstein, W., *Am. Mineral.* **39:** 92 (1954).
72. Matsuoka, O., Clementi, E., Yoshimine, M., *J. Chem. Phys.* **64:** 1351 (1976).
73. Mayorga, G.D., Peterson, D.L., *J. Phys. Chem.* **76:** 1641 (1972).
74. McQuarrie, D.A., *Statistical Mechanics*. New York: Harper & Row, 1976.
75. Meier, W.M., Olson, D.H., *Atlas of Zeolite Structure Types*. Pittsburgh; International Zeolite Association, 1978.
76. Meier, W.M., Meier, R., Gramlich, V., *Zeit. F. Krystallog.* **147:** 329 (1978).
77. Meier, W.M., Mock, H.J., *J. Solid State Chem.* **27:** 349 (1979).

78. Melchior, M.T., Vaughan, D.E.W., Jarman, R.H., Jacobson, A.J., *Nature* **298:** 455 (1982).
79. Metropolis, N., Rosenbluth, A.W., Rosenbluth, M.N., Teller, A.H., Teller, E., *J. Chem. Phys.* **21:** 1087 (1953).
80. Mezei, M., Harrison, S.W., Ravishanker, G., Beveridge, D.L., *Israel J. Chem.* **27:** 163 (1986).
81. Mortier, W.J., Sauer, J., Lercher, J.A., Noller, H., *J. Phys. Chem.* **88:** 905 (1984).
82. Mulla, D.J., *J. Coll. Int. Sci.* **100:** 576 (1984).
83. Mulla, D.J., Cushman, J.H., Low, P.F., *Water Resources Research* **20:** 619 (1984).
84. Müller, A., *Proc. Roy. Soc. London Ser. A* **154:** 624 (1936).
85. Nowak, A.K., Cheetham, A.K., Pickett, S.D., Ramdas, S., *Molecular Simulation* **1:** 67 (1987).
86. Olson, D.H. Kokotailo, G.T., Lawton, S.L., Meier, W.M., *J. Phys. Chem.* **85:** 2238 (1981).
87. Pettitt, B.M., Karplus, M., Rossky, P.J., *J. Phys. Chem.* **90:** 6335 (1986).
88. Rahman, A., *Phys. Rev.* **136:** A405 (1964).
89. Ravishanker, G., Mezei, M., Beveridge, D.L., *J. Comput. Chem.* **7:** 345 (1986).
90. Rhee, C.H., Metzger, R.M., Wiygul, F.M., *J. Chem. Phys.* **77:** 899 (1982).
91. Salem, L., *Mol. Phys.* **3:** 441 (1960).
92. Sauer, J., Hozba, P., Zahradnik, R., *J. Phys. Chem.* **84:** 3318 (1980).
93. Sauer, J., Engelhardt, G., *Z. Naturforsch.* **37a:** 277 (1982).
94. Sauer, J., Zahradnik, R., *Int. J. Quantum Chem.* **26:** 793 (1984).
95. Sauer, J., *Acta Phys. Chem.* **31:** 19 (1985).
96. Shin, J.M., No, K.T., Jhon, M.S., *J. Chem. Phys.* **92:** 4533 (1988).
97. Singh, U.C., Brown, F.K., Bash, P.A., Kollman, P.A., *J. Amer. Chem. Soc.* **109:** 1607 (1987).
98. Smith, J.V., Bailey, S.W., *Acta Cryst.* **16:** 801 (1963).
99. Sonnenschein, R., Laaksonen, A., Clementi, E., *J. Comput. Chem.* **7,** 645 (1986).
100. Soto, J.L., Myers, A.L., *Mol. Phys.* **42:** 971 (1981).
101. Stroud, H.J.F., Richards, E., Limcharoen, P., Parsonage, N.G., *J. Chem. Soc., Faraday Trans. I* **72:** 942 (1976).
102. Swamy, K.N., Clementi, E., *BIOMOL a Molecular Dynamics Simulation Package for Nucleic Acid Hydration.* Kingston, NY: IBM Corp. Dept 48B MS 428, 1984.
103. Takaishi, T., *Pure & Appl. Chem.* **58:** 1375 (1986).
104. Thomas, J.-M., Fyfe, C.A., Ramdas, S., Klinowski, J., Gobbi, C.G., *J. Phys. Chem.* **86:** 3061 (1982).
105. Valleau, J.P., Whittington, S.G., *Statistical Mechanics Part A,* (B.J. Berne, ed.). New York: Plenum Press, 1977, Vol. 5, p. 137.
106. Vaughan, P.A., *Acta Crystallogr.* **21:** 983 (1966).
107. Verlet, L., *Phys. Rev.* **159:** 98 (1967).
108. Wojcik, M., Corongiu, G., Detrich, J., Mansour, M.M., Clementi, E., Lie, G.C., *Supercomputer Simulations in Chemistry, Lectures Notes in Chemistry* (M. Dupuis, ed.). Berlin: Springer-Verlag, 1986, Vol. 44.
109. Wong, C.F., McCammon, J.A., *Israel J. Chem.* **27:** 211 (1986), and references therein.

2
Conceptual Background for the Conversion of Hydrocarbons on Heterogeneous Acid Catalysts

J.A. Martens and P.A. Jacobs

INTRODUCTION

It is generally assumed that carbocations are reaction intermediates in the conversion of hydrocarbons over heterogeneous acid catalysts.[1] *Direct* evidence for the presence of carbocations on the surface of a working catalyst has, however, never been provided. The arguments in favor of the occurrence of cationic intermediates have been that the reaction pathways of hydrocarbons can be predicted by considering the stability of the carbocations that may be formed and the rearrangement pathways these cations might follow.[1,2]

The thermochemistry of gaseous cations has become experimentally accessible by the technique of mass spectrometry, according to which neutral molecules are ionized and the molecular or fragment ions or both are recognized according to their mass/charge ratio and kinetic energy.[3] Quantification of substituent effects on the stability of carbocations and quantitative prediction of the chemical reactivity became possible. Long-living alkylcarbenium ions can be generated in superacid solutions at moderate temperatures and their structure and rearrangements studied with ^{13}C and ^{1}H nuclear magnetic resonance (NMR).[4] A typical preparation method of alkylcarbenium ions consists of introducing an alkyl fluoride (or chloride) in a mixture of SbF_5 and SO_2ClF:

$$R_1\text{–CHF–}R_2 + SbF_5 \rightarrow R_1\text{–CH}^+\text{–}R_2 + SbF_6^- \qquad (2.1)$$

The data obtained in the gas phase have proven to be very useful for the rationalization of the carbocation behavior in solution.[5] Likewise it will be illustrated subsequently that it is indispensable to have a background in carbocation chemistry in order to be able to understand the working of heterogeneous acid catalysts. The use of the concept of carbocation intermediates will be illustrated for the conversion of alkanes over bifunctional catalysts.

NOMENCLATURE AND REPRESENTATION OF ALKYLCARBENIUM AND ALKYLCARBONIUM IONS RELEVANT TO ALKANE CONVERSION

Alkylcarbenium Ions

Alkylcarbenium ions are ionic species that contain a positively charged carbon atom, bound either to one alkyl group and two hydrogen atoms (**primary** ions), or to two alkyl groups and one hydrogen atom (**secondary** ions), or to three alkyl groups (**tertiary** ions). The electron-deficient carbon atom shows sp^2 hybridization, giving rise to a planar structure. The stability of the alkylcarbenium ions increases along the following cation sequence:

$$\text{primary} < \text{secondary} < \text{tertiary} \qquad (2.2)$$

A convenient way to denote specific alkylcarbenium ions, which does not violate IUPAC rules,[6] is to add the word "cation" to the radical name. Examples given in Eqs. (2.3) and (2.4) illustrate this nomenclature:

2-methyl-2-propyl cation
or tertiobutyl cation (2.3)

2,3-dimethyl-6-ethyl-5-octyl cation (2.4)

For the sake of clarity in the discussion on the scission reactions of acyclic alkylcarbenium ions, the positions of branchings in alkylcarbenium ions will be denoted by Greek letters. They will be used to determine the distance of a given carbon atom in the main chain from the positively charged carbon atom, which for formal reasons has to occupy the α-position [Eq. (2.5)]. Positions between the positive charge and the end carbon atom of the main carbon chain are denoted with β', γ', δ', etc., to distinguish them from the corresponding positions in the cation between the positive charge and the initial carbon atom. Within the scope of this chapter, it will not be necessary to introduce notations for carbon atom positions in the chain branchings.

δ β β' δ'
... + ... (2.5)
γ α γ'

The main chain can be selected using the IUPAC rules for alkane notation,[6] provided the chain contains the positively charged carbon atom.

This procedure allows one to denote in a unique way specific configurations of side chains and to relate them to the position of the positively charged carbon atom without specifying the carbon number of the alkylcarbenium ion. For example, the alkylcarbenium ion shown in Eq. (2.6) will be denoted as a $\alpha,\gamma,\gamma,\beta'$-tetramethyl-branched cation.

(2.6)

$\alpha,\gamma,\gamma,\beta'$-tetrabranched alkylcarbenium ion

Alkylcarbonium Ions

An **alkylcarbonium ion** is a pentacoordinated carbocation. Alkylcarbenium and alkylcarbonium ions are chemically different. Whereas, in general, alkylcarbenium ions can be represented by the formula CR_3^+, in which R stands for a hydrogen atom or an alkyl group, the general chemical formula for alkylcarbonium ions is CR_5^+. The alkylcarbonium ions of interest to the isomerization of acyclic alkylcarbenium ions are secondary, tertiary, or quaternary, depending on the number of alkyl substituents. In Figure 2.1 possible representations for the CHR_4^+ alkylcarbonium ion are given. This carbocation can be regarded as (i) a σ complex between the carbenium ion CR_3^+ and a C–H bond in a HR molecule (Fig. 2.1A), (ii) a σ complex between a proton and a C–H or C–C bond of the CR_4 molecule (Fig. 2.1B), or (iii) a molecule containing a three-center bond between two carbon atoms and a hydrogen atom or two hydrogen atoms and one carbon atom (Fig. 2.1C).[7] Alkylcarbonium ions can be represented by a positively charged carbon atom with five substituents (Fig. 2.1D), keeping in mind, however, that these five bonds only contain eight electrons.

Alkylcarbonium ions invoked as potential intermediates in the isomerization mechanisms of acyclic alkylcarbenium ions are **substituted pro-**

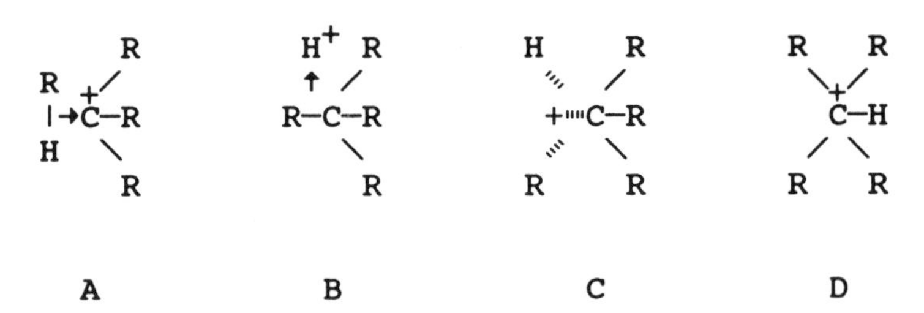

Figure 2.1. Representations of CHR_4^+ alkylcarbonium ions. R stands for a hydrogen or an alkyl group.

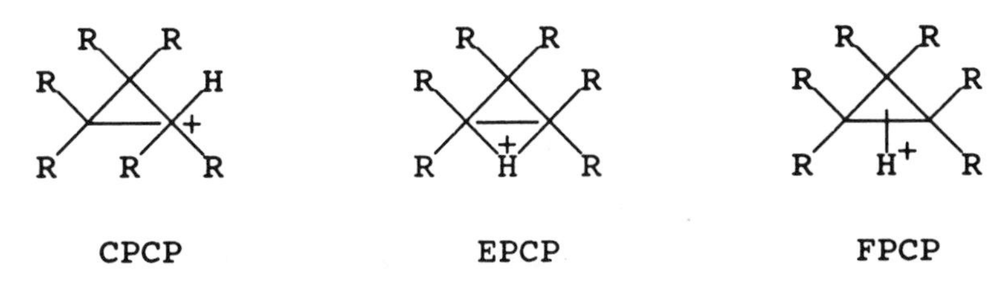

Figure 2.2. Representation of substituted corner (CPCP), and edge (EPCP) and face (FPCP) protonated cyclopropanes. R stands for a hydrogen or an alkyl group.

tonated cycloalkanes.[7] **Substituted protonated cyclopropanes (PCP)** contain the smallest possible rings among these structures. PCP exists in three different forms, as shown in Figure 2.2. Differentiation among the different types of PCP is based on the number of bonds that are present between the proton and the carbon atoms of the cyclopropane ring. In a **substituted corner protonated cyclopropane** (CPCP), the proton is bound to only one carbon atom of the cyclopropane ring (Fig. 2.2). If the proton is bound to two carbon atoms of the cyclopropane ring, the PCP ion is denoted as **substituted edge protonated cyclopropane** (EPCP). In **substituted face protonated cyclopropanes** (FPCP) the proton is bound to all three cyclopropane carbon atoms.

For large reaction schemes it is convenient to use a simplified representation of substituted corner protonated cycloalkanes, in which the formal representation of the C–H bonds is omitted. An example of the simplified representation of a CPCP is shown in Eq. (2.7):

$$\equiv \tag{2.7}$$

Within the scope of this chapter, it is necessary to introduce a nomenclature for substituted corner protonated cycloalkanes. It involves numbering of the ring carbon atoms by arabic numbers. Carbon atom 1 carries the proton. The direction for numbering of the other ring carbon atoms is chosen to give the smallest numbers possible to the substituent positions. Examples given in Eqs. (2.8) and (2.9) illustrate this nomenclature.

1-protonated
1,2,2-trimethyl-3-ethylcyclopropane (2.8)

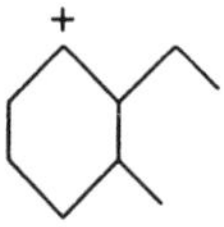

1-protonated 2-ethyl-3-methylcyclohexane (2.9)

RELATIVE STABILITIES OF ALKYLCARBENIUM AND ALKYLCARBONIUM IONS

The relative stability of an alkylcarbenium ion is governed by the inductive effects of its substituents. Since alkyl groups behave as electron donor substituents on the positively charged carbon atom, the replacement of a hydrogen atom by an alkyl group will result in a considerable stabilization.

The relative stability of a few alkylcarbenium ions in the gas phase is given in Table 2.1. Among the $C_4H_9^+$ carbenium ions the primary cations (1-butyl and 2-methyl-1-propyl cation) are 60–70 kJ mol^{-1} less stable than the secondary 2-butyl cation. The latter is another 67 kJ mol^{-1} less stable than the tertiary tertiobutyl cation. The stabilizing effect of an alkyl substituent depends mainly on the number of C–C bonds in the β position to the positively charged carbon atom.[5] A methyl group substituted to the positive carbon atom does not contain β C–C bonds, while ethyl and larger *n*-alkyl groups contain such a bond. The replacement of a methyl group with an ethyl group results in stabilization, as can be seen from Table 2.1, when the $\Delta H°$ values for 2-methyl-2-propyl (no β C–C bond), 2-methyl-2-butyl (one β C–C bond), and 3-methyl-3-pentyl cation (two β C–C bonds) are compared. The influence of the nature of the substituents

Table 2.1. Enthalpy differences between alkylcarbenium ions in the gas phase (after Refs. 12 and 13).

	$\Delta H°$ (kJ mol^{-1})[a]
1-Butyl cation	+138
2-Methyl-1-propyl cation	+130
2-Butyl cation	+67
2-Methyl-2-propyl cation	0
2-Methyl-2-butyl cation	−12
2-Methyl-2-pentyl cation	−19
3-Methyl-3-pentyl cation	−19
2,3-Dimethyl-2-butyl cation	−21
2-Heptyl cation	+6.3
3-Heptyl cation	+2.5
4-Heptyl cation	0.0

[a] At 298 K and 0.1 MPa.

on the cation stability is, however, small for tertiary ions with the same carbon number. This is illustrated in Table 2.1 when the ΔH° values for 2-methyl-2-pentyl, 3-methyl-3-pentyl, and 2,3-dimethyl-2-butyl cations are compared. In the respective cations, the substituents of the positively charged carbon atom consist either of two methyl groups and a propyl group, two ethyl groups and a methyl group, or two methyl groups and one isopropyl group. Table 2.1 further shows that 2-heptyl cation has the highest energy among the different secondary *n*-heptyl cations and that the energy differences between 3-heptyl cation and 4-heptyl cation are small. Therefore, all secondary *n*-alkylcarbenium ions with the same carbon number should have a comparable stability, unless one of the two substituents of the positively charged carbon atom is a methyl group.

PCP carbonium ions are not stable in superacids. Their existence and properties have been inferred from studies of the products of reactions of alkylcarbenium ions.[7] Their energies have been estimated from activation energies of alkylcarbenium ion rearrangements in superacids[7] and from molecular orbital calculations.[8–11]

Analogous to the behavior of the alkylcarbenium ions, the stability of alkylcarbonium ions increases along the following cation sequence:

$$\text{primary} < \text{secondary} < \text{tertiary} < \text{quaternary} \tag{2.10}$$

The alkyl substituent effects on the stability of alkylcarbonium ions are, however, much less pronounced than on alkylcarbenium ions, as a result of the larger number of substituents and the concomitantly larger charge delocalization.[14] For example, the stabilizing effect of a methyl substituent on a CPCP corresponds to only one-third of the stabilization energy effected by the same substituent on an alkylcarbenium ion.[15]

Molecular orbital calculations indicate that corner and edge protonated cyclopropanes are much more stable compared to face protonated cyclopropanes.[8–11] The stability of corner protonated cyclopropane is situated between that of the 1-propyl cation and 2-propyl cation.[15] The stability of substituted CPCP ions increases faster with increasing carbon number than that of secondary alkylcarbenium ions. Consequently, energy equilization for substituted CPCP ions and secondary alkylcarbenium ions occurs for ions with carbon number 5 and 6.[15,16]

REARRANGEMENTS OF ACYCLIC ALKYLCARBENIUM IONS IN SUPERACIDS

Isomerization Mechanisms

From NMR investigations on deuterium (D) and ^{13}C-labeled molecules, the mechanisms of the isomerization of acyclic alkylcarbenium ions with carbon number less than nine have been studied at temperatures in the range from 173 to 373 K.[17] Many rearrangements of the small alkylcarbenium ions with less than six carbon atoms result in **intramolecular scrambling** of the carbon and hydrogen atoms and do not give rise to chemically distinguishable molecules.

The isomerization reactions of acyclic alkylcarbenium ions in superacids formally belong to one of the following two categories:

- **Isomerization reactions of type A,** which change the position of a side chain, without changing the degree of branching.
- **Isomerization reactions of type B,** which increase or reduce the degree of branching.

Isomerization of type A is illustrated in Eq. (2.11) for the isomerization of a 2-methyl-2-pentyl into a 3-methyl-3-pentyl cation.

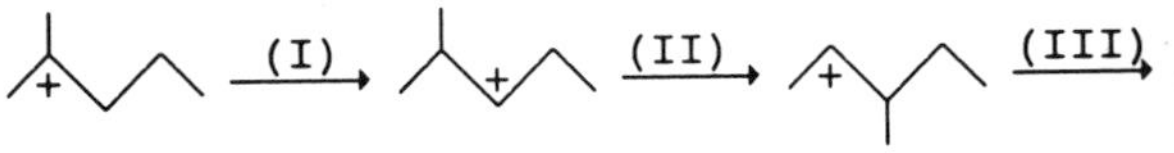

(I) (II) (III) (2.11)

It consists of an **1,2-hydride shift** (I), followed by an **1,2-alkyl shift** (II) and another **1,2-hydride shift** (III), in order to restore a tertiary alkylcarbenium ion. The 1,2-alkyl shift shown in step II of Eq. (2.11) is not necessarily an elementary step but, according to generally accepted principles in superacid chemistry, may proceed as shown in Eq. (2.12) through cyclization of the alkylcarbenium ion into an intermediate CPCP structure followed by reopening of the cyclopropane ring in the latter ion:

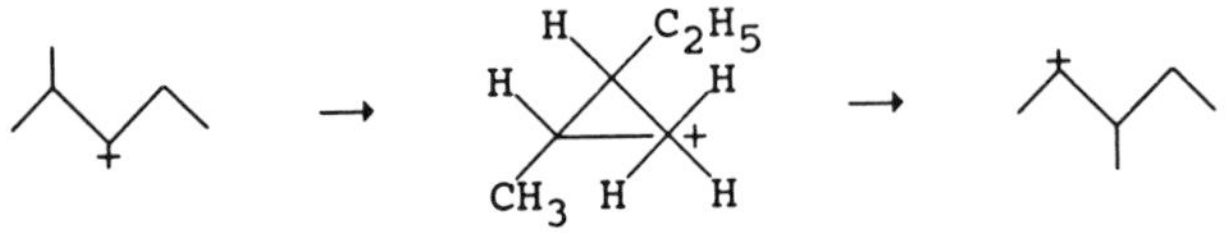

H C_2H_5 H H + CH_3 H H (2.12)

The mechanism, currently accepted in superacid catalysis for branching rearrangements (type B isomerization), also involves the formation of CPCP cations as intermediates. Branching of an alkyl cation occurs only when prior to opening of the cyclopropane ring of the CPCP intermediate a **corner-to-corner migration of hydrogen** has occurred. The main difference between type A and B rearrangements is the occurrence of such a corner-to-corner hydrogen shift in the latter case.

In fact, the rearrangement illustrated in Eq. (2.13) and denoted in literature[17] as **corner-to-corner hydrogen shift** comprises a **proton jump** on the cyclopropane ring. It should be stressed that such a proton jump is distinctly different from the 1,2-hydride shift mentioned earlier, an example of which is shown in Eq. (2.11).

(2.13)

In order for a branching to be generated in a *n*-alkylcarbenium ion, the corner-to-corner proton jump should involve a displacement of the positive charge toward a corner carbon atom of the cyclopropane ring that is free of alkyl substituents. After ring opening of such CPCP ion, an increase of the degree of branching of the carbon skeleton is observed. It is evident that the opposite reaction, involving a decrease in the degree of branching, should start from a CPCP ion with the positive charge located on a carbon atom free of alkyl substituents. This mechanism further implies that only methyl branchings can form and vanish.

In Figure 2.3 the isomerization mechanism via formation of CPCP intermediates, followed by a corner-to-corner proton jump, is illustrated. It provides an energetically favorable reaction path for the branching of $C_5H_{11}^+$ and heavier ions, since it avoids the formation of primary alkylcar-

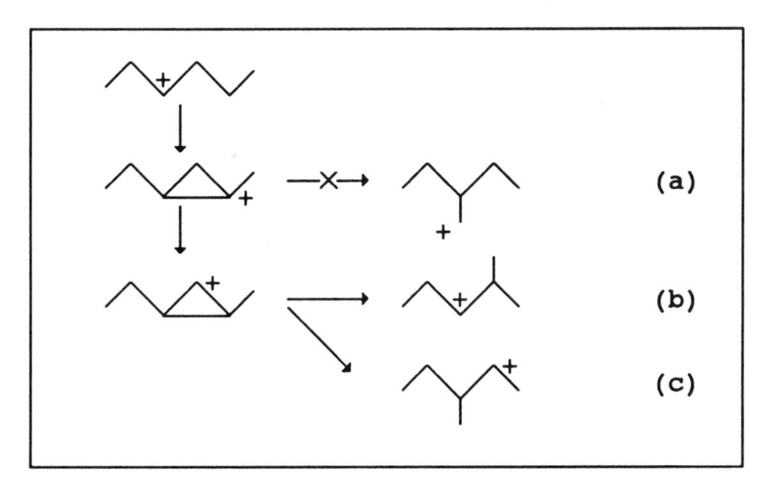

Figure 2.3. Possible type *B* isomerization of 3-hexyl cation without (a) and with (b and c) consecutive corner-to-corner proton jump.

benium ions. It cannot interconvert, however, 2-butyl and tertiobutyl cation. In this case the opening of the protonated methylcyclopropane ring inevitably generates a primary cation in order to change the degree of branching [Eq. (2.14)].

(2.14)

The mechanisms of type A and B isomerization of $C_6H_{13}^+$ alkylcarbenium ions are illustrated Figure 2.4. The figure shows two examples of type A rearrangement:

- 2-Methyl-2-pentyl cation (U) is converted to 3-methyl-3-pentyl cation (R) and vice versa.

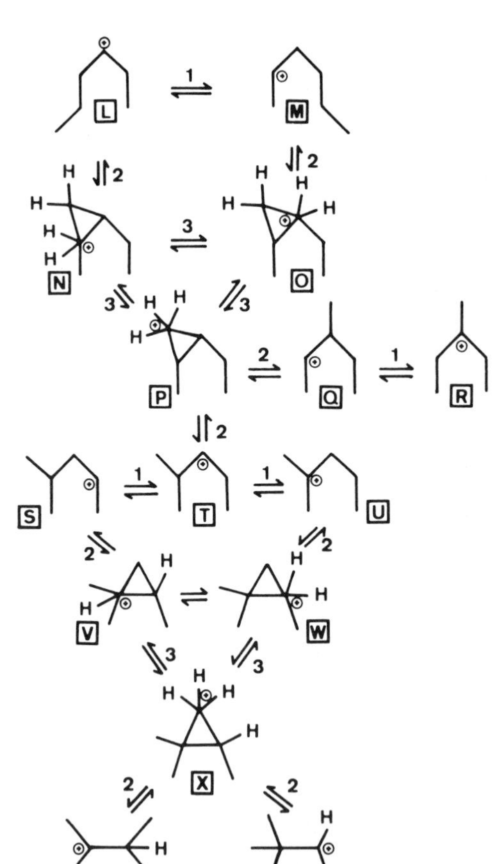

Figure 2.4. Types *A* and *B* rearrangements of $C_6H_{13}^+$ alkylcarbenium ions via CPCP intermediates in superacids, after Ref. 13; 1, 1,2-hydride shift; 2, ring opening and closure; 3, corner-to-corner proton jump. The solid lines represent C–C bonds, while –H stands for a C–H bond.

- 2,3-Dimethyl-2-butyl cation (Y) to 3,3-dimethyl-2-butyl cation (Z) and vice versa.

As an example of the conversion of unbranched to monobranched ions, Figure 2.4 shows the isomerization of 2- and 3-hexyl cation (M and L) to 2-methyl-2-pentyl (U) and 3-methyl-3-pentyl cation (R). Examples of the conversion from mono- to dibranched ions are found for the isomerization of the monobranched 2-methyl-2-pentyl cation (U) toward the dibranched 2,3-dimethyl-2-butyl cation (Y) and 3,3-dimethyl-2-butyl cation (Z). All corner-to-corner proton jumps involved in these type B isomerization reactions are illustrated in the interconversion of the cations, denoted as N, O, and P, and V, W, and X.

In superacids type B rearrangements are always slower than type A ones and the energy barriers of processes involving an overall change of the degree of branching (type B) are about 12 kJ mol^{-1} higher than those rearrangements not involving such change (type A).[18,19] Figure 2.5 shows the mechanism of 1,2-methyl shift and unbranching on 2,3-dimethyl-3-pentyl cation (A) into 2,4-dimethyl-2-pentyl cation (E) and 2-methyl-2-hexyl cation (H), respectively. The first two steps, viz., 1,2-hydride shift (A into B) and cyclization (B into C) are common for both rearrangements. Cation F is formed from the secondary CPCP alkylcarbonium ion C via the unbranching pathway and is more stable, since it is of tertiary nature. CPCP ring opening via the 1,2-methyl shift and the unbranching pathway bear the secondary alkylcarbenium ions D and G, respectively. The latter ions should be close in energy (Table 2.1). D and G are transformed into the more stable tertiary alkylcarbenium ions E and H, respectively.

The intermediates written in Figure 2.5 do not provide an explanation for the rate difference between type A and B isomerizations. The corner-to-corner proton migration on the CPCP ring (C to F in Fig. 2.5), which is an unavoidable step in the type B rearrangement, should therefore proceed over a high energy barrier and account for the rate difference.

A study by Saunders et al.[18] on the rearrangements of the 1,1,1-trideutero-2-^{13}C-propyl cation (D_3C-$^{13}C^+H$-CH_3) in SbF_5/SO_2ClF solution provided experimental evidence that CPCP ring closure and opening is more rapid compared to corner-to-corner proton jump. With this probe, carbon atom as well as hydrogen atom interchange can be detected. At a temperature of 213 K, hydrogen (deuterium) shifts with and without concomitant carbon atom interchange were observed, the latter ones being faster. Processes involving carbon atom interchange necessarily proceed via CPCP intermediates. Figure 2.6 shows the possible conversion pathways of D_3C-$^{13}C^+H$-CH_3 via CPCP intermediates. Most of the conversion reactions of Figure 2.6 result in carbon together with hydrogen (deute-

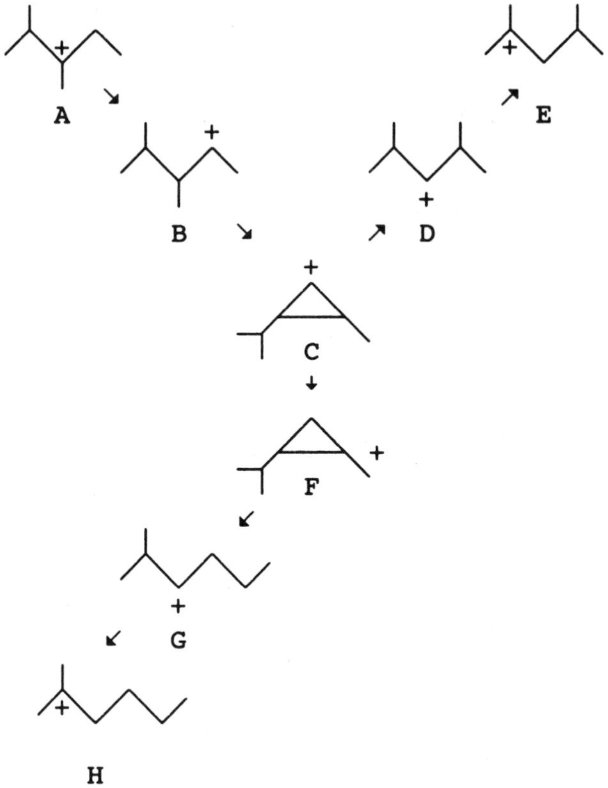

Figure 2.5. Types *A* and *B* isomerization of 2,3-dimethyl-4-pentyl cation via the 1,2-methyl shift mechanism and the unbranching pathway, respectively.

rium) shifts. Figure 2.6 further shows that the expected products in the presence or absence of a corner-to-corner proton jump are different. The product distributions found by Saunders et al. allow one to conclude that at a temperature of 213 K, the pathways involving corner-to-corner proton jumps on CPCP intermediates are much slower than those involving CPCP ring opening and closure only.[18]

The corner-to-corner proton jump could proceed through the formation of an edge-protonated cyclopropane (EPCP) intermediate or transition state,[18] as shown in Figure 2.7. The experiment by Saunders et al.[18] suggests that CPCP is more stable than EPCP, in agreement with the results of molecular orbital calculations.[20]

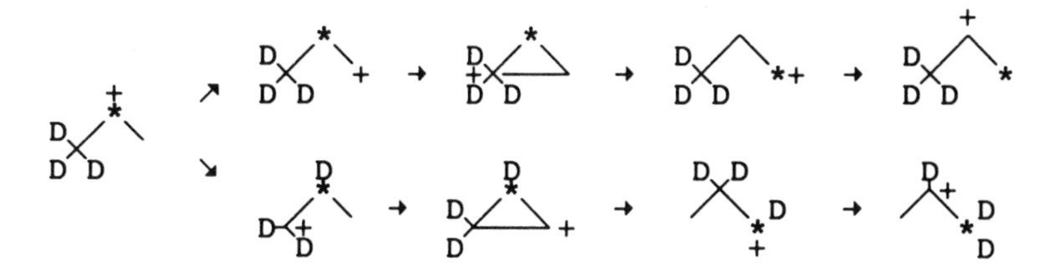

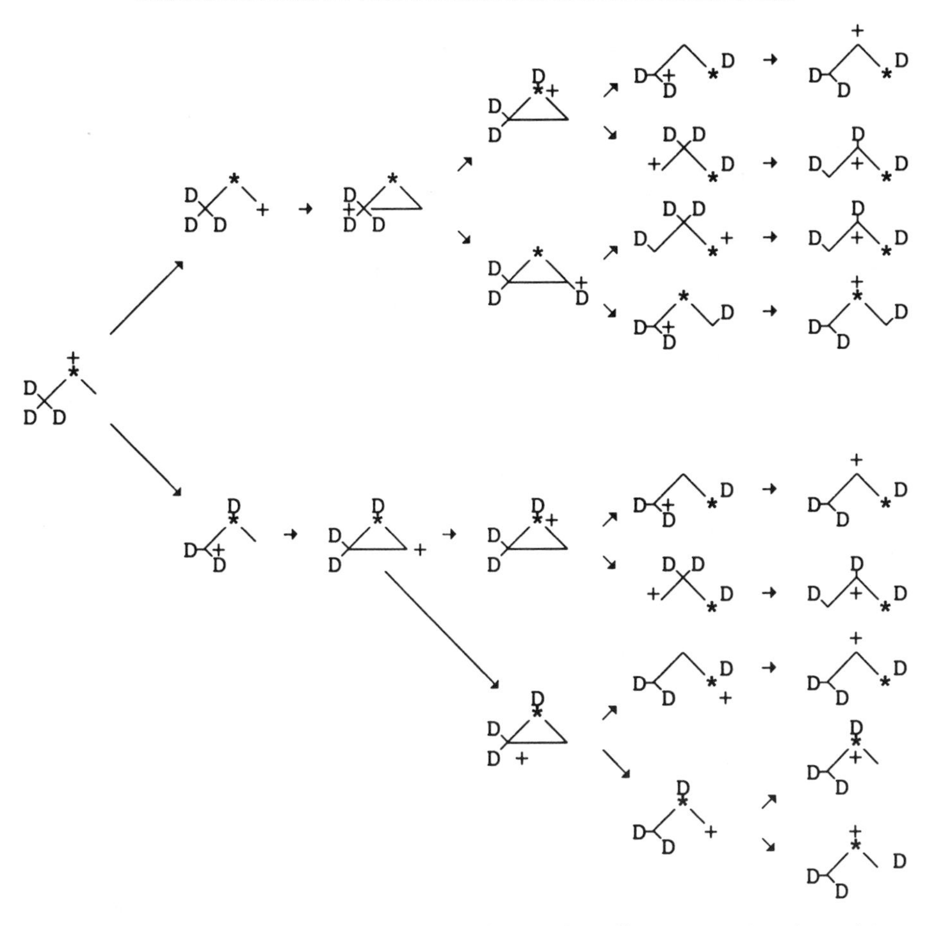

Figure 2.6. Rearrangements of 1,1,1-trideuterio-2-^{13}C-2-propyl cation with and without corner-to-corner proton (deuterium) jump according to mechanisms involving CPCP intermediates.

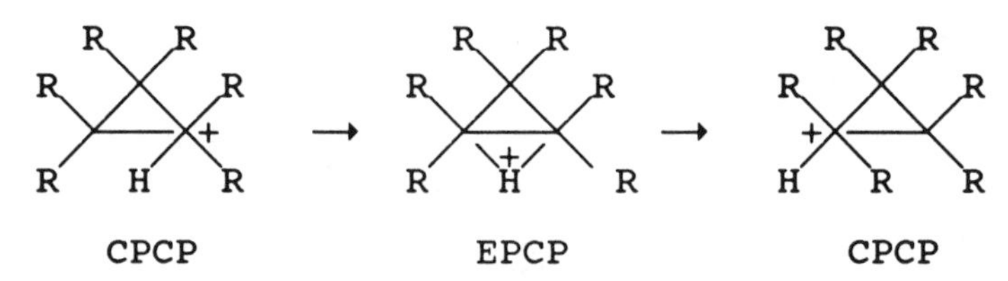

Figure 2.7. Corner-to-corner proton jump on a CPCP through the formation of an EPCP intermediate.

β-Scission

Fragmentation of alkylcarbenium ions in superacids proceeds via β-scission.[21] β-Scission involves the migration of two electrons of the C–C bond in the β position of the positively charged carbon atom toward the C–C bond in the α position. Thus, after the scission this α C–C bond becomes olefinic, while the carbon atom, originally in the γ position, ends up as the electron-deficient carbon atom of a smaller alkylcarbenium ion. Depending on the branching configuration of the original carbenium ion, there are five modes of β-scission reactions of secondary and tertiary alkylcarbenium ions possible.[22] These modes are shown in Figure 2.8 and are de-

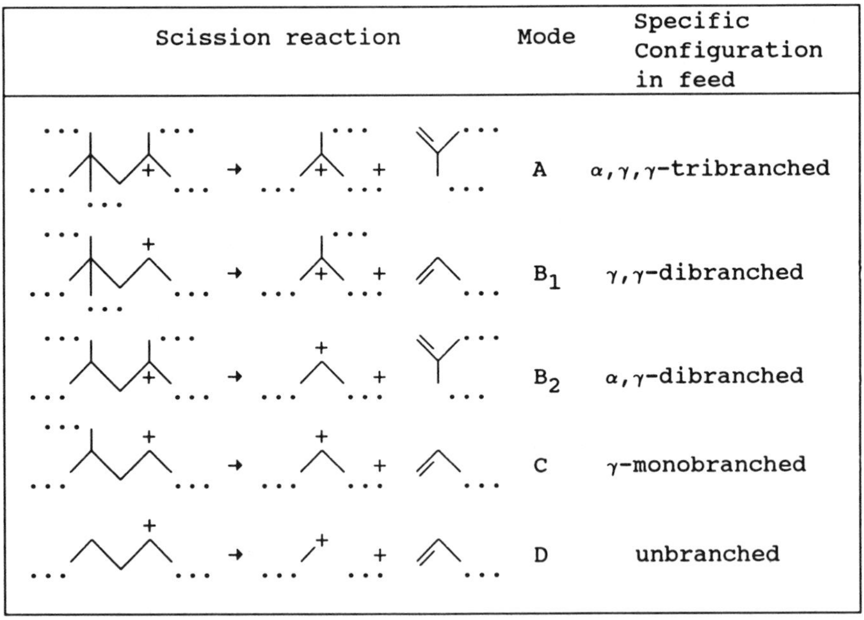

Scission reaction	Mode	Specific Configuration in feed
	A	α,γ,γ-tribranched
	B_1	γ,γ-dibranched
	B_2	α,γ-dibranched
	C	γ-monobranched
	D	unbranched

Figure 2.8. Specific branching configurations and β-scission modes of acyclic alkylcarbenium ions. The dashed lines stand for *n*-alkyl groups.

noted as A, B_1, B_2, C, and D types of β-scission. Table 2.2 summarizes the characteristics of the various β-scission modes of alkylcarbenium ions.

The distinction between the different β-scission mechanisms is based on the minimum number of side chains and their specific positions on the main hydrocarbon chain with respect to the position of the positively charged carbon atom. This way, each scission mode is characterized by the following **specific branching configurations** (SBC): α,γ,γ; γ,γ; α,γ; and γ for the scission modes A, B_1, B_2, and C, respectively. It is obvious that the positions of the side chains of immediate interest in these different types of β-scission cannot be farther away from each other than three carbon atoms. The presence of such specific branching arrangements will allow cracking to occur, irrespective of the degree of branching elsewhere in the chain.

Mechanism A requires the presence of three side chains, positioned in α,γ,γ. The smallest cation in which such configuration is possible has eight carbon atoms (Table 2.2); $C_7H_{15}^+$ and smaller alkylcarbenium ions are not susceptible to conversion by this mechanism. Accordingly, it is necessary to subdivide the alkylcarbenium ions into **short-chain** ions, having carbon numbers lower than eight, and **long-chain** ions, containing eight or more carbon atoms. Only long-chain alkylcarbenium ions are susceptible to β-scission according to the five mechanisms.

Mechanism A generates exclusively methylbranched fragments from tribranched parent ions, unless the substituents at the quaternary carbon atom in the parent ion are both ethyl or larger groups. As seen in Figure 2.8, mechanism B_1 acts on dibranched cations with a quaternary carbon atom in the γ-position to the positively charged carbon atom. It causes the formation of an equal number of normal and monobranched fragments from dibranched parent ions. The monobranched ions formed are also

Table 2.2. Characteristics of the β-scission modes of alkylcarbenium ions with specific branching configurations (SBC) only.

Scission Mode	A	B_1	B_2	C	D
Minimum carbon number required	8	7	7	6	4
Ions involved[a] as feed	t	s	t	s	s
as products	t	t	s	s	p
Minimum branching required[b]	T	D	D	M	U
Positions of the branchings (SBC)	α, γ, γ	γ, γ	α, γ	γ	—
Branching of products[c]	a + m	a + u	m + u	u + u	u + u

[a] t = tertiary, s = secondary, p = primary alkylcarbenium ion.
[b] T = tri-, D = di-, M = mono, and U = unbranched alkylcarbenium ion.
[c] a = alkylbranched; m = methylbranched; u = unbranched.

methylbranched, unless ethyl or larger groups constitute the two branchings in the parent ion (Fig. 2.8). The other β-scission mode involving at least dibranched ions is denoted as B_2. It can be distinguished from the B_1 route by the presence of a different configuration of the branchings in the parent ion. Figure 2.8 shows that according to this mechanism, α,γ-dibranched ions are converted into an equal number of monobranched and linear fragments. A B_2 β-scission starts from a tertiary ion and generates a secondary ion, while the opposite occurs for a scission of the B_1 type. From dibranched ions, the formation of an ethylbranched fragment together with a normal fragment is only possible with the B_1 but not with the B_2 mechanism (Fig. 2.8). In contrast to type A cracking, the cracking modes B_1 and B_2 can already operate on $C_7H_{15}^+$ ions.

Figure 2.8 further shows that for the occurrence of type C β-scission, only one single γ-positioned branching is required. In this scission mode a secondary ion is converted into a n-alkene and a smaller secondary n-alkylcarbenium ion. This cracking route is already possible with $C_6H_{13}^+$ ions (Table 2.2). Unbranched fragments are obtained via type C β-scission. For the alkylcarbenium ions $C_4H_9^+$ and $C_5H_{11}^+$, mechanism D is the only β-scission mechanism available. It involves the formation of primary cations (Fig. 2.8).

ALKYLCARBENIUM IONS ON THE SURFACE OF HETEROGENEOUS CATALYSTS

Alkylcarbenium Ion Formation from Alkenes on Catalysts with Brønsted Acidity

Formally, the formation of an alkylcarbenium ion from an alkene can be visualized as follows. The attack of a proton of the acid catalyst on the π electrons of an alkene leads to the formation of a σ bond with one of the carbon atoms of the double bond, while the second carbon atom of the double bond becomes positively charged:

$$R_2C{=}CR_2 + H^+Z^- \longrightarrow R_2CH{-}C^+R_2 + Z^- \qquad (2.15)$$

in which Z^- represents the catalyst.

A bifunctional catalyst consists of a metal phase, which is dispersed on a support with Brønsted acidity. Platinum and palladium are currently used metals. Zeolites and silica–alumina are examples of acidic supports. When the metal phase dehydrogenates the alkanes into alkenes, formally the generation of alkylcarbenium ions can readily occur through protonation of the double bond in the alkenes on Brønsted acid sites. This classical reaction scheme, which has been proposed by Weisz[23] and Coonradt

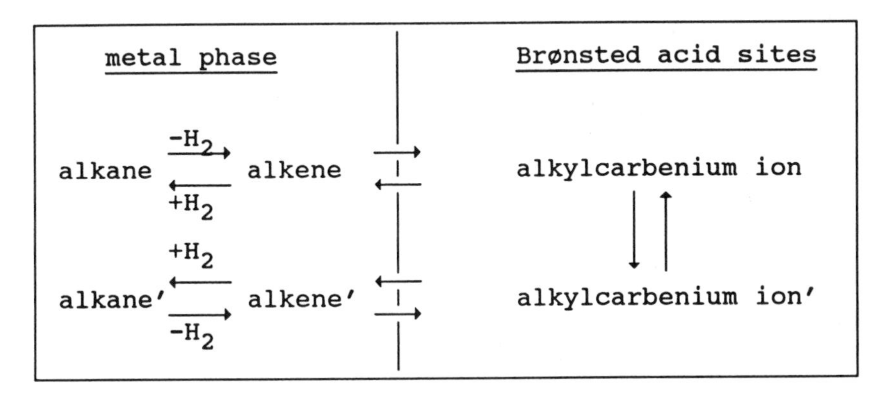

Figure 2.9. Classical bifunctional reaction scheme.

and Garwood,[24] is shown in Figure 2.9. Provided that hydrogenation–dehydrogenation and protonation–deprotonation are fast enough, the rearrangements of the alkylcarbenium ions become the rate-limiting steps of the reaction scheme.

Evidence for Free Alkylcarbenium Ions in Superacids and Zeolites

The properties of alkylcarbenium ions in superacids vary little over a range of media, indicating that these carbocations are well-established entities and that the anion and the solvent play a nonspecific role.[16,25] Evidence from the literature will now be presented showing that carbocations with a similar degree of freedom occur as reaction intermediates during the conversion of hydrocarbons in the intracrystalline void volumes of large-pore acid zeolites.

Simple intramolecular reactions of alkylcarbenium ions (type *A* and *B* isomerization) have an entropy of activation, $\Delta S\ddagger$, that is practically zero.[21] Hence, the free energy of activation

$$\Delta G\ddagger = \Delta H\ddagger - T\,\Delta S\ddagger \tag{2.16}$$

is practically temperature independent and the Arrhenius activation energy E_a becomes

$$E_a = \Delta G\ddagger + RT \tag{2.17}$$

Consequently, an increase of the temperature by 100 K results in an increase of E_a of only 0.8 kJ mol^{-1}. It can therefore be expected that the activation energy does not vary too much with temperature.

A study on the isomerization of 2-methyl-2-pentene and 2-methyl-1-pentene over an ultrastable zeolite Y (USY) showed a close correspondence between the apparent activation energies for methyl migrations, catalyzed by zeolites at temperatures in the range from 323 to 473 K and by superacids at temperatures below 273 K.[26] A value of 40–58 kJ mol^{-1} was found with the zeolite, compared to 46 kJ mol^{-1} with the superacid.

Other support for the occurrence of free alkylcarbenium ions on zeolite USY at temperatures of 453–493 K has been provided by Froment.[27] The isomerization of octane into the three monomethylheptanes over bifunctional zeolite USY could be modeled using the classical bifunctional reaction scheme, and rearrangements of alkylcarbenium ions by 1,2-hydride shift, 1,2-methyl shift, and branching via PCP. In Table 2.3 the estimated activation energies from that model are compared to activation energies of corresponding rearrangements of C_4 to C_8 alkylcarbenium ions in superacids. Again, the correspondence between the two sets of data is striking. It can be concluded that the extrapolation of the low-temperature kinetic parameters for the rearrangements of carbocation species in superacids to the operation temperature of heterogeneous catalysts should be extremely relevant.

Some rate constants of alkylcarbenium ion rearrangements, measured in superacid at a temperature of 195 K, were extrapolated in Table 2.4 to temperatures of 400 and 500 K using the Eyring equation:

$$k = \left(\frac{k^*T}{h}\right) \exp\left(-\frac{\Delta G\ddagger}{RT}\right) \qquad (2.18)$$

in which k^* and h are the Boltzmann and Planck constants, respectively. Table 2.4 shows that at 195 K the rate of a 1,2-methyl shift in 2-methyl-2-pentyl cation is of the same order of magnitude as that of an 1,2-ethyl shift

Table 2.3. Activation energy (kJ mol^{-1}) of rearrangements of C_8 alkylcarbenium ions on Pt/USY zeolite[27] and C_4 to C_8 alkylcarbenium ions in superacid.[16,21]

	Activation Energy	
	Pt/USY (at 453–493 K)	Superacid (at 143–196 K)
1,2-Hydride shift		
secondary into secondary cation	11	<13
tertiary into secondary cation	59	46
1,2-Methyl shift		
secondary into secondary cation	25	21

Table 2.4. Rate constants (k) of some alkylcarbenium ion rearrangements at temperatures of 400 and 500 K, obtained by extrapolation of kinetic data from Ref. 21 measured in superacid at 195 K.

	$k_{195\ K}$ ($10^{-7}\ s^{-1}$)	$\Delta G\ddagger$ (kJ mol^{-1})	$k_{400\ K}$ ($10^{3}\ s^{-1}$)	$k_{500\ K}$ ($10^{5}\ s^{-1}$)
1. 1,2-Methyl shift				
C—C$^+$(C)—C—C—C → C—C—C$^+$(C)—C—C	8,000	60	200	100
2. 1,2-Ethyl shift				
C—C—C$^+$(C—C)—C—C → C—C—C—C$^+$(C—C)—C	5,000	59	100	50
3. Di- to mono unbranching				
C—C(C)—C$^+$(C)—C → C—C$^+$(C)—C—C—C	7	70	5	4
4. Mono- to *n*-unbranching				
C—C$^+$(C)—C—C → C—C$^+$—C—C—C	0.2	77	0.9	1
5. Type *A* β-scission				
C—C(C)(C)—C—C$^+$(C)—C → C—C$^+$(C)(C) + C—C(C)=C	5,000[a]	—		

[a] At 200 K.

in 3-ethyl-3-pentyl cation. As both reactions show practically the same activation energy, this similarity is maintained at higher temperatures. At a temperature of 195 K, type *A* isomerization of $C_6H_{13}^+$ ions is approximately 1000 times faster than their type *B* conversion from di- to monobranched ions and 40,000 times faster than unbranching of the 2-methyl-2-butyl cation (Table 2.4). As type *B* compared to type *A* rearrangements have substantially higher activation energies, extrapolation to 500 K allows one to conclude that the rate of unbranching of di- to monobranched $C_6H_{13}^+$ cation is only 25 times slower than their isomerization via methyl migration. Under these conditions, the rate constant of unbranching of di- to monobranched $C_6H_{13}^+$ cation remains only four times higher than that for unbranching of the monobranched $C_5H_{11}^+$ cation.

Table 2.4 also shows that in superacids and at 200 K, the rate of type *A*

β-scission of 2,4,4-trimethyl-2-pentyl cation is comparable to that of type *A* isomerization and much faster than the rate of type *B* isomerization. Under these conditions the B_1, B_2, and C β-scission pathway involving the cleavage of linear, mono-, di-, or even tribranched cations is very slow, even compared to type *B* isomerization.[21] Unfortunately, no values for the activation energy of the different β-cleavage reactions in superacids are available for extrapolation to elevated temperatures.

CARBOCATION CHEMISTRY AND BIFUNCTIONAL CONVERSION OF SHORT-CHAIN ALKANES

A study of the conversion of several C_4 to C_6 alkanes on silica–alumina (SA), containing 1.2 wt% of platinum metal (1.2 Pt/SA), revealed that their selective isomerization was possible.[28] Relative values for the apparent rate constants of the isomerization of C_4, C_5 and C_6 alkanes over the bifunctional catalyst mentioned are given in Table 2.5. These data allow us to make the following general statements.

Table 2.5. Comparison of relative isomerization rates of short-chain alkanes in superacids and on a bifunctional catalyst. *A*, apparent rate constant[a] for isomerization over 1.2 Pt/SA at 573 K (after Ref. 27); *B*, true rate constant of the corresponding rearrangements of alkylcarbenium ions in superacids (after Ref. 21), extrapolated to 573 K.

Rearrangement	*A*	*B*
Butane → isobutane	0.02	
[1-^{13}C]Butane → [2-^{13}C]butane	2.7	
Pentane → isopentane	7.6	
Isopentane → pentane	3.2	1.8
Hexane → 2-methylpentane	13.2	
2-Methylpentane → hexane	6.5	
Hexane → 3-methylpentane	9.7	
3-Methylpentane → hexane	7.5	
2-Methylpentane → 3-methylpentane	100.0	100.0
3-Methylpentane → 2-methylpentane	152.0	
2,3-Dimethylbutane → 2-methylpentane	1.6	5.8

[a] Defined by the authors as the product of the true rate constant times the dehydrogenation equilibrium constant.

- The rate constant of ^{13}C scrambling in butane is much higher than that of its skeletal isomerization and is even of the same order of magnitude as the rate for isomerization of pentane and hexane. These data suggest the occurrence of the type *B* isomerization mechanisms, involving CPCP intermediates.
- The isomerizations of 2- to 3-methylpentane and vice versa are the fastest reactions among those listed in Table 2.5. In these reactions the degree of branching does not change, while it does in all the other reactions. This observation can readily be explained by the difference in rate that exists between types *A* and *B* isomerization reactions of alkylcarbenium ion intermediates.

In Table 2.5, the rate data for isomerization of light alkanes on the 1.2 Pt/SA catalyst are compared with the relative rate constants of the corresponding alkylcarbenium rearrangements in superacids, extrapolated to the same temperature. The direct comparison of the two sets of data, listed in Table 2.5, requires some caution. Indeed, the apparent isomerization rate constants cannot be compared directly with the rate constants of alkylcarbenium ion rearrangements. Strictly speaking, the rates of alkane isomerization over a heterogeneous catalyst are the product of a rate constant and a surface concentration. The data of Table 2.5 are only comparable if the coverage of the active sites is the same in all instances. Although no evidence is available that this is true, the similarity of the two sets of data is surprisingly good.

At a temperature of 573 K, the bifunctional conversion of an alkane involving a change in the degree of branching is one to two orders of magnitude slower than for conversions that do not cause such changes (Table 2.5). This was predicted in Table 2.4 by extrapolation of the rate constants of type *A* and *B* rearrangements of alkylcarbenium ions in superacids to such temperatures.

The similarities between isomerization of light alkanes over a bifunctional heterogeneous catalyst and of alkylcarbenium ions in superacids are also strong arguments in favor of the occurrence of superacid chemistry and the existence of carbocations. Their generation must occur via the classical bifunctional reaction scheme, and free intermediate olefins must be formed.

CARBOCATION CHEMISTRY AND BIFUNCTIONAL CONVERSION OF LONG-CHAIN ALKANES

Studies by ^{13}C and ^{1}H NMR on the rearrangements of long-chain alkylcarbenium ions in superacid solutions are lacking in the literature, possibly because of the high degree of complexity they may involve. It will be shown in this section that the features of the products obtained from the conversion of long-chain alkanes on heterogeneous catalysts can be explained perfectly, based on the occurrence of alkylcarbenium ions as unstable, transient intermediates and using the general rules of their rearrangements as determined for short-chain cations in superacids. In addition, isomerization and hydrocracking of higher alkanes over bifunctional catalysts wil turn out to be powerful tools for studying the rearrangements of long-chain alkylcarbenium ions.

Overall Conversion Processes of Long-Chain *n*-Alkanes

When a long-chain *n*-alkane, such as decane, is reacted in the presence of hydrogen over Pt/USY, no products other than alkanes are formed. The formation of mono-, di-, and tribranched isodecanes and cracked products follows a typical pattern when plotted against the degree of conversion of the feedstock (Fig. 2.10), indicating that the monobranched isomers

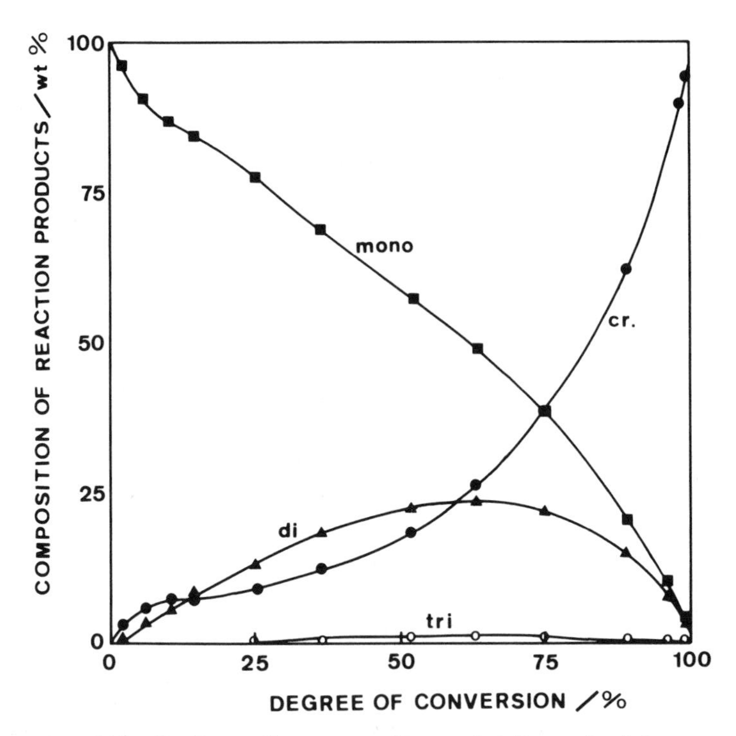

Figure 2.10. Distribution of mono-, di-, and tribranched isomers and cracked products (in weight%) from decane against its degree of conversion (%) using a Pt/USY catalyst. The total reaction pressure was 0.35 MPa, the H_2/decane molar ratio was 100, the molar flow rate at the reactor inlet (F_0/W) was 0.4 mmol kg^{-1} s^{-1}. The conversion was changed by increasing the reaction temperature from 400 to 500 K.

are primary products from decane, while the di- and tribranched isodecanes as well as the cracked products are at best of secondary nature. In this case cracked products refer to alkanes with a carbon number lower than 10.

Branching isomerization of long-chain *n*-alkanes over a Pt/USY bifunctional catalyst clearly is a step-by-step reaction. Indeed, Figure 2.10 shows that mono-, di-, and tribranched feed isomers are obtained in consecutive reactions. This sequence of branching has also been observed with octane,[29,30] decane and undecane,[31,32] and tridecane,[29] using Pt/USY[30,32] as well as Pt/CaY[29,31] as bifunctional catalysts. It can therefore be considered as a general feature in isomerization of long-chain *n*-alkanes.

Formation of Monobranched Isomers from *n*-Alkanes

All possible monobranched isodecanes are detected in the stream of reaction products formed from decane over the Pt/USY catalyst. The monobranched isodecanes encompass the methylbranched isomers (2-, 3-, 4-, and 5-methylnonane), as well as the ethylbranched and propylbranched ones (3-ethyloctane, 4-ethyloctane, and 4-propylheptane). A peculiarity in the distribution of the methylbranched isomers from decane is the reduced formation of 2-methylnonane, mainly at low levels of conversion (Fig. 2.11). As a result, under these conditions all other methylnonanes are found in amounts exceeding their equilibrium value. This hindrance in the formation of methylbranchings in 2-position should therefore be of kinetic origin.

In Figure 2.11 it is shown that the composition of the methylnonanes gradually changes toward a constant composition, which should reflect their thermodynamic equilibrium. Figure 2.11 further indicates that the literature data on this equilibrium are inaccurate and that the present experimental method of determining thermodynamic equilibrium data is preferable.

In Figure 2.12 the distribution of the monobranched feed isomers is shown against the degree of conversion of decane over a Pt/USY catalyst. Next to the methylnonanes, the ethyloctanes and perhaps also 4-propylheptane could be primary products.

Branching via Substituted Protonated Cycloalkanes Larger than Cyclopropane

By analogy to the methylbranching mechanism that involves CPCP intermediates, a mechanism via substituted corner protonated cyclobutanes (CPCB) could be invoked to explain the formation of ethyloctanes as primary products. The formation of longer side chains, i.e., propyl and

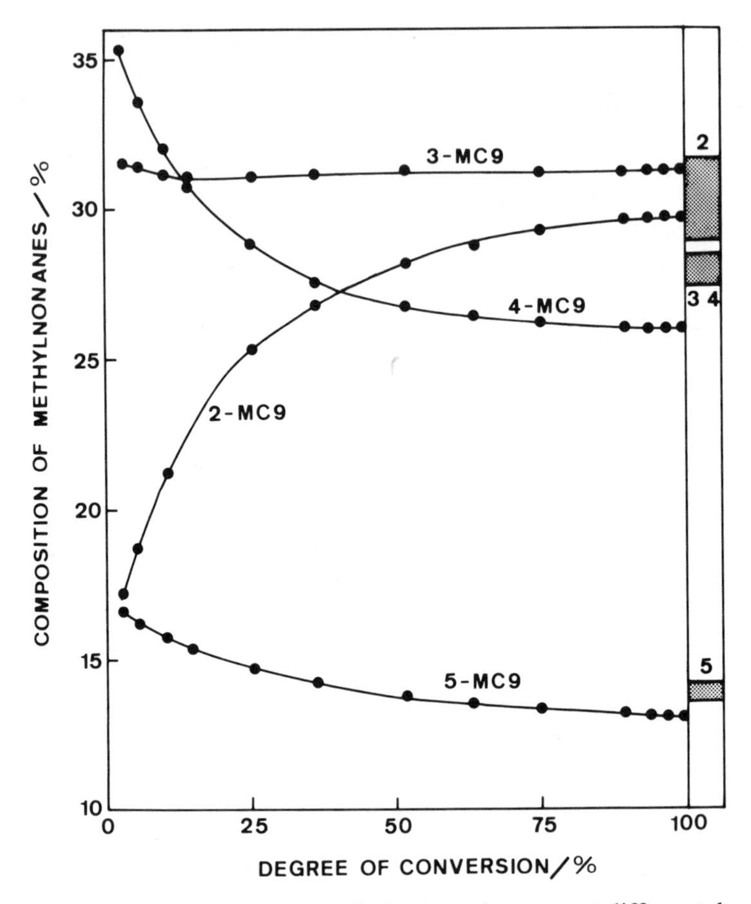

Figure 2.11. Distribution of the methylnonane isomers at different degrees of conversion of decane over Pt/USY, using the reaction conditions described in Figure 2.10. The hatched areas at the right side refer to their calculated thermodynamic equilibrium levels (after Ref. 33).

butyl branchings, would then proceed through substituted corner protonated cyclopentane (CPCPe) and substituted corner protonated cyclohexane (CPCH) intermediates, respectively. Examples of ethylbranching, propylbranching, and butylbranching mechanisms of tridecyl cations, involving large substituted protonated rings, are shown in Figure 2.13. By analogy to the branching mechanism via CPCP intermediates, a corner-to-corner migration of a proton on the CPCB, CPCPe, and CPCH intermediates has to take place before ring opening can occur. In contrast with the ring opening of CPCP intermediates, for which there are always two

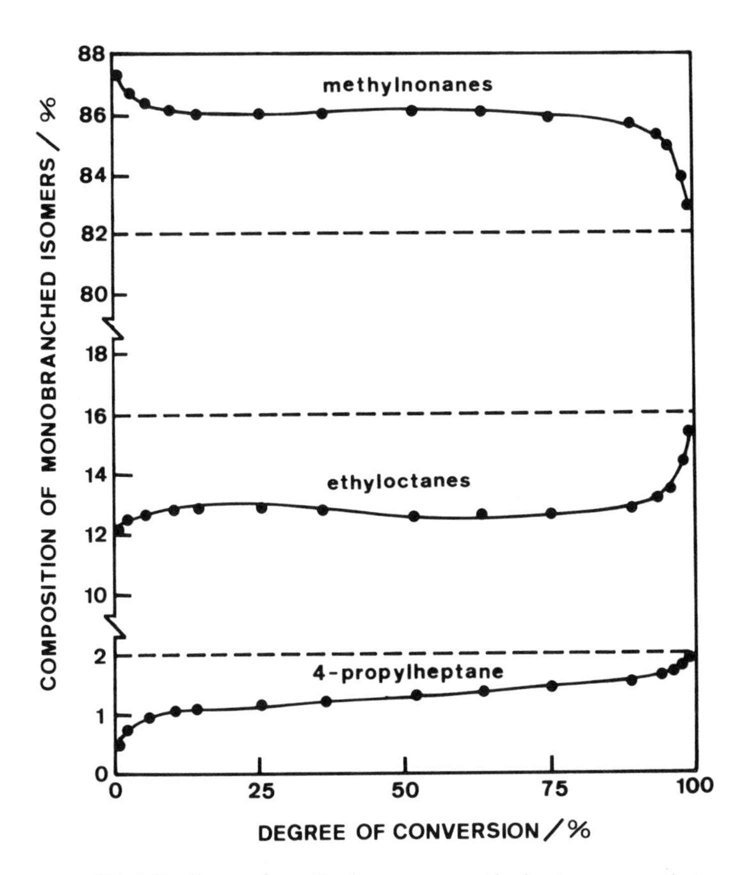

Figure 2.12. Distribution of methylnonanes, ethyloctanes, and 4-propylheptane in the monobranched isomerization products against the degree of conversion of decane over Pt/USY. The reaction conditions are described in Figure 2.10. The dashed lines correspond to the thermodynamic equilibrium at the temperature range of interest (after Ref. 33).

competing pathways, it can be seen in Figure 2.13 that, with larger rings in order to avoid primary alkylcarbenium ions, there is only one ring opening mode possible after the occurrence of a single corner-to-corner proton jump.

For the branching rearrangements of short-chain alkylcarbenium ions in superacids, there is no need to invoke mechanisms via cyclic alkylcarbonium ions, larger than PCP. Protonated cyclobutane (PCB) is 130 kJ mol^{-1} less stable than protonated methylcyclopropane in the gas phase.[34] On this basis, it has been suggested that in superacids carbon atom scrambling in 2-butyl cation should proceed through a protonated methylcyclo-

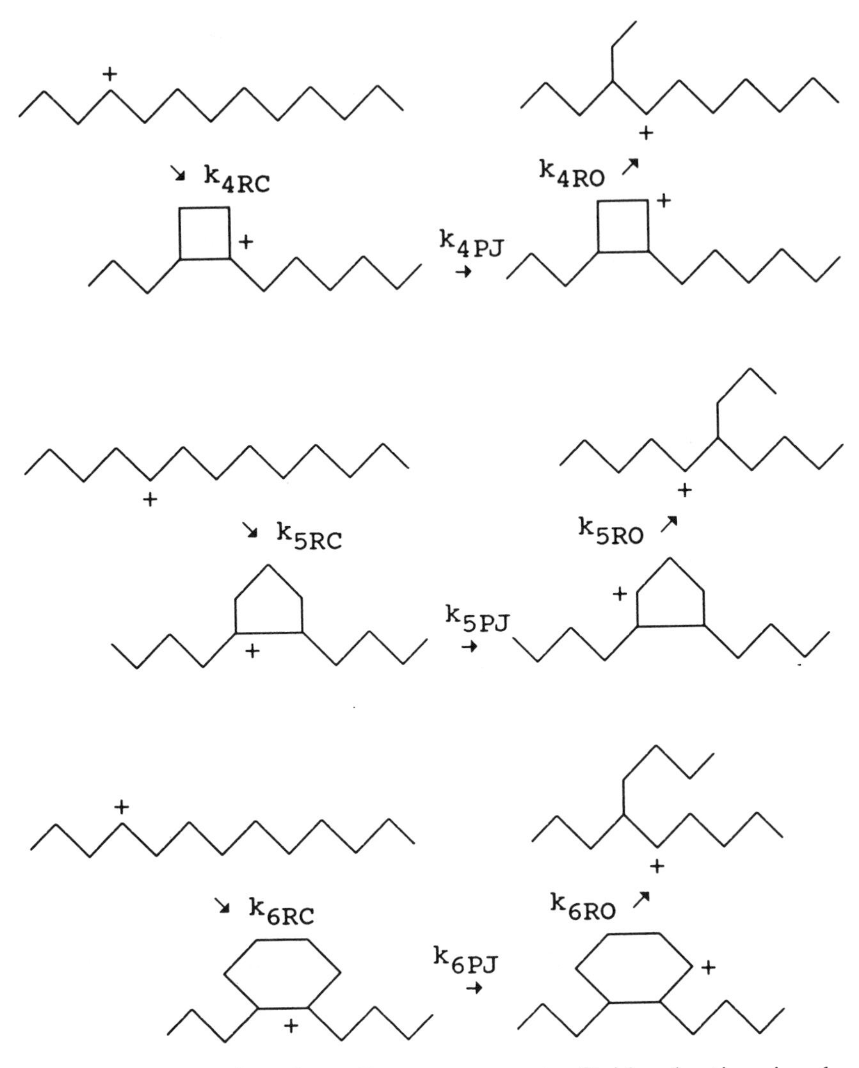

Figure 2.13. Examples of type *B* rearrangements of tridecyl cations involving CPCB, CPCPe, and CPCH intermediates. The indices RC, PJ, and RO of the rate constants refer to ring closure, proton jump, and ring opening, respectively.

propane and not through a protonated cyclobutane intermediate.[35] All this, however, does not preclude that in the bifunctional conversion of long-chain *n*-alkanes on heterogeneous catalysts, CPCB and other alkylcarbonium ions containing larger rings can play an important role, since reactions with high activation energies may become important at higher temperatures.

Side-Chain Elongation via Type *A* Isomerization

An alternative mechanism for the formation of larger than methyl side chains is through consecutive shifts of ethyl and bulkier alkyl groups on methylbranched cations. This mechanism is shown in detail in Figure 2.14 for the consecutive transformation of the 5-methyl-5-nonyl cation into 4-ethyl-4-octyl and 4-propyl-4-heptyl cation. The rearrangements correspond to a step-by-step elongation of the side chain, viz., a methyl side chain is transformed into an ethylbranching, which in its turn can be converted to a propyl group. In this mechanism, however, a direct interconversion of methyl and propyl side chains is precluded. Formally, the elongation of a methyl side chain can be considered to proceed through the 1,2-shift of an alkyl group, which is larger than a methyl group. For example, the conversion of 4-methyl-3-pentyl cation into 3-ethyl-2-pentyl

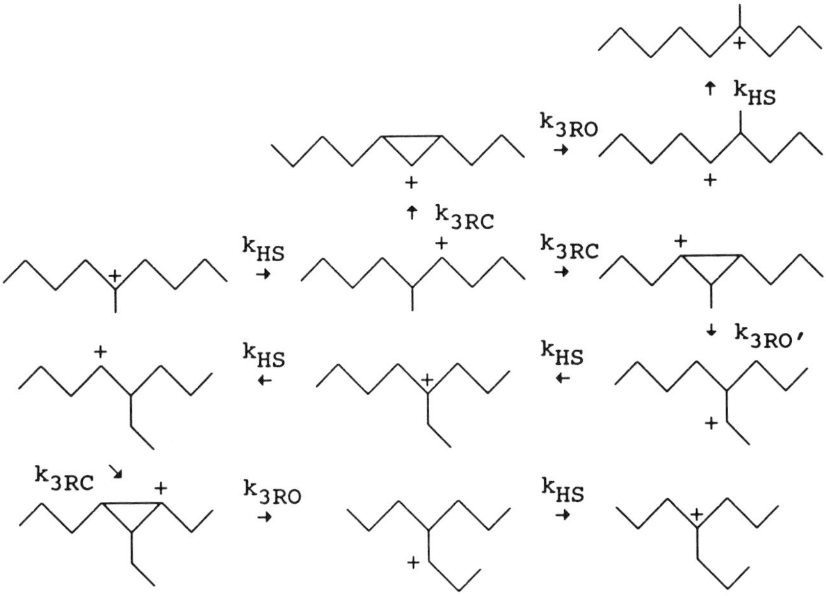

k_{HS} k_{3RO} k_{3RC} k_{HS} k_{3RC} $k_{3RO'}$ k_{HS} k_{HS} k_{3RC} k_{3RO} k_{HS}

Figure 2.14. 1,2-Methyl shift and one-carbon-at-a-time elongation of the methyl side chain in 5-methyl-5-nonyl cation via CPCP intermediates. The index HS of the rate constant refers to 1,2-hydride shift.

cation corresponds to the shift of a propyl group as shown in the following:

$$\text{(skeletal structure)}^{+} \rightarrow \text{(skeletal structure)}^{+} \qquad (2.19)$$

For the formation of monobranched isodecanes out of decane, according to the side chain elongation mechanism via alkyl shifts the following conversion sequence is expected:

```
            2-methylnonane
          ↗
            3-methylnonane  ↘
        ↗                     3-ethyloctane
decane  ↘                   ↗
            4-methylnonane                      ↗ 4-propylheptane
          ↘                 ↘
                              4-ethyloctane
                            ↗
            5-methylnonane
```

(2.20)

According to this mechanism, all monobranched isomers with larger than methyl side chain are at best of secondary nature. The larger the side chain of a monobranched alkane, the later this isomer is formed when starting from the *n*-alkane.

Compositions of monobranched feed isomers, obtained at less than 5% conversion of *n*-alkanes in the range from hexane to pentadecane over bifunctional faujasites, have been collected in Table 2.6. An experimental difficulty in determining initial compositions of the monobranched isomerization products from *n*-alkanes is that the commercially available feedstocks are always contaminated with a few tenths of a percent of methylbranched isomers. The initial compositions of monobranched isomerization products, listed in Table 2.6, are product distributions, which have been corrected assuming no conversion of the methylbranched isoalkane impurities. As the latter is certainly not true, a substantial error is always introduced. Nevertheless, the data of Table 2.6 indicate that ethyl-, propyl-, and butylbranchings are formed at these low levels of conversion and suggest that monobranched isomers with bulky *n*-alkyl side chains are primary isomerization products from long-chain *n*-alkanes.

These experimental observations are therefore in contrast with the occurrence of a mechanism involving a one-carbon-atom-at-the-time elongation of side chains. Moreover, it is difficult to rationalize that at very low levels of conversion of decane, step-by-step elongation of the methyl side chain of methylnonane products accounts for the formation of ethyloc-

Table 2.6. Distribution (%) of the monobranched isomers obtained from hexane through pentadecane *n*-alkanes over a Pt/CaY (after Ref. 31) and a Pt/USY catalyst at less than 5% conversion and prediction according to the branching mechanism via CPCP intermediates.[a]

	Pt/USY	Pt/CaY	Hyp[b]	CPCP[c]
Hexane				
2-methylpentane	57.8	59.5	57.8	50
3-methylpentane	42.2	40.5	42.2	50
Heptane				
2-methylhexane	42.2 (44.1)	44.6 (46)	40.0	33.3
3-methylhexane	53.5 (55.9)	52.3 (54)	56.6	66.6
3-ethylpentane	3.4	3.1	3.4	
Octane				
2-methylheptane	27.5 (29.1)	30.2 (30.6)	27.5	25
3-methylheptane	46.5 (49.2)	49.0[d] (49.7)	39.4	50
4-methylheptane	20.6 (21.8)	19.4 (19.7)	27.5	25
3-ethylhexane	5.5	1.4[d]	5.5	
Nonane				
2-methyloctane	19.0 (20.9)	20.3 (22.4)	19.0	20
3-methyloctane	35.1 (38.6)	34.8 (38.4)	32.5	40
4-methyloctane	36.9 (40.5)	35.5 (39.2)	39.5	40
3-ethylheptane	4.9	5.1	4.9	
4-ethylheptane	4.1	4.3	4.1	
Decane				
2-methylnonane	15.0 (17.0)	15.8 (17.8)	15.0	16.7
3-methylnonane	27.6 (31.3)	27.0 (30.4)	26.5	33.3
4-methylnonane	31.0 (35.2)	30.4 (34.2)	31.2	33.3
5-methylnonane	14.5 (16.5)	15.6 (17.6)	15.4	16.7
3-ethyloctane	4.9	4.5	4.9	
4-ethyloctane	6.5	6.4	6.5	
4-propylheptane	0.5	0.3	0.5	
Undecane				
2-methyldecane	12.7 (14.3)	11.6 (12.9)	12.7	14.3
3-methyldecane	23.1 (26.0)	22.3 (24.8)	24.0	28.6
4-methyldecane	25.8 (29.1)	25.9 (28.8)	25.9	28.6
5-methyldecane	27.1 (30.6)	30.0 (33.4)	26.2	28.6
3-ethylnonane	3.0	2.5	3.0	
4-ethylnonane	4.7	4.5	4.7	
5-ethylnonane	2.3	2.1	2.3	
4-propyloctane	1.4	1.1	1.4	
Dodecane				
2-methylundecane	11.4 (14.3)	9.4 (10.8)	11.4	12.5
3-methylundecane	21.7 (24.9)	17.8 (20.4)	17.8	25
4-methylundecane	20.8 (23.9)	22.2 (25.4)	23.1	25
5-methylundecane	22.5 (25.8)	24.7 (28.3)	23.2	25
6-methylundecane	10.7 (12.3)	13.2 (15.1)	11.7	12.5
3-ethyldecane	3.1	2.8	3.1	
4-ethyldecane	4.0	4.0	4.0	
5-ethyldecane	4.2	4.2	4.2	
4-propylnonane	0.8	0.8	0.8	
5-propylnonane	0.8	0.7	0.8	

Table 2.6. (Continued).

	Pt/USY	Pt/CaY	Hyp[b]	CPCP[c]
Tridecane				
2-methyldodecane	9.3 (11.0)	7.0 (8.1)	9.3	11.1
3-methyldodecane	17.9 (21.3)	15.1 (17.4)	17.7	22.2
4-methyldodecane	18.0 (21.4)	18.4 (21.2)	19.0	22.2
5-methyldodecane	19.6 (23.3)	22.8 (26.3)	19.1	22.2
6-methyldodecane	19.4 (23.0)	23.4 (27.0)	19.1	22.2
3-ethylundecane	2.7	1.9	2.7	
4-ethylundecane	3.4	3.4	3.4	
5-ethylundecane	3.9	3.6	3.9	
6-ethylundecane	2.2	2.0	2.2	
4-propyldecane	1.0	0.7	1.0	
5-propyldecane	1.8	1.4	1.8	
5-butylnonane	0.7	0.3	0.7	
Tetradecane				
2-methyltridecane		5.8 (6.8)		10
3-methyltridecane		12.1 (14.3)		20
4-methyltridecane		15.5 (18.3)		20
5-methyltridecane		20.4 (24.1)		20
6-methyltridecane } 7-methyltridecane }		31.0 (36.6)		20 10
3-ethyldodecane		1.5		
4-ethyldodecane		3.2		
5-ethyldodecane		3.7		
6-ethyldodecane		3.6		
4-propylundecane		0.7		
5-propylundecane		1.5		
6-propylundecane		0.6		
5-butyldecane		0.4		
Pentadecane				
2-methyltetradecane		6.4 (7.7)		9.1
3-methyltetradecane		12.0 (14.4)		18.2
4-methyltetradecane		13.5 (16.2)		18.2
5-methyltetradecane		16.9 (20.3)		18.2
6-methyltetradecane		17.2 (20.6)		18.2
7-methyltetradecane		17.4 (20.9)		18.2
3-ethyltridecane		1.5		
4-ethyltridecane		3.1		
5-ethyltridecane		3.3		
6-ethyltridecane		3.1[d]		
7-ethyltridecane		1.6[d]		
4-propyldodecane		0.8		
5-propyldodecane		1.6		
6-propyldodecane		1.6		
5-butylundecane } 6-butylundecane }		1.0		

[a] The compositions of the methylbranched isomers alone are given within brackets. The origin of the hypothetical (Hyp) composition is explained in the text.
[b] Hypothetical composition.
[c] According to the mechanism via CPCP intermediates.
[d] Estimated percentages.

tanes and 4-propylheptane, which represent some 12% of the products (Fig. 2.12), while at the same moment the initial composition of the methylnonanes is not much perturbed by methyl shifts (Fig. 2.11). Evidence exists, indeed, that for long-chain alkylcarbenium ions,[36] side chain elongation is **not faster** than shifts of methyl groups, in agreement with the reaction scheme shown in Figure 2.14. At 4.9% conversion of 5-methylnonane over Pt/CaY the reaction products contained 52.7% of 4-methylnonane and only 28.0% of 4-ethyloctane. Therefore, side chain elongation via alkyl shifts **cannot** be the most important pathway for the formation of ethyl or bulkier linear side chains during isomerization of a *n*-alkane.

Branching of Decyl Cations via CPCP Intermediates

Figure 2.15 shows the reaction scheme for branching of the secondary decyl cations via CPCP intermediates. This scheme has been developed using the mechanism for branching of short-chain alkylcarbenium ions. The CPCP intermediates formed upon ring closure are denoted as $CPCP_i^*$, where i refers to the position of the ring in the carbon skeleton. The position of the ring is formally represented as follows:

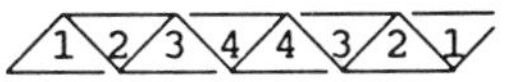

$$1 \quad 2 \quad 3 \quad 4 \quad 4 \quad 3 \quad 2 \quad 1 \tag{2.21}$$

The CPCP intermediates obtained after the corner-to-corner proton jump are denoted "without asterisk."

Only $CPCP_2^*$ can be formed out of the 2-decyl cation, while out of the other secondary cations two $CPCP_i^*$ intermediates can be formed. Corner-to-corner proton jumps rearrange the primarily obtained $CPCP_i^*$ ions into $CPCP_i$ ions in which the positive charge is now located on a ring carbon atom, free of alkyl side chains. $CPCP_2$ can thus be obtained from 2-decyl cation, $CPCP_1$ and $CPCP_3$ from 3-decyl cation, $CPCP_2$ and $CPCP_4$ from 4-decyl cation, and $CPCP_3$ and $CPCP_4$ from 5-decyl cation. The formation of $CPCP_1$ does not lead to the formation of a methylbranching upon opening of the cyclopropane ring, unless a primary cation is formed. Therefore, this reaction path is considered to be very unlikely. Upon ring opening of $CPCP_2$, $CPCP_3$, or $CPCP_4$, a secondary methylnonyl cation is obtained. In each case one supplementary hydride shift is sufficient to obtain a tertiary methylnonyl cation. Upon ring opening of $CPCP_2$, $CPCP_3$, and $CPCP_4$, two different positions for the methylbranching can be obtained.

As CPCP formation is fast with respect to the corner-to-corner proton jumps, the $CPCP_i^*$ structures should be equilibrated through ring opening and closure reactions and hydride shifts on the secondary alkylcarbenium ions. The distribution of the CPCP ions before the occurrence of corner-

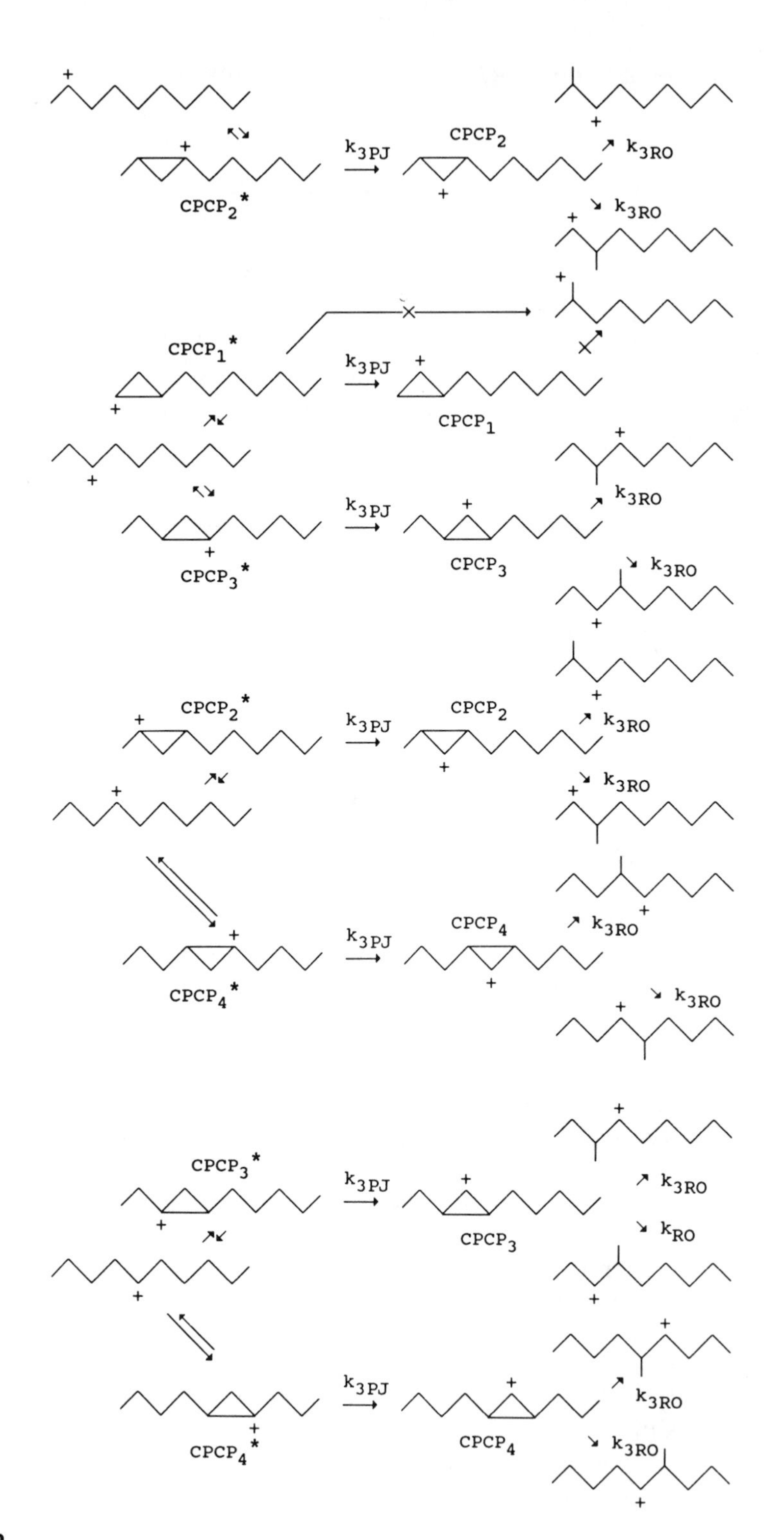
CPCP$_2$
k_{3PJ}
k_{3RO}
k_{3RO}
CPCP$_2$*
CPCP$_1$*
k_{3PJ}
CPCP$_1$
k_{3PJ}
k_{3RO}
CPCP$_3$
k_{3RO}
CPCP$_3$*
CPCP$_2$*
k_{3PJ}
CPCP$_2$
k_{3RO}
k_{3RO}
CPCP$_4$
k_{3PJ}
k_{3RO}
k_{3RO}
CPCP$_4$*
CPCP$_3$*
k_{3PJ}
k_{3RO}
k_{RO}
CPCP$_3$
k_{3PJ}
k_{3RO}
CPCP$_4$*
CPCP$_4$
k_{3RO}

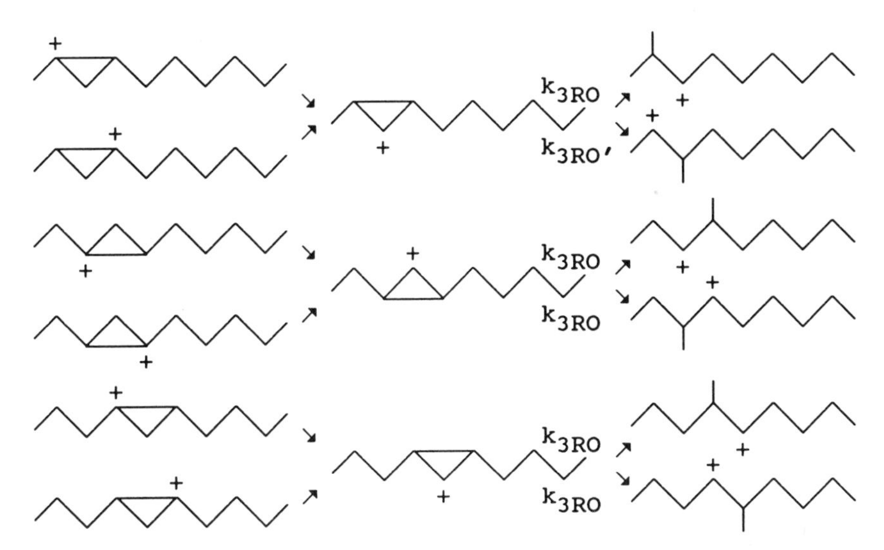

Figure 2.16. Simplified scheme for branching of decyl ions via CPCP intermediates.

to-corner proton jump, should, therefore, reflect their relative stabilities. As the stabilities of CPCP ions with the same carbon number are very close, it is reasonable to assume that their surface coverage is identical.

Reaction schemes for branching via CPCP alkylcarbonium ion intermediates can now be simplified. It is sufficient to consider all different CPCP* structures that can be formed by ring closure and to assume that their surface coverage is the same. If, subsequently, one corner-to-corner proton jump is allowed, CPCP intermediates are obtained that upon ring opening result in a branched cation. The relative surface coverages of the CPCP intermediates correspond then to the number of adequate corner-to-corner proton jumps resulting in their formation. The final step to be considered is ring opening. This simplified scheme is shown in Figures 2.16 and 2.17 for CPCP isomerization of decyl and undecyl cations, respectively. From the schemes it appears that for decyl cation branching

$$\theta_{CPCP_2} = \theta_{CPCP_3} = \theta_{CPCP_4} \tag{2.22}$$

and for undecyl branching

$$\theta_{CPCP_2} = \theta_{CPCP_3} = \theta_{CPCP_4} = 2\theta_{CPCP_5} \tag{2.23}$$

Figure 2.15. Scheme for branching of *n*-decyl cations via CPCP intermediates.

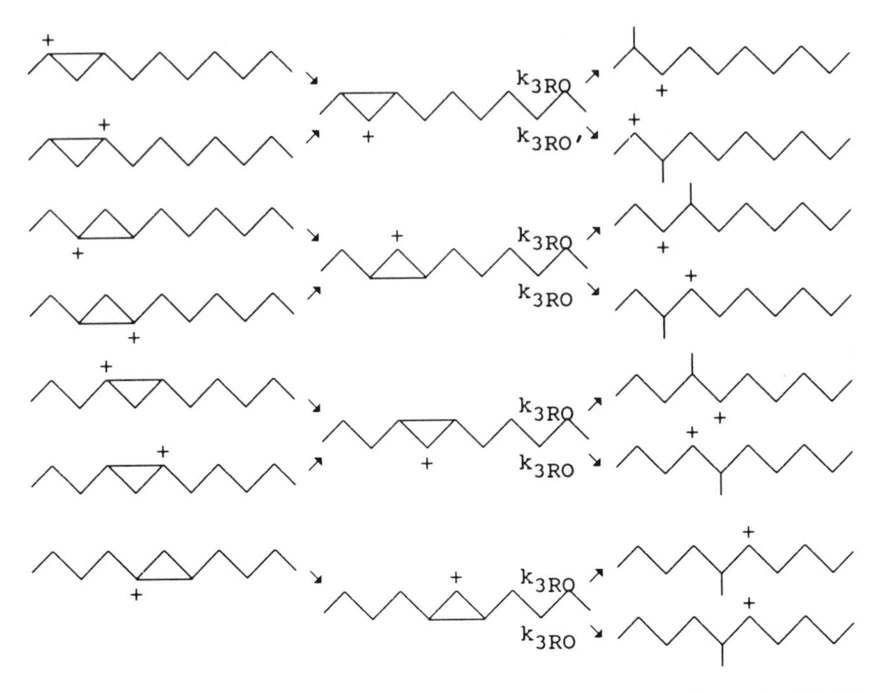

Figure 2.17. Simplified scheme for branching of undecyl ions via CPCP intermediates.

Therefore, for odd carbon numbers ($n = 2m + 1$), the number of $CPCP_i$ intermediates with $i = m$ and thus with the cyclopropane ring in the central position of the carbon chain is equal to half of that of the other $CPCP_i$.

In one of the ring opening modes of a $CPCP_2$ intermediate, a 3-methyl-2-alkylcarbenium ion is formed. This is illustrated for 1-protonated-2-ethyl-3-methyl-cyclopropane in Figure 2.18. The opening of the cyclopropyl ring gives either 3-methyl-2-pentyl cation and leads to methylbranching at position 3, or 2-methyl-3-pentyl cation, resulting in methylbranching at position 2. As the stability of the 3-methyl-2-pentyl ion is lower than of the 2-methyl-3-pentyl cation (the former cation contains a methyl substituent group on the positively charged carbon atom), the formation of 2-methyl-3-pentyl cation should be favored. In Figure 2.18 a different notation has therefore been used for these two ring opening modes. The rate constant denoted with a "prime" is for the $CPCP_2$ ring opening into an ion with branching at position 3.

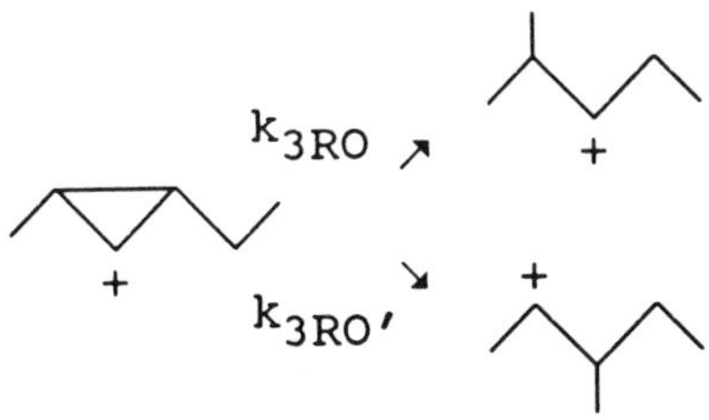

Figure 2.18. Rival ring opening modes of a 1-protonated-2-ethyl-3-methyl-cyclopropane intermediate.

Relative Rates of Type *B* Rearrangements of Decyl Ions

The detailed distributions of the monobranched feed isomers, obtained at low levels of conversion of *n*-alkanes over bifunctional faujasite catalysts, collected in Table 2.6, provide excellent means for determining the mechanism of branching rearrangements of *n*-alkylcarbenium ions. In light of the preceding argument, the following qualitative statements on the branching mechanisms can be made:

1. Next to CPCP intermediates, CPCB and cyclic alkylcarbonium ions with larger rings have to be considered. CPCX will be used as notation for the set of substituted protonated rings with ring size X. In principle, the largest substituted protonated ring that can be formed for an *n*-alkane with carbon number *n*, and which gives rise to the formation of a side chain upon ring opening, is an $(n - 2)$-membered ring.
2. The rate-determining step in branching isomerization via CPCP is a corner-to-corner proton jump. It may be expected that this holds for any of the $CPCX_i^*$ intermediates. It is assumed that the rate constant for corner-to-corner proton jump is the same for cylic alkylcarbonium ions with the same ring size. This rate constant will be denoted as k_{XPJ}, in which X is the number of carbon atoms in the ring.
3. For each value of X, the surface coverage of all cyclic alkylcarbonium ions, $CPCX_i^*$, is the same. Corner-to-corner proton jumps on $CPCX_i^*$ intermediates transform them in a unique way into $CPCX_i$ isomers, in which the ring can open according to one single mode only, if X is larger than 3. This is illustrated in Figure 2.19 for the branching of decyl cations via CPCB intermediates, for which X equals 4. Relative values for the surface coverages of the $CPCX_i$ intermediates with larger than cyclopropane rings can thus be determined and the reaction schemes simplified. It is sufficient to consider the ring opening reactions of the different $CPCX_i$ intermediates only.
4. Ring opening of a $CPCX_2$ occurs at a lower rate when it results in the formation of a secondary alkylcarbenium ion, charged at carbon atom 2.

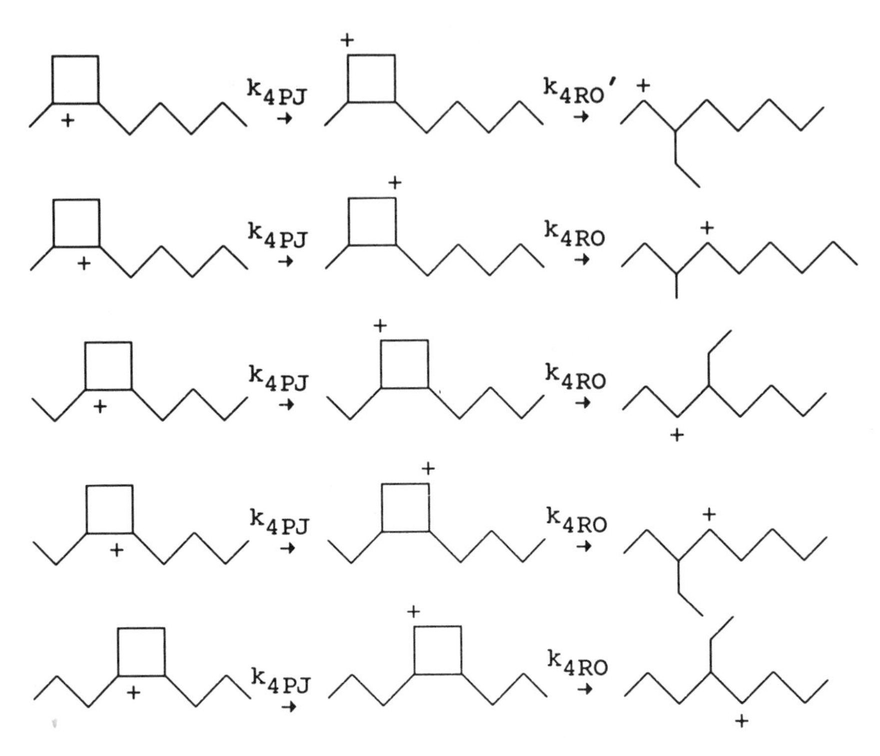

Figure 2.19. Scheme for branching isomerization of decyl cations via CPCB intermediates.

All other ring openings of $CPCX_i$ cations with the same carbon number occur at the same rate. The rate constants for ring opening, k_{XRO}, will be denoted with a "prime" when the positive charge becomes located on a carbon atom at position 2 in the resulting alkylcarbenium ion.

Based on these rules, the branching mechanism of decane can be outlined as in Table 2.7.

Based on the kinetic treatment, given in Table 2.7, 11 rate constants (k_{3RO}, $k_{3RO'}$, k_{4RO}, $k_{4RO'}$, k_{5RO}, $k_{5RO'}$, k_{6RO}, $k_{6RO'}$, k_{7RO}, $k_{7RO'}$, and $k_{8RO'}$) have to be considered for the branching of decane. Since there are only seven monobranched isomers from decane, it is impossible to determine all these parameters. The surface coverages of the $CPCX_i$ intermediates can be considered as fractions, f_i, of the total coverage of CPCX:

$$\theta_{CPCX} = \sum_i \theta_{CPCX_i} \tag{2.24}$$

$$\theta_{CPCX_i} = f_i \theta_{CPCX} \tag{2.25}$$

Table 2.7. Scheme for branching of decyl cations via CPCX intermediates.

Intermediate	Distribution	Ring Opening Reaction
	0.33	k_{3RO} ↗ 2-Methyl-3-nonyl; $k_{3RO'}$ ↘ 3-Methyl-2-nonyl
	0.33	k_{3RO} ↗ 3-Methyl-4-nonyl; k_{3RO} ↘ 4-Methyl-3-nonyl
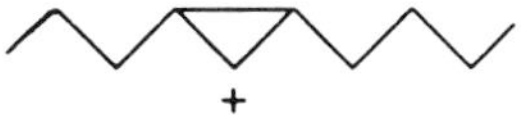	0.33	k_{3RO} ↗ 4-Methyl-5-nonyl; k_{3RO} ↘ 5-Methyl-4-nonyl
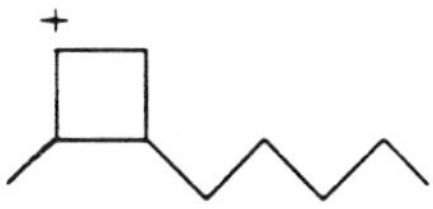	0.20	$\xrightarrow{k_{4RO'}}$ 3-Ethyl-2-octyl
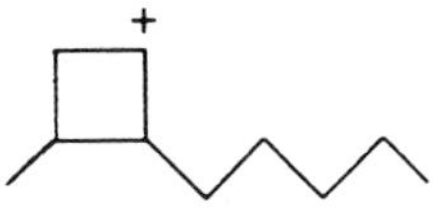	0.20	$\xrightarrow{k_{4RO}}$ 3-Methyl-4-nonyl
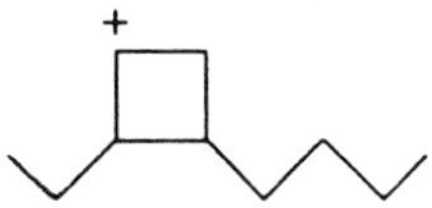	0.20	$\xrightarrow{k_{4RO}}$ 4-Ethyl-3-octyl
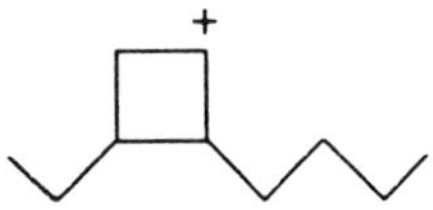	0.20	$\xrightarrow{k_{4RO}}$ 3-Ethyl-4-octyl
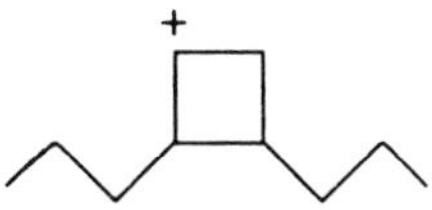	0.20	$\xrightarrow{k_{4RO}}$ 4-Ethyl-5-octyl

Table 2.7. (Continued).

Intermediate	Distribution	Ring Opening Reaction
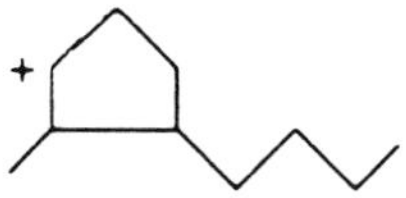	0.25	$\xrightarrow{k_{5RO'}}$ 3-Propyl-2-heptyl
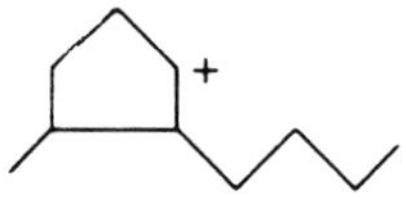	0.25	$\xrightarrow{k_{5RO}}$ 4-Methyl-5-nonyl
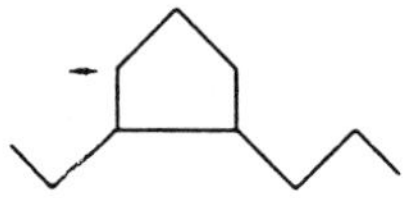	0.25	$\xrightarrow{k_{5RO}}$ 4-Propyl-3-heptyl
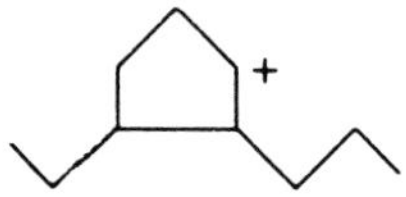	0.25	$\xrightarrow{k_{5RO}}$ 4-Ethyl-5-octyl
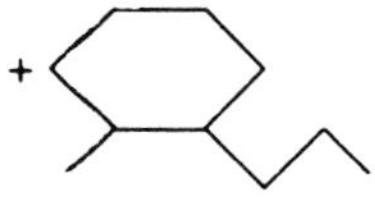	0.33	$\xrightarrow{k_{6RO'}}$ 3-Propyl-2-heptyl
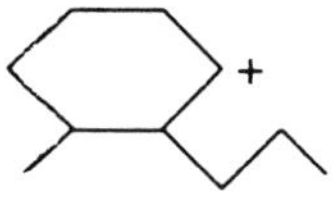	0.33	$\xrightarrow{k_{6RO}}$ 5-Methyl-4-nonyl
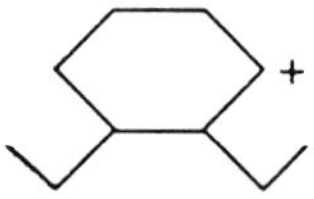	0.33	$\xrightarrow{k_{6RO}}$ 4-Ethyl-3-octyl
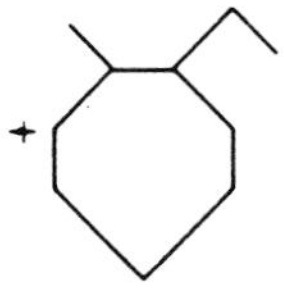	0.50	$\xrightarrow{k_{7RO'}}$ 3-Ethyl-2-octyl
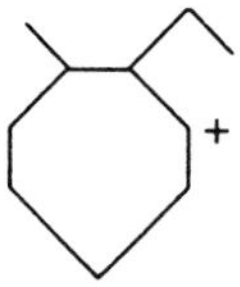	0.50	$\xrightarrow{k_{7RO}}$ 4-Methyl-3-nonyl
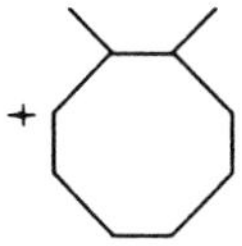	1.00	$\xrightarrow{k_{8RO'}}$ 3-Methyl-2-nonyl

Expressions for the relative rates of the formation of monobranched isomers from decane can then be derived based on the ring opening modes shown in Table 2.7:

$$r_{2\text{-MC9}} = 0.33k_{3RO}\theta_{CPCP} \tag{2.26}$$

$$r_{3\text{-MC9}} = 0.33(k_{3RO} + k_{3RO'})\theta_{CPCP} + 0.20k_{4RO}\theta_{CPCB} + 1.00k_{8RO'}\theta_{CPCO} \tag{2.27}$$

$$r_{4\text{-MC9}} = 0.67k_{3RO}\theta_{CPCP} + 0.25k_{5RO}\theta_{CPCPe} + 0.50k_{7RO}\theta_{CPCHe} \tag{2.28}$$

$$r_{5\text{-MC9}} = 0.33k_{3RO}\theta_{CPCP} + 0.33k_{6RO}\theta_{CPCH} \tag{2.29}$$

$$r_{3\text{-EC8}} = 0.20(k_{4RO'} + k_{4RO})\theta_{CPCB} + 0.50k_{7RO'}\theta_{CPCHe} \tag{2.30}$$

$$r_{4\text{-EC8}} = 0.40k_{4RO}\theta_{CPCB} + 0.25(k_{5RO} + k_{5RO'})\theta_{CPCPe} + 0.33(k_{6RO} + k_{6RO'})\theta_{CPCH} \tag{2.31}$$

$$r_{4\text{-PC7}} = 0.25k_{5RO}\theta_{CPCPe} \tag{2.32}$$

This set of equations cannot be solved, since there are more parameters than equations. However, it is possible to determine relative rates, R_{CPCX}, of the branching modes via the different CPCX intermediates in the following way:

$$k_{3RO}\theta_{CPCP} = R_{CPCP} \tag{2.33}$$

$$k_{3RO'}\theta_{CPCP} = R_{CPCP'} \tag{2.34}$$

$$k_{4RO}\theta_{CPCB} = R_{CPCB} \tag{2.35}$$

$$k_{4RO'}\theta_{CPCB} = R_{CPCB'} \tag{2.36}$$

$$k_{5RO}\theta_{CPCPe} = R_{CPCPe} \tag{2.37}$$

$$k_{6RO}\theta_{CPCH} = R_{CPCH} \tag{2.38}$$

$$k_{7RO}\theta_{CPCHe} = R_{CPCHe} \tag{2.39}$$

$$k_{8RO'}\theta_{CPCO} = R_{CPCO'} \tag{2.40}$$

Equations (2.26)–(2.32) then become

$$r_{2\text{-MC9}} = 0.33R_{CPCP} \tag{2.41}$$

$$r_{3\text{-MC9}} = 0.33R_{CPCP} + 0.33R_{CPCP'} + 0.20R_{CPCB} + 1.00R_{CPCO'} \tag{2.42}$$

$$r_{4\text{-MC9}} = 0.67R_{\text{CPCP}} + 0.25R_{\text{CPCPe}} + 0.50R_{\text{CPCHe}} \quad (2.43)$$

$$r_{5\text{-MC9}} = 0.33R_{\text{CPCP}} + 0.33R_{\text{CPCH}} \quad (2.44)$$

$$r_{3\text{-EC8}} = 0.20R_{\text{CPCB}} + 0.20R_{\text{CPCB}'} + 0.50R_{\text{CPCHe}} \quad (2.45)$$

$$r_{4\text{-EC8}} = 0.40R_{\text{CPCB}} + 0.25R_{\text{CPCPe}} + 0.25R_{\text{CPCPe}'} + 0.33R_{\text{CPCH}} + 0.33R_{\text{CPCH}'} \quad (2.46)$$

$$r_{4\text{-PC7}} = 0.25R_{\text{CPCPe}} \quad (2.47)$$

Equation (2.42) contains two parameters, $R_{\text{CPCP}'}$ and $R_{\text{CPCO}'}$, that do not occur in the other equations. $R_{\text{CPCO}'}$ was considered to be zero. It will subsequently become clear why this is a reasonable assumption. With the initial distribution of the monobranched isomers from decane, given in Table 2.6, the resolution of Eqs. (2.41)–(2.47) leads to negative values of some of the R_{CPCX} values. A hypothetical composition for the monobranched isomerization products from decane was derived in the following way, based on the distribution of the monobranched isomers obtained with the Pt/USY catalyst. As the decane feedstock used in the catalytic experiment with the Pt/USY catalyst was not contaminated with 2-methylnonane, 3-ethyloctane, 4-ethyloctane, and 4-propylheptane, the experimental distribution of these reaction products was considered to be exact. The amounts of 3-, 4-, and 5-methylnonane are obtained by solving Eqs. (2.41)–(2.47).

The hypothetical initial composition of the monobranched isodecanes thus calculated is given in Table 2.6. The calculated composition of the methylnonanes does not deviate substantially from the experimentally obtained values, supporting the model.

When R_{CPCP} is taken as the standard and a value of 100 is given to it, the following relative rates for the different branching modes of decane are found:

$$R_{\text{CPCP}} = 100 \quad (2.48)$$

$$R_{\text{CPCP}'} = 61 \quad (2.49)$$

$$R_{\text{CPCB}} = 26 \quad (2.50)$$

$$R_{\text{CPCB}'} = 24 \quad (2.51)$$

$$R_{\text{CPCPe}} = 4.4 \quad (2.52)$$

$$R_{\text{CPCH}} = 2.6 \quad (2.53)$$

$$R_{\text{CPCHe}} = 1.8 \quad (2.54)$$

$$R_{\text{CPCO}'} = 0.0 \quad (2.55)$$

From these results it is concluded that during branching of decane on bifunctional faujasite catalysts a reaction pathway involving CPCX intermediates becomes less important for larger values of X. Thus the relative rate of branching via CPCHe intermediates is very low. Consequently, it seems to be acceptable that the assumption made earlier is true: the rate of isomerization via formation of substituted corner protonated cyclooctanes with ring opening into a monobranched isodecyl cation, charged at position 2, $R_{CPCO'}$, equals zero.

As expected, $R_{CPCP'}$ and $R_{CPCB'}$ are found to be smaller than R_{CPCP} and R_{CPCB}, respectively.

The predicted distribution of methylnonanes obtained according to a branching mechanism involving exclusively CPCP intermediates is given in Table 2.6 for comparison. In contrast to the mechanism involving CPCP exclusively, the present mechanism also involving larger CPCX intermediates provides a rationale for the higher amounts of 4-methylnonane with respect to 3-methylnonane initially formed (Table 2.6).

Effect of Carbon Number on the Relative Rates of the Different Branching Pathways Involving CPCX Intermediates

The kinetic approach followed for the branching of decane has been applied to the branching of hexane through pentadecane. Among the feedstocks used for determining the distribution of monobranched isomers from *n*-alkanes, listed in Table 2.6, only hexane was very pure and had a purity of 99.98%[31] and 99.95% in the conversion over Pt/CaY and Pt/USY, respectively. In that case, the experimentally determined distribution of isomers can be considered to be exact. For octane and larger *n*-alkanes, a hypothetical distribution of their monobranched isomers was derived similar to the method used for decane. The distribution of the isomers with ethyl or bulkier side chains, the ratio of the isomers with bulky side chains to those with methylbranching, as well as the ratio of the 2-methylbranched isomers to those with side chains bulkier than methyl, were taken from the experiments over the Pt/USY catalyst. The composition of the other methylbranched isomers was then derived from a best fit with the kinetic equations. For heptane, from the experimental data only, the ratio of 3-ethylpentane to methylhexanes was used.

The following equations can be derived as was done for decane. It gives the rate of formation of the monobranched isomers as a function of the rates of the different branching pathways:

for hexane:

$$r_{2\text{-MC5}} = 1.0R_{CPCP} \tag{2.56}$$

$$r_{3\text{-MC5}} = 1.0R_{CPCP'} + 1.0R_{CPCB'} \tag{2.57}$$

for heptane:

$$r_{2\text{-MC6}} = 0.67R_{\text{CPCP}} \tag{2.58}$$

$$r_{3\text{-MC6}} = 0.67R_{\text{CPCP}} + 0.67R_{\text{CPCP}'} + 0.50R_{\text{CPCB}} + 1.0R_{\text{CPCPe}'} \tag{2.59}$$

$$r_{3\text{-EC5}} = 0.50R_{\text{CPCB}'} \tag{2.60}$$

for octane:

$$r_{2\text{-MC7}} = 0.50R_{\text{CPCP}} \tag{2.61}$$

$$r_{3\text{-MC7}} = 0.50R_{\text{CPCP}} + 0.50R_{\text{CPCP}'} + 0.33R_{\text{CPCB}} + 1.0R_{\text{CPCH}'} \tag{2.62}$$

$$r_{4\text{-MC7}} = 0.50R_{\text{CPCP}} + 0.50R_{\text{CPCPe}} \tag{2.63}$$

$$r_{3\text{-EC6}} = 0.67R_{\text{CPCB}'} + 0.50R_{\text{CPCPe}} \tag{2.64}$$

for nonane:

$$r_{2\text{-MC8}} = 0.40R_{\text{CPCP}} \tag{2.65}$$

$$r_{3\text{-MC8}} = 0.40R_{\text{CPCP}} + 0.40R_{\text{CPCP}'} + 0.25R_{\text{CPCB}} + 1.0R_{\text{CPCHe}'} \tag{2.66}$$

$$r_{4\text{-MC8}} = 0.80R_{\text{CPCP}} + 0.33R_{\text{CPCPe}} + 0.50R_{\text{CPCH}} \tag{2.67}$$

$$r_{3\text{-EC7}} = 0.25R_{\text{CPCB}} + 0.25R_{\text{CPCB}'} + 0.50R_{\text{CPCH}'} \tag{2.68}$$

$$r_{4\text{-EC7}} = 0.25R_{\text{CPCB}} + 0.67R_{\text{CPCH}} \tag{2.69}$$

for undecane:

$$r_{2\text{-MC10}} = 0.29R_{\text{CPCP}} \tag{2.70}$$

$$r_{3\text{-MC10}} = 0.29R_{\text{CPCP}} + 0.29R_{\text{CPCP}'} + 0.17R_{\text{CPCB}} + 1.0R_{\text{CPCN}'} \tag{2.71}$$

$$r_{4\text{-MC10}} = 0.57R_{\text{CPCP}} + 0.20R_{\text{CPCPe}} + 0.50R_{\text{CPCO}} \tag{2.72}$$

$$r_{5\text{-MC10}} = 0.57R_{\text{CPCP}} + 0.25R_{\text{CPCH}} + 0.33R_{\text{CPCHe}} \tag{2.73}$$

$$r_{3\text{-EC9}} = 0.33R_{\text{CPCB}} + 0.50R_{\text{CPCO}} \tag{2.74}$$

$$r_{4\text{-EC9}} = 0.33R_{\text{CPCB}} + 0.40R_{\text{CPCPe}} + 0.67R_{\text{CPCHe}} \tag{2.75}$$

$$r_{5\text{-EC9}} = 0.17R_{\text{CPCB}} + 0.50R_{\text{CPCH}} \tag{2.76}$$

$$r_{4\text{-PC8}} = 0.40R_{\text{CPCPe}} + 0.25R_{\text{CPCH}} \tag{2.77}$$

for dodecane:

$$r_{2\text{-MC11}} = 0.25R_{\text{CPCP}} \tag{2.78}$$

$$r_{3\text{-MC11}} = 0.25R_{\text{CPCP}} + 0.25R_{\text{CPCP}'} + 0.14R_{\text{CPCB}} + 1.0R_{\text{CPCD}'} \tag{2.79}$$

$$r_{4\text{-MC11}} = 0.50R_{\text{CPCP}} + 0.17R_{\text{CPCPe}} + 0.50R_{\text{CPCN}} \tag{2.80}$$

$$r_{5\text{-MC11}} = 0.50R_{\text{CPCP}} + 0.20R_{\text{CPCH}} + 0.33R_{\text{CPCO}} \tag{2.81}$$

$$r_{6\text{-MC11}} = 0.25R_{\text{CPCP}} + 0.25R_{\text{CPCHe}} \tag{2.82}$$

$$r_{3\text{-EC10}} = 0.14R_{\text{CPCB}} + 0.14R_{\text{CPCB}'} + 0.50R_{\text{CPCN}} \tag{2.83}$$

$$r_{4\text{-EC10}} = 0.29R_{\text{CPCB}} + 0.33R_{\text{CPCPe}} + 0.67R_{\text{CPCO}} \tag{2.84}$$

$$r_{5\text{-EC10}} = 0.29R_{\text{CPCB}} + 0.40R_{\text{CPCH}} + 0.50R_{\text{CPCHe}} \tag{2.85}$$

$$r_{4\text{-PC9}} = 0.33R_{\text{CPCPe}} + 0.25R_{\text{CPCHe}} \tag{2.86}$$

$$r_{5\text{-PC9}} = 0.40R_{\text{CPCH}} + 0.17R_{\text{CPCPe}} \tag{2.87}$$

for tridecane:

$$r_{2\text{-MC12}} = 0.22R_{\text{CPCP}} \tag{2.88}$$

$$r_{3\text{-MC12}} = 0.22R_{\text{CPCP}} + 0.22R_{\text{CPCP}'} + 0.13R_{\text{CPCB}} + 1.00R_{\text{CPCU}'} \tag{2.89}$$

$$r_{4\text{-MC12}} = 0.44R_{\text{CPCP}} + 0.14R_{\text{CPCPe}} + 0.50R_{\text{CPCD}} \tag{2.90}$$

$$r_{5\text{-MC12}} = 0.44R_{\text{CPCP}} + 0.17R_{\text{CPCH}} + 0.33R_{\text{CPCN}} \tag{2.91}$$

$$r_{6\text{-MC12}} = 0.44R_{\text{CPCP}} + 0.20R_{\text{CPCHe}} + 0.25R_{\text{CPCO}} \tag{2.92}$$

$$r_{3\text{-EC11}} = 0.13R_{\text{CPCB}} + 0.13R_{\text{CPCB}'} + 0.50R_{\text{CPCD}} \tag{2.93}$$

$$r_{4\text{-EC11}} = 0.25R_{\text{CPCB}} + 0.14R_{\text{CPCPe}} + 0.67R_{\text{CPCN}} + 0.14R_{\text{CPCPe}} \tag{2.94}$$

$$r_{5\text{-EC11}} = 0.25R_{\text{CPCB}} + 0.33R_{\text{CPCH}} + 0.50R_{\text{CPCO}} \tag{2.95}$$

$$r_{6\text{-EC11}} = 0.13R_{\text{CPCB}} + 0.40R_{\text{CPCHe}} \tag{2.96}$$

$$r_{4\text{-PC10}} = 0.29R_{\text{CPCPe}} + 0.25R_{\text{CPCO}} \tag{2.97}$$

$$r_{5\text{-PC10}} = 0.29R_{\text{CPCPe}} + 0.17R_{\text{CPCH}} + 0.40R_{\text{CPCHe}} \tag{2.98}$$

$$r_{5\text{-BC9}} = 0.33R_{\text{CPCH}} \tag{2.99}$$

The relative rates of the different types of B isomerization of hexane through tridecane, calculated from Eqs. (2.56)–(2.99), using the hypothetical compositions of the monobranched isomers listed in Table 2.6, are given in Table 2.8. For the isomerization of an n-alkane with carbon number n it was assumed that $R_{\text{CPCX}'} = 0$ with $\text{X} = n - 2$. As the number

Table 2.8. Relative rates of type *B* isomerization pathways of *n*-alkanes involving CPCX intermediates.

CN[a]	6	7	8	9	10	11	12	13
R_{CPCP}	100	100	100	100	100	100	100	100
$R_{CPCP'}$	73	33	33	59	61	77	43	76
R_{CPCB}	—	11	15	18	26	20	24	26
$R_{CPCB'}$	—	—	—	—	24	—	20	25
R_{CPCPe}	—	—	0	6.1	4.4	5.6	3.3	6.3
$R_{CPCPe'}$	—	—	—	—	—	—	—	5.0
R_{CPCH}	—	—	—	2.3	2.6	3.6	3.0	5.0
R_{CPCHe}	—	—	—	—	1.8	3.4	2.6	5.0
R_{CPCO}	—	—	—	—	—	0.0	0.9	2.3
R_{CPCN}	—	—	—	—	—	—	0.2	1.1
R_{CPCD}	—	—	—	—	—	—	—	0.1

[a] Carbon number of *n*-alkane.

of rates that could be determined directly using the experimental data was always restricted, the R_{CPCX} values were always derived using as many $R_{CPCX'}$ parameters as possible. In the latter category, those rates with smallest ring sizes were preferentially retained in the equations. For those with large ring size it was assumed that $R_{CPCX'}$ equals R_{CPCX}. The rates of the different mechanisms listed in Table 2.8 to a certain extent might be inaccurate because of the uncertainties on the experimentally determined compositions of the monobranched isomerization products. Experiments with *n*-alkanes with very high purity are required for refining the present data. For tetradecane and pentadecane, the calculations were not made, given the uncertainty in the experimental data on the distribution of monobranched isomers with large side chains (Table 2.6).

Nevertheless, the rates of the different types of *B* isomerization, calculated from the data, clearly indicate that the proposed mechanism is essentially correct and that next to substituted protonated cyclopropanes, substituted protonated cyclobutanes, cyclopentanes, etc., contribute to the formation of the first branching in a linear hydrocarbon skeleton. The following relations between the rates of the different type *B* isomerization pathways for *n*-alkanes ranging from hexane through tridecane were found:

$$R_{CPCP} > R_{CPCP'};\ R_{CPCB} > R_{CPCB'};\ R_{CPCPe} > R_{CPCPe'} \qquad (2.100)$$

$$R_{CPCP} > R_{CPCB} > R_{CPCPe} > R_{CPCH} > \cdots \qquad (2.101)$$

Most probably they are also valid for larger *n*-alkanes.

The relative rates of the different reactions seem to be independent of carbon number (Table 2.8). As the number of rings that can be formed increases with increasing carbon number, the contribution of larger than cyclopropane rings in the branching event increases. This could have been predicted, since the amount of isomers with larger than methyl branchings formed by isomerization of an *n*-alkane increases with increasing carbon number (Table 2.6).

The predicted distributions of monomethylbranched isomers obtained according to a branching mechanism involving CPCP intermediates exclusively are given in Table 2.6 for comparison. Table 2.6 shows that systematic deviations between predicted and experimental data are encountered. When hexane and heptane are converted over Pt/USY and Pt/CaY, the amount of the 2-methylbranched isomer is always in excess of that predicted by the CPCP mechanism. With hexane the formation of 2-methylpentane is favored over 3-methylpentane. The same observation was made when hexane was reacted over Pt/silica–alumina.[28] With the latter catalyst, the initial composition of the monobranched isomers was 57.8% of 2-methylpentane and 42.2% of 3-methylpentane. These results are similar to those obtained with the faujasite-based catalysts given in Table 2.6. When heptane is converted over either Pt/USY or Pt/CaY, initially much more 2-methylhexane is formed than predicted by the CPCP mechanism (Table 2.6). In another study[37] with a Pt/USY catalyst initially 40% 2-methylhexane and 60% 3-methylhexane were found.

For *n*-alkanes with carbon numbers higher than 12, an opposite effect is encountered. The formation of 2-methylbranched isomers is now lower than predicted by the CPCP mechanism. Whereas equal amounts of the other methylbranched isomers are expected, it is experimentally found that branching at centrally located carbon atoms is favored.

Table 2.6 shows that these deviations vanish when, next to CPCP intermediates, substituted protonated cycloalkanes with larger than three-rings are also considered. The hypothetical compositions of the monobranched isomerization products, given in Table 2.6, are in fair agreement with the experimental results.

Formation of the Second Branching

Even with most advanced high-resolution gas–liquid chromatography (GLC) it is very difficult to analyze the composition of the dibranched isomerization products from a long-chain *n*-alkane below 10% conversion of the feedstock, since they are numerous and are formed in low yields. The distribution of the dibranched isodecanes, formed from decane over Pt/USY, is plotted against the degree of conversion of the feed in Figure 2.20. All dibranched isodecanes seem to be formed simultaneously. From

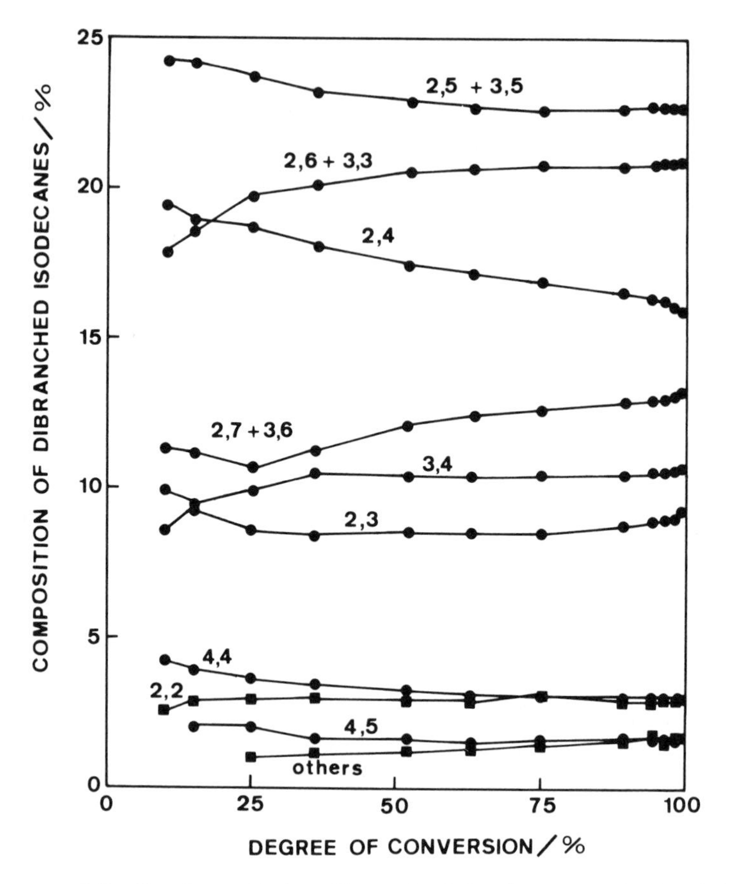

Figure 2.20. Distribution of the dibranched isodecanes obtained from decane at increasing levels of conversion. The figures on the curves indicate the positions of the methylbranchings. "Others" stand for methylethylbranched isodecanes. The reaction conditions are given in Figure 2.10.

the data of Figure 2.20 it would be erroneously concluded that methylethylheptanes are formed only in trace amounts. Actually only few of these isomers can be resolved from the larger chromatographic peaks of the more abundant dimethyloctanes.

A study of the second branching step is hampered by the observation that the monobranched feed molecule is converted quickly into its positional isomers owing to the occurrence of alkyl shifts, which compete with the branching reactions to be studied.[36,40] To illustrate this, the yield of the *n*-decane and mono-, di-, and tribranched isomers and cracked products obtained at increasing levels of conversion of 2-methylnonane over Pt/USY is shown in Figure 2.21. From these data it can be deduced that type *A* isomerization of 2-methylnonane into its monobranched isomers is some seven times faster than type *B* isomerization into dibranched isomers.

There is evidence for the second branching step occurring through the same mechanism as the first one. Dibranched isoalkanes with a quaternary carbon atom are readily formed in branching reactions; neopentane, which is never observed among the C_5 products obtained over bifunctional catalysts, is the exception. This can readily be understood if a branching mechanism involving CPCP intermediates occurs. It can be seen in Eq. (2.102) that the formation of neopentyl cation from 2-methylpentyl cation via a CPCP intermediate should be very slow, since it necessitates the formation of a primary ion upon ring opening. The reaction is in this way analogous to the conversion of butyl to isobutyl cation, shown in Eq. (2.14).

(2.102)

In Figure 2.22 the simplified reaction scheme for methyl shift and branching via CPCP intermediates of 2-methyl-2-nonyl cations has been developed. The first hydride shift and cyclization step are not shown for clarity. Starting from 2-methyl-2-nonyl cation, one single type *A* rearrangement is possible, yielding 3-methyl-3-nonyl cation. On the other hand, from the secondary 2-methyl-2-nonyl cations, a whole set of dibranched isomers can be formed directly by branching via CPCP interme-

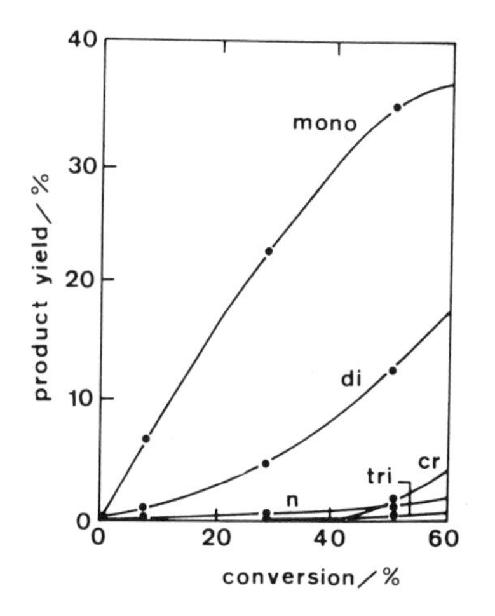

Figure 2.21. Yield of *n*-decane, and mono-, di-, and tribranched isomers and cracked products (cr) at increasing levels of conversion from 2-methylnonane over Pt/USY. The total pressure was 0.1 MPa, the molar H_2/2-methylnonane ratio was 120 and F_0/W was 0.75 mmol kg^{-1} s^{-1}. The reaction temperature was varied between 400 and 430 K.

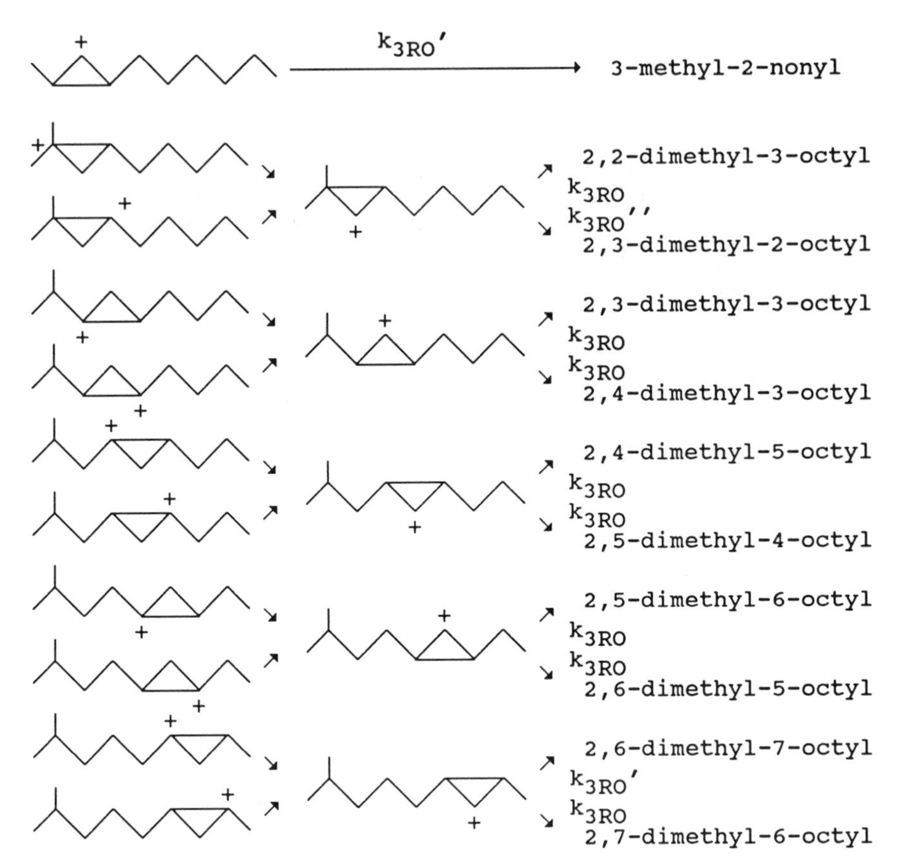

Figure 2.22. Simplified scheme for 1,2-methyl shift and branching via CPCP intermediates of 2-methylnonyl cations.

diates. It is assumed that corner-to-corner proton jumps are again the rate-limiting steps and that, therefore, the surface coverages of the different CPCP isomers formed after ring closure are equal. The reaction scheme shows that there are 10 CPCP intermediates that can generate a second branching, whereas there is only one intermediate (1-protonated 2-methyl-3-hexyl-cyclopropane) which is responsible for a 1,2-methyl shift. From Figure 2.22 it can be seen that 2,2-, 2,3-, 2,4-, 2,5-, 2,6-, and 2,7-dimethyloctyl cations can be generated from 2-methylnonane via a mechanism involving CPCP intermediates.

Methyl shift from 2- to 3-methylnonane is seven times faster than branching of 2-methylnonane into dibranched isomers (Fig. 2.21). Roughly, the rate of methyl shift should be 70 times faster than branching of mono- into dimethylbranched isomers. This ratio between the rates of type *A* and *B* rearrangements is of the same order of magnitude as predicted from the kinetic data on rearrangements of $C_6H_{13}^+$ alkylcarbenium ions in superacids. When the rates of both rearrangements were extrapolated to 400 K, a two orders of magnitude difference was obtained (see Table 2.4).

The simplified scheme for branching of 2-methyloctyl cations is shown in Figure 2.23. From this scheme it can be expected that the relative amounts of the dimethylheptane (DMC_7) products must follow the sequence:

$$\begin{matrix} & & & & 3,3\text{-DMC}_7 \\ 2,3\text{-DMC}_7 & & & & \\ & & 2,2\text{-DMC}_7 & & 3,4\text{-DMC}_7 \\ 2,4\text{-DMC}_7 & > & & > & \\ & & 2,6\text{-DMC}_7 & & 3,5\text{-DMC}_7 \\ 2,5\text{-DMC}_7 & & & & \\ & & & & 4,4\text{-DMC}_7 \end{matrix} \tag{2.103}$$

Weitkamp et al.[36] reported product distributions from 2-methyloctane, converted over Pt/CaY catalyst that are in agreement with this sequence.

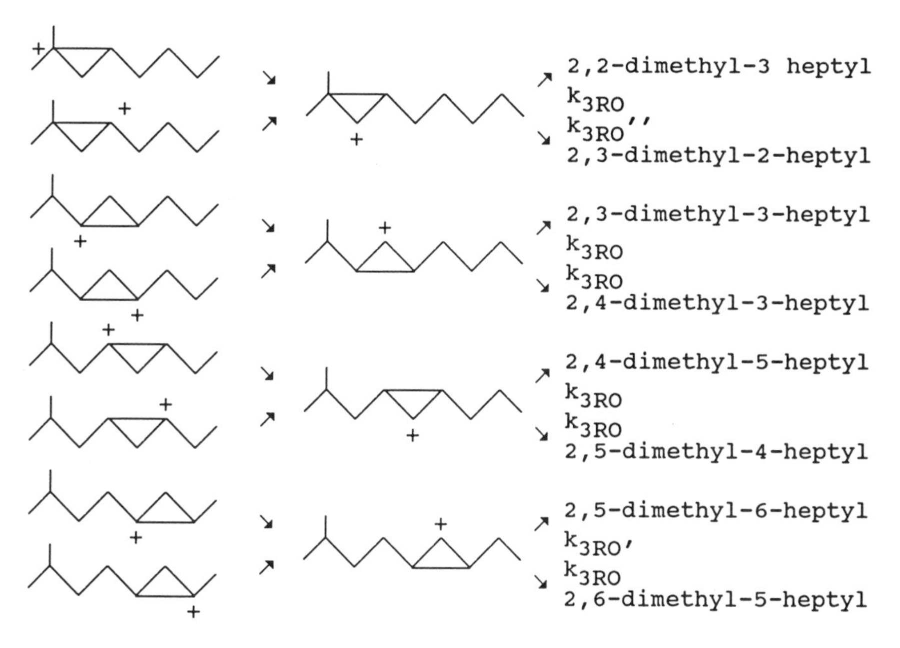

Figure 2.23. Simplified scheme for branching via CPCP intermediates of 2-methyloctyl cations.

By analogy with the branching of *n*-alkanes, the formation of methylethylbranched isomers from methylbranched feeds should proceed mainly via CPCB intermediates. From the data of Table 2.8 it can be seen that during branching of *n*-alkanes with carbon number *n*, the relative rates of the branching pathways involving CPCX intermediates with (*n*-2)- and (*n*-3)-membered rings are very slow. It can, therefore, be expected that the contribution of branching pathways involving CPCX intermediates with (*n*-3)- and (*n*-4)-membered rings will also be low during branching of a monomethylbranched alkane with carbon number *n*, as the main chain is shortened by one carbon atom.

1,2-Alkyl Shift

Weitkamp et al.[36] revealed that for long-chain methylbranched alkanes, the shift of the methyl group along the main chain is a stepwise event. Subtle rate differences were found between the different positional 1,2-methyl shifts. When two alternative 1,2-shifts of a methyl group are possible, the preferred shift is the one toward the center of the main chain. From the distribution of the methylbranched isomers obtained from methylnonanes over the Pt/CaY catalyst, the following relative rates for the different displacements of methyl groups can be obtained:

$$\text{3-methylnonane} \begin{array}{l} \nearrow^{1.0} \ \text{2-methylnonane} \\ \searrow_{1.5} \ \text{4-methylnonane} \end{array} \qquad (2.104)$$

$$\text{4-methylnonane} \begin{array}{l} \nearrow^{1.0} \ \text{3-methylnonane} \\ \searrow_{1.1} \ \text{5-methylnonane} \end{array} \qquad (2.105)$$

The conversion of 3-methylnonane into 4-methylnonane is faster than into 2-methylnonane. It can easily be verified that the reaction pathway leading to 2-methylnonane involves the formation of an intermediate secondary alkylcarbenium ion with the positive charge at carbon atom 2. It is known that this cation is less stable than cations with the positive charge in other positions. The conversion of 4-methylnonane into 3- and 5-methylnonane occurs at comparable rates, which is in agreement with the stabilities of the intermediate secondary alkylcarbenium ions.

In the section on the branching mechanisms, the primary nature of methylnonanes, ethyloctanes, and 4-propylheptane as reaction products from decane was discussed. Once formed, these products might be the subject of interconversions via type *A* isomerization. In the reaction of decane, the distribution of the methyl-, ethyl- and propylbranched isomerization products among the monobranched isomers is constant over a very wide range of conversions, going from 5 to 95% (Fig. 2.24). At conversions higher than 95%, there is a sudden conversion toward the thermodynamic equilibrium composition, which is richer in ethyloctanes and 4-propylheptane. This behavior indicates that the elongation of the side chain (from methyl over ethyl to propyl branching) is not fast enough to establish the thermodynamic equilibrium among the monobranched isomers at medium levels of conversion.

In contrast to this, the methylnonanes (Fig. 2.11) and the ethyloctanes (Fig. 2.12) reach their respective thermodynamic equilibrium values starting at a certain level of conversion of decane. The ethyloctanes are already at thermodynamic equilibrium from 5% conversion of decane. The thermodynamic equilibrium composition of the methylnonanes is only reached for feed conversions higher than 50%. This difference can readily be explained by the features of the type *A* isomerization mechanism. For the ethyloctanes, equilibration can proceed through the shift of an ethyl group over one single position, viz., the conversion of the excess 3-ethyloctane into 4-ethyloctane, without involving alkylcarbenium ions, which are charged at carbon atom 2. For the equilibration of the methyl-

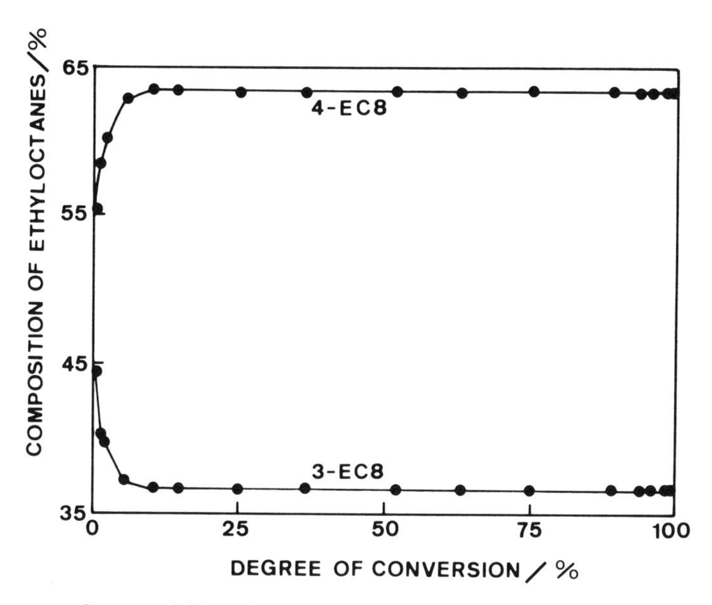

Figure 2.24. Composition of the ethyloctanes in function of the degree of conversion of decane over Pt/USY. The reaction conditions are those of Figure 2.10.

nonanes, an excess of 4- and 5-methylnonane has to be transformed into 2-methylnonane, requiring multiple shifts of the methylbranching toward the end of the hydrocarbon chain, which is an unfavorable reaction. The one-carbon-atom-at-a-time elongation of a side chain is an unfavorable reaction too, since it proceeds through alkylcarbenium ions, which are charged at carbon atom 2 (see Fig. 2.14).

The composition of the dibranched isomers of decane remains essentially constant between 25% and 99% conversion (Fig. 2.20). As no major changes of the composition are observed at very high levels of conversion, these compounds are probably at thermodynamic equilibrium, indicating that, as with the ethyloctanes, a few alkyl shifts are sufficient to establish the thermodynamic equilibrium composition of the dibranched isomers, immediately after their formation via branching reactions.

Hydrocracking Mechanisms

Figure 2.25 shows the conversion scheme of $C_8H_{17}^+$ cations in superacid at 200 K. The cracking of any normal, and mono-, di-, or tribranched $C_8H_{17}^+$

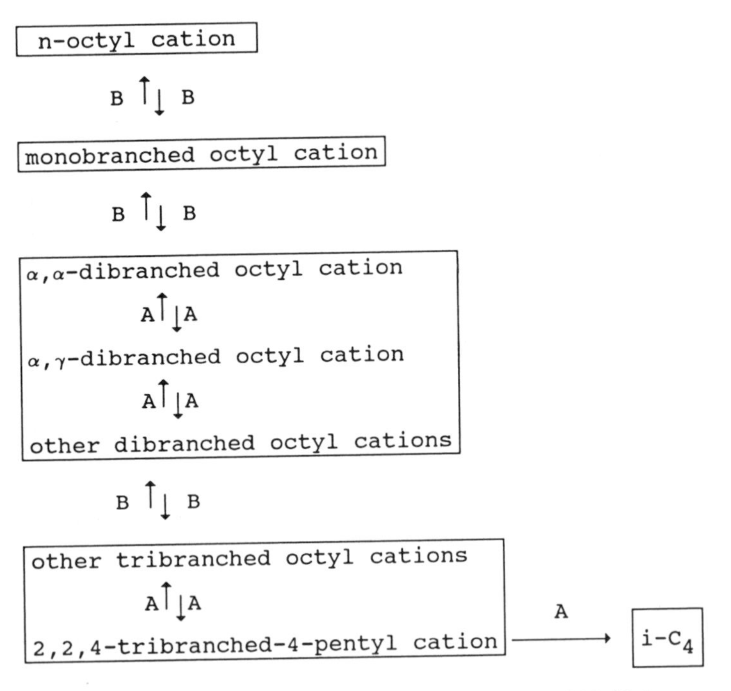

Figure 2.25. Conversion pathways of octyl cations at 200 K in superacid (from Ref. 21).

cation under these conditions results in exclusive formation of isobutene and tertiobutyl cation.[21] These fragments result from type *A* β-scission of 2,4,4-trimethyl-2-pentyl cation. Figure 2.25 shows that before the octyl ions can be ultimately consumed via the fastest β-scission mechanism, they will be equilibrated via several consecutive type *A* and *B* isomerization steps to form the α,γ,γ-tribranched ion. During this isomerization process, mono-, di-, and tribranched cations are formed that are susceptible to β-scission according to the mechanisms B_1, B_2, and *C* leading to different fragments. At the low temperatures applied (200 K), these β-scissions are, however, substantially slower than types *A* and *B* isomerization and type *A* β-scission.[21] The situation might, however, be changed at higher temperatures. Table 2.5 shows, for example, that the difference in rate between types *A* and *B* isomerization is reduced by one to two orders of magnitude by a temperature increase of 200 K. Similarly, the rate difference between *A* and B_1, B_2, and *C* type β-scission may become smaller at higher temperature.

When octane is converted over a Pt/USY catalyst at 469 K, not only isobutane, but also propane, butane, pentane, and isopentane are found as cracked products.[39] It has been possible to simulate the generation of the **individual** cracked products of octane by considering all possible β-scissions of branched octyl cations via *A*, B_1, B_2, and *C*.[39,40] An exhaustive list of β-scission reactions is given in Table 2.9. In that work[39,40] it was assumed that the gas-phase composition of the isooctane reaction products can be used to estimate the relative carbenium ion concentrations in the zeolite pores. The relative apparent rate constants for types B_1, B_2, and *C* hydrocracking of dibranched isooctanes, isononanes, and isodecanes obtained with this approach are given in Table 2.10. With respect to B_2, the apparent rate constant of B_1 hydrocracking seems to increase with increasing carbon number, since, strictly speaking, the rate constants, taken at slightly different temperatures, are not directly comparable. The relative rate constant of type C hydrocracking is smaller than that of B_1 and B_2.

In Table 2.10, some rate data for branching isomerization, type *A* hydrocracking, and 1,2-methyl shift are given as well. The rate constant for formation of a second or third branching of a C_8 to C_{10} alkane is lower than for type B_2 hydrocracking. For isooctanes and isononanes, the formation of a third branching proceeds slower than that of the second one. For isodecanes, the formation of the second and third branching is equally fast. Type *A* hydrocracking is by far the fastest reaction (Table 2.10). It is much faster than isomerization via 1,2-methyl shifts, which in its turn is substantially faster than types B_1, B_2, and *C* hydrocracking. A comparison of the data in Tables 2.5 and 2.10 shows that the increase in temperature from 195 K to 405 K reverses the rate order of type *A* isomerization and type *A* β-scission.

Table 2.9. β-Scissions of octyl cations.

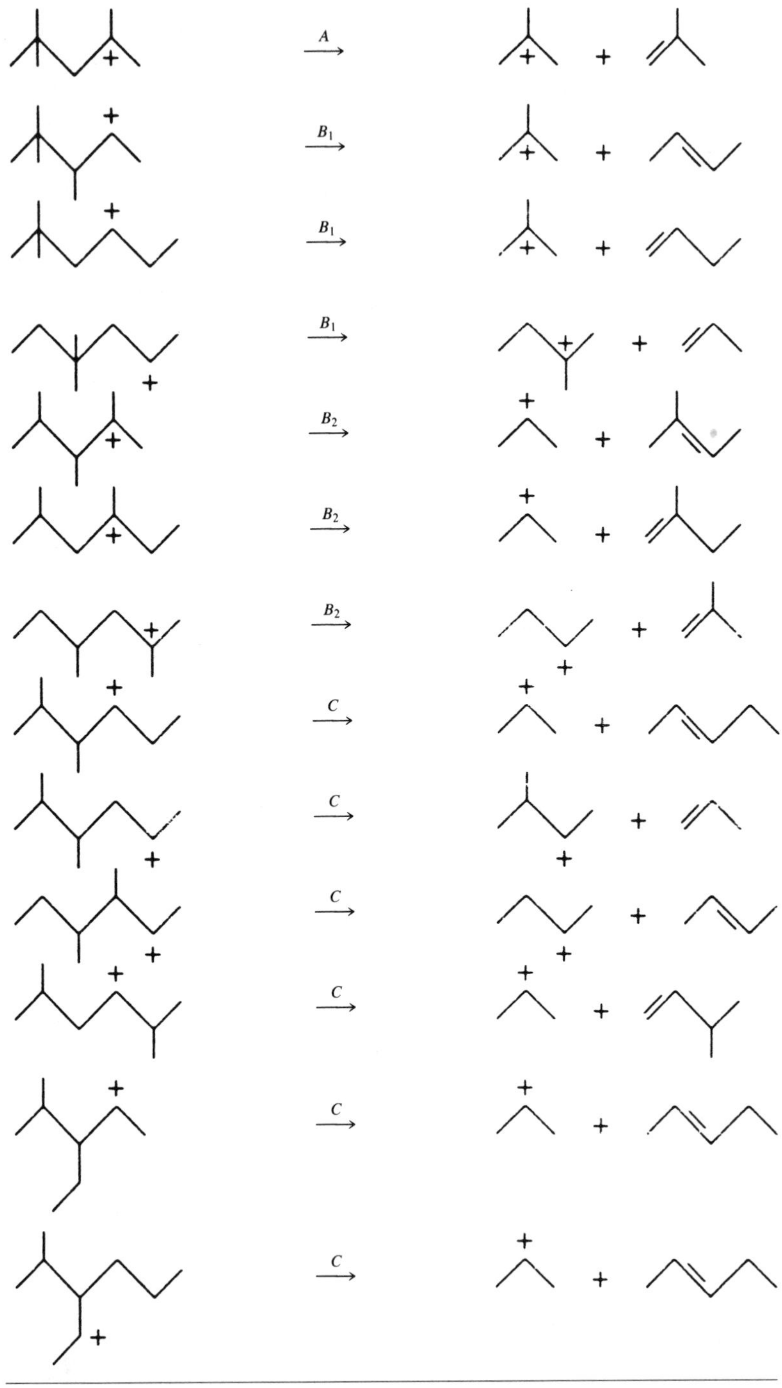

	$\xrightarrow{A}$	+
	$\xrightarrow{B_1}$	+
	$\xrightarrow{B_1}$	+
	$\xrightarrow{B_1}$	+
	$\xrightarrow{B_2}$	+
	$\xrightarrow{B_2}$	+
	$\xrightarrow{B_2}$	+
	$\xrightarrow{C}$	+
	$\xrightarrow{C}$	+
	$\xrightarrow{C}$	+
	$\xrightarrow{C}$	+
	$\xrightarrow{C}$	+
	$\xrightarrow{C}$	+

Table 2.10. Relative apparent rate constants of types A, B_1, B_2, and C hydrocracking, 1,2-methyl shift, and branching of isooctanes, isononanes, and isodecanes with B_2 as standard over the Pt/USY catalyst (data adapted from Refs. 38 and 39).

	Temperature (K)			
	469 Isooctanes	439 Isononanes	434 Isodecanes	405
Hydrocracking				
B_1	0.34	1.5	2.8	
B_2	1.0	1.0	1.0	
C	0.34	0.77	0.40	
A				1050
Type B isomerization				
Mono- → dibranched	0.34	0.85	0.80	0.80[a]
Di- → tribranched	0.28	0.77	0.80	
Type A isomerization				
1,2-Methyl shift				56

[a] Taken as standard = 0.80.

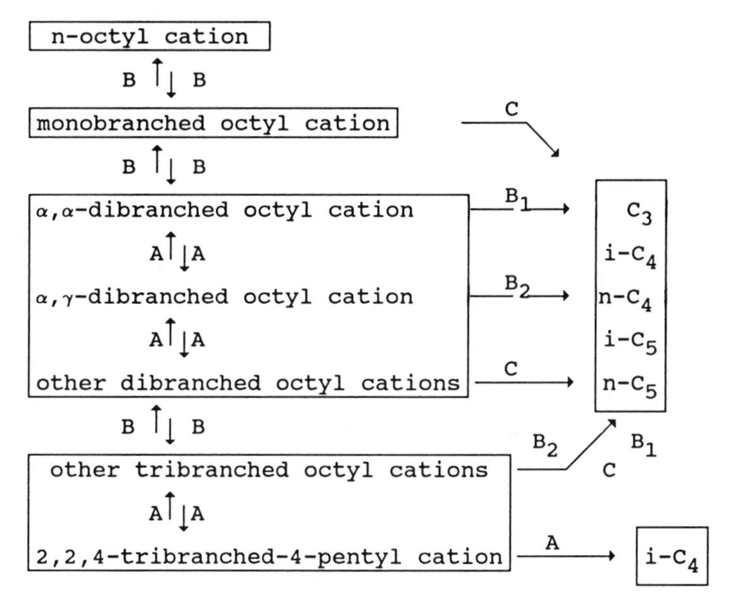

Figure 2.26. Conversion pathways of octyl cations at 469 K in Pt/USY zeolite (from Ref. 39).

The conversion pathways of octyl cations on the Pt/USY catalyst at 469 K are shown in Figure 2.26. The difference with the pattern at 200 K in superacids (Fig. 2.25) is the occurrence of cracking via B_1, B_2, and C routes.

GENERAL CONCLUSIONS

The kinetic parameters of alkylcarbenium ion rearrangements measured directly in superacid solutions are very useful for rationalizing product selectivities observed in heterogeneous catalysis. The Arrhenius activation energy of intramolecular reactions of alkylcarbenium ions in heterogeneous catalysis and superacid catalysis is very similar, as for such conversions it is practically independent of temperature. A close correspondence between the apparent activation energy of 1,2-hydride shift and 1,2-methyl shift catalyzed by zeolites at elevated temperatures and superacids at moderate temperatures is observed experimentally.

Alkylcarbenium ions are rearranged in superacids via type *B* and type *A* isomerization, depending on whether the degree of branching of the carbon skeleton is changed or not. Both types of rearrangements involve the formation of an intermediate substituted corner protonated cyclopropane (CPCP). In type *B* isomerization, a corner-to-corner hydrogen jump has to occur before opening of the cyclopropane ring of such CPCP. Since the corner-to-corner proton jump proceeds over an important energy barrier, rearrangements of type *B* are always slower than of type *A*. The mechanism involving CPCP intermediates provides an energetically favorable reaction path for branching of $C_5H_{11}^+$ and larger cations, as it avoids the formation of primary cations.

The conversion pathways of alkanes over bifunctional catalysts can be understood if the classical bifunctional reaction scheme is assumed and if in that scheme the rearrangements of alkylcarbenium ions are rate limiting. In this way it can be understood why ^{13}C scrambling in butane is much faster than its skeletal isomerization and equally fast as pentane isomerization. As predicted by an extrapolation of kinetic data on types *A* and *B* rearrangements of alkylcarbenium ions in superacids to temperatures typical for bifunctional heterogeneous catalysis, a bifunctional conversion of an alkane involving a change in the degree of branching is one to two orders of magnitude slower than a conversion that does not cause such change.

Mechanistic data on the rearrangements in superacids of alkylcarbenium ions larger than $C_7H_{13}^+$ are lacking in the literature. The product distributions obtained from the conversion of long-chain alkanes over bifunctional catalysts suggest that new conversion pathways of alkylcarbenium ions become available when the cation contains eight or more carbon atoms. Accordingly, it is necessary to subdivide the alkylcarbenium ions into short-chain and long-chain ions and, likewise, the alkanes into short-chain and long-chain alkanes.

Branching isomerization of a long-chain *n*-alkane over a bifunctional faujasite catalyst is a step-by-step reaction. Mono-, di-, and tribranched feed isomers are obtained in consecutive reactions. When a *n*-alkane is converted, all possible monobranched isoalkanes with the same carbon number are formed. Next to the methylbranched isoalkanes, at least part of the ethyl- and larger *n*-alkylbranched isoalkanes are also primary products from *n*-alkanes.

Type *B* isomerization via substituted protonated cycloalkanes with cyclopropane and larger rings explains satisfactorily the formation of monobranched isomers with all possible *n*-alkyl side chains. Corner-to-corner hydrogen jumps on the cycloalkane rings are the rate-limiting reaction steps. Reaction schemes for the branching via CPCX ions of long-chain *n*-alkyl cations are very simple, since for a given ring size, X, all CPCX intermediates are equally stable. It is sufficient to consider the number of CPCX ions which can be formed and their ring opening modes. Ring opening does not proceed at the same rate for all CPCX intermediates. As the stability of a secondary *n*-alkyl cation, charged at position 2, is lower than for its isomers, CPCX ring opening into such an ion is slower. With this model the distribution of all monobranched reaction products from long-chain *n*-alkanes can be predicted. It was found that next to cyclopropane intermediates to a lower extent cyclobutanes, cyclopentanes, and substituted corner protonated cycloalkanes with larger rings contribute to the formation of the first branching in a linear hydrocarbon skeleton.

There is evidence that the formation of the second branching occurs through the same mechanism as the first one. A study of the second branching step is, however, hampered by the occurrence of alkyl shifts, which compete with the branching reactions to be studied.

With long-chain alkanes, 1,2-alkyl shifts are responsible for the displacement of side chains along the main chain and for step-by-step elongation of side chains.

Hydrocracking of alkanes on bifunctional catalysts occurs through β-scission of alkylcarbenium ions. There are five modes of β-scission, denoted as A, B_1, B_2, C, and D. The distinction between the different mechanisms is based on the specific branching configuration required in the parent cation for β-scission to occur. These specific configurations are α,γ,γ-, γ,γ-, α,γ-, and γ-branching for β-scission of types A, B_1, B_2, and C, respectively. Type D β-scission operates on unbranched cations. Types A, B_1, B_2, C, and D hydrocracking modes are transformation reactions for alkanes, in which the corresponding β-scission modes are rate determining.

Mechanism *A* requires the presence of three side chains, positioned in α,γ,γ. The smallest cation in which such a configuration is possible has eight carbon atoms. During the cracking of any normal, and mono-, di-, or tribranched $C_8H_{17}^+$ cation in superacid solution, the parent cations are rearranged via several consecutive types *A* and *B* isomerization steps into 2,4,4-trimethyl-2-pentyl cation, which is ultimately consumed via the fast β-scission mechanism *A*. On bifunctional zeolite catalysts hydrocracking via the B_1, B_2, and *C* routes is superimposed on mechanism *A*. On zeolites, type *A* hydrocracking is by far the fastest reaction. It is much faster than type *A* isomerization, which in its turn is substantially faster than types B_1 and B_2 hydrocracking, type *B* isomerization, and type *C* hydrocracking. An increase of the reaction temperatures from those typical of superacid catalysis, to those typical of heterogeneous bifunctional catalysis reverses the rate order of type *A* isomerization and type *A* β-scission.

ACKNOWLEDGMENTS

J.A.M. and P.A.J. acknowledge the Flemish NFWO for a Research Position and the Belgian Ministry for Science Policy for a research grant.

REFERENCES

1. Pines, H., *The Chemistry of Catalytic Hydrocarbon Conversions*. New York: Academic Press, 1981.
2. Wojciechowski, B.W., Corma, A., *Catalytic Cracking, Catalysts, Chemistry and Kinetics,* New York, Basel: Marcel Dekker, 1986.
3. Levsen, K., *Fundamental Aspects of Organic Mass Spectrometry*. Weinheim, New York: Verlag Chemie, 1978.
4. Olah, G.A., von R. Schleyer, P., *Carbonium Ions*. New York: Wiley, 1968, Vol. 1.
5. Vogel, P., *Carbocation Chemistry*. Amsterdam, Oxford, New York, Tokyo: Elsevier, 1985.
6. Rigaudy, J., Klesney, S.P., *Nomenclature of Organic Chemistry, IUPAC, Organic Chemistry Division*. Pergamon Press, New York, 1979.
7. Reference 5, p. 62.
8. Radom, L., Poppinger, D., Haddon, R.C., in *Carbonium Ions* (G.A. Olah and P.v.R. Schleyer, eds.). New York: Wiley-Interscience, 1976, No. 5, p. 2303.
9. Al-Khowaiter, S.H., Wellington, C.A., *Tetrahedron* **33:** 2843 (1977).
10. Pakkanen, T., Whitten, J.L., *J. Am. Chem. Soc*. **97:** 6337 (1978).
11. Schwarz, H., Franke, W., Chandrasekhar, Schleyer, P.v.R., *Tetrahedron* **35:** 1969 (1979).
12. Reference 5, p. 162.
13. Corma, A., Lopez Agudo, A., Nebot, I., Tomas, F., *J. Catal*. **77:** 159 (1982).
14. Reference 5, p. 167.
15. Reference 5, p. 342.
16. Brouwer, D.M., Hogeveen, H., in *Progress in Physical Organic Chemistry* (A. Streitwieser Jr. and R.W. Taft, eds.). New York: Wiley-Interscience, 1972, Vol. 9, p. 179.
17. For a review see Ref. 5, p. 323.

18. Saunders, M., Vogel, P., Hagen, E.L., Rosenfeld, J.C., *Acc. Chem. Res.* **6:** 53 (1973).
19. Saunders, M., Budiansky, S.P., *Tetrahedron* **35:** 929 (1979).
20. Radom, L., Pople, J.A., Buss, V., Schleyer, P.v.R., *J. Amer. Chem. Soc.* **93:** 1813 (1971).
21. Brouwer, D.M., in *Chemistry and Chemical Engineering of Catalytic Processes,* NATO ASI Ser. E, No. 39 (R. Prins, and G.C.A. Schuit, eds.), Rockville: Sijthoff and Noordhoff, 1980, p. 137.
22. Weitkamp, J., Jacobs, P.A., Martens, J.A., *Applied Catal.* **8:** 123 (1983).
23. Weisz, P.B., *Adv. Catal.* **13:** 137 (1962).
24. Coonradt, M.L., Garwood, W.E., *Ind. Eng. Chem. Prod. Res. Dev.* **3:** 38 (1964).
25. Brouwer, D.M., *Rec. Trav. Chim.* **88:** 9 (1969).
26. Kramer, G.M., McVicker, G.B., Ziemiak, J.J., *J. Catal.* **92:** 355 (1985).
27. Froment, G.F., *Catal. Today* **1**(4): 455 (1987).
28. Chevalier, F., Guisnet, M., Maurel, R., in *Proceedings of the 6th International Conference on Catalysis* (G.C. Bond, P.B. Wells, and F.C. Tompkins, eds.). London: The Chemical Society, 1977, p. 478.
29. Weitkamp, J., *ACS Symp. Ser.* **20:** 1 (1975).
30. Vansina, H., Baltanas, M.A., Froment, G., *Ind. Eng. Chem. Prod. Res. Dev.* **22:** 526 (1983).
31. Weitkamp, J., *Ind. Eng. Prod. Res. Dev.* **21:** 550 (1982).
32. Steijns, M., Froment, G., Jacobs, P., Uytterhoeven, J., Weitkamp, J., *Ind. Eng. Chem. Prod. Res. Dev.* **20:** 654 (1981).
33. Stull, R.D., Westrum, E.F., Sinke, G.C., *The Thermodynamics of Organic Compounds.* New York: Wiley, 1969.
34. Fiaux, A., Smith, D.L., Futrell, J.H., *Int. J. Mass Spectrom. Ion Phys.* **25:** 281 (1977).
35. Lee, C.C., Reichle, R., Weber, U., *Can. J. Chem.* **56:** 658 (1978).
36. Weitkamp, J., Farag, H., *Acta Phys. Chem., Szeged.* **2:** 327 (1978).
37. Guisnet, M., Alvarez, F., Giannetto, G., Perot, G., *Catal. Today* **1:** 415 (1987).
38. Martens, J.A., Tielen, M., Jacobs, P.A., *Catal. Today* **1:** 435 (1987).
39. Martens, J.A., Jacobs, P.A., Weitkamp, *J. Applied Catal.* **20:** 239 (1986).
40. Martens, J.A., Jacobs, P.A., Weitkamp, *J. Applied Catal.* **20:** 283 (1986).

3
The Role of Next Nearest Neighbors in Zeolite Acidity and Activity

WILLIAM A. WACHTER

INTRODUCTION

Knowledge of the acid chemistry and catalysis in zeolites has grown dramatically in the last 20 years. Early work described zeolites in terms of "strong" acidity and "superactivity." Today, better understanding has made the amazing commonplace, and the chemistry has been placed in the context of well-established solution chemistry.

Early titration studies with organic bases and indicators showed that aluminosilicates, especially zeolites, were capable of protonating dyes that in solvating media required acids with the strength of strong sulfuric acid $H_0 < -7$.[1-3] In addition, reaction studies demonstrated that zeolites cracked paraffins up to four orders of magnitude faster than amorphous silica–alumina.[4]

The correlation of the chemical properties of zeolites with their catalytic performance is not straightforward. Paraffin cracking is linear in Al content for as-synthesized high-silica zeolites such as ZSM-5[5,6] and even for zeolites that have been converted to high Si forms such as high-silica Y zeolite.[7-9] However, isooctane cracking rates are not linear in protonated Al "T" sites generated by the removal of basic charge-compensating cations.[10]

Acidity in zeolites has been described either as a discrete phenomenon with sites of varying acid "strength" or as a continuous phenomenon in which interacting sites have an "activity coefficient" associated with the whole. These aspects of acid behavior can be shown to be parts of the same phenomenon by modeling the spatial relationships of acid sites and charge-compensating cations. In particular, cations that compensate for charge deficiencies in the zeolite framework associated with tetrahedral aluminate ions behave similarly in zeolites and in solution. Zeolites with low silica/alumina ratios have high compensating cation concentrations and low activity coefficients. Zeolites with high silica–alumina ratios have low compensating cation concentrations with high activity coefficients. Description of zeolite acidity must therefore distinguish between the intrinsic activity of an acid site and the expression of that acidity as buffered by its environment.

THE ALUMINOSILICATE ACTIVE SITE

Zeolites and Clays

The most surprising aspect of aluminosilicate chemistry is that a chemical combination of silicon dioxide (which is relatively inert catalytically) and aluminum oxide (which has only fair activity when compared with zeolites) can produce both good catalysts such as zeolites and poor catalysts such as clays. Zeolites typically have surface areas of more than 300 m^2/g,[11,12] while clays have areas less than 100 m^2/g.[13] Clays are more stable in a hydrothermal environment than zeolites. For example, steam rehydrates metakaolin to kaolinite after kaolinite has been transformed into metakaolin by calcination at 850°C.[14] On the other hand, the zeolite NaA is 50% destroyed after 16 hr at 755°C, and the presence of steam accelerates this destruction. NH_4A decomposes at 140°C even though its final decomposition product has roughly the same chemical composition as kaolin.[15] Since catalytic activity implies a metastable entity that reacts with feed but not with products, it is not surprising that zeolites are more readily decomposed than clays in the catalytic cracking environment. To understand the source of this metastability and catalytic activity, an even closer look at the structural differences between clays and zeolites is useful.

Silicon and Aluminum Coordination in Oxide Systems

Silicon and aluminum each have only one oxidation state readily accessible in zeolites and clays—+4 for Si, +3 for Al. Thus, intrinsic acidity depends to a first approximation on their coordination geometry in the oxide system that they share. Silicon coordination with oxygen is almost exclusively fourfold. The exceptions are in a high-pressure allomorph, stishovite[16]; SiP_2O_7(AIII)[17,18]; and an ammonium catecholate.[19] None of the circumstances that occasion the formation of octahedral Si in the preceding compounds occur during the synthesis of zeolites or clays. Silicon can therefore be considered to have an exclusively fourfold coordination geometry in zeolites and clays. Aluminum, on the other hand, has an amphoteric chemistry, which is reflected in its variable coordination geometry. In general, aluminum has a tetrahedral coordination geometry in aluminosilicates when associated with counterions with small charge/radius ratios, i.e., in basic systems such as zeolites, and an octahedral coordination geometry when the counterions have high charge/radius ratios, i.e., in acid systems such as clays.[20] Zeolites are synthesized in basic media, and only on ammonium ion exchange or calcination or both do they become acidic. Clays are synthesized in acidic or neutral media. Acidic clays such as montmorillonite are synthesized at a pH above 7 and have been used as cracking catalysts in their own right. Clays synthesized below pH 7, such as kaolin, are relatively inert.[21] Thus, tetrahedrally coordinated aluminum in solid acids results from basic syntheses. Its association with strong acidity in zeolites has been recapitulated in recent reviews.[22,23] When stressed in a hydrothermal environment, acidic tetrahedral aluminum migrates to nonframework octahedral sites in the zeolite.[24–26] An ^{27}Al MASNMR peak lying between those for octahedral (sixfold) and tetrahedral (fourfold) coordination grows into the spectrum upon steaming so-called "ultrastable Y." This has been attributed to the formation of five-coordinate Al.[27] Examples of five-coordinate Al in crystalline systems are rare, one example being andalusite[28–30]; however, increasing evidence for this species exists in hydrothermally decomposed aluminosilicates.

The dependence of acid performance on the metastable tetrahedral coordination geometry of aluminum was recognized nearly 20 years ago for amorphous aluminosilicate gels. After hydrothermal deactivation, these could be "tetrahedralized or (re)activated *in situ*" by treatment with base.[31] Shortly thereafter, reinsertion of Al into tetrahedral sites of zeolites using base was also accomplished.[32] Recently, the reinsertion of Al into tetrahedral sites in zeolites has been corroborated by NMR studies.[33,34]

Formal Descriptions of Aluminosilicate Acidity

Acidity in zeolites can be described formally at least two ways. In one, tricoordinate aluminum, a strong Lewis acid, polarizes and enhances the acidity of an adjacent silanol group. In the other, the charge-compensating proton associated with the tetrahedral aluminum in the framework (an Al "T" site) occupies a larger free volume (an octahedral hole) than it would otherwise. This free volume enhances the ability of the proton to associate with polarizable entities other than the Al "T" site. The ultimate acid strength of a proton may be fixed by this free volume since a proton in a vacuum can protonate even Xe and Kr.[35–38]

BUFFERED BEHAVIOR AND ZEOLITE STRUCTURE

The dependence of buffered behavior on the separation of ionizable sites is well known in inorganic and organic chemistry.[39] If zeolites behave like other condensed-phase acids, then the expression of the strong intrinsic acidity in zeolites should depend not only on the presence of a charge-compensating proton at an Al "T" site but also on the absence of any other monovalent cations (except protons) within four bonds of the proton. For example, in phosphoric acid, the first and second ionization constants differ by five orders of magnitude and the ionized sites are separated by two P–O bonds. Furthermore, although all the sites in phosphoric acid are identical, the system behaves as though sites of three different strengths are present during titration. In other words, the measurement (titration) modifies the property measured (the strength and number of acid sites). In pyrophosphoric acid, the first and second ionization constants differ by only one order of magnitude and ionized sites are separated by four P–O bonds. The ionization of the second proton in pyrophosphoric acid is not buffered by the ionization of the first. Hexabasic $H_6(Co(II)W_{12}O_{40})$ is perhaps the ultimate example of an isolated but compact acid site system and gives only one endpoint on titration with sodium hydroxide in water.[40]

The "T" sites in zeolites are framework cations (Si^{4+}, Al^{3+}) in tetrahedral coordination with oxygen. Each Al "T" site has four Si "T" sites as nearest neighbors. This is known as "Lowenstein's Rule."[41] The four Si "T" sites are connected to between 8 and 12 other "T" sites depending on the zeolite structure, and these sites are the next nearest neighbors of the original Al "T" site. Each Al "T" site can therefore be described in terms of its environment: an Al "T" site with no Al next nearest neighbors is a 0NNN site, an isolated pair of Al sites are two Al 1NNN sites, an isolated 6-ring with alternate Al and Si "T" sites contains three Al 2NNN sites, etc.

Since 4-rings are the smallest structural units in zeolites, a site can be classified according to the number of 4-rings containing it. A site contained in "*n*" 4-rings is a (12-*n*)NNN site. All sites in faujasites are contained in three 4-rings and are thus all 9NNN sites.[42,43] An excellent summary of the state of knowledge about the local environment of Si and Al in zeolite systems has recently been published.[44] In general, zeolites synthesized in more basic environments have Si/Al ratios less than 5 and are 9NNN systems, while those synthesized in less basic environments have Si/Al ratios greater than or equal to 5 and are 10 and higher NNN systems.

Zeolites are described here using the convention of Meier and Olson,[45,46] where the different structural types are denoted by three letters. Thus FAU is the structural type of X, Y, and faujasite, LTL is Linde type L, MFI is ZSM-5, OFF is offretite, etc.

The Measurement and Modeling of FAU Acidity

Historically, the identification of acid behavior with composition was occasioned by a very thorough experimental study of the *n*-butylamine titration of the large-pore FAU structure at Si/Al ratios between 1.5 and 3.7.[47–49] These studies framed the results in terms of solution chemistry and noted the low activity coefficients of the low Si/Al system.

The use of the *n*-butylamine titration technique to determine these acidities has been closely questioned.[50,51] However, calorimetric data on *n*-butylamine adsorption show essential agreement with the indicator dye technique for faujasites.[52,53]

A number of models have been developed to explain and correlate these data. Beaumont and Barthomeuf, the original investigators,[54–56] preferred to use activity coefficients to describe the "self-neutralization effect." However, other authors have chosen to use a topological approach to acidity, and one model explicitly tries to model both the topology of Al "T" sites and the role of the charge-compensating cations. In fact, one can speak of a European school favoring the "activity coefficient" approach, an English school favoring the topological approach, and an American school favoring a computational approach in a topological framework.

The activity coefficient approach has been used by Mortier and co-workers,[57,58] Jacobs,[59] Hocevar and Drzaj,[60] as well as Beaumont and Barthomeuf. Sanderson electronegativities were used to calculate acid strengths by Mortier, and this work was elaborated on by Hocevar and Drzaj, and reviewed by Jacobs.

However, other workers have used the topology of faujasite to describe Al site types. Mikovsky and Marshall[61–63] discussed Al occupancy in the three 4-rings about an Al "T" site. They proposed only four site-types (0NNN, 1NNN, 2NNN, 3NNN) and used the rate of ammonia evolution from an ammonium Y zeolite to corroborate their model. Their work did not take into account the role played by charge-compensating cations. Dempsey[64–66] outlined the necessary requirements for symmetry to be preserved with this approach and assigned strong acidity to 0NNN and 1NNN sites.

Peters[67,68] was more interested in the arrangement of "T" sites in faujasite and used a random distribution of sites bound by the constraints of Lowenstein's rule (i.e., no two Al "T" sites can be neighbors). Mikovsky[69] also showed that the ^{29}Si MASNMR for the FAU structure was consistent with Lowenstein's rule and a more or less random distribution of Al. Melchior,[70] on the other hand, was able to show from a series of FAU zeolites of varying Si/Al ratios that Al "T" sites actually preferred to site themselves during synthesis so that they avoided interaction even at the next-nearest-neighbor level. For example, FAU synthesis can be described by an assembly process in which Al "T" sites maximize their separation from each other across the double six-ring. This is referred to in this chapter as the D6R model. Dwyer and co-workers[71,72] showed that a computer model that randomly placed Al into the FAU structure under the constraints of Lowenstein's rule had increasing difficulty as the Al "T" site concentration rose from $\frac{1}{3}$ to $\frac{2}{5}$ of the total number of "T" sites present in the system. Dwyer also calculated the probability that each of the nine types of NNN sites was an Al "T" site for a range of Si/Al ratios. Wachter[73] used a computational model with a less sophisticated Al distribution than that of Dwyer to estimate the effect of Al "T" sites associated with basic charge-compensating cations on neighboring protonated Al "T" sites. The following sections show that using the distribution of Al "T" sites proposed by Melchior in this computational model fits the original data of Beaumont and Barthomeuf[47] better than other distributions.

The Next-Nearest-Neighbor Model

The Next-Nearest-Neighbor (NNN) Model is an algorithm that "titrates" acidity in zeolite structures by:

1. first titrating the 0NNN site acidity and summing as "strong" acidity;
2. then titrating the acidity at 1NNN sites, summing "strong" acidity as appropriate; then summing basic monovalent cations and "titrated" acids as basic "buffering" charge-compensating cations impacting on the ability of remaining protons to be strong acids;
3. then titrating the acidity of 2NNN sites, etc.

The probability an acid site is a "strong" acid site is proportional to the probability that the Al NNN present do not have basic cations or titrated protons as charge-compensating cations. This probability is calculated using

$$P = (1 - \mathrm{C/Al}) \sum_{n=0}^{m} \sum_{c=1}^{c_{\max}} \Delta \tag{3.1}$$

where

$\Delta = n\mathrm{NNN}/c_{\max} \times \{1 - \eta[1/(1 - 0\mathrm{NNN})][(1 - 0\mathrm{NNN})\mathrm{C/Al} + \mathrm{B/Al}]\}^n$

n = the number of Al next nearest neighbors

$B = 0$ for $n = 0$; and $B = \Sigma\Sigma\Delta$ for $n > 0$.

B/Al = the ratio of titrated protons on Al "T" sites to Al "T" sites

C/Al = the ratio of basic charge-compensating cations to Al "T" sites

nNNN = the probability a given "T" site will have n Al "T" NNN sites

η = the cation buffering efficiency and accounts for the "average" influence of cations
c_{max} = the number of iterations for a given nNNN Al "T" site
m = the last site type with a significant probability of occurring
P = the strong acid probability.

In order to calculate the strong acid titer using the NNN model, the distribution of the Al site types—0NNN, 1NNN, 2NNN, etc.—is needed as well as the initial concentration of basic (monovalent, large charge/radius ratio) charge-compensating cations at these sites. In addition, the buffering efficiency of the basic charge-compensating cations must be calculated.

NNN Site Type Distributions. The original work with the NNN model[70] used a binomial distribution of Al site types given by

$$y\text{NNN} = n!/[(n - y)!y!][1/(1 + R)]^y[1 - 1/(1 + R)]^{(n-y)} \qquad (3.2)$$

where
yNNN = the probability an Al "T" site will have y Al "T" sites as next nearest neighbors
n = the total number of possible next nearest neighbors (set by the structure and between 8 and 12 in zeolites)
R = the Si/Al ratio.

This distribution of sites cannot be correct, since, for example, at a Si/Al ratio of 1 in a 9NNN system such as FAU, all Al "T" sites must be 9NNN sites, that is, have nine Al "T" sites as next nearest neighbors. The binomial distribution ignores Lowenstein's rule and assigns half the sites to the 4NNN and 5NNN site types. Table 3.1 contains calculations for binomial and other distributions, which are explained subsequently in the text.

An improvement on this distribution can be obtained by using the random Lowenstein distribution calculated by Beagley et al.[72] In this case, the probability each of the 9NNN sites is occupied by an Al is given by calculation and not by the Si/Al ratio. A weighted average of these probabilities was used in Eq. (3.3) to calculate Al "T" site populations and the results for Si/Al ratios between 1.18 and 2.42 are also contained in Table 3.1:

$$y\text{NNN} = n!/[(n - y)!y!](RL)^y(1 - RL)^{(n-y)} \qquad (3.3)$$

Table 3.1. 9NNN site distributions.

	Site Type									
	0NNN	1NNN	2NNN	3NNN	4NNN	5NNN	6NNN	7NNN	8NNN	9NNN
Si/Al = 1.0										
Lowenstein	0.00	0.00	0.00	0.00	0.00	0.00	0.00	0.00	0.00	1.00
Binomial	0.00	0.02	0.07	0.16	0.25	0.25	0.16	0.07	0.02	0.00
Si/Al = 1.18										
Random Lowenstein	0.00	0.00	0.00	0.00	0.00	0.00	0.03	0.14	0.38	0.44
Si/Al = 1.40										
Random Lowenstein	0.00	0.00	0.00	0.00	0.01	0.05	0.15	0.29	0.33	0.17
Si/Al = 1.67										
Random Lowenstein	0.00	0.00	0.00	0.02	0.06	0.15	0.26	0.28	0.18	0.05
Si/Al = 2.00										
Random Lowenstein	0.00	0.02	0.08	0.17	0.25	0.25	0.16	0.06	0.00	0.00
Si/Al = 2.42										
Random Lowenstein	0.00	0.03	0.10	0.20	0.26	0.22	0.13	0.05	0.01	0.00
D6R	0.00	0.29	0.39	0.32	0.00	0.00	0.00	0.00	0.00	0.00
Si/Al = 3.00										
D6R	0.04	0.50	0.38	0.08	0.00	0.00	0.00	0.00	0.00	0.00
Si/Al = 4.00										
D6R	0.22	0.63	0.15	0.08	0.00	0.00	0.00	0.00	0.00	0.00
Si/Al = 5.00										
D6R	0.57	0.43	0.00	0.00	0.00	0.00	0.00	0.00	0.00	0.00

where

yNNN = the probability an Al "T" site will have y Al "T" sites as next nearest neighbors

n = the total number of possible next nearest neighbors (set by the structure and between 8 and 12 in zeolites)

RL = the weighted average of probabilities that a site will be occupied by Al under the constraints of Lowenstein's rule.

Many more distributions of Al "T" site types are possible. The D6R distribution of Melchior[70] sites Al in "T" sites to avoid NNN neighbor interactions where possible. Table 3.1 also contains D6R distributions of Al "T" site types for Si/Al ratios between 2.42 and 5.

Buffering Efficiencies of Charge-Compensating Cations. If one assigns a buffering efficiency of 1 to basic charge-compensating cations, all protons at Al "T" sites that have an Al "T" site as a NNN with a basic charge-compensating cation would have their acidity buffered. Figure 3.1 illustrates the situation for a simple linear case, where an Al "T" site associated with a basic charge-compensating cation buffers the acidity of two protonated Al "T" site next nearest neighbors. This simple approach overestimates the buffering capacity of the basic charge-compensating cation. Since the basic charge-compensating cation can buffer only one of the protons, the appropriate buffering efficiency is 0.5 for this case.

The FAU structure has 9NNN sites associated with any given acid site, and these are associated with $9 \times 4 = 36$ anionic sites. Figure 3.2 shows that 12 of these anionic sites are in a position to buffer the acidity of the acid site. The other 24 anionic oxygens associated with Al NNN sites are also shared with Si nearest neighbors of still other Al "T" sites, so we must account in some way for the impact of the charge-compensating cations at these Al NNN sites on the potential acidity of these other Al "T" sites. If these cations are basic, the influence should be negligible, since the correlation of these ions should have a minimal impact on the acidity of the system. If the cations are protons, then the buffering efficiency of the charge-compensating cation at the original Al "T" site will be higher. Since the other 24 anionic oxygens are also shared with $\frac{12}{36}$ of the anionic oxygens shared with other Si nearest neighbors of still other Al "T" sites and could alternately buffer those sites, the total buffering

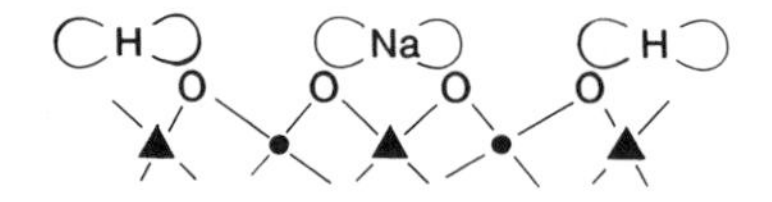

Figure 3.1. Buffering efficiency in a linear system. ▲ = Al; ● = Si.

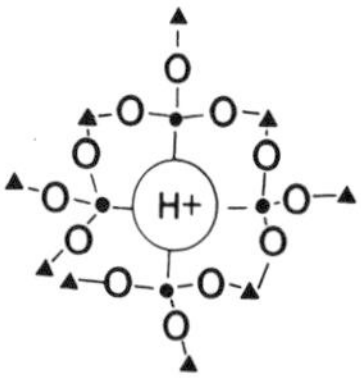

Figure 3.2. Buffering anionic sites in a 9NNN system. ▲ = Al; ● = Si; ○ = buffering anionic sites.

capacity of a monovalent charge-compensating cation on an Al NNN site is $\frac{12}{36} + \frac{24}{36} \times \frac{12}{36} = \frac{5}{9}$. Figure 3.2 illustrates the position of buffering anionic sites in a 9NNN system for the calculation of buffering efficiencies. The general formula for buffering efficiency, η, as a function of the NNN structure type is given by

$$\eta = 12/(\#\text{NNN} \times 4) + [12 \times (\#\text{NNN} \times 4 - 12)/(\#\text{NNN} \times 4)^2] \quad (3.4)$$

where #NNN is the number of next nearest neighbors for each "T" site. The results of these calculations are contained in Table 3.2.

The buffering efficiency of charge-compensating cations must rise as the Si/Al ratio rises. For example, if all Al "T" sites are pairs of 1NNN sites, then the buffering efficiency of all basic charge-compensating cations would be 1.0. To account for this effect, a weighted buffering efficiency, η_w, has been calculated for the NNN model using

$$\eta_w = 1.0 \times (1\text{NNN})^2/(1 - 0\text{NNN}) + \tfrac{5}{9}\{1 - [(1\text{NNN})^2/(1 - 0\text{NNN})]\} \quad (3.5)$$

This formula gives all sites a weighted buffering efficiency where paired Al 1NNN "T" sites contribute a buffering efficiency of 1.0 and Al "T" sites with higher Al "T" site NNN contribute a buffering efficiency dependent on the structure and calculated previously.

Table 3.2. Buffering efficiencies for different structure types.

NNN	Buffering Efficiency η	Maximum Strong Acid Titer/10000 "T" Sites
9	0.556	172
10	0.510	171
11	0.471	170
12	0.438	169

Charge-Compensating Cation Distributions over NNN Site Types. Once the Al "T" site distribution and basic charge-compensating cation buffering efficiencies are chosen, both the Al "T" sites and the charge-compensating cations must be distributed in the structure prior to carrying out the titration. There are three simple ways to distribute basic charge-compensating cations over the Al "T" site types. The first is to distribute them globally, that is, to assign to each site type the same proportion of basic charge-compensating cations as are found globally. The second is successively to fill Al "T" sites with increasing numbers of Al NNN with the basic charge-compensating cations before titration. The third is successively to fill Al "T" sites with decreasing numbers of Al NNN with the basic cations before titration.

Figure 3.3 contains the results of using the NNN model with random Lowenstein and D6R distributions at Si/Al ratios between 1.0 and 5.0 with an initial global distribution of basic charge-compensating cations and weighted buffering efficiencies for a 9NNN structure such as faujasite. The results clearly show the major inhibition of the expression of strong acidity in systems where the Si/Al ratio is less than 2.4. Site isolation has a major effect on the appearance of strong acidity, since the random Lowenstein site distribution gives 25% lower strong acid titers

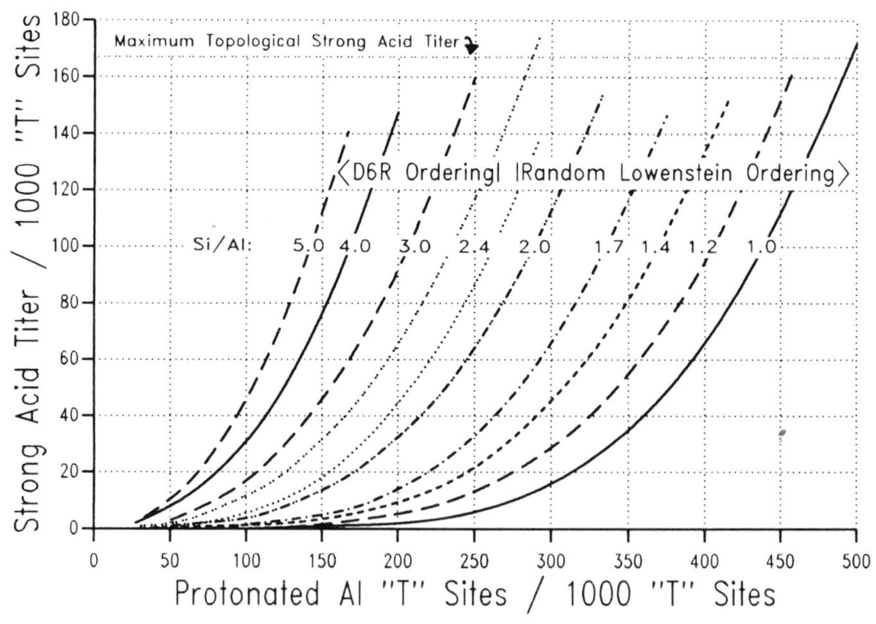

Figure 3.3. NNN model predictions for the 9NNN system.

than the D6R distribution at a Si/Al ratio of 2.4. The maximum strong acid titer calculated for each of the systems approaches but never exceeds roughly $\frac{1}{6}$ of the total "T" sites in the system.

The topology of the faujasite system is such that no more than 32 out of 192 or $\frac{1}{6}$ of the Al "T" sites can be 0NNN sites in FAU.[74] Therefore, no calculation procedure should give more than $32/192 \times 1000 = 167$ strong acid sites per 1000 "T" sites for a 9NNN system. The NNN model returns a number of maximum acid titers, none of which exceed the value of 167 by more than 4%. The failure of real zeolite systems to achieve strong acid titers near the 2.8 meq/g this value represents testifies to the effects of acidity on the structural stability of zeolites. In effect, zeolites with low Si/Al ratios decompose long before they can reach their maximum strong acid titer. To evaluate the precision of the NNN Model calculations for the maximum acid titer, we note that the strong acid titer at a Si/Al ratio of 5 with the D6R distribution can be no more than $(0.57 + 0.43/2) \times 1000/6 = 131/1000$ "T" sites. The NNN Model returns a value of 141/1000, indicating that further improvements in the NNN model are warranted.

Figure 3.4 contains the results for different distributions of basic charge-compensating cations over the Al site types for the random Lo-

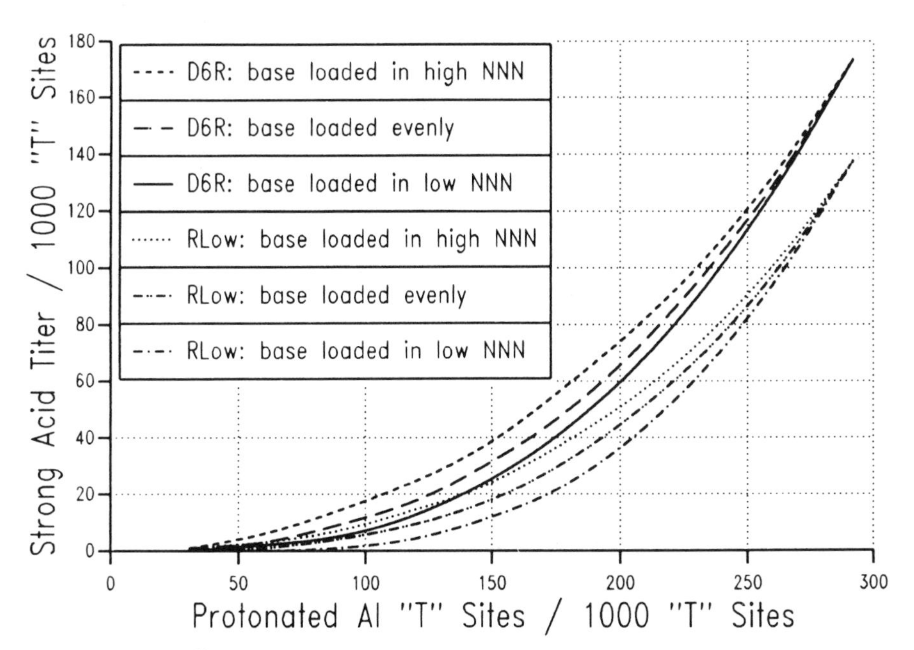

Figure 3.4. Effect of basic charge-compensating cation distributions on strong acid titer for Si/Al = 2.42.

wenstein (RLow) and D6R site distributions at a Si/Al ratio of 2.42. As one might expect, preloading low NNN sites with basic cations before titration results in a lower strong acid titer than the global distribution. The global distribution in turn gives a smaller strong acid titer than that obtained by preloading basic cations into high NNN sites.

Figure 3.5 contains the classic *n*-butylamine titration data for $H_0 < -7$ of Beaumont and Barthomeuf surperimposed on the NNN model calculations of Figure 3.3. The model fits data from ion-exhanged as-synthesized zeolites (Si/Al = 1.2, 2.4) best. Modified zeolites give higher titers than projected, suggesting site isolation is effected by extraction procedures. The D6R NNN site distribution does a better job of matching acid titers for Si/Al = 2.42 than does a random Lowenstein model. The *n*-butylamine titers can be checked against values reported for isosteric heats of adsorption of ammonia on the FAU structure of 2.2–2.7 meq/g of ammonia adsorbed with heats of adsorption greater than 100 kJ/mol.[61,75,76] These strong acid titers translate into 126 and 167 strong acid sites/1000 "T" sites, a result consistent with predictions of the NNN Model. The 100–110 kJ cutoff for characterizing strong acidity seems reasonable in light of the work by Klyachko[77] on high silica zeolites.

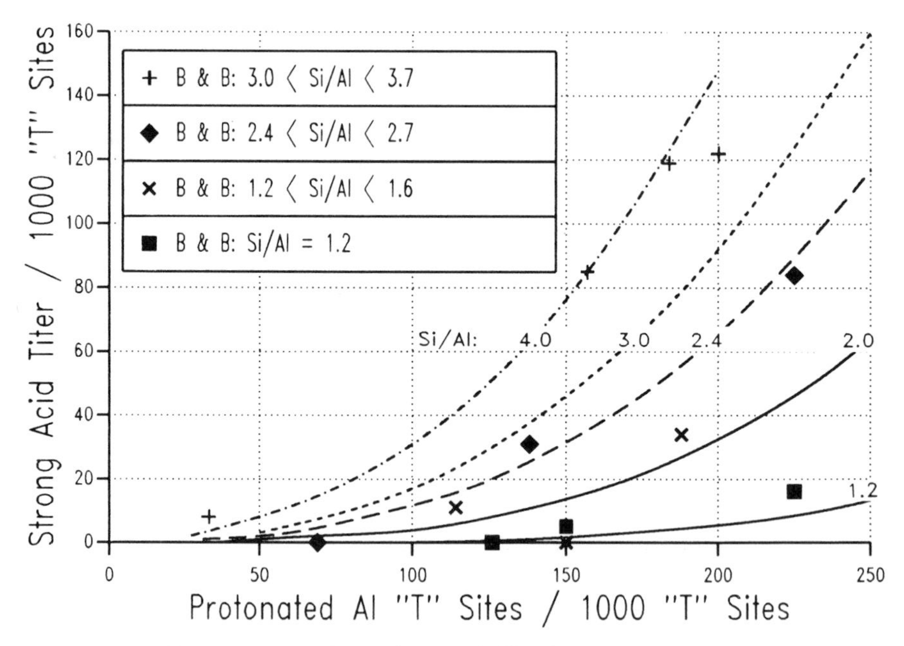

Figure 3.5. Beaumont and Barthomeuf titration results superimposed on 9NNN model predictions.

Acidity in Zeolites other than FAU

The study of the acidity of zeolites other than 3-D, large-pore zeolites such as FAU using *n*-butylamine runs the risk of underestimating the acidity present because access is limited into the zeolite.

Table 3.3 shows how poorly *n*-butylamine measures the acid titer in a single-channel large-pore zeolite, LTL.[78] The total titer measured by *n*-butylamine is only $\frac{1}{10}$ that expected on the basis of the elemental analysis (assuming all Al are in "T" sites and associated with whatever Na is present). If the *n*-butylamine technique measures a representative portion of the LTL acidity, however, the proportion of acidity that is strong should be representative as well. The proportion of strong acidity measured is nearly twice as high as that predicted, however. This proportion occurs at a Si/Al ratio too low to allow complete site isolation implied by the high proportion of strong acidity. It is likely a measurement artifact.

Al "T" Site Pairing: Acidity Data. Four studies on the single-channel, large-pore zeolite, MOR, using ammonia as an acidity probe[77,79–81] allow one to assess the extent to which the limitations of pore structure affect acidity measurement results with *n*-butylamine on LTL. In all these studies, ammonia with a heat of adsorption greater than 110 kJ was considered to titrate strong acidity. These results are summarized in Table 3.4.

In these cases, the strong acid titer as measured by ammonia seems to agree well at lower total acid titers as projected from elemental analyses. This indicates the inadequacy of the *n*-butylamine technique for any zeolite but FAU.

However, the strong acid titer as indicated by the adsorption by ammonia with an isosteric heat of adsorption greater than 110 kJ/mol is nowhere near that expected if all the Al "T" sites were isolated. In almost all

Table 3.3. LTL *n*-butylamine titration data.

Si/Al	Elemental Analysis Projection of Total Acid Titer (Measured)	Ratio of Strong Acid to Total Acid Measured	NNN Model AlOH/1000 NNN Model	Ratio of Strong Acid to Total Acid Predicted
3.3	147 (14.4)	0.37	150	0.22
4.4	100 (11.4)	0.85	114	0.32
4.4	114 (9.2)	0.81		
5.4	64 (14.7)	0.64	67	0.28
5.4	86 (11.4)	0.93	100	0.46

Table 3.4. MOR ammonia adsorption data.

Si/Al	Na/Al	Total Acid Titer/1000 "T" Sites Projected from Elemental Analysis	NH_3 Titer with ΔH_{ads} > 110 kJ/1000 "T" Sites	Ratio of NH_3 > 110 kJ/Total Titer (Ref.)
5.04	0.67	55	4	0.07 (75)
5.04	0.54	77	20	0.26 (75)
5.04	0.17	139	75	0.54 (75)
5.04	0.02	164	90	0.55 (75)
5.0	0.54	85	51	0.60 (73)
5.0	0.10	150	89	0.60 (73)
6.3	0.00	137	38	0.28 (73)
5.0	0.00	167	30	0.18 (76)
6.5	0.00	167	90	0.54 (76)
5.77	0.39	89	43	0.48 (77)
5.77	0.29	114	49	0.43 (77)
5.77	0.00	148	86	0.58 (77)

cases, the strong acid titer is one-half the total acid titer even though the Si/Al ratio is high enough that the sites could be isolated. Paired Al "T" sites are limited to one strong acid site per pair by the buffered behavior of the titrating base. MASNMR results to be discussed subsequently indicate that indeed Al "T" site pairing does occur in a number of zeolites with Si/Al ratios greater than 5. Results where a roughly 1/1 ratio between strong acid sites and weaker acid sites are found not only in MOR but also in MFI and are found in Table 3.5.[82] Table 3.5 also contains microcalorimetric results on a variety of MFI samples in which one-half of the potential acid sites exhibit strong acid behavior.[83]

Al "T" Site Pairing: ^{29}Si MASNMR Data. The body of zeolite synthesis data indicates that zeolites which are made in a higher pH regimen at lower Si/Al ratios tend to be structures that can both obey Lowenstein's rule and have a Si/Al ratio of 1. Those made at higher Si/Al ratios have 5-rings present in the structure, which precludes meeting both these conditions. Thus structural types MOR and MFI with Si/Al ratios at or above 5 are comprised of 6-, 5-, and 4-rings, while LTA and FAU with Si/Al ratios below 5 are comprised of 6- and 4-rings. At present ^{29}Si MASNMR studies are being used to delineate (albeit indirectly) the effect of structure type on Al siting. The data to date indicate that Al tends to occupy 6-rings in FAU, a 9NNN system (in this system all sites are topologically equiva-

Table 3.5. MFI ammonia adsorption data.

Si/Al	AlOH/1000[a]	"Medium" H^+/1000	"Strong" H^+/1000
42.6	22.7	10.4	8.8
42.6	21.2	10.4	7.9
42.6	11.4	8.3	4.2
42.6	11.4	1.9	0.0

Reference 82

Formula/Unit Cell	Total AlOH/Unit Cell	NH_3 Titer for $\Delta H_{ads} > 110$ kJ
$Na_{0.03}H_{2.67}Al_{2.7}Si_{93}$	2.7	1.8
$Na_{0.2}H_{3.2}Al_{3.4}Si_{93}$	3.2	1.7
$Na_{0.96}H_{3.4}Al_{4.4}Si_{92}$	3.4	2.1
$Na_{0.05}H_{4.75}Al_{4.8}Si_{91}$	4.8	2.2
$Na_{0.10}H_{6.3}Al_{6.4}Si_{90}$	6.3	3.0

Reference 83

[a] Calculated by assuming the ratio of water lost to acid sites present is constant and that the number of acid sites present is equal to the initial Al concentration.

lent, but Al "T" site placement is based on a 6-ring subunit)[70]; 6-rings in MAZ, a 10NNN system[84–86]; 6-rings in FER, a 12NNN system[87]; 6-rings in MTN, a 12NNN system[87]; 4-rings in MOR, an 11/12NNN system[88,89]; and 5-rings in MFI, an 11/12NNN system.[87]

The siting studies provide an experimental benchmark on which *ab initio* calculations have been tested for MFI, MOR, and FER.[90–92] An *ab initio* calculation on TON indicates that "T" sites with longer bond lengths and smaller bond angles are preferred by Al. This is consistent with a preference for Al siting in larger rings but does not fit the MOR data.[93]

Data supporting the preferential paired siting of Al "T" sites in 4-rings in mordenites is found in the analysis of ^{29}Si MASNMR.[89] Water enhances MFI activity for paraffin cracking, and this has been interpreted to mean that Al "T" sites are paired in this structure until the concentration of the Al "T" sites drops below 1/unit cell.[94] The MASNMR spectra for MFI have been interpreted to mean that Al "T" sites are not in the 4-rings as they are in MOR, however. Since the number of sites containing 4-rings is limited in MFI, and the Si/Al ratio is generally above 10, Al "T" siting in 4-rings would seem to be preferred. Further investigation seems warranted in this area.

STABILITY CRITERIA AND THE NEXT-NEAREST-NEIGHBOR MODEL

In FAU, the concentration of "T" sites is 12.7 T/nm^3. All Al "T" sites can be isolated at Si/Al = 5, so the maximum isolated site concentration is 32/unit cell or 167/1000 "T" sites or 2.12 "T" sites/nm^3. If all these Al "T" sites are protonated, the acid concentration is 3.5 *M*! Thus, the acid concentration that can be created on heating an NH_4HY zeolite is greater than the pH at which the hydrated Al^{3+} cation becomes the most stable species (ca pH 3).[95]

When the protonated Al "T" site concentration rises above 32/unit cell in FAU, protonated Al "T" sites must become next nearest neighbors,[74] forcing protonation of two siloxy groups attached to one framework Al site. This may lead to a sharp decrease in the coordinative saturation of the Al site and its subsequent excision from the structure. Kuehl and Schweizer[96] discovered that if the (Si + NaAl)/AlOH ratio drops below 5 in the FAU structure, i.e., more than 32 protonated Al "T" sites/unit cell are present, the structure is unstable. Bolton and Lanewala[97] showed that the rate of dehydration of ammonium Y (FAU, Si/Al = 2.5) slowed as the hydroxyl content dropped below 32/unit cell, consistent with acid sites being isolated below this level.

The stability criterion used for FAU, a 9NNN system, should also be valid with a little variation for systems such as LTL (which has 24 "T" sites with 9NNN and 12 "T" sites with 10NNN in a unit cell). In an LTL study,[78] the measured acidity began to drop when the (Si + KAl)/AlOH ratio fell below 7 even though fewer basic cations were present in the structure.

Mirodatus and Barthomeuf[98] note that OFF samples with a (Si + KAl)/AlOH ratio at or above (14.4 + 0.6)/2.5 = 6 "have good crystallinity." A sample with a ratio of (14.4 + 0.4)/2.7 = 5.6 did not. The OFF structure has 12 9NNN "T" sites and 6 10NNN "T" sites, and thus should have stability more like that of LTL than FAU.

The NNN model when tested with buffering efficiencies that mirror the relative oxygen anion concentrations of Al NNN sites and thus their direct and indirect buffering capacity indicates that the criterion for protonated Al "T" site isolation is the same in 9NNN, 10NNN, 11NNN, and 12NNN systems if all sites are topologically equivalent and equally populated. That is, as long as the protonated Al "T" sites have equivalent environments and their concentration does not rise above 167/1000 "T" sites, the structure should be stable.

Why does the NNN model fail to predict the instability of the LTL and OFF structures? The initial assumption that all "T" sites have equal probability of being Al "T" sites is known to be wrong on the basis of

^{29}Si MASNMR spectra. Apparently if Al "T" sites cluster, the fabric of the structure in which they reside is more likely to rip. If, for example, Al sites were concentrated in the 9NNN sites in OFF and LTL, then the protonated Al "T" site concentration would be reduced by 2/3, so the non-AlOH "T" site/AlOH "T" site ratio would need to be $(5 + 1/3)/(2/3) = 8$ for stability.

PARAFFIN CRACKING ACTIVITY

Paraffin Cracking in Unbuffered Environments

n-Paraffin cracking activity has been shown to be a linear function of the protonated Al content of the zeolite as long as the Si/Al ratio is greater than 5. Since this is the Si/Al ratio at which Al "T" sites can be isolated, this is not surprising. This linear relationship between paraffin cracking activity has been observed for both the MFI[5,6] and FAU[7–9] structures. These span the likely range of NNN systems,[9–12] and their pore structures are not likely to become blocked, thereby obscuring activity with deactivation.

Figure 3.6 contains results for the cracking of isobutane,[9] hexane,[8] and hexadecane[7] as a function of the framework AlOH content in FAU. Noth-

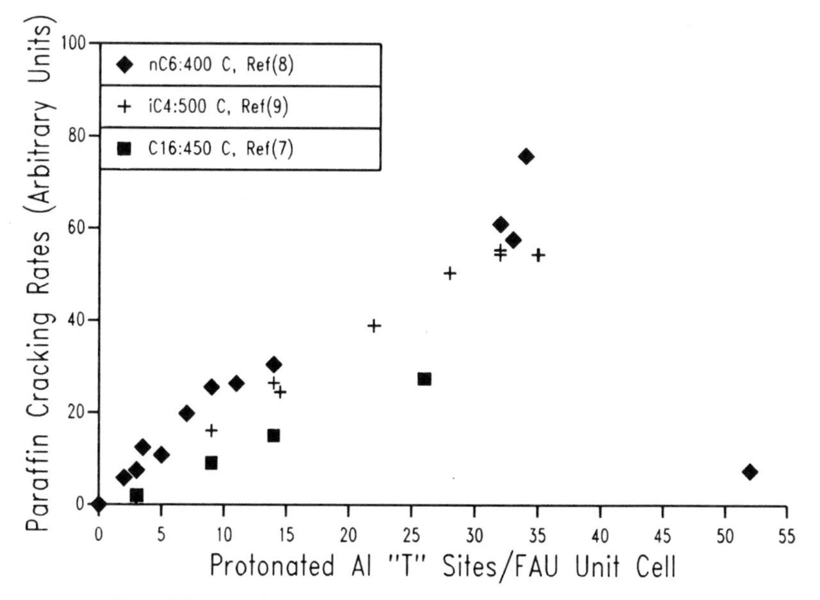

Figure 3.6. Paraffin cracking dependence on protonated Al "T" sites in unbuffered environments.

ing is so desired or boring as a straight line in plotting data, and all the data up to 32 protonated Al "T" sites in FAU fall on straight lines. These data have been interpreted to mean that paraffin cracking depends on the presence of individual isolated protonated Al "T" sites, but it can also be interpreted to mean that paraffin cracking rates rise with a site corresponding to a certain Al "T" site configuration, such as a pair of NNN Al "T" sites.

For FAU zeolites with between 11 and 55 protonated Al "T" sites/unit cell, the hexane cracking activity rises steadily as the framework Al content increases to 32 Al "T" sites/unit cell. At this point, two distinct behaviors occur. Catalysts with somewhat more than 32 Al "T" sites/unit cell have moderately higher activity; however, an ammonium Y that starts with 55 Al "T" sites becomes quite inactive. The source of the accentuated activity for the catalysts with Al "T" site concentrations greater than 32/unit cell has been ascribed to the presence of extraframework Al. This would be surprising, since the instability of the Al in the framework caused it to migrate to a more thermodynamically favored position out of the framework. How could a more stable material be more active? A more satisfying rationale is that an incremental number of paired protonated Al "T" sites enhance catalytic activity for those sites. The exceedingly low activity of the ammonium Y has a number of possible sources. The first is that it is partially decomposed. The second is that the zeolite rapidly deactivates. Bremer's study with pulsed hexadecane[7] shown in Figure 3.6 also indicated that the deactivation function, $[1/k(h)] - [1/k(0)]$, where $k(h)$ is the activity after h amount of hydrocarbon has been converted by the catalyst and $k(0)$ is the activity for the first pulse of hydrocarbon, is linear in the number of framework Al sites. Thus, a zeolite with twice as many protonated Al "T" sites would convert twice as much material and deactivate twice as fast on the material it converted to have a final activity of one-quarter that of its less active counterpart.

Paraffin Cracking in Buffered Environments

Paraffin cracking in a buffered environment differs greatly from that in an unbuffered environment. The results of Beaumont and Barthomeuf on isooctane cracking are presented in Figure 3.7 together with the data of Figure 3.6 for comparison. In this case, activity for cracking does not rise linearly with protonated Al "T" site content. Instead, it first rises when there are fewer than 32 sodium ions/unit cell (or more than 24 protonated Al "T" sites) in NaHY. At this point protonated Al "T" sites can first exist without being buffered by Na.

A second inflection point occurs at the point where less than 15 Na^+ ions/unit cell remain or 41 protonated Al "T" sites/unit cell are exposed.

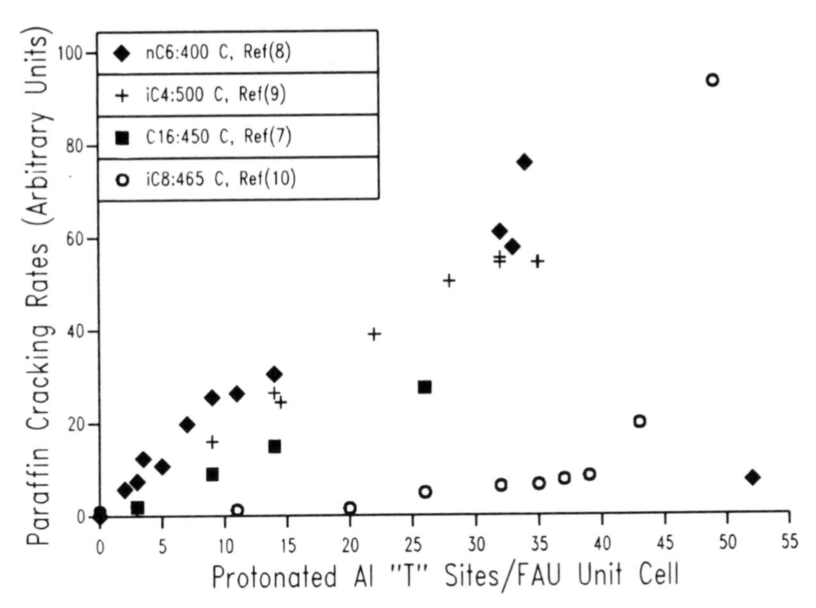

Figure 3.7. Paraffin cracking dependence on protonated Al "T" sites in buffered and unbuffered environments.

At this point, the D6R Model indicates that 1NNN Al sites must become protonated, suggesting that higher activity attends the existence of protonated 1NNN Al "T" sites than protonated 2NNN and 3NNN Al "T" sites.

Na contents greater than 8/unit cell suppress the carbonium ion activity of NaHY zeolites with Si/Al = 4.5.[9] In this case, the D6R Model places 15 Al "T" sites/unit cell in pairs of 1NNN sites. Na contents less than 7.5/unit cell must force these 1NNN sites to exist as protonated pairs, possibly accentuating carbonium ion activity.[9] Thus, forced exposure of paired protonated 1NNN Al "T" sites accounts for accelerated activity in FAU structures with Si/Al ratios at or below 5 in at least two cases.

The reason for the importance of paired sites is readily apparent. Paraffin cracking has been shown to be first order in olefin as well as first order in paraffin,[99,100] a result consistent with the necessity of having two adjacent protonated Al "T" sites. The carbonium ion formed by the reaction of a proton with an olefin of one of the Al "T" sites can serve as a hydride acceptor from a relatively inert paraffin; the carbonium ion thus formed from this paraffin can site at the other Al "T" site while the hydride acceptor desorbs as a paraffin. With one site containing a carbonium ion and the other a proton, the carbonium ion can split to form two protonated olefins at the two adjacent Al "T" sites. An alternative formulation of this statement is that a carbonium ion and a proton can react to form two carbonium ions. The kinetic implications of having a reaction in which a product is a participant is that the reaction is autocatalytic as long as the olefin does not coke up the surface.

CUMENE CRACKING ACTIVITY

Cumene cracking is also linear with protonated Al "T" site concentrations below 32 sites/FAU unit cell at 290°C as is shown in Figure 3.8.[101] Above 55 sites/FAU unit cell, the cumene cracking rate at 300°C declines steadily.[102] Since FAU zeolites with more than 32 protonated Al "T" sites/unit cell have the potential to decompose, the steady decline in activity above 55 sites/FAU unit cell probably reflects the loss in structural integrity of the zeolite as the Si/Al ratio falls.

SUMMARY AND CONCLUSIONS

Reasonable estimates of the acidic behavior of zeolites can be made with the following information: the Si/Al ratio, the framework Al content, the relative crystallinity, the basic charge-compensating cation content (Na/

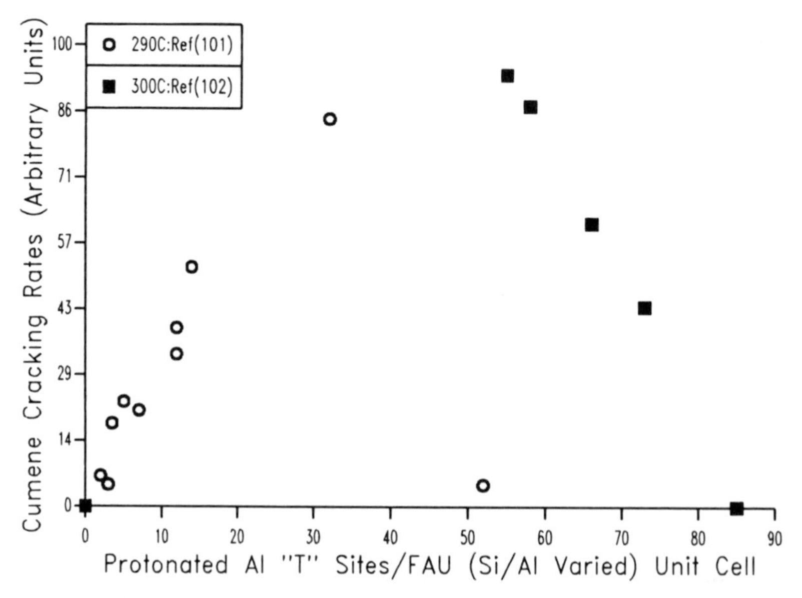

Figure 3.8. Cumene cracking dependence on protonated Al "T" sites in unbuffered environments.

Al, K/Al), the distribution of Al among the various next-nearest-neighbor site types, and the distribution of their associated charge-compensating cations. The same information will provide a reasonable estimate of the tendency of the zeolite to decompose when in its ammonium ion-exchanged form. Estimating the catalytic activity of the zeolite for model compound conversions is not so straightforward. If the rate-limiting step is protonation, as it seems to be with cumene cracking, then the variation in the initial activity will be determined by the total acidity of the zeolite. If the rate-limiting step is the creation of an activated hydrocarbon/acid site as it seems to be with paraffin cracking, then the activity will depend on the relative distribution of Al site types and their associated charge-compensating cations.

ACKNOWLEDGMENTS

The author thanks the management of the Exxon Research and Development Lab for permission to publish this review. He also appreciates the assistance of M.T. Melchior and stimulating discussions with W.L. Schuette, L.A. Pine, M.A. Freeman, P.J. Maher, and G. McVicker.

REFERENCES

1. Benesi, H.A., Winquist, B.H.C., *Adv. Catal.* **27:** 97–182 (1978).
2. Atkinson, D., Curthoys, G., *Chem. Soc. Revs.* **8:** 475–497 (1979).
3. Forni, F., *Catal. Rev.* **8:** 69 (1973).
4. Miale, J.N., Chen, N.Y., Weisz, P.B., *J. Catal.* **6:** 278–287 (1966).
5. Olson, D.H., Lago, R.M., Haag, W.O., *J. Catal.* **61:** 390 (1980).
6. Namba, S., Sato, K., Fujita, K. Kim, J.H., Yashima, T., *Studies in Surface Science and Catalysis,* Proceedings of the 7th International Zeolite Conference, Tokyo, Aug. 17–22, 1986. New York: Elsevier, 1986. Vol. 28, pp. 661–668.
7. Bremer, H., Wendlandt, K.-P., Chuong, T.K., Lohse, U., Stach, H., Becker, K., *Proceedings of the Vth International Symposium on Heterogeneous Catalysis, Part I.* Varna, 1983. pp. 435–439.
8. Sohn, J.R., DeCanio, S.J., Fritz, P.O., Lunsford, J.H., *J. Phys. Chem.* **90:** 4847–4851 (1986).
9. Beyerlein, R.A., McVicker, G.B., Yacullo, L.N., Ziemiak, J.J., *J. Phys. Chem.* **92:** 1967–1970 (1988).
10. Beaumont, R., Barthomeuf, D., *J. Catal.* **30:** 288–297 (1973).
11. Breck, D.W., *Zeolite Molecular Sieves: Structure, Chemistry and Use.* New York: John Wiley and Sons, 1974.
12. Barrer, R.M., *Zeolites and Clay Minerals as Sorbents and Molecular Sieves.* New York: Academic Press, 1978.
13. van Olphen, H., Fripiat, J.J., *Data Handbook for Clay Materials and Other Non-metallic Minerals.* New York: Pergamon, 1979.
14. Roy, R., Brindley, G.W., *Natl. Acad. Sci. Publ.* **456:** 125–132 (1956).
15. Breck, D.W., *Zeolite Molecular Sieves: Structure, Chemistry and Use.* New York: John Wiley and Sons, 1974, p. 495.

16. Stishov, S.M., Popova, S.V., *Geokhimiya* **10:** 837 (1961).
17. Liebau, F., Bissert, G., Koppen, N., *Z. Anorg. Allgem. Chem.* **359:** 113–134 (1968).
18. Bissert, G., Liebau, F., *Acta Cryst.* **26:** 233–240 (1970).
19. Rosenheim, A., Raibmann, B., Schendel, G., *Z. Anorg. Allgem. Chem.* **196:** 160–176 (1931).
20. Wells, A.F., *Structural Inorganic Chemistry,* 5th ed. Oxford: Clarendon Press, 1984.
21. Barrer, R.M., *Hydrothermal Chemistry of Zeolites.* New York: Academic Press, 1982, p. 79.
22. Weisz, P.B., *Ind. Eng. Chem. Fundam.* **25:** 53–58 (1986).
23. Haag, W.O., Lago, R.M., Weisz, P.B., *Nature* **309:** 589–591 (1984).
24. Klinowski, J., Thomas, J.M., Fyfe, C.A., Gobbi, G.C., *Nature* **296:** 533 (1982).
25. Klinowski, J., Thomas, J.M., Fyfe, C.A., Gobbi, G.C., Hartman, J.S., *Inorg. Chem.* **22:** 63 (1983).
26. Fyfe, C.A., Thomas, J.M., Klinowski, J., Gobbi, G.C., *Angew. Chemie, Int. Edn.* **22:** 259 (1983).
27. Gilson, J.P., Edwards, G.C., Peters, A.W., Rajagopalan, K., Wormsbecher, R.F., Roberie, T.G., Shatlock, M.P., *J. Chem. Soc., Chem. Commun.,* 91–92 (1987).
28. Hey, J.S., Taylor, W.H., *Zeit. Krist.* **80:** 428 (1931).
29. Vaughn, M.T., Weidner, D.J., *Phys. Chem. Minerals* **3:** 133–144 (1978).
30. Cruickshank, M.C., Dent-Glasser, L.S., Barri, S.A.I., Poplett, I.J.F., *J. Chem. Soc., Chem. Commun.,* 23–24 (1986).
31. Maher, P.K., Baker, R.W., McDaniel, C.V., U.S. Patent 3,423,332 (January 21, 1969).
32. Eberly, P.E., U.S. Patent 3,631,120 (December 28, 1971).
33. Liu, X., Klinowski, J., Thomas, J.M., *J. Chem. Soc., Chem. Commun.,* 582–584 (1986).
34. Man, P., Klinowski, J., *Chem. Phys. Lett.* **147:** 581–584 (1988).
35. Field, F.H., Head, H.H., Franklin, J.L., *J. Amer. Chem. Soc.* **84:** 1118 (1962).
36. Hertel, G.R., Koski, W.S., *J. Amer. Chem. Soc.* **87:** 1686 (1965).
37. Wexler, S., *J. Amer. Chem. Soc.* **85:** 272 (1963).
38. Olah, G.A., Surya Prakash, G.K., Sommer, J., *Superacids.* New York: John Wiley and Sons, 1985.
39. Bruckenstein, S., Kolthoff, I.M., *Treatise on Analytical Chemistry* (I.M. Kolthoff and P.J. Elving, eds.). New York: John Wiley and Sons, 1964. Part I, Vol. 5, pp. 2707–2838.
40. Simmons, V.E., Ph.D. Dissertation, Boston University, 1963.
41. Lowenstein, W., *Am. Mineral.* **39:** 92 (1954).
42. (a) Meier, W.M., Moeck, H.J., *J. Solid State Chem.* **27:** 349–355 (1979). (b) Brunner, G.O., *J. Solid State Chem.* **29:** 41–45 (1979). (c) Brunner, G.O., *Nature* **337:** 146–147 (1989).
43. Sato, M., Ogura, T., *Anal. Chim. Acta* **133:** 759–764 (1981).
44. Engelhardt, G., Michel, D., *High Resolution Solid State NMR of Silicates and Zeolites.* New York: John Wiley and Sons, 1987, pp. 222–238.
45. Meier, W.M., Olson, D.H., *Atlas of Zeolite Structure Types,* 2nd ed. Stoneham: Butterworths, 1988.
46. Smith, J.V., *Chem. Rev.* **88:** 149–182 (1988).
47. Beaumont, R., Barthomeuf, D., *J. Catal.* **26:** 218–225 (1972).
48. Beaumont, R., Barthomeuf, D., *J. Catal.* **27:** 45–51 (1972).
49. Beaumont, R., Barthomeuf, D., *J. Catal.* **30:** 288–297 (1973).
50. Drushel, H.V., Sommers, A.L., *Anal. Chem.* **38:** 1723–1731 (1966).
51. Deeba, M., Hall, W.K., *J. Catal.* **60:** 417–429 (1979).
52. Bezman, R., *J. Catal.* **68:** 242–244 (1981).

53. Ghosh, A.K., Curthoys, G., *J. Chem. Soc., Faraday Trans.* **1** (79): 147–153 (1983).
54. Beaumont, R., Barthomeuf, D., Trambouze, Y., "Molecular Sieve Zeolites," *ACS Advances in Chemistry Series, 101* (E.M. Flanigen and L.B. Sand, eds.). Washington, D.C.: American Chemical Society, 1971, pp. 327–336.
55. Barthomeuf, D., *J. Phys. Chem.* **83:** 249–256 (1979).
56. Barthomeuf, D. *Materials Chemistry and Physics* **17:** 49 (1987).
57. Mortier, W.J., *J. Catal.* **55:** 138–145 (1978).
58. Van Genechten, K.A., Mortier, W.J., *Zeolites* **8:** 273–283 (1988).
59. Jacobs, P.H., *Catal. Rev.-Sci. Eng.* **24:** 415–440 (1982).
60. Hocevar, S., Drzaj, B., *J. Catal.* **73:** 205–215 (1982).
61. Mikovsky, R.J., Marshall, J.F., *J. Catal.* **44:** 170–173 (1976).
62. Mikovsky, R.J., Marshall, J.F., *J. Catal.* **49:** 120–121 (1977).
63. Mikovsky, R.J., Marshall, J.F., Burgess, W.P., *J. Catal.* **58:** 489–492 (1979).
64. Dempsey, E., *J. Catal.* **33:** 497–499 (1974).
65. Dempsey, E., *J. Catal.* **39:** 155–157 (1975).
66. Dempsey, E., *J. Catal.* **49:** 115–119 (1977).
67. Peters, A.W., *Symposium on Advances in Zeolite Chemistry, ACS Las Vegas Meeting,* March 28–April 2, 1982, pp. 482–486.
68. Peters, A.W., *J. Phys. Chem.* **86:** 3489–3491 (1982).
69. Mikovsky, R.J., *Zeolites* **3:** 90–92 (1983).
70. Melchior, M.T., *ACS Symp. Ser. 218, Intrazeolite Chem.* 243–265 (1983).
71. Dwyer, J., Fitch, F.R., Nkang, E.E., *J. Phys. Chem.* **87:** 5402–5404 (1983).
72. Beagley, B., Dwyer, J., Fitch, F.R., Mann, R., Walters, J., *J. Phys. Chem.* **88:** 1744–1751 (1984).
73. Wachter, W.A., *Proceedings of the 6th International Zeolite Conference, Reno, July 1983*. England: Butterworths, 1983. pp. 141–150.
74. Ramdas, S., Thomas, J.M., Klinowski, J., Fyffe, C.A., Hartman, J.S., *Nature* **292:** 228–230 (1981).
75. Cattanach, J., Wu, E.L., Venuto, P.B., *J. Catal.* **11:** 342–347 (1968).
76. (a) Masuda, T., Taniguchi, H., Tsutsumi, K., Takahishi, H., *Bull. Chem. Soc. Japan* **52:** 2849–2852 (1979). (b) Stach, H., Wendt, R., Lohse, U., Jaenchen, J., Spindler, H. *Catalysis Today* **3:** 431–436 (1988). (c) Magnoux, P., Cartraud, P., Mignard, S., Guisnet, M., *J. Catal.* **106,** 235–241 (1987).
77. Klyachko, A.L., Kapustin, G.I., Brueva, T.R., Rubinstein, A.M., *Zeolites* **7:** 119–122 (1987).
78. Parra, C.F., Ballivet, D., Barthomeuf, D., *J. Catal.* **40:** 52–60 (1975).
79. Bankos, I., Valyon, J., Kapustin, G.I., Kallo, D., Klyachko, A.L., Brueva, T.R., *Zeolites* **8:** 189–195 (1988).
80. Kiovsky, J.R., Goyette, W.J., Notermann, T.M., *J. Catal.* **52:** 25–31 (1978).
81. Ratnaswamy, P., Sivasankar, S., Vishnoi, S., *J. Catal.* **69:** 428–433 (1981).
82. Bapu, G.P., Hegde, S.G., Kulkarni, S.B., Ratnasamy, P., *J. Catal.* **81:** 471–477 (1983).
83. Aurox, A., Gravelle, P.C., Vedrine, J.C., Rekas, M., in *Proceedings of the 5th International Conference on Catalysis, Naples, 1980.* (L.V. Rees, ed.). London: Heyden, 1980, pp. 433–439.
84. Fyfe, C.A., et al., *Zeolites* **5:** 179 (1985).
85. Klinowski, J., Anderson, M.W., *J. Chem. Soc., Faraday Trans.* **1** (82): 569 (1986).
86. Massiani, P., Fajula, F., Figueras, F., Sanz, J., *Zeolites* **8:** 332–337 (1988).
87. Gabelica, Z., Nagy, J.B., Bodart, P., Debras, G., Derouane, E.G., Jacobs, P.A., in *Zeolites: Science and Technology,* F.R. Ribeiro, A.E. Rodrigues, L.D. Rollmann, and C. Naccache, eds.). NATO ASI Series, Series E: Applied Sciences, No. 80, 1984, pp. 193–210.

88. Debras, G., Nagy, J.B., Gabelica, Z., Bodart, P., Jacobs, P.A., *Chem. Lett.*, 199 (1983).
89. Itabashi, K., Okada, T., Igawa, K., *Studies in Surface Science and Catalysis, Proceedings of the 7th International Zeolite Conference*, Tokyo, Aug. 17–22, 1986. New York: Elsevier, 1986. Vol. 28, pp. 369–376.
90. Derouane, E.G., Fripiat, J.G., *Zeolites* **5:** 165 (1985).
91. Derouane, E.G., Fripiat, J.G., *Proceedings of the 6th International Zeolite Conference,* Reno, July 1983. England: Butterworths, 1983, pp. 717–726.
92. Fripiat, J.G., et al., *J. Phys. Chem.* **89:** 1932 (1985).
93. O'Malley, P.J., Dwyer, J., *Zeolites* **8:** 317–321 (1988).
94. Lago, R.M., Haag, W.O., Mikovsky, R.J., Olson, D.H., Hellring, S.D., Schmitt, K.D., Kerr, G.T., *Studies in Surface Science and Catalysis, Proceedings of the 7th International Zeolite Conference,* Tokyo, Aug. 17–22, 1986. New York: Elsevier, 1986, Vol. 28, pp. 677–684.
95. Baes, C.F., Mesmer, R.E., *The Hydrolysis of Cations*. New York: John Wiley and Sons, 1976.
96. Kuehl, G.H., Schweizer, A.E., *J. Catal.* **38:** 469–476 (1975).
97. Bolton, A.P., Lanewala, M.A., *J. Catal.* **18:** 154–163 (1970).
98. Mirodatus, C., Barthomeuf, D., *J. Catal.* **57:** 136–146 (1979).
99. Anufriev, D.M., Kuznetsov, P.N., Ione, K.G., *React. Kinet. Catal. Lett.* **9:** 297–302 (1978).
100. Anufriev, D.M., Kuznetsov, P.N., Ione, K.G., *J. Catal.* **65:** 221–226 (1980).
101. DeCanio, S.J., Sohn, J.R., Fritz, P.O., Lunsford, J.H., *J. Catal.* **101:** 132–141 (1986).
102. Tsutsumi, K., Takahashi, H., *J. Catal.* **24:** 1–7 (1972).

4
Electronegativity Equalization, Solid-State Chemistry, and Molecular Interactions

W.J. Mortier

INTRODUCTION

Catalytic conversions in heterogeneous catalysis originate from the molecular interactions occurring at the gas–solid and liquid–solid interface. The study of these is a complex problem for the theoretical chemist, while the experimentalist usually has no problems in systematizing her or his findings. The vast number of industrial applications and successes certainly gives credit to the empirical approach. The experimentalist, however, may soon realize that in the absence of firm guidelines and scientific understanding progress is costly and requires a significant effort in the screening of various materials. On the other hand, theoretical progress is mostly slow and lags behind the experimental results, and little time is taken to translate the discoveries into comprehensible language.

A major difficulty in studying heterogeneous catalysis is the lack of direct evidence about the "active complex." The many different ways for explaining catalytic phenomena are rather related to the methods of study, all seeing supposedly different things and explaining them in different ways, than to the object of study. Theory is related to all of these and constitutes a common basis: "We have to do the mathematics right before we can think about the physics." Electrons and nuclei, atoms and molecules are brought together, and there should be no other rules governing their interactions than laid down in the principles of physics. Rules of thumb need to be derived from these which in a simple way predict the outcome of these interactions and which therefore may serve as a guideline to ensure rapid experimental progress.

There are several sources for attaining this information. There is of course quantum chemistry as the classic resource for fundamental understanding in chemistry. Then there are the experimental measures of molecular and crystalline properties, which also contain indirect evidence of the interaction between atoms and molecules. We will rely on the principles of density functional theory for constructing a bridge between experiment and theory.

Density functional theory has recently been found to rationalize long-standing concepts in chemistry.[1,2] Especially, the concept of electronegativity[3] was demonstrated to be a fundamental constant of a system of electrons and nuclei, where before it merely reflected chemical intuition.[4] Electronegativity equalization, first postulated by Sanderson,[5] also naturally follows from it for ground-state systems.[3] It will become clear that this will allow us to formulate some general but very simple expressions, stated in the proper language of the chemist, for qualitatively as well as quantitatively describing molecular interactions, charge density reorganizations (easier to imagine than perturbations of the molecular orbitals, but equally fundamental) and, because of experimental evidence, variations in bond strength (the onset of catalytic conversions). Molecules as well as solids can be described by the same formalism. Applications will be sought in the field of zeolite and molecular sieve sciences, but the rules of thumb should be sufficiently general for applying to other solid–fluid interactions.

ELECTRONEGATIVITY OF AN ATOM IN A MOLECULE

"The power of an atom in a molecule to attract electrons to itself," which is the first definition of electronegativity by Pauling,[4] is sufficiently intuitive and general that it can be found in many regularities of the periodic table. Many of these properties have been used as a basis for assigning a "degree of electronegativity" to atoms. Several authors recently reviewed the electronegativity concept and the possible applications to chemistry[6] such that there is no need here to detail the historical evolution of the concept. A rigorous mathematical definition, however, was not given before 1978, when R.G. Parr identified electronegativity with the negative of the "chemical potential" of the electrons in a system of electrons and nuclei described in density functional theory.[3] It also was demonstrated that the electronegativity of all orbitals in an atom or a molecule in the ground state was equalized, whereby Sanderson's electronegativity equalization principle[5] was validated. Because this will form the basis of our discussions, a general outline (recalled from Ref. 7) of the way in which the effective electronegativity concept is derived and what its relation is to density functional theory are given here again.

Two theorems by Hohenberg and Kohn[8] certify that the electron density distribution function $\rho(\mathbf{r})$ determines all ground-state properties of the molecule. The ground-state energy can therefore be written as a functional of the electron density $E[\rho]$. A similar functional can be written for the energy in quantum mechanics: $E[\Psi]$, where Ψ now fully describes the system. Any trial function ρ or Ψ can give us an approximation to the energy (but always above the exact value). The best choice is the one that gives the lowest energy. Only normalized functions are acceptable, i.e., $\langle\Psi|\Psi\rangle = 1$ and $N[\rho] = \int\rho(\mathbf{r})d\mathbf{r} = N$, with N being the number of electrons. We can find the minimizing function by applying the Euler equation[9] associated with this variational problem. (For an introductory text on the calculus of variations, see Ref. 10.) The normalization conditions are considered by using a Lagrangian multiplier (λ and μ): $\{E[\Psi] - \lambda\langle\Psi|\Psi\rangle\}$ and $\{E[\rho] - \mu N[\rho]\}$ are minimized instead of $E[\Psi]$ or $E[\rho]$. The conditions for a minimum energy are (i) in quantum mechanics, $H\Psi = \lambda\Psi$, from which it is apparent that $\lambda = E$, the system's energy[8]; and (ii) in density functional theory (at constant external potential v, such as generated by the nuclei),

$$\mu = \left(\frac{\delta E[\rho]}{\delta\rho(\mathbf{r})}\right)_v \tag{4.1}$$

which can also be identified[3] with

$$\mu = \left(\frac{\partial E}{\partial N}\right)_v = -\chi \tag{4.2}$$

whereby the Lagrange multiplier μ is given its physical meaning[3]: minus the electronegativity χ as defined by Iczkowski and Margrave[11] [Eq. (4.2)].

For an excellent discussion of Eq. (4.2), and why this reflects the electronegativity, we refer to the same paper. Briefly (referring to Fig. 4.1): if we bring together two atoms for which the slope of the $E(N)$ curve at the origin (neutral atoms) is different, electrons will be transferred from the

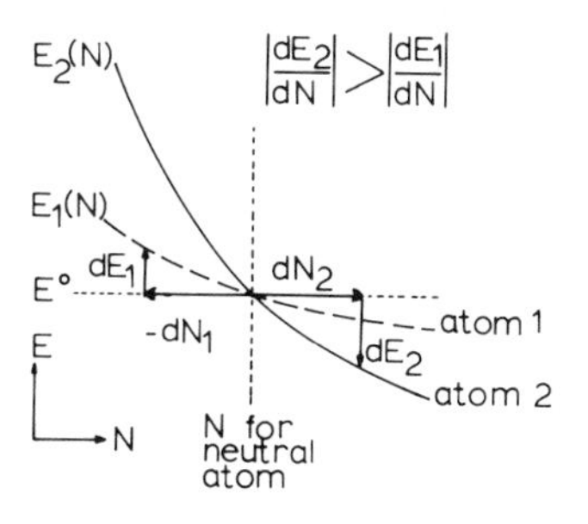

Figure 4.1. The electronegativity concept[11] ($\chi = -dE/dN$) illustrated by considering an energy lowering for a combined system of two atoms upon electron transfer (dN): an atom with a higher slope of $E(N)$ vs N will consistently take away electrons from the atom having a smaller slope.

atom with the smallest slope to the atom with the largest slope. This necessarily results in an energy lowering for the combined system: the energy invested in taking away electrons from the atom with the smallest slope is consistently smaller than the energy gained by transferring these to the atom with the largest slope. This slope therefore simulates a tendency to attract electrons, and because this slope is always negative, the electronegativity is defined as $\chi = -(dE/dN) = -\mu$. Because of the analogy with thermodynamics, μ can be defined as the "chemical potential" of the electrons. μ is a fundamental constant of a system of electrons and nuclei, in the same way as E (which is also a Lagrange multiplier). Equation (4.1) also asserts the constancy of the electronegativity throughout the system.

Provided that we carry all the terms, any function can be expanded without loss of information. A Taylor expansion of the energy as a function of the number of electrons (N) around the neutral atom ($N = Z$, the nuclear charge) is most convenient and immediately allows the application of Eq. (4.2). Using for

$$E(N - Z) = E^0 + \mu^0(N - Z) + \eta^0(N - Z)^2 \tag{4.3}$$

$E(-1) = E^0 + I$ (ionization potential) and $E(+1) = E^0 - A$ (electron affinity), $-\mu^0 = \chi^0 = (I + A)/2$, i.e., Mulliken's[12] definition of electronegativity, and $\eta^0 = (I - A)/2$, which Parr and Pearson[13] defined as the hardness. From this [Eqs. (4.2) and (4.3)] also immediately follows an expression for the variation of the electronegativity of the free atom with charge $q = -(N - Z)$, viz.,

$$\chi = \chi^0 + 2\eta^0 q \tag{4.4}$$

Politzer and Weinstein[14] showed that for a molecule in the ground state, for which the energy is expressed as a function of the number of electrons associated with the different atoms ($N_\alpha, N_\beta, \ldots$), the nuclear charges ($Z_\alpha, Z_\beta, \ldots$), and the interatomic distances ($R_{\alpha\beta}, \ldots$), we have

$$\left(\frac{\partial E}{\partial N_\alpha}\right)_{N_\beta, \ldots, R_{\alpha\beta}, \ldots} = \left(\frac{\partial E}{\partial N_\beta}\right)_{N_\alpha, \ldots, R_{\alpha\beta}, \ldots} = \cdots = \mu_\alpha = \mu_\beta = \cdots \tag{4.5}$$

In analogy with Eq. (4.2), we may now adopt Eq. (4.5) as the mathematical formulation of the chemical potential of the electrons (electronegativity) for an "atom in a molecule." Their electronegativities are then equal in the ground state at equilibrium.

EXPLICIT EXPRESSION FOR THE EFFECTIVE ELECTRONEGATIVITY

It now becomes clear that isolated-atom electronegativities [Eq. (4.4)] cannot become equalized in a molecule: the energy E in expression (4.5) is the total molecular energy. The isolated-atom energy will be supplemented by the interaction energy of the electrons with all other nuclei and all other electrons in the molecule (and even outside). Moreover, the intraatomic terms (i.e., kinetic energy terms and nuclear–electron and electron–electron interaction energies) must also change upon bond formation because of changes in the size and shape of the electron cloud occurring during bond formation.[7,15,16] It is however still possible to write the equivalent of Eq. (4.3) for the intraatomic parts of the total energy, but with different expansion coefficients: μ^* and η^*. Using a spherical-atom approximation, it is also possible to write the interatomic terms in a simplified way, and the total molecular energy can then be written as a sum of intraatomic and interatomic contributions, viz.,[16]

$$E = \sum_{\alpha} \Bigg(E_\alpha^* + \mu_\alpha^*(N_\alpha - Z_\alpha) + \eta_\alpha^*(N_\alpha - Z_\alpha)^2 \qquad \text{(intraatomic)}$$

$$\qquad\qquad\qquad\qquad\qquad\qquad\qquad\qquad\qquad\qquad\qquad\qquad (4.6)$$

$$- \underbrace{N_\alpha \sum_{\beta \neq \alpha} \frac{Z_\beta}{R_{\alpha\beta}}}_{(V_{ne})} + \underbrace{\tfrac{1}{2} N_\alpha \sum_{\beta \neq \alpha} \frac{N_\beta}{R_{\alpha\beta}}}_{(V_{ee})} + \underbrace{\tfrac{1}{2} \sum_{\beta \neq \alpha} \frac{Z_\alpha Z_\beta}{R_{\alpha\beta}}}_{(V_{nn})} \Bigg) \qquad \text{(interatomic)}$$

From this expression for the molecular energy, the effective electronegativity of an atom in a molecule, as it was defined by Eq. (4.5), can be directly evaluated as[16]

$$\chi_\alpha = \chi_\alpha^* + 2\eta_\alpha^* q_\alpha + \sum_{\beta \neq \alpha} \frac{q_\beta}{R_{\alpha\beta}} \tag{4.7}$$

THE ELECTRONEGATIVITY EQUALIZATION METHOD (EEM)

If, for all atom types, χ^* and η^* in Eq. (4.7) are known, the molecular charge distribution and the average electronegativity can be calculated from a set of n equations (for an n-atom molecule) of type (4.7) and one supplementary equation that fixes the total molecular charge (Σq_α = constant). This method is referred to as the Electronegativity Equalization Method (EEM).[16,17] If the expansion coefficients are not known, they can be calibrated to a set of known charges for different molecules by the methods of the multiple regression.[16] There is by now overwhelming evidence that the expansion coefficients can be transferred for most atom

Table 4.1. χ^* and η^*, calibrated to STO-3G *ab initio* charges (Mulliken population analysis), relative to $\chi_0^* = 8.5$ eV.[18]

Atom Type	χ_α^*	η_α^*
H (δ+)	4.40877	13.77324
H (δ−)	3.17392	9.91710
C	5.68045	9.05058
N	10.59916	13.18623
O	8.5	11.08287
F	32.42105	90.00488
Al	−2.23952	7.67245
Si	1.33182	6.49259
P	2.90541	6.29415

types into a wide variety of environments. Atomic charges are very accurately reproduced. A consistent set was determined by Uytterhoeven et al.,[18] using a Mulliken population analysis[19] on STO-3G *ab initio* wavefunctions for a limited number of atoms. These are given in Table 4.1. (There is an arbitrary constant in χ^*; its value for oxygen was fixed to 8.5 eV.)

In the atomic regions, $\rho(\mathbf{r})$ integrates to a certain number of electrons. The atomic charge distribution, which is accurately calculated by the EEM method, therefore constitutes a finite-difference approach to the electron density distribution function. Because the latter contains all information,[8] the atomic charge distribution should also contain important chemical information. There is also a second property of $\rho(\mathbf{r})$, i.e., the compactness of the electron cloud. This can be related to the average electronegativity.[20] Atomic charges and average electronegativities should therefore not be confused: they have an entirely different meaning.

THE SOLID STATE[21,22]

Equation (4.7) is equally applicable to the solid state, provided that the external potential is calculated for the whole crystal. Fortunately, the calculation of the electrostatic interaction energy is straightforward using Madelung-type summations,[23] and the potential (V) generated by all surrounding charges at all atomic sites (i) is easily derived from this. The total electrostatic interaction energy E per unit cell is given by

$$E = \frac{1}{2} \sum_{i=1}^{N} \sum_{j \neq i}^{\infty} \frac{q_i q_j}{r_{ij}} \tag{4.8a}$$

and the potential is

$$V_i = \left(\frac{\partial E}{\partial q_i}\right) = \sum_{j \neq i}^{\infty} \frac{q_j}{r_{ij}} \tag{4.8b}$$

where the summation of i runs over all atoms of the unit cell and j over all atoms in the crystal. It is obvious that all crystallographically distinct positions will carry a different charge and also that the average electronegativity will be different for different structure types.

MOLECULAR INTERACTIONS

Whenever the environment of a molecule is disturbed, e.g., by adsorption or by molecular interactions, the charges must be redistributed: the external potential in Eq. (4.7) is supplemented by the extra contribution of all surrounding charges. Qualitatively, a positive $\Sigma q/R$ in Eq. (4.7) will result in an increased electronegativity and a higher electron density; the reverse is true if this sum is negative. A charge redistribution must occur in all interacting species, and this also includes the surface itself in the case of adsorption. The power of the EEM method for predicting these charge shifts was illustrated for water dimers[24] in Ref. 16. Moreover, the charge shifts within the molecules are much more important than the charge transfer between molecules. Charge shifts upon interaction are reproduced within a few thousandths of an electron by the EEM method (see Fig. 4.2) and are the direct consequence of the changing electrostatic potential at each atom site.

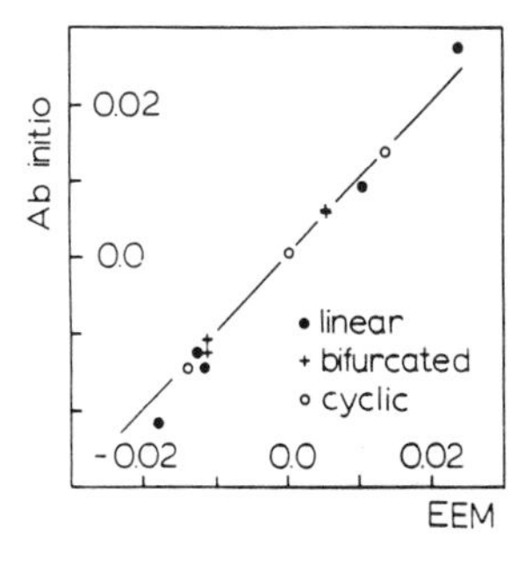

Figure 4.2. *Ab initio* charge shifts[24] in interacting water molecules (dimers) versus charge shifts calculated with the EEM method.[16]

Charge shifts upon molecular interactions were already rationalized and put into simple rules by Gutmann.[25] We consider a schematic representation of two interacting molecules as follows:

$$>C \overset{\text{bond lengthening}}{\overline{\qquad\qquad}} \overset{\delta-}{D} \Longrightarrow \overset{\delta+}{A} \overset{\text{bond lengthening}}{\overline{\qquad\qquad}} B<$$

	Donor	Acceptor
Electron density	INCREASE	DECREASE
	Pileup effect	Spillover effect

At the donor site ($\delta-$), the charges are further accumulated (pile-up effect: larger increase in electron density than electrons transferred in the donor–acceptor interaction) because of the proximity of the acceptor site ($\delta+$), which induces an increased effective electronegativity. By the same mechanism, the electron density must decrease at the acceptor site (spillover effect: larger decrease of electron density than electrons received). The charge density rearrangement is however not confined to the site of interaction, but the entire molecule is affected, which is clearly understood in terms of Eq. (4.7). This charge density rearrangement is also directly related to changes in bond strengths (the onset of catalytic conversions). An increase in bond ionicity is generally accompanied by a bond weakening (lengthening; there is an exception for T–O bonds where an increased T–O–T angle results in a decreased T–O bond length and an increased negative charge on oxygen[26]). Although there is no proof of this, it agrees with experimental findings. (For an example of the relation between bond length, ionicity, and reactivity of the C—O bond in R_1–O–R_2 compounds, see Refs. 27 and 28.) Gutmann's bond-length variation rules[25] explicitly state that (i) the bonds adjacent to the site of interaction are lengthened (varying, of course, according to the donor–acceptor interaction strength); (ii) for charge shifts farther in the molecule (direction of the charge shift determined by the donor $\rightarrow$ acceptor interaction) from a more electronegative to a less electronegative atom, bonds are shortened, and for an electron shift in the molecule from a less electronegative to a more electronegative atom, bonds are lengthened.

FRAMEWORK: INTRINSIC PROPERTIES

There are two main features of the compounds that we will always have to consider: the charge distribution and the average electronegativity. On the other hand, if we think about catalysis, we must realize that the interactions at the surface form only part of a picture, which also involves diffusion and dynamic aspects.

We will focus our attention first on the intrinsic properties of the solid surface, and in a following paragraph on the interactions of this surface with molecules. Acidity, because of its importance, will be treated separately. In each discussion, theoretical aspects as well examples of applications will be given.

Properties Related to the Average Electronegativity

The compactness of the electron cloud, of course, increases with the average electronegativity of the framework: the binding of the electrons will be tighter. More energy will be needed to remove an electron from the system and/or the electron affinity of the system will also increase, both having the effect of increasing the slope of the $E(N)$ curve (Fig. 4.1) and therefore also of $\bar{\chi}$. By virtue of Eq. (4.7), the organization of the building elements will directly influence the average electronegativity. An analysis of several zeolite structure types with (hypothetical) SiO_2 composition[29] together with a number of silica polymorphs revealed that the average framework electronegativity increases with decreasing framework density (Fig. 4.3). Also, since the electronegativity directly relates to the strength with which the electrons are held in place, a near-perfect correlation with the refractive index was found[21,22] (Fig. 4.4).

Electronegativity cannot be directly measured. The variation of the refractive index is only indirectly related to it. The definition given in Eq. (4.3) and its explicit dependence on the ionization potential and electron affinity (Mulliken's[12] definition) is of course valid for any system (and there should also be only one electronegativity value for every system, as there is only one ionization potential and one electron affinity). Fortunately, variations in the binding energies of the outer orbitals are also reflected in the variations of the binding energies of the inner orbitals. In that case, X-ray electron spectroscopy should be the tool of choice for investigating the electronegativity variations in molecules or crystals. Indeed, in the ground-state potential model, the ESCA shifts are calculated[30] in a remarkably similar way to that given in Eq. (4.7) for the effective electronegativity: the atomic charge, the Madelung potential,

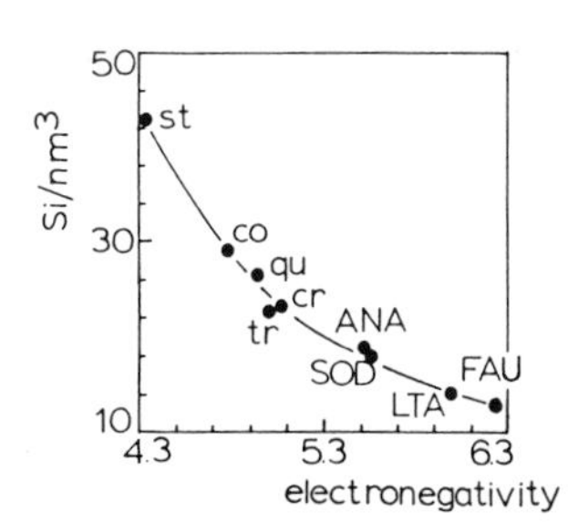

Figure 4.3. Variation of the average framework electronegativity with the framework density (number of Si atoms per nm³) for stishovite (st), coesite (co), quartz (qu), tridymite (tr), and cristobalite (cr). Zeolite-type frameworks of the analcime-, sodalite-, zeolite A-, and faujasite-type with hypothetical SiO_2 composition are also shown.

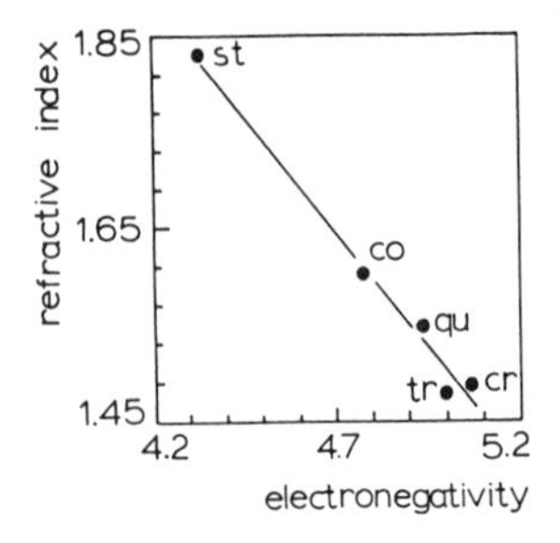

Figure 4.4. Variation of the average framework electronegativity with the refractive index n at $\lambda = 589\ \mu m$ for stishovite (st), coesite (co), quartz (qu), tridymite (tr), and cristobalite (cr).

and a constant are required in the expression for the binding energy. X-ray photoelectron spectroscopy was applied to heterogeneous catalysts by Vinek et al.[31] These authors postulated that variations in the binding energies of the inner orbitals should be indicative of the binding energies of the valence orbitals, and these should then reflect the donor or acceptor strength of the active sites. For a series of Mg compounds (MgO, $MgSO_4$, $MgHPO_4$), it was shown that Mg and oxygen undergo the same variations: acidic (electron pair acceptor strength) or basic (electron pair donor strength) character should be associated with the compound as a whole rather than with only one of its sites. This is of course in agreement with electronegativity equalization. At the same time, this means that there cannot be a unique relation between the atomic charge and the acid/base site strength, also by virtue of Eq. (4.7).

In an X-ray photoelectron spectroscopic study of silicate minerals, Seyama and Soma[32] came to the same conclusion: the shifts in binding energies of Si, O, and tetrahedrally coordinated Al in the framework are highly correlated. These decrease with increasing formal negative charge on the framework (increasing substitution of Si^{4+} by Al^{3+}), which is indicative of a decrease in average electronegativity of these systems. The octahedrally coordinated Al ions and the Na ions are much less sensitive. This can be explained readily by the fact that for these the effective electronegativity is mainly determined by the charge contribution [Eq. (4.7)], which agrees with the longer bond lengths in the coordination polyhedron. The "extra-framework" cations will also directly influence the average electronegativity, even when these can be considered as carrying their formal charge: their distances to the framework directly influence the electrostatic potential at the coordinating oxygens. (This is also true for all other framework atoms.)

Properties Related to the Charge Distribution

Structure Type. The charge distribution will also be influenced directly by the structure type. For a series of hypothetical zeolite structure types,

Van Genechten et al.[29] found a correlation between the average T–O–T bonding angle in a silicate tetrahedron and the charge on Si, which increases with increasing angle. This is of course in agreement with theoretical findings (see, e.g., Ref. 26) and with ^{29}Si NMR.[33,34] A striking example of the influence of the structure on the charge distribution is found for tridymite versus cristobalite.[22] These silica polymorphs differ in a similar way as wurtzite and sphalerite (ZnS), respectively. The results indicate that the wurtzite-type structure intrinsically favors a higher ionicity. A series of sulfides, selenides, telurides, . . . , crystallize[35] in either one of these polymorphic forms; some are diamorphic. Analyzing these, there seems to be a tendency of the more ionic materials to adopt the wurtzite-type structure.

Composition. The influence of the composition on the charge distribution is more pronounced, however, than changes in the structure type. The replacement of an element with a different electronegativity (characterized by χ^* and η^*) immediately influences the average electronegativity and the local charge distribution. The ^{29}Si MASNMR gives an example: different chemical shifts are obtained in framework aluminosilicates depending on the number of Al neighbors.[36] There is of course a range of chemical shifts observed: all surrounding charges do influence the local electron density distribution. For an example of the influence of Ca ions in dehydrated faujasite-type zeolites on the ^{29}Si chemical shift see, e.g., Ref. 37.

Because of the prevailing influence of the composition, several empirical formalisms based on electronegativity equalization principles were also successfully applied for quantitatively estimating its effect on the physicochemical properties of zeolites. A review of these and a critical evaluation were made in Refs. 7 and 15. It is obvious that an equalization of isolated-atom electronegativities [Eq. (4.4)] cannot be correct. "Atomic charges" are often calculated in this way, but these cannot be connectivity dependent (in acetone, e.g., the carbonyl C and the methyl C must have the same charge). Moreover, there is no more information in these charges than in the average electronegativity. There is, however, one important advantage: in the intermediate composition ranges, for homologous series of compounds, these numbers do correlate with the charges or the average *effective* electronegativity.[7,15]

Experimental evidence for the variation of the intrinsic framework properties with charge for zeolite frameworks can further be found in IR spectroscopic investigations. A theoretical study, combined with IR experiments, was made by Datka et al.[38] For a series of dehydrated faujasite-type and ZSM-5-type zeolites with different Al content and exchanged with monovalent or divalent cations, or in the H-form, it was found that within a homologous series (e.g., varying the Al content only, or the

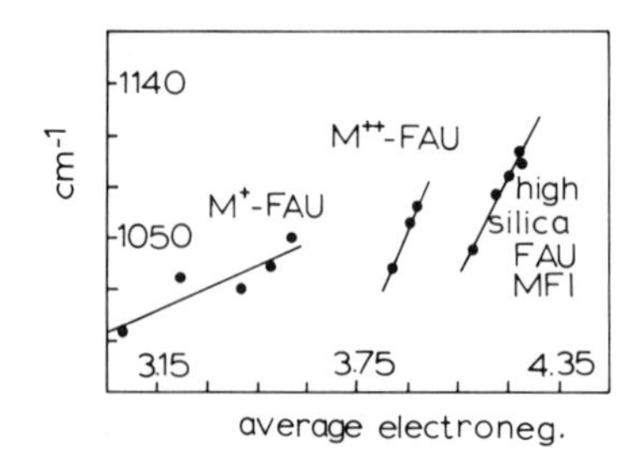

Figure 4.5. Variation of the TO asymmetric stretching internal vibration frequency (cm^{-1}) for monovalent and divalent cation-exchanged faujasite-type zeolites and high silica H-faujasite and H-ZSM-5-type zeolites with average electronegativity.[38]

cation type, or the structure type), the frequency of several framework stretching bands varied linearly with the average compound electronegativity; see Figure 4.5. The average compound electronegativity was calculated as the geometric average of the isolated-atom electronegativity of all atoms in the compound.[39] The variation of the individual bond strength with chemical composition can be understood by applying Gutmann's second rule: for an increased electronegativity of the substituents of a T–O–T moiety, the T–O bonds should be strengthened, according to the following scheme which explains the direction of the IR shift (increased bond strength with increased average electronegativity):

⟸ T T ⟹ (T–O–T, with dotted arrows toward O)

⟹ Increased substituent electronegativity

⋯⋯▸ Bond shortening

FRAMEWORK: MOLECULAR INTERACTIONS

Whenever there is some degree of ionicity in the framework as well as in the molecules interacting with the solid surface, the EEM method will be well suited for predicting the charge shifts and the concomitant change in bond strength. Both the surface and the adsorbed molecules must be affected. For most cases, it is sufficient to consider the molecule in the field of the solid surface (and vice versa) without accounting for charge transfer, which is usually negligible with respect to the charge density reorganization within the molecules themselves (see, e.g., the case of interacting water molecules[24]), because the following conditions must be fulfilled[40] to have chemical bonds formed:

1. The energies of the binding orbitals in their respective atoms should be of comparable magnitude.
2. The orbitals should overlap as much as possible.
3. The orbitals should have the same symmetry relative to the bond axis.

The chemical natures of the surface of the solid catalysts and of the interacting molecules are most often so different (e.g., hydrocarbons adsorbed on an oxide or zeolite surface) that it certainly suffices to consider the polarization in the molecules and in the adsorbent surface only.

The generalized donor–acceptor approach by Gutmann,[25] together with the effective electronegativity equalization concept, prove to be very convenient tools for discussing molecular interactions and catalysis. At any "ionic" surface, as well as in any molecule, donor ($\delta-$) and acceptor ($\delta+$) sites can be found. An excellent example of the power of this approach for explaining heterogeneous catalysis, specifically elimination reactions, was presented by Vinek et al.[31] For the elimination of X^- and H^+ from a hydrocarbon, the following adsorption structure may be drawn:

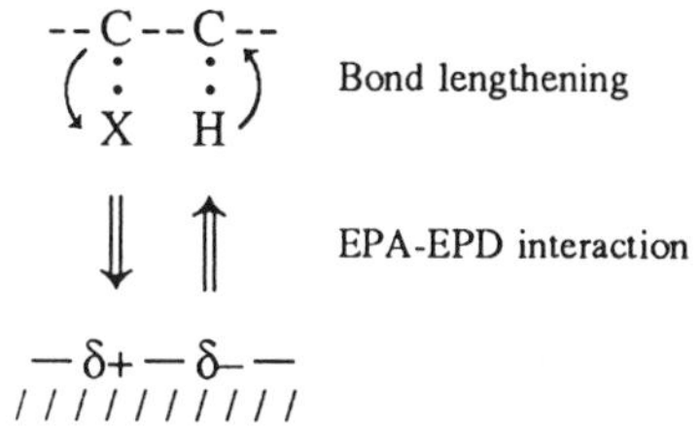

```
--C--C--
 (:  :)      Bond lengthening
  X  H

  ⇓  ⇑       EPA-EPD interaction

—δ+—δ−—
//////////
```

Both the C–X and the C–H bonds should be weakened according to Gutmann's first rule,[25] or because we can expect a higher ionicity due to the extra positive (negative) charge in the immediate vicinity of X (H), resulting in a higher (lower) electronegativity of this atom than before adsorption. Three possible mechanisms can be distinguished for elimination of HX, depending on the relative strength of the donor–acceptor interactions: E1, E2 and E1cB, which differ in the timing of the bond rupture and which may give different product distributions:

$$\text{X–C–C–H} \left| \begin{array}{l} \xrightarrow{\text{E1}} \text{X}^- + {}^+\text{C–C–H} \longrightarrow \\ \xrightarrow{\text{E2}} \text{X}^{\delta-} \;.\;.\;.\; \text{C–C} \;.\;.\;.\; \text{H}^{\delta+} \longrightarrow \\ \xrightarrow{\text{E1cB}} \text{X–C–C}^- + \text{H}^+ \longrightarrow \end{array} \right| \text{C=C} + \text{HX}$$

As was previously mentioned, the binding energies of the inner orbitals—since these can be experimentally determined by X-ray photoelectron spectroscopy—should correlate with the EPA and EPD strength (and the effective electronegativity) of the surface atoms of the catalyst. We may compare the donor strength of the oxygens in different catalysts, as given in Table 4.2 (data from Ref. 31). The lower the binding energy, the more pronounced the donor strength of the oxygen, while the acceptor strength must increase with the binding energy. The catalysts are arranged according to the EPD strength of the oxygens in basic and acidic

Table 4.2. O(1*s*) and Mg(2*p*) binding energies (data from Ref. 31).

Catalyst	Binding Energies (eV)	
	Mg(2*p*)	O(1*s*)
Basic		
La_2O_3		529.0
MgO	48.0	530.2
Acidic		
$MgSO_4$	49.1	531.1
$MgHPO_4$	49.8	531.8
Al_2O_3		531.8
SiO_2		533.1
Al_2O_3-SiO_2		532.1
Zeolites		
MgNaX (49.5%)	50.2	530.6
MgNaY (50.7%)	49.4	530.9

oxides, with the zeolites classified separately because of their high EPD strength of the oxygens, together with a remarkably high EPA strength of the cations, which [as exemplified by the Mg(2*p*) binding energy] is higher than in any other compound. For zeolites, the elimination mechanism has been found to be E2 or E1, indicating the importance of the cation–X interaction. The low O(1*s*) binding energy for the basic oxides favors an E1cB mechanism. On $MgSO_4$ and $MgHPO_4$, an E2 mechanism is again consistent with the high acceptor strength of the cations. For a further detailed analysis along these lines, see Ref. 31.

Theoretical support for a two-site adsorption model in the dehydration of 2-butanol was recently provided by Sedlacek.[41] Two interacting acid–base pairs consisting of a zeolite hydroxyl group and a zeolite oxygen, on the one hand, and the butanol OH group and the hydrogen atom on the adjacent carbon, on the other hand, are considered to initiate the dehydration reaction. The conclusions of this CNDO study are in complete agreement with the preceding discussion. In agreement also with Eq. (4.7), the absolute values of both negative and positive charges on all atoms of the two interacting systems (zeolite and butanol) increase. This again stresses the importance of the increased polarization of the bonds during adsorption. The concomitant weakening of the bonds is most pronounced in the closest vicinity of the adsorption sites. Experimental evidence for charge reorganizations in the framework occurring during adsorption is given by NMR.[42]

A direct probing of the negative surface charge of the framework oxygens is possible by monitoring the bond strength changes in adsorbed molecules interacting with the oxygens. Barthomeuf[43] investigated the wavenumber shift of the IR N . . . H stretching vibration of pyrrole adsorbed on zeolites with different composition and structure (faujasite, zeolite L, and mordenite). In accordance with Gutmann's rules,[25] we may expect an N . . . H bond weakening, which should correlate with the donor (the zeolite oxygen)–acceptor (pyrrole H) interaction strength. The oxygen basicity (negative charge) measured in this way is primarily a function of the composition and correlates well with Sanderson's average electronegativity: the higher the Al content (i.e., the higher the residual negative charge originating from the isomorphous substitution in the framework of Si^{4+} by Al^{3+}), the larger the bathochromic shift of the NH stretching frequency. The framework Al content is however not the only parameter: the electronegativity of the exchangeable cations and the structure type (as can be inferred from the discussion on the intrinsic properties of the framework) are also important. The ionicity of the mordenite-type framework is predicted to be higher than for zeolite L and faujasite. Indeed, Van Genechten et al.[29,44] calculated for these zeolite structure types with theoretical SiO_2 composition a significantly higher average silicon charge for mordenite (2.04) than for zeolite L (1.90) and faujasite (1.86).

Further evidence for the interaction of molecules with the zeolitic surface is found in the adsorption equilibria. The lower adsorption of ethane and ethene[45] with rising degree of dealumination is another example of the decreasing negative charge on the oxygens with decreasing degree of isomorphous substitution. At the same time, the structure becomes more and more important. This is reflected in the higher adsorption in those zeolites with the smaller pores.[45] The influence of the pore size on the molecular interaction between zeolite surface and guest molecules is also evident from the ^{13}C chemical shift in NMR for tetramethylammonium ions trapped in zeolite pores during synthesis[46] (shift to lower field as the pore size decreases).

While hydrocarbons preferentially interact with the zeolite surface oxygens through their partially positively charged hydrogens, once more electronegative atoms such as oxygen and nitrogen are incorporated interaction with the zeolite occurs, specifically at the cationic centers. NMR techniques have therefore also been used extensively for probing the positive centers (cations, Lewis-acid sites) using such molecules as CO, NH_3, amines, pyridine, and acetonitrile. It is obvious from the effective electronegativity considerations [Eq. (4.7)] that the charge distribution in these molecules will also be perturbed by adsorption on cations in cation-exchange sites. This is particularly true for interactions with polyvalent cations in cation–exchange sites such as the extra-framework Al^{3+} formed in the activation step of the catalysis. Details of these studies are beyond the scope of this chapter, and the reader is referred to Refs. 47–50.

BRÖNSTED ACIDITY: INTRINSIC PROPERTIES

Considerable attention has been devoted to explaining Brönsted acidity in zeolites. Protons can be incorporated into the zeolite framework by (i) ion exchange in acid medium if the zeolite is stable under these conditions (high-silica zeolites only), (ii) exchange with ammonium ions followed by an activation step whereby ammonia is expelled, and (iii) during dehydration of multivalent cation-exchanged zeolites whereby water hydrolysis occurs. The surface Brönsted acidity will hence be determined by the number of sites *and* by their individual strengths. For the applications, both these parameters will have to be balanced for maximum performance. The intrinsic strength of the individual OH groups will be the subject of this section, and their interaction with molecules will be considered in the next section. Again, although this is focused on zeolites, the ideas put forward are sufficiently general for application to other systems that are characterized by a considerable charge delocalization. (The zeolite framework is indeed highly covalent in nature; the anionic framework must be considered as a soft base.[51]) For an excellent recent review on quantum-chemical calculations using cluster models of acid–base sites in oxide and zeolite catalysts, see Ref. 52. For reviews on the acidity concept in zeolites, see Refs. 53–55.

A schematic representation of the two possible configurations in which OH groups can occur in zeolites is given below: i.e., hydroxyl groups of the terminal type [i.e., at the termination of the framework at crystal boundaries or stacking faults (Si–OH), and in polyhydroxy cations] and of the bridged type:

$$M^{n+}\text{--}O{-}H \qquad (T{-}O)_3{\equiv}Si{-}\overset{\overset{\displaystyle H}{\bullet}}{O}{-}Al{\equiv}(O{-}Si)_3$$

It is immediately obvious that the electronic properties of the OH group will strongly depend on the nature (electronegativity) of the atoms in the proximity (data on chemisorption and catalysis advocate a local approach[52]; perturbations are very quickly attenuated because of the rapidly increasing number of bonds and atoms with increasing topological distance). For the bridging OH this means that each T-atom (tetrahedral cation) can be either Si or Al (according to Löwenstein's Al avoidance rule,[56] the nearest neighbors of Al will be Si consistently). The same will be true for the nature of the M^{n+} cation and its nearest neighbors in the terminal OH case. For a higher electronegativity of the surrounding atoms, an electron shift from the less electronegative H to the more electronegative O will be accompanied by a weakening of the O–H bond (Gutmann's rules[25]). It is also obvious that the terminal OH bond will be much stronger (and therefore also less acidic) than the bridging OH bond.[57] The latter also directly follows from Gutmann's rules (two-coordinated versus three-coordinated oxygen in the terminal and bridging OH, respectively, results in longer bonds for the higher coordination).

The OH bond strength can be directly measured in the IR by its stretching frequency, which has traditionally been used as a means of characterizing the intrinsic acid strength of zeolites. A band at 3745 cm^{-1}, characteristic for terminal Si–OH groups, will not be considered here. For the bridging OH groups, usually two bands are detected for faujasite-type zeolites: a sharp high-frequency (HF) band at 3660 cm^{-1} and a broad low-frequency (LF) band around 3556 cm^{-1}. There is quite some variation in these frequencies, depending on the composition and also on the structure type. Jacobs et al.[58,59] assigned the HF band to OH groups freely vibrating into the large cavities or channels of zeolites. For OH groups located in framework six-rings or eight-rings, the proximity of supplementary negatively charged oxygens around the proton will cause a further decrease of its electronegativity [by virtue of Eq. (4.7)] and therefore also increase its positive charge and acidity. For OH groups in six-rings, the bathochromic shift amounts to about 100 cm^{-1} with respect to the HF band, and for eight-rings to about 30 cm^{-1}.

There have been several attempts to model the influence of the composition on the bridging OH stretching frequency. Obviously, application of the EEM method cannot be envisaged because of the necessity to dispose of all crystallographic details. Jacobs et al.[58,59] found an excellent correlation (Fig. 4.6) between the HF stretching frequency and the geometric

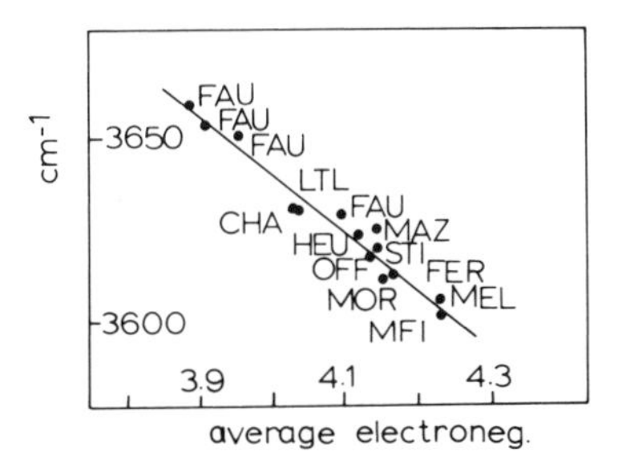

Figure 4.6. Variation of the OH stretching frequency (HF band) for various cation-exchanged H-zeolites with different structure type and framework composition with the average electronegativity.[59]

average compound electronegativity calculated according to Sanderson,[39] summarizing in this way influences of Al content and cation loading, irrespective of the structure type: the composition seems to be the most important parameter. In agreement with the previous discussion, the frequency decreases (increased acidity) with increased compound electronegativity.

The foregoing approach does not have any structural input, and no structure-dependent information can therefore be expected. For zeolites with high cation loading, the predictions no longer apply owing to a direct interaction of the exchangeable cations with the protons.[7]

Models based on a local cluster approach consider the occupancy of the framework tetrahedra surrounding the hydroxyl group (making abstraction of the influence of the exchangeable cations and possible perturbations by nearby framework oxygens). Following the previous schematic representation, we may distinguish in the second "coordination shell" (the first being consistently Si and Al) four cases ranging from all T atoms being Si to all being Al. This has been the approach by Kazansky.[52,60] An increasing number of Al atoms will of course decrease the acidity.

A refinement of the local approach is possible by looking at the next nearest neighbors as well.[54,55] These will obviously contribute less because of the longer topological distance from the acid site, but these effects will certainly become apparent for high silica zeolites, where it was experimentally found that beyond a Si/Al ratio of about 10, the OH stretching frequency no longer changes.[61] The highest acidity will have been reached for no Al in next-nearest-neighbor sites. Note that for these compositions, the influence of exchangeable cations will also be minimal, such that these local models will largely suffice. We refer to Refs. 54, 55, and 62–65 for the details of the models and discussions of the acidic properties in relation to this approach.

Not only the composition but also the local geometry will be of importance for the acid strength of the surface hydroxyl groups. The electronegativity of the oxygen depends on the hybridization state of its bonding orbitals, which is directly influenced by the bond angle.[66] The larger the bonding angle, the more pronounced the *s* character, and a higher electronegativity of the oxygen hybrid orbitals is to be expected.[26] According to Pelmenshchikov et al.,[67] the influence of the chemical composition should be negligibly small with respect to these local geometric effects. However, this needs some more extensive testing, certainly in view of the experimental evidence and recent work by Senchenya et al.[68] where it was demonstrated that the structural effect was less important than the composition. In a semiempirical study, Mortier and Geerlings[69] found that after protonation of an Al–O–Si bridge, the influence of the bonding angle on the OH bond strength became negligible. In view of the covalent character of the framework, it is hard to understand that the composition has a minor effect. However, the two effects cannot be entirely separated: new equilibrium geometries will be established for every compositional change.

A more detailed analysis reveals that not only the broad LF band but also the HF band,[54,70–72] is composed of several peaks. It is therefore obvious that all parameters previously discussed are of importance, and will contribute to the fine details. Constructing a model of the actual situation is impossible. However, the principles outlined here are general, and it should be possible without too much difficulty to apply them to other systems such as silica–alumina catalysts. For some recent calculations on alumina, see Refs. 73 and 74.

BRÖNSTED ACIDITY: MOLECULAR INTERACTIONS

Two questions will be addressed (again limiting ourselves to the surface hydroxyls in zeolites but keeping in mind that the same discussion can easily be extended to other systems): (i) to what extent does the difference between bridging and terminal OH change the interaction characteristics with molecules, and (ii) how does a compositional variation influence this. An extended discussion can be found in the Refs. 28, 75, and 76.

The interaction between donor molecules and bridging and terminal hydroxyls is schematically represented below following Gutmann's[25] notation (the arrow points in the direction of the electron shift; a solid line indicates a bond lengthening, and a dashed line a bond shortening):

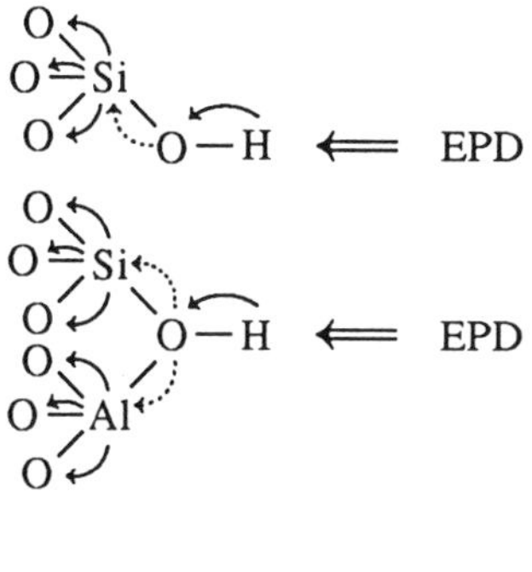

O
O ⇆ Si
O   O—H ⟸ EPD

O
O ⇆ Si
O   O—H ⟸ EPD
O
O ⇆ Al
O

Data from *ab initio* calculations[75] are presented in Table 4.3. From these, it is immediately obvious that:

1. The OH bond weakening, the OH stretching force constant, and concomitant decrease in frequency are progressively more impor-

Table 4.3. Overall characteristics[75] of the complexes of terminal hydroxyls (type I) and bridging hydroxyls (type II) with CO, NH_3, and H_2O: stabilization energy ΔE, equilibrium O–H bond length r_{OH} and change relative to the free OH Δr_{OH}, force constant and change f_{OH} and Δf_{OH}, and stretching frequency ν_{OH} (and scaled in parentheses) and the change relative to the free OH $\Delta \nu_{OH}$.

		ΔE (kJ mol^{-1})	r_{OH} (Å)	Δr_{OH}	f_{OH} (N m^{-1})	Δf_{OH}	ν_{OH} (cm^{-1})	$\Delta \nu_{OH}$
I	CO	15.5	0.960	0.001	885	−7	3980 (3797)	15 (14)
	H_2O	55.7	0.975	0.016	741	−151	3642 (3474)	353 (336)
	NH_3	56.2	0.983	0.024	706	−186	3555 (3392)	440 (419)
II	CO	27.1	0.972	0.005	785	−78	3748 (3576)	183 (174)
	H_2O	91.1	1.004	0.037	558	−305	3160 (3015)	771 (735)
	NH_3	101.2	1.029	0.062	432	−431	2781 (2653)	1150 (1097)

tant with increasing donor properties of the adsorbed molecules: CO < H_2O < NH_3. This is true for the bridging as well as for the terminal OH groups.

2. The effect on the bridging hydroxyls largely exceeds the effect of the same donor molecules on the terminal hydroxyls.

The shift of the hydroxyl bands interacting with different donor molecules can be observed experimentally, and should be a better measure for the acidity than the stretching frequency itself, which is indicative only of the intrinsic properties. This was demonstrated for bridging and terminal OH groups (NaHY zeolites and aerosil, respectively) by Paukshtis and Yurchenko[77] (Fig. 4.7) and for terminal OH (aerosil) by Horill and Noller.[78]

The second question relates to the influence of the framework composition on the O–H bond weakening upon interaction with adsorbed molecules. A higher electronegativity of the framework will draw away more electrons from the OH group and the bond weakening upon interaction with donor molecules increases with increasing electronegativity of the framework. This was theoretically established in Refs. 28 and 76 and

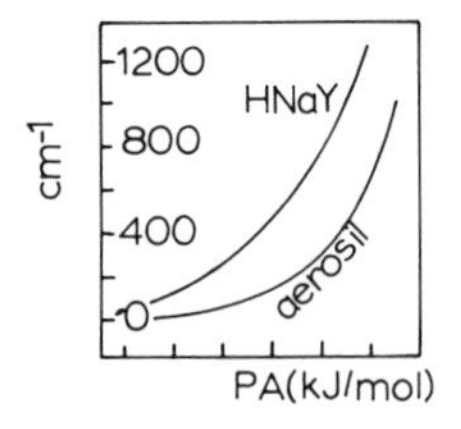

Figure 4.7. Variation of the batochromic shift ($\Delta\nu$, cm^{-1}) of the terminal OH in aerosil and the bridging OH in HNaY zeolites upon adsorption of molecules with varying proton affinity (PA).[77]

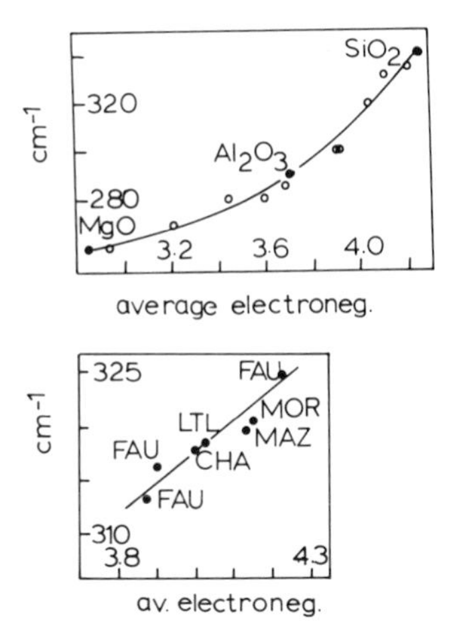

Figure 4.8. *Top:* Batochromic shift ($\Delta\nu$, cm^{-1}) of the terminal OH in mixed oxides upon adsorption of acetone[79] versus the average compound composition of the oxide. *Bottom:* Batochromic shift ($\Delta\nu$, cm^{-1}) of the bridging OH in H-zeolites upon adsorption of benzene[80] versus the average electronegativity of the zeolite.

experimentally demonstrated for terminal OH groups by Lercher and Noller[79] and for bridging OH groups by Jacobs.[80] Note (Fig. 4.8) that the batochromic shift of the OH stretching frequency is of the same order of magnitude (about 300 cm^{-1}) for acetone interacting with terminal OH as for the much weaker donor molecule benzene interacting with bridging hydroxyls. It was then empirically found[79,80] that the framework composition could best be summarized by again using the average compound electronegativity. The donor properties of the molecules are best quantitatively given by Gutmann's donor number.[25,78] Also note that the framework properties should be affected by the interactions of molecules with the surface OH groups.

We can generally derive from the preceding discussion that the weaker bond will be increasingly sensitive to perturbations (interactions with adsorbed molecules or changes in the composition). Recent theoretical calculations addressing various aspects of molecular interactions with solid surfaces can be found in Refs. 81 and 82, from which the prior literature can be searched.

THE ACTIVE SITE

Although much of the perturbations in the molecules and in the surface can now be qualitatively and quantitatively understood, the preceding picture is by no means complete enough to give a full account of "catalysis." The foregoing "equilibrium" phenomena may have a different bearing under "catalytic conditions," where diffusion and temperature are two major parameters neglected. Reaction schemes may be easy to write, but a proof of validity is most often very difficult to give. More often than not, the active surface takes no part in the proposed mechanisms. Acidity is not just Brönsted acidity, but Lewis sites (cationic sites) may have a more drastic effect on molecules. Radical mechanisms were also not considered, while active participation of, e.g., framework species (such as the oxygens[83]), cannot be excluded either.

However, despite all these shortcomings, understanding the initial step of activating the molecule is a must if we are to design future catalysts in a more efficient way. We again cannot refrain from referring to the first principles of physics. This is important enough as a final reminder. In quantum chemistry, a system of electrons and nuclei is completely determined as soon as we are in a position to write the Hamiltonian operator $\hat{H}$. There only then remains to solve the Schroedinger equation $\hat{H}\Psi = E\Psi$, a purely mathematical exercise, in order to obtain the wavefunction Ψ and the corresponding energy E. One of the characteristics of this operator is that for every system of N electrons, the terms for the kinetic energy $\hat{T}$ and for the electron–electron repulsion $\hat{V}_{ee}$ are identical. Two systems of N electrons can only differ in the "external potential" (i.e., the potential felt by the electrons, primarily due to the nuclear charge, but also all other charges in the surroundings) $\hat{V}_{ne}$. Using atomic units and summation indexes involving all electrons (μ,ν) and all nuclei (α) with charge Z_α, we may write

$$\hat{H} = \underbrace{-\frac{1}{2}\sum_{\mu=1}^{N} \nabla_\mu^2}_{\hat{T}} + \underbrace{\sum_{\mu<\nu}^{N} \frac{1}{r_{\mu\nu}}}_{\hat{V}_{ee}} + \underbrace{\sum_{\mu=1}^{N} v(\mu)}_{\hat{V}_{ne}} \qquad \left(v(\mu) = -\sum_{\alpha} \frac{Z_\alpha}{r_{\mu\alpha}}\right) \tag{4.9}$$

It is exactly this property that was used by Hohenberg and Kohn[8] for proving their two theorems. Perturbing a molecule therefore ultimately means interfering with the number of electrons or with the external potential.

ACKNOWLEDGMENTS

The author acknowledges a permanent position with the Belgian National Fund for Scientific Research (N.F.W.O.) as Research Director (Onderzoeksdirekteur).

REFERENCES

1. Parr, R.G. in *Electron Distributions and The Chemical Bond* (P. Coppens and M.B. Hall, eds). New York: Plenum, 1982, p. 95.
2. Parr, R.G., *Ann. Rev. Phys. Chem.* **34:** 631 (1983).

3. Parr, R.G., Donnelly, R.A., Levy, M., Palke, W.E. *J. Chem. Phys.* **68:** 3801–3807 (1978).
4. Pauling, L., *J. Am. Chem. Soc.* **54:** 3570–3582 (1932).
5. Sanderson, R.T., *Science* **114:** 670–672 (1951).
6. Sen, K.D., Jørgensen, C.K. (eds.), *Structure and Bonding, Vol. 66.* Berlin: Springer-Verlag, 1987.
7. Mortier, W.J., *Studies of Surface Science and Catalysis* **37:** 253–268 (1988).
8. Hohenberg, P., Kohn, W., *Phys. Rev. Sec.* **136B:** 864–871 (1964).
9. If it is desired to find a stationary value for an integral $\int_a^b I(x,y,y')dx$, for which the integrand I depends on the choice of the function y as well as of $y' = dy/dx$, satisfying the Euler equation $\delta I/\delta y - d(\delta I/\delta y')/dx = 0$ will ensure that our choice is correct, i.e., that y is an extremal. Note that also $E[\Psi]$ and $E[\rho]$ are integrals which we like to minimize by varying Ψ or ρ.
10. Margenau, H., Murphy, G.M., *The Mathematics of Physics and Chemistry,* 2nd ed. Princeton NJ: D. Van Nostrand Co., 1956, pp. 198–215.
11. Iczkowski, R.P., Margrave, J.L., *J. Am. Chem. Soc.* **83:** 3547–3551 (1961).
12. Mulliken, R.S., *J. Chem. Phys.* **2:** 782–793 (1934).
13. Parr, R.G., Pearson, R.G., *J. Am. Chem. Soc.* **105:** 7512–7516 (1983).
14. Politzer, P., Weinstein, H., *J. Chem. Phys.* **71:** 4218–4220 (1979).
15. Mortier, W.J., *Structure and Bonding* **66:** 125–143 (1987).
16. Mortier, W.J., Ghosh, S.K., Shankar, S., *J. Am. Chem. Soc.* **108:** 4315–4320 (1986).
17. Uytterhoeven, L., Mortier, W.J., EEM program (FORTRAN) (1986). This program solves the set of equations and allows the calibration of the expansion coefficients. Copies can be obtained upon request.
18. Uytterhoeven, L., Lievens, J., Van Genechten, K., Mortier, W.J., Geerlings, P., *Preprints conf. Eberswalde (D.D.R.),* 1987.
19. Mulliken, R.S., *J. Chem. Phys.* **23:** 1833–1840 (1955).
20. Mortier, W.J., in *New Developments in Zeolite Science and Technology* (Y. Murakami, A. Iijima, and J.W. Ward, eds.). Amsterdam: Elsevier, 1986, pp. 423–428.
21. Van Genechten, K., Mortier, W.J., Geerlings, P., *J. Chem. Soc., Chem. Comm.* 1278–1279 (1986).
22. Van Genechten, K., Mortier, W.J., Geerlings, P., *J. Chem. Phys.* **86:** 5063–5071 (1987).
23. Bertaut, F., *J. Phys. Radium* **13:** 499–505 (1952).
24. Kollman, P.A., Allen, L.C., *J. Chem. Phys.* **51:** 3286–3293 (1969).
25. Gutmann, V., *The Donor-Acceptor Approach to Molecular Interactions.* New York: Plenum Press, 1978.
26. Newton, M.D., in *Structure and Bonding in Crystals* (M. O'Keeffe and A. Navrotsky, eds). New York: Academic Press, 1981, Vol. I, pp. 175–193.
27. Allen, F.H., Kirby, A.J., *J. Am. Chem. Soc.* **106:** 6197–6200 (1984).
28. Jones, P.G., Kirby, A.J., *J. Am. Chem. Soc.* **106:** 6207–6212 (1984).
29. Van Genechten, K., *Ph. D. Thesis #160.* K.U. Leuven, Fac. Agronomy, Nov. 1987.
30. Siegbahn, K., *Molecular Spectroscopy.* London: Heyden and Son, 1977, pp. 227–312.
31. Vinek, H., Noller, H., Ebel, M., Schwartz, K., *J. Chem. Soc., Faraday Trans. I* **73:** 734–746 (1977).
32. Seyama, H., Soma, M., *J. Chem. Soc., Faraday Trans. I* **81:** 485–495 (1985).
33. Smith, J.V., Blackwell, C.S., *Nature* **303:** 223–224 (1983).
34. Engelhardt, G., Radeglia, R., *Chem. Phys. Lett.* **108:** 271–274 (1984).
35. Adams, D.M., *Inorganic Solids.* London: Wiley, 1974, p. 60.
36. Klinowski, J., *Progress in NMR Spectroscopy* **16:** 237–309 (1984).
37. Grobet, P.J., Mortier, W.J., Van Genechten, K., *Chem. Phys. Lett.* **119:** 361–364 (1985).

38. Datka, J., Geerlings, P., Mortier, W.J., Jacobs, P., *J. Phys. Chem.* **89:** 3483–3488 (1985).
39. Sanderson, R.T., *Polar Covalence.* New York: Academic Press, 1983.
40. McWeeny, R., *Coulson's Valence,* 3rd Ed. Oxford: Oxford University Press, 1979, p. 81.
41. Sedlacek, J., *J. Mol. Catalysis* **39:** 169–179 (1987).
42. Fyfe, C.A., Kennedy, G.J., De Schutter, C.T., Kokotailo, G.T., *J. Chem. Soc., Chem. Commun.* 541–542 (1984).
43. Barthomeuf, D., *J. Phys. Chem.* **88:** 42–45 (1984).
44. Van Genechten, K., Mortier, W.J., *Zeolites* **8:** 273–283 (1988).
45. Stach, H., Lohse, U., Thamm, H., Schirmer, W., *Zeolites* **6:** 74–90 (1986).
46. Hayashi, S., Suzuki, K., Shin, S., Hayamizu, K., Yamamoto, O., *Chem. Phys. Lett.* **113:** 368–371 (1985).
47. Michel, D., Germanus, A., Pfeifer, H., *J. Chem. Soc. Faraday Trans. I* **78:** 237–254 (1982).
48. Bosacek, V., Freude, D., Gründer, W., Meiler, W., Pfeifer, H., *Z. Phys., Chemie (Leipzig)* **265:** 241–249 (1984).
49. Pfeifer, H., Freude, D., Hunger, M., *Zeolites* **5:** 274–286 (1985).
50. Michael, A., Meiler, W., Michel, D., Pfeifer, H., Hoppach, D., Delmau, J., *J. Chem. Soc., Faraday Trans. I* **82:** 3053–3067 (1986).
51. Derouane, E.G., Fripiat, J.G., *J. Phys. Chem.* **91:** 145–148 (1987).
52. Zhidomirov, G.M., Kazansky, V.B., *Advances in Catalysis* **34:** 131–202 (1986).
53. Mortier, W.J., Schoonheydt, R.A., *Progress in Solid State Chemistry* **16:** 1–126 (1985).
54. Dwyer, J., *Studies of Surface Science and Catalysis* **37:** 333–354 (1988).
55. Barthomeuf, D., *Materials Chemistry and Physics* **17:** 49 (1987).
56. Löwenstein, N., *Amer. Mineral.* **39:** 93 (1954).
57. Mortier, W.J., Sauer, J., Lercher, J.A., Noller, H., *J. Phys. Chem.* **88:** 905–912 (1984).
58. Jacobs, P.A., Mortier, W.J., Uytterhoeven, J.B., *J. Inorg. Nucl. Chem.* **40:** 1919–1923 (1978).
59. Jacobs, P.A., Mortier, W.J., *Zeolites* **2:** 226–230 (1982).
60. Kazansky, V.B., *Proc. Fourth Nat. Symp. Cat. (Cat. Soc. India)* Bombay Dec. 2–4, 1978, pp. 14–26.
61. Freude, D., Hunger, M., Pfeifer, H., *Chem. Phys. Lett.* **128:** 62–66 (1986).
62. Barthomeuf, D., *J. Phys. Chem.* **83:** 249–256 (1979).
63. Dwyer, J., Fitch, F.R., Nkang, E.E., *J. Phys. Chem.* **87:** 5402–5404 (1983).
64. Beagley, B., Dwyer, J., Fitch, F.R., Mann, R., Walters, J., *J. Phys. Chem.* **88:** 1744–1751 (1984).
65. Wachter, W.A., *Proc. 6th Int Conf. Zeol.* (D.H. Olson and A. Bisio, eds.). Guildford, England: Butterworths, 1984), pp. 141–150.
66. Hinze, J., Jaffé, H.H., *J. Chem. Phys.* **84:** 540 (1964).
67. Pelmenshchikov, A.G., Pavlov, V.I., Zhidomirov, G.M., Beran, S., *J. Phys. Chem.* **91:** 3325–3327 (1987).
68. Senchenya, I.N., Kazansky, V.B., Beran, S., *J. Phys. Chem.* **90:** 4857–4859 (1986).
69. Mortier, W.J., Geerlings, P., *J. Phys. Chem.* **84:** 1982–1986 (1980).
70. Dombrowski, D., Hoffmann, J., Fruwert, J., Stock, T., *J. Chem. Soc., Faraday Trans. I* **81:** 2257–2271 (1985).
71. Dźwigaj, S., Haber, J., Romotowski, T., *Zeolites* **3:** 134–138 (1983).
72. Kustov, L.M., Borovkov, V. Yu., Kazansky, V.B., *J. Catalysis* **72:** 149–159 (1981).
73. Kawakami, H., Yoshida, S., *J. Chem. Soc., Faraday Trans. I* **81:** 1117–1127 (1985).
74. Senchenya, I.N., Pelmenshchikov, A.G., Zhidomirov, G.M., Kazanskii, V.B., *React. Kinet. Catal. Lett.* **31:** 101–105 (1986).

75. Geerlings, P., Tariel, N., Botrel, A., Lissillour, R., Mortier, W.J., *J. Phys. Chem.* **88:** 5752–5759 (1984).
76. Datka, J., Geerlings, P., Mortier, W.J., Jacobs, P.A., *J. Phys. Chem.* **89:** 3488–3493 (1985).
77. Paukshtis, E.A., Yurchenko, E.N., *React. Kinet. Catal. Lett.* **16:** 131–135 (1981).
78. Horill, P., Noller, H., *Zeit. Phys. Chemie NF* **100:** 155 (1976).
79. Lercher, J.A., Noller, H., *J. Catalysis* **77:** 152–158 (1982).
80. Jacobs, P.A., *Catal. Rev.-Sci. Eng.* **24:** 415–440 (1982).
81. Choumakos, B.C., Gibbs, G.V., *J. Phys. Chem.* **90:** 996–998 (1986).
82. Sauer, J., *J. Phys. Chem.* **91:** 2315–2319 (1987).
83. Takaishi, T., Endoh, A., *J. Chem. Soc., Faraday Trans. I* **83:** 411–424 (1987).

5
Quantum-Chemical Studies of Zeolites

STANISLAV BERAN†

INTRODUCTION

The study of zeolite properties continues to attract considerable interest because of the ever increasing possibilities of practical applications of these substances both as catalysts in various transformations of hydrocarbons used in the chemistry of crude oil processing and potentially also in C1 chemical processes and as adsorbents in various separation processes. The unusual structure and consequent physicochemical and catalytic properties of zeolites have led to their being among the most important acid catalysts and adsorbents used in the chemical industry.[1-4]

From a chemical point of view, zeolites are crystalline aluminosilicates with the composition $M^{n+}_{x/n}(AlO_2)_x(SiO_2)_y \cdot zH_2O$, where M is a metal cation or proton. The basic structural unit in the zeolite skeleton is the TO_4 tetrahedron (where T = Si or Al) forming the aluminosilicate rings $(TO_3)_k$ with various sizes. The combination of several aluminosilicate rings then leads to the formation of structural channels or cavities connected by channels. The diameter of these channels or cavities depends on the zeolite structural type and, depending on the size of aluminosilicate rings, attains values of units of angstroms. Consequently, the internal zeolite structure is available only for molecules with certain dimensions (the molecular sieve effect of zeolites), which plays an important role in their utilization as catalysts and adsorbents—the shape selectivity effect. As a result of the presence of Al, the zeolite skeleton exhibits a negative charge that can be compensated either by protons bonded to the skeletal O atoms and forming bridging OH groups or by metal cations coordinated in the cation positions to several skeletal O atoms. While the bridging OH groups are responsible for the zeolite Brönsted acidity, metal cations can act as donors or, more especially, as acceptors of electron density and correspond to the zeolite Lewis sites. The terminal ≡SiOH groups also exhibit Brönsted acidity. Basic sites are formed by the skeletal O atoms, especially of the ≡SiOAl≡ type. In addition to these acid–base sites, vari-

† Deceased

ous types of zeolite treatment (such as thermal dehydroxylation) can lead to the formation of so-far unspecified, coordinationally unsaturated sites exhibiting variously strong Lewis acidity. A further factor that has also often been considered in connection with the catalytic activity of zeolites is the large electrostatic field operating in their structural channels.

This brief review of the structural properties of zeolites indicates that, in addition to their practical importance, zeolites are chemically interesting systems with defined active sites and structure. Consequently, they constitute an excellent model system for the study of heterogeneous catalytic processes. This has led to intensive study of the physicochemical characteristics of zeolites, as well as of their interaction complexes with molecules, employing various experimental (especially spectroscopic) techniques. The utilization of these experimental methods then led to the need to obtain further information on zeolite properties necessary for the interpretation of the results obtained. One of the ways of obtaining this information consists of carrying out quantum-chemical calculations of the properties of models of zeolite sites and their interaction complexes with molecules. The development in computer technology then permitted since the middle of the 1970s quantum-chemical calculations being increasingly used as a source of this complementary and, gradually, independent information on the properties of zeolite systems. At present, several hundred original quantum-chemical studies of the properties of aluminosilicates and silicates and a number of reviews[5–11] covering larger or smaller regions in this field can be found in the literature.

This chapter was written in order to illustrate the capabilities of quantum-chemical calculations of cluster models to yield information on the properties of zeolites and their interaction complexes with molecules. The first part is devoted to problems connected with the selection of a suitable cluster model for zeolites, its size and termination, the effect of the electrostatic field, and quantum-chemical methods used to describe these models. The second part describes and discusses results of calculations of zeolite geometric and electronic structure, the properties of the cationic and Lewis sites, and various types of OH groups in zeolites and the stability of the zeolite skeleton. The third part then demonstrates the capabilities of quantum-chemical calculations to describe interactions of molecules (H_2O, CO, NH_3, and hydrocarbons) with various zeolite sites.

ZEOLITE MODELS AND QUANTUM-CHEMICAL METHODS

The Cluster Approximation

Study of the properties of a solid phase by quantum-chemical methods requires the use of models of this solid phase consisting of a finite number of atoms. The solid phase or its active sites are thus represented by a fragment of a crystal—a cluster—which should describe all the characteristics of the studied system. The formation of this fragment is connected with the breaking of the fragment–crystal bonds, resulting in the formation of artificial surface states (located in the forbidden zone), disturbing of the charge distribution in the model, etc. However, as will be demonstrated, this problem, which represents one of the most important shortcomings of the cluster approximation, can be satisfactorily solved for zeolite cluster models. A further important consideration is the necessary size of the cluster model required for a sufficiently good description of the properties of the studied sites. The cluster approximation can be used only to describe the properties of the solid phase that have local character, i.e., properties determined by the interactions of the atoms used to form the model. Fortunately, it has been found that most of the properties of polyatomic crystals of dielectric or semiconductor substances (and this also includes zeolites) are local in character and can thus be described successfully in terms of the cluster model. The electrostatic field of zeolite crystals constitutes an exception, since its magnitude can be correctly described only by an infinite or at least large number of atoms forming the crystal. One of the great advantages of the cluster approximation is the fact that, in contrast to the procedures used in solid-phase physics, it permits construction and description of molecular models of arbitrary structurally and chemically ordered sites, accenting the chemical nature of these systems. Thus, experience and analogies in molecular chemistry can be used to understand the physicochemical properties of active sites in the solid phase.

Termination of Cluster Models

The formation of artificial surface states in cluster models as well as other consequences of artificial scission of bonds resulting from the formation of cluster models can be removed in a number of ways. One of the most widely used approaches involves saturation of dangling bonds formed by breaking of the model fragment away from the crystal by using monovalent atoms (usually H atoms).[5–12] The properties of these monovalent atoms (their electronegativity) are, however, generally different from those of the remainder of the crystal that they model. This leads to a poor description of the cluster atoms to which these monovalent atoms are bonded, resulting, for example, in incorrect charge values on these atoms. Thus, the next step in the development of this procedure was termination of the cluster with monovalent pseudoatoms (usually atoms with an *s* orbital), whose properties are selected to describe as well as possible the properties of the remainder of the crystal, which is not included in the model.[13–16] A formally analogous method of cluster termination is the use

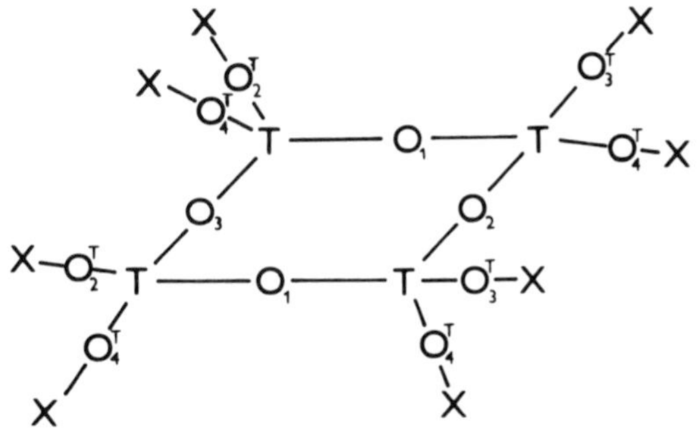

Figure 5.1. Schematic depiction of the $Si_4O_4(OX)_8$, $Si_3AlO_4(OX)_{12}H$, and $Si_2Al_2O_4(OX)_8H_2$ cluster models of four-member windows in zeolites with indication of the types of O atoms. Protons forming bridging OH groups are bonded to O_1, O_2, and O_3 atoms of the ring. T = Si or Al and X stands for pseudoatoms, pseudoions, or H.

of boundary pseudoions[17,18] (point charges) compensating the charge of selected ionic models of the lattice and localized in the boundary atom positions. This approach is analogous to the Watson spheres employed in X_α calculations.

These procedures are illustrated on the example of the CNDO/2 calculations[16] of the charges on atoms in the model of a four-member zeolite window represented by the cyclic $Si_4O_4(O^TX)_8$ cluster (see Fig. 5.1), where X is the H atom, a pseudoatom with various electronegativities $[\frac{1}{2}(I + A)]$ or a positive charge (see Table 5.1). These results demonstrate that, while the charge on the O^T atoms changes with the properties of the boundary atom X, the charge on the Si and O atoms located in the second coordination sphere of boundary atom X is practically independent of its properties. This fact documents the local character of interactions in simi-

Table 5.1. Dependence of charges on atoms of the $Si_4O_4(O^TX)_8$ cluster on its termination.

X	$\frac{1}{2}(I + A)$ (eV)	q_{Si}	q_{O_1}	q_{O_2}	q_{O_3}	$q_{O_2^T}$	$q_{O_3^T}$	$q_{O_4^T}$
Hydrogens	7.18	1.63	−0.73	−0.74	−0.74	−0.58	−0.57	−0.58
Pseudoatoms	3.0	1.64	−0.74	−0.75	−0.75	−0.68	−0.68	−0.68
Pseudoatoms	2.0	1.63	−0.75	−0.76	−0.75	−0.70	−0.69	−0.68
Pseudoatoms	1.0	1.62	−0.75	−0.76	−0.75	−0.73	−0.72	−0.73
Pseudoatoms	0.5	1.48	−0.74	−0.75	−0.75	−0.79	−0.78	−0.80
Pseudoions	—	1.51	−0.71	−0.72	−0.72	−1.42	−1.42	−1.38

lar systems and confirms the justifiability of the use of the cluster model for quantum-chemical study of zeolites. The value of parameter $\frac{1}{2}(I + A) = 1$ eV (see Table 5.1) for which the charge on the terminal O^T atom is practically identical with the charge on the O atoms forming the window then corresponds to the optimized parameter for the pseudoatom. Similarly, optimal parameters were found for pseudoatoms modeling Si for the CNDO/BW and MINDO/3 methods.[13,14] In this case, optimal parameters were sought for pseudoatom X in the volume fragment $Si(OX)_4$, so that the electroneutrality condition $q_{Si} = -2q_O$ is fulfilled. In conclusion it can be stated that the model zeolitic cluster can be terminated by H atoms without danger of qualitative errors, although the use of optimized pseudoatoms is certainly physically more correct and yields better results. The justifiability of this approach was confirmed both by the large number of qualitatively or semiqualitatively successful studies of models terminating by H atoms and by the fact that the electronegativity of the H atom is not very different from that of Si or Al atoms. A definite advantage of this approach is the possibility of using standard unmodified quantum-chemical programs, as well as the fact that terminal OH groups simultaneously model actual OH groups terminating the zeolite crystal. On the other hand, however, it is understandable that the results of the calculation of some characteristics (e.g., dissociation of OH bonds, interactions of the models with molecules, and the charge transfer connected with this interaction) are dependent on the electronegativity of the boundary atoms (see following discussion) and are only qualitative in nature.

The importance of boundary pseudoatoms is not, however, only in the possibility of forming a physically more correct model of the solid phase. Changes in the electronegativity of pseudoatoms can be used to describe the effect of changes in the chemical environment around a given site on its properties in the framework of very simple models. Examples of this utilization of pseudoatoms can be found in CNDO/BW calculations[19] of the dependence of the dissociation energy of bridging OH groups, modeled by the $(XO)_3Si\overset{H}{O}Al(OX)_3$ cluster, on the value of the valence orbital ionization potential (VOIP) of one of the pseudoatoms X. It follows from these calculations that an increase in VOIP of X from 5 to 20 eV results in a decrease in the dissociation energy of the OH group from 1496 to 1351 kJ mol^{-1}. This result qualitatively illustrates the large changes in the properties of bridging OH groups that can occur from any change in their third coordination sphere, expressible in terms of the electronegativity (e.g., changes in the chemical composition). Simultaneously, it demonstrates how the properties of the OH groups depend on the electronegativity of the atoms terminating the cluster.

The Size of Zeolite Models

There are no general rules for the selection of a model of the solid phase and its necessary size, except for the trivial rule that an increase in the size generally leads to an improvement in the quality of the results. Obviously, the necessary size of the model depends on the properties of the solid phase to be described. Experience in the quantum-chemical description of zeolites has indicated that relatively good quality description of the structural characteristics of zeolite fragments can be achieved using very simple cluster models, e.g., of the $Si(OH)_4$ or H_3SiOTH_3 type.[6,7] Similarly, basic qualitative differences in the acidities of terminal and bridging OH groups can be described using very simple models such as H_3SiOH and $H_3Si\overset{H}{O}AlH_3$. The latter model can also be used to describe qualitatively the effect of zeolite structural characteristics (the length of TO bonds and TOT angles) on the properties (acidity and vibrational frequencies) of the bridging OH groups. A large model is necessary for description of the effect of the chemical environment on the properties of bridging OH groups. For example, the effect of the Si : Al ratio can be modeled using clusters of the $Si_{4-n}Al_nO_4(OH)_8H_n$ or $Si_{6-n}Al_n(OH)_{12}H_n$ type (cf. Fig. 5.2) or can be simulated qualitatively using X pseudoatoms (whose electronegativity corresponds to that of Si or Al atoms) with a cluster model of the $(XO)_3Si\overset{H}{O}Al(OX)_3$ type (see following discussion). Large models are essential for correct description of the properties of metal cations coordinated by three or more O atoms in the zeolite skeleton, forming one or more zeolite windows. It is then necessary to use a model including at least this zeolite window, i.e., clusters of the $T_6O_6H_{12}Cat$ type or, preferably, $T_6O_6(OH)_{12}Cat$.

Electrostatic Fields

Polyatomic crystals have a positive or negative charge localized on their atoms, leading to the formation of an electrostatic field. The magnitude of this field is determined by a large number of atoms forming the crystal and can attain values of units of volts per angstrom.[20,21] Obviously, this electrostatic field can be approximated only very inadequately in the cluster approximation and often the approximation does not even exhibit correct symmetry. The effect of the electrostatic field on the characteristics calculated using cluster models can be estimated by including this field in calculations by using a sufficiently large number of atoms of the crystal lattice in the form of point charges. The Hamiltonian of quantum-chemical methods then contains terms describing the interactions of electrons

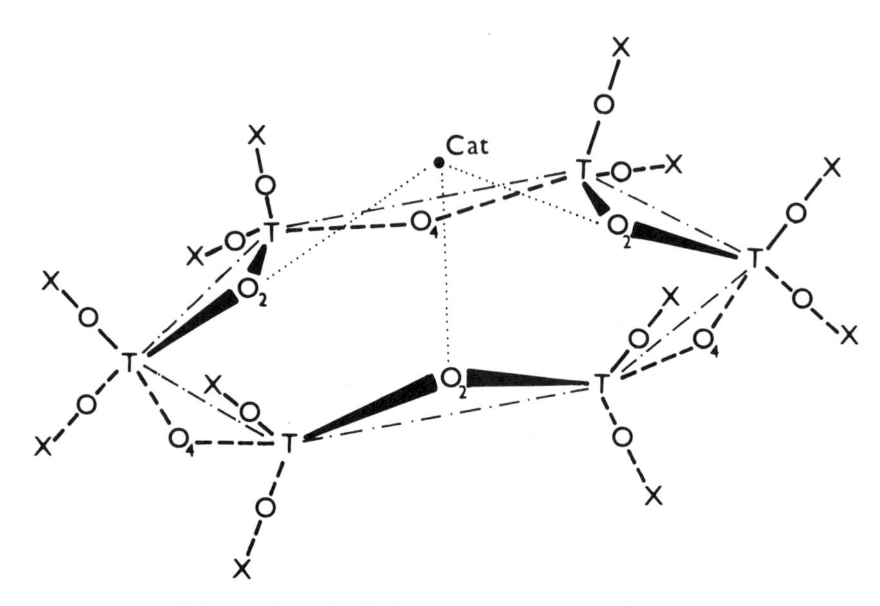

Figure 5.2. Schematic depiction of the $Si_{6-n}Al_nO_6(OX)_{12}Cat$ and $Si_{6-n}Al_nO_6(OX)_{12}H_n$ models of six-member windows in zeolites with indication of the types of O atoms. Protons forming bridging OH groups are bonded to O_2 or O_4 atoms of the ring. T = Si or Al, and X stands for pseudoatoms or H atoms.

and atomic nuclei forming the cluster with these point charges.[9,16,22] The results of these calculations[16,22] for models of cationic and H forms of zeolite, however, have demonstrated that the inclusion of an additional electrostatic field does not greatly affect the calculated characteristics of the zeolite skeleton. A substantial increase in the charge as a result of the inclusion of an additional electrostatic field was found only for cations localized in the zeolite cationic positions and for protons of bridging OH groups.[16,22] It can thus be summarized that the poor description of the electrostatic field of the crystal by cluster models has no great effect on the quality of the calculated characteristics. However, on the other hand, it has been found that the electrostatic field in the pores of the zeolite can greatly affect the physicochemical characteristics of the molecules diffusing through the zeolite channels. For example, model INDO calculations[23] of the effect of the electrostatic field on the methanol molecule have demonstrated that fields with value of units of volts per angstrom can even lead to bond dissociation in this molecule. The effect of the electrostatic field on the properties of water molecules located in various positions inside of the faujasite structure can be seen in Fig. 5.3.

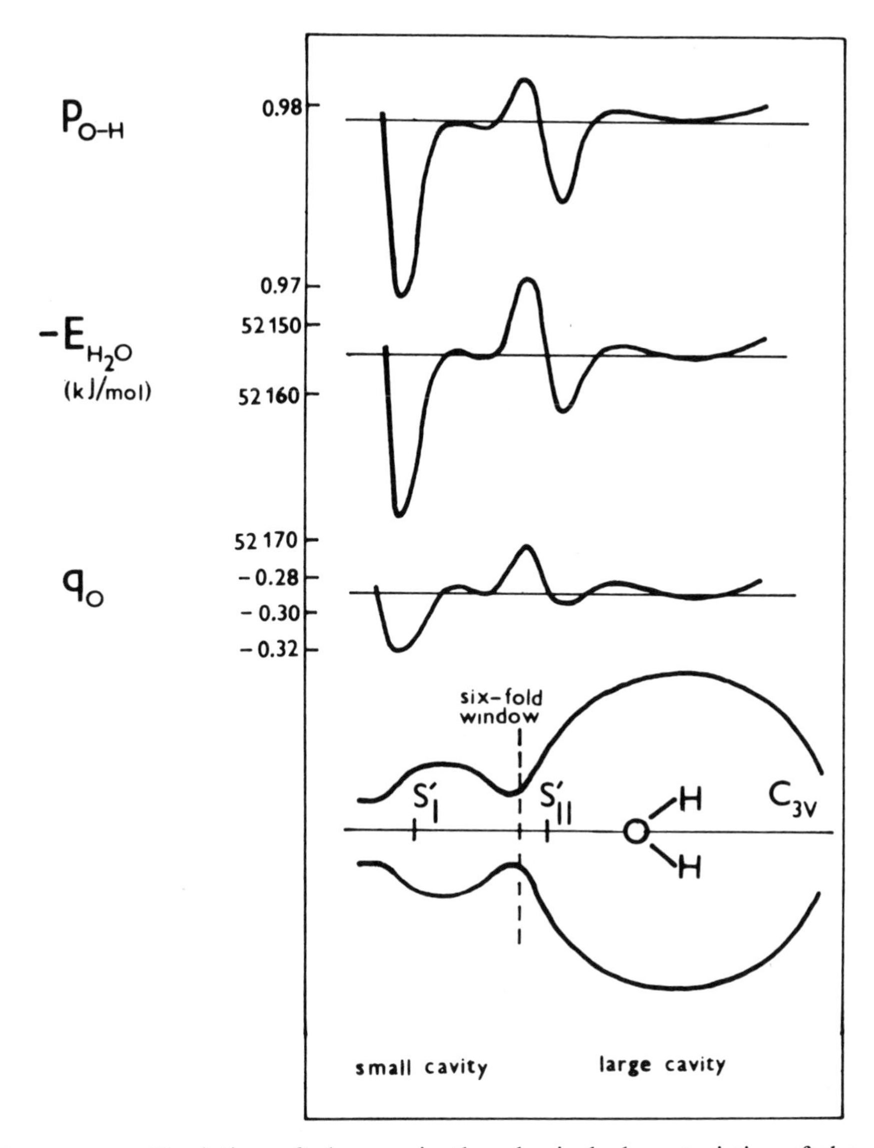

Figure 5.3. Depiction of changes in the physical characteristics of the water molecule (overall energy, bond orders, and charge on O) caused by the zeolite electrostatic field depending on the position of its O atom on threefold axis of faujasites without cations. [From Beran, S., CNDO/2 calculations on cluster models of zeolite crystals including their electrostatic field, *Chem. Phys. Lett.* **91:** 86–90 (1982). Courtesy of Elsevier, Amsterdam.]

Quantum-Chemical Methods

The quality of the results of quantum-chemical description of the properties of the solid phase or its interactions with molecules is determined by two factors: the quality of the model of the solid phase and the quality of the quantum-chemical method used. These two factors are then limited by calculation capabilities that are available. The quantum-chemical calculation of zeolite cluster models can be carried out using (and have been used[5–11]) variously approximative methods from empirical (of the EHT type)[24] through semiempirical (CNDO/2, CNDO/BW, INDO, MINDO/3)[25–27] to nonempirical methods with bases from minimal STO-3G to 6-31G.[28] Detailed analysis of the quality of the information yielded by the individual methods is outside the scope of this chapter and has been carried out elsewhere.[24–30] Here, we will limit ourselves to stating that the quality and reliability of the characteristics yielded by these methods increases from empirical to nonempirical methods. While semiempirical and especially empirical methods yield only qualitative information on the properties of the studied systems and are sometimes even qualitatively unreliable, except for a few exceptions, nonempirical methods yield semiquantitatively to quantitatively correct results, depending on the base used and the degree of inclusion of the correlation energy. However, the transition from semiempirical to nonempirical methods is connected with a sharp increase in the computer time requirements. Most of the quantum-chemical studies carried out so far, especially older works, used semiempirical methods and yielded qualitatively correct conclusions, many of which were later confirmed nonempirically. As computer techniques advance, the range of models that can be used for nonempirical description is rapidly expanding, leading to a constant increase in the number of calculations on zeolites using nonempirical methods. However, where it is necessary to use large cluster models or test model systems using a large number of calculations, semiempirical methods are still used extensively.

ZEOLITE PROPERTIES

Structure

The structure and properties of ionic crystals and thus also of zeolites can be explained and understood on the basis of traditional ideas of theoretical chemistry on the building of these systems included under various empirical concepts, such as Pauling's rule,[31] the Brown[32] and Gutmann[33] rules, or Sanderson's concept of the average electronegativity.[34] These empirical methods of theoretical chemistry (employing the concept of ionic radii, electronegativity, electrostatic valence, etc.) can contribute to the understanding of crystal bonding in many instances, but permit formation of only very qualitative conclusions about the details of building up the lattice with complex compositions. On the other hand, these rules can be viewed as a consequence of a more general rule of crystal structure: the process of lattice formation must proceed so that the energy of the local interactions in the crystal is minimal. It is thus apparent that the use of quantum-chemical methods considerably broadens (and quantifies) the possibilities of utilizing a "local" view of chemical interactions in crystals. To begin with, the formation of even very small lattice fragments must fulfill the requirements of minimal energy for local interactions in these fragments, thus revealing a possibility to calculate their optimal geometry using very small molecular analogs. Second, association in a given set of structural elements should also correspond to a minimum in the sum of the energies of their mutual interactions. It then becomes possible to find the energetically most probable arrangement of these elements in the crystal lattice.

The development of efficient automatic techniques for the optimization of molecular geometry and of computer techniques themselves then permits intensive and systematic study of both the optimal geometric characteristics of molecular systems modeling various aluminosilicates and the energetically most probable association of the individual fragments in the crystal.

Geometric Characteristics. The comparison of the calculated optimal geometric characteristics for even relatively small molecular systems with the structural characteristics of real crystals revealed unexpectedly good agreement. For example, the simplest model (molecular analog) of one of the basic fragments of the aluminosilicates tetrahedron SiO_4 is the orthosilicic acid molecule, $Si(OH)_4$, for which the STO-3G calculation[35] yielded the following optimal geometric parameters: $d(SiO) = 1.659$ Å, $d(OH) = 0.981$ Å, $<SiOSi = 107.1°$ or $114.2°$ and $<SiOH = 108.9°$. The corresponding values found for hydrated silicates are as follows: 1.63–1.70 Å, 1.06 Å, 106–116°, and 108–125°. Considering the great simplicity of the model, the agreement between the calculated and experimental values is very good. The results of the STO-3G calculations[6,7,35–38] of the optimal geometric characteristics of larger cluster models of the $(HO)_3SiOT(OH)_3$ and H_3SiOTH_3 types indicate that this agreement is not simply fortuitous. In addition to yielding the lengths of the SiO and AlO bonds comparable with the average experimentally observed values, these calculations predict variations in the lengths of these bonds as well as the corresponding force constants and values of the SiOAl angle.[6,7,35,36] It has been found, for example, that the increasing *s*-character of the hybrid orbitals forming the SiO and AlO bonds with increasing SiOAl angle leads to a shortening of

these bonds and an increase in the values of their stretching force constants.[6,7,35,36] The dependence found corresponds well with the experimentally observed correlation between the lengths of the SiO bonds and the SiOSi angles. Very important is the discovery that the overall energy of the system changes very little with a change in the SiOT angle in the region from 120° to 180°, as follows from calculations of the dependences of the properties of the ≡SiOT≡ systems on the SiOT angle. In addition, it was found that the Boltzmann weighted SiOSi angle distribution function found on the basis of the total energy calculated for various SiOSi angles conforms with the experimentally observed distribution of SiOSi angles.[6,7,35,38] This conclusion is very important for understanding the broad variability in aluminosilicate structures, as well as the variability in SiOT angles in a given structural type.

The situation is very different if a proton is bonded to the bridging O atoms in the ≡SiOAl≡ fragment forming a bridging OH group. STO-3G calculations[37] for the $(HO)_3Si\overset{H}{O}Al(OH)_3$ cluster have revealed that variation in the SiOAl angle leads to much greater changes in the overall energy of the system compared to the $(HO)_3SiOAl(OH)_3^-$ system (see Fig. 5.4. This indicates much greater rigidity of the SiOAl angles in protonated fragments. Consequently, the SiOAl angle for skeletal atoms forming OH groups should have values close to optimal ones.

Thus, the effect of the coordination of protons and cations to the ≡SiOAl≡ fragment on its geometry characteristics has been studied[39] in detail in the framework of the STO-3G theory using the $H_3SiOAlH_3^-$ and $H_3Si\overset{M}{O}AlH_3$ cluster models (M = H, Li, or Na). First, the optimal geometry for these systems was not affected by any deformation of the ≡SiOAl≡ and $\equiv Si\overset{M}{O}Al\equiv$ fragments by the zeolite structure (see Table 5.2). The sterically deformed ≡SiOAl≡ fragment was then modeled by the $H_3SiOAlH_3^-$ system, whose SiOAl angle was set at 150° during geometry optimization (i.e., the usual value for zeolites). As has already been mentioned, this deformation leads to a relatively small increase in the overall energy of the system: 17 kJ mol^{-1}. The rigidity of the zeolite lattice preventing free relaxation of the ≡SiOAl≡ fragment after coordination of ions was then modeled by fixing the terminal H atoms in the $H_3Si\overset{M}{O}AlH_3$ system in the positions found for the completely optimized ($<$SiOAl = 122°) and deformed ($<$SiOAl = 150°) $H_3SiOAlH_3^-$ systems, while the positions of the other atoms were optimized. Although the condition used to limit relaxation of the geometry of the ≡SiOAl≡ fragments is apparently stronger than the limitation of the relaxation of these fragments in the zeolite lattice, it was found that the optimized lengths of

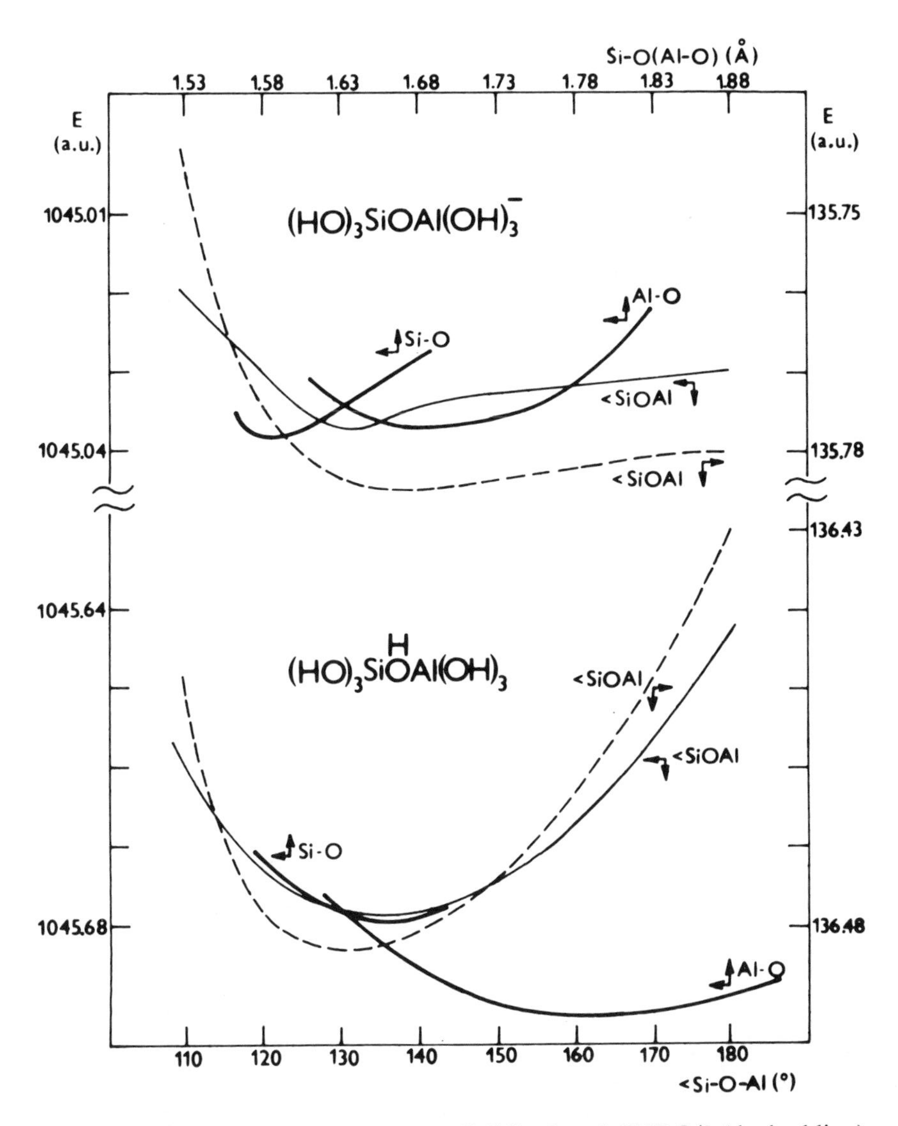

Figure 5.4. Plots of the total *ab initio* (full line) and CNDO/2 (dashed line) energies of the $(HO)_3SiO(H)Al(OH)_3$ and $(HO)_3SiOAl(OH)_3^-$ clusters against the SiOAl angle or the lengths of the SiO and AlO bonds. [From I.N. Senchenya et al., Quantum chemical study of the effect of the structural characteristics of zeolites on the properties of their OH groups. 2, *J. Phys. Chem.* **90:** 4857–4859 (1986). Courtesy of Amer. Chem. Soc., Washington, DC.]

Table 5.2. Optimized to (T = Si or Al) bridging bond lengths, *d* (Å), and TOT bond angles for various model systems.

Model	*d* (SiO)	*d* (AlO)	<TOT°
$H_3SiOSiH_3$ [a]	1.655		126
$H_3SiOAlH_3^-$ [b]	1.631	1.748	122
$H_3Si\overset{H}{O}AlH_3$ [b]	1.698	1.867	130
$H_3Si\overset{Li}{O}AlH_3$ [b]	1.676	1.798	125
$H_3Si\overset{Na}{O}AlH_3$ [b]	1.651	1.775	125
$H_3SiONaH_3SiH_3^{2-}$ [a]	1.640		136
$H_3SiOMgH_3SiH_3^-$ [a]	1.670		123
$(HO)_3SiOSi(OH)_3$ [a]	1.600		140
$(HO)_3SiOAl(OH)_3^-$ [a]	1.590	1.700	130
$(HO)_3Si\overset{H}{O}Al(OH)_3$ [a]	1.670	1.795	133

[a] Taken from Ref. 6.
[b] Taken from Ref. 39.

the SiO and AlO bonds in both these models and in the completely optimized $H_3Si\overset{M}{O}AlH_3$ system differ by a maximum of 0.02 Å. The difference in the SiOAl angles are somewhat larger; e.g., the SiOAl angle in the completely optimized $H_3Si\overset{H}{O}AlH_3$ model is 130°, while the values for identical systems with limited geometry relaxation (with original SiOAl angles of 122° and 150°) equal 126° and 144°.

Therefore, nonempirical calculations have demonstrated that the optimal lengths of the SiO and AlO bonds in the ≡SiOAl≡ and $\equiv Si\overset{H}{O}Al\equiv$ fragments differ by at least 0.1 Å, similar to the optimal lengths of the TO bonds for the protonated and unprotonated ≡SiOT≡ fragment (cf. Table 5.2). Although structural deformation of these fragments can lead to changes in these geometric parameters of up to 0.02 Å, the differences in the individual types of TO bonds (e.g., in Ref. 40 values of 1.590 and 1.715 Å are recommended for the lengths of SiO and AlO bonds in unprotonated fragments and 1.68 and 1.84 Å in protonated fragments, respectively) are sufficiently large to permit utilization in interpretation of X-ray data for zeolites. The structural information obtained from X-ray data yielding average lengths of the T_iO_j bonds between Si and Al atoms in the T_i position and protonated and unprotonated oxygen atoms of the O_j type depends, of course, on the relative occupation of T_i positions by Si and Al atoms, as well as on the ratio of protonated to unprotonated oxygen atoms of the O_j type. The lengths of the T_iO_j bonds to T_i atoms, obtained from X-ray data, or their average qualitatively indicate the relative contents of Si and Al in the T_i positions. Similarly, the length of the T_iO_j bonds to O_j atoms depends on the ratio of SiO_j and AlO_j bonds as well as on the ratio of protonated to unprotonated O_j atoms. For example, in the H-forms of faujasites (one type of T and four types of O atoms), X-ray data have yielded[41] lengths of the TO_1, TO_2, TO_3, and TO_4 bonds of 1.65, 1.63, 1.66, and 1.63 Å, respectively, indicating that the protons will be localized especially on the O_3 and O_1 atoms.[42,43] This conclusion, which is also supported by the values found for the TOT angles, is in agreement with experimental findings.[44] X-ray data[45] for ZSM-5 zeolites (with 12 types of T positions) have indicated that the largest average length of the T_iO bonds is that in the T_{12} and T_2 positions, indicating that Al atoms will be particularly localized in these positions.

On the other hand, it is apparent that the experimentally determined geometry of the ≡SiOAl≡ fragments of faujasites for O_1 and O_3 atoms is closer to the optimal geometry of the protonated systems than that for the O_2 and O_4 atoms. Thus, quantum-chemical calculations for clusters with geometry derived from X-ray data predict that localization of protons on the O_3 and O_1 bridging atoms is energetically favorable.[46,47] The results of quantum-chemical calculations[48–52] on localization of Al in the ZSM-5 and ZSM-11 zeolites, mordenite, and ferrierite can be interpreted similarly. However, these calculations are not limited to determination of the probable localization of Al in the individual T positions, but also predict the energetically most probable pairing of Al atoms in the zeolite skeleton, which cannot be estimated directly from X-ray data. For example, STO-3G calculations[51] have indicated that, because of the topology of the ZSM-5 zeolite, random siting of Al in its T_2 and T_{12} sites leading to formation of Al pairs will favor *cis* pairing where the Al atoms are located in the same channel.

Ordering of Fragments in a Lattice. In addition to determining the geometric characteristics of the elementary structural units (fragments) of zeolites, quantum-chemical calculations of optimized cluster models can be used to predict the energetically most favorable linkage of these fragments in the crystals.

This approximation has been used to interpret the mutual arrangements of SiO_4 and AlO_4 tetrahedra in the zeolite skeleton. It followed from STO-3G calculations[53,54] on cluster models of the $T_4O_4H_8$ and $(HO)_3TOT(OH)_3$ types with various arrangements of Si and Al atoms that, for example, the reactions

$$2(HO)_3SiOAl(OH)_3^- \rightarrow (HO)_3SiOSi(OH)_3 + (HO)_3AlOAl(OH)_3^{2-}, \quad E = 487 \text{ kJ mol}^{-1}$$

$$2Si_3AlO_4H_8^- \rightarrow Si_4O_4H_8 + SiAlSiAlO_4H_8^{2-}, \quad E = 370 \text{ kJ mol}^{-1}$$

$$2Si_3AlO_4H_8^- \rightarrow Si_4O_4H_8 + SiSiAlAlO_4H_8^{2-}, \quad E = 540 \text{ kJ mol}^{-1}$$

are strongly endothermal, indicating that the formation of ≡SiOAl≡ fragments in the zeolite lattice is energetically more favorable than ≡AlOAl≡ linkages (Loewenstein rule).

Similarly, the MINDO/3 method has been used to study[55] the mutual combination of AlO_3 and SiO_4 elements in aluminosilicates for systems with the compositions $2SiO_2 \cdot Al_2O_3 \cdot 7H_2O$ and $2SiO_2 \cdot Al_2O_3 \cdot Na_2O \cdot 6H_2O$. The relative energies (in kJ mol^{-1}) of possible combinations of structural elements in the $2SiO_2 \cdot Al_2O_3 \cdot 7H_2O$ system

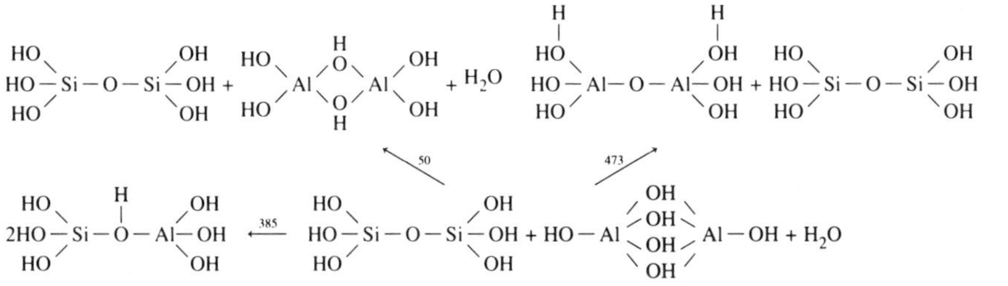

$(HO)_3Si{-}O{-}Si(OH)_3 + (HO)_2Al(\mu\text{-}OH)_2Al(OH)_2 + H_2O$ ⟵ 50 ⟶ $(HO)_3Si{-}O{-}Si(OH)_3 + HO{-}Al(\mu\text{-}OH)_4Al{-}OH + H_2O$ ⟶ 473 ⟶ $(HO)_3Al(OH)\!-\!O\!-\!Al(OH)_3(H) + (HO)_3Si{-}O{-}Si(OH)_3$

$(HO)_3Si{-}O{-}Si(OH)_3 + HO{-}Al(\mu\text{-}OH)_4Al{-}OH + H_2O$ ⟶ 385 ⟶ $2(HO)_3Si{-}O(H){-}Al(OH)_3$

indicated that the mutual coordination of AlO_3 elements leading to

$$>Al\langle {}^{O}_{O} \rangle Al< \qquad -Al\langle {}^{O}_{O}{}^{O}_{O} \rangle Al-$$

structures is energetically most favorable. It thus becomes clear why these types of combinations of AlO_3 elements are observed in solid systems with the composition of kaolinite and diktite. However, when the system contains Na ions, the combination of the AlO_3 and SiO_4 elements occurs differently, as can be seen from comparison of the energies for identical combination of elements in the system with the $2SiO_2 \cdot Al_2O_3 \cdot Na_2O \cdot 6H_2O$ composition:

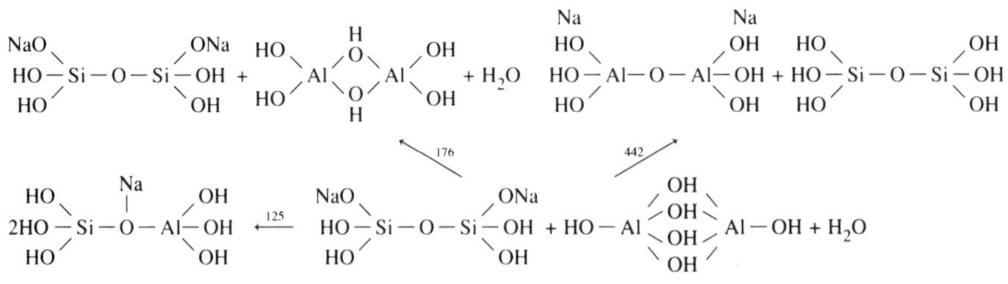

$(NaO)(HO)_2Si{-}O{-}Si(ONa)(OH)_2 + (HO)_2Al(\mu\text{-}OH)_2Al(OH)_2 + H_2O$ ⟵ 176 ⟶ $(NaO)(HO)_2Si{-}O{-}Si(ONa)(OH)_2 + HO{-}Al(\mu\text{-}OH)_4Al{-}OH + H_2O$ ⟶ 442 ⟶ $(HO)_3Al(Na){-}O{-}Al(OH)_3(Na) + (HO)_3Si{-}O{-}Si(OH)_3$

$(NaO)(HO)_2Si{-}O{-}Si(ONa)(OH)_2 + HO{-}Al(\mu\text{-}OH)_4Al{-}OH + H_2O$ ⟶ 125 ⟶ $2(HO)_3Si{-}O(Na){-}Al(OH)_3$

Then, the formation of $\equiv Si-\overset{Na}{O}-Al\equiv$ fragments, typical for zeolites, becomes energetically the most favorable; i.e., systems with the composition $nSiO_2 \cdot m(Al_2O_3 \cdot Me_2O) \cdot kH_2O$. The formation of $\equiv Al-O-Al\equiv$ fragments is energetically unfavorable, in agreement with the Loewenstein rule. These results also explain the higher stability of the zeolite cationic forms compared with the H-forms, which correspond to metastable states.

Electronic Structure

Nonempirical and semiempirical calculations of the electronic structures of cluster models of zeolites have demonstrated that the occupied and unoccupied molecular orbitals (MO) are separated by a gap with a width of 10–14 eV (see Fig. 5.5), depending on the method used.[56–59] The properties (energy and localization) of the frontier MO [highest occupied molecular orbital (HOMO) and lowest unoccupied molecular orbital (LUMO)], which can act as electron donors or acceptors, are most important for the interactions of zeolites with molecules. In zeolite skeletons without metal cations and protons, the HOMOs correspond to lone electron pairs on the skeletal oxygen atoms, while the LUMOs are localized mainly on the Si atoms.[60]

If the zeolite contains protons or cations of nontransition metals coordinated to the skeletal O atoms, the HOMOs once again correspond to the lone electron pairs of the O atoms, which do not participate in the coordination of metal cations or protons. The LUMO is then formed mainly by the AOs of the metal cation and, depending on the type of cation, has lower energy than the LUMO of the skeleton without the cation.[61–64]

If a transition metal cation is coordinated to the zeolite skeleton, then both frontier MOs are formed mostly by the AOs of the metal cation. Their energies depend on the type and oxidation state of the metal cation and often lie in the gap between the occupied and unoccupied bands (cf. Table 5.3 and Fig. 5.5).

Cationic Forms

As mentioned previously, the negative charge on the zeolite skeleton can be compensated by the metal cations, which are coordinated in the cationic positions by several skeletal O atoms. For the use of the zeolite as a catalyst, the presence of metal cations in the zeolite is important for two reasons: the metal cations affect the acidity and number of Brönsted sites and can also act as catalytically important sites capable of accepting or donating electron density. Their positive effect on the stability of the zeolite structure is also important.

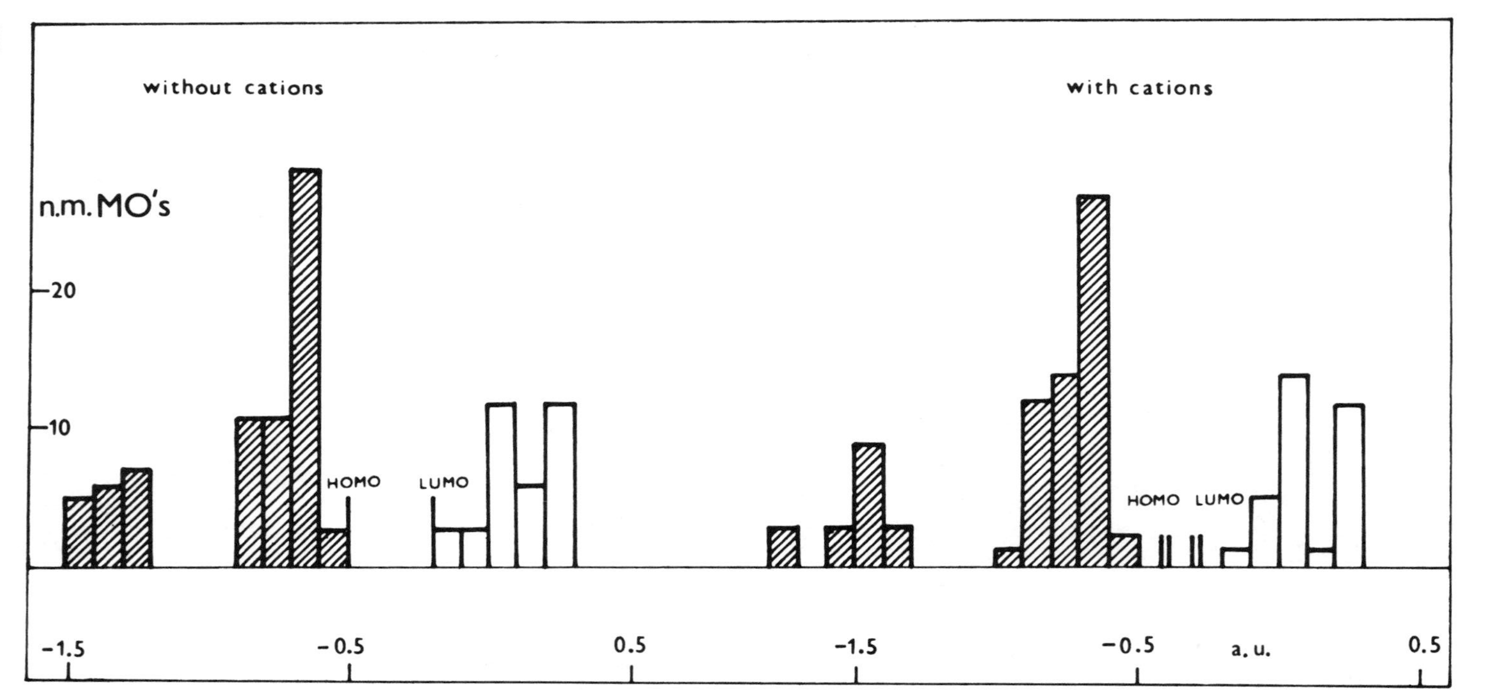

Figure 5.5. Representation of the band structure of the six-fold windows in zeolites without and with cations by a histogram of the number of molecular orbitals versus the energy at 0.1 a.u. intervals. The shaded areas represent occupied molecular orbitals.

Table 5.3. CNDO/2 charges on atoms, q, Wiberg bond orders, p, and energies of HOMO, E_{HOMO} (eV), or LUMO, E_{LUMO} (eV), calculated for clusters $T_6O_6(OH)_{12}Cat$ with various cations[a]

	Li^+	Na^+	K^+	Mg^{2+}	Ca^{2+}	Al^{3+}	Cu^+	Cu^{2+}	Cr^{2+}	Cr^{3+}	Fe^{2+}	Fe^{3+}	Co^{2+}	Co^{3+}	Ni^+	Ni^{2+}
Si: Al = 1																
q_{Si}	1.47	1.50	1.38	1.50	1.43	1.56	1.54	1.59	1.61	1.58	1.58	1.59	1.57	1.55		1.58
q_{Al}	1.39	1.37	1.36	1.38	1.37	1.34	1.35	1.38	1.38	1.36	1.38	1.38	1.37	1.34		1.38
q_{O2}	−0.65	−0.72	−0.71	−0.75	−0.73	−0.59	−0.63	−0.67	−0.67	−0.68	−0.67	−0.66	−0.67	−0.70		−0.65
q_{Cat}	0.03	0.23	0.56	0.98	1.30	0.84	0.09	0.43	1.00	1.20	0.39	1.28	0.46	1.27		0.31
p_{SiO2}	0.86	0.93	0.98	0.83	0.91	0.62	0.72	0.71	0.77	0.67	0.72	0.69	0.73	0.67		0.69
p_{AlO2}	0.53	0.56	0.60	0.52	0.56	0.42	0.47	0.46	0.47	0.44	0.47	0.48	0.48	0.45		0.46
$p_{Cat\text{-}O2}$	0.38	0.17	0.05	0.32	0.12	0.66	0.55	0.60	0.57	0.64	0.54	0.59	0.40	0.58		0.57
Si: Al = ∞																
q_{Si}	1.65	1.61	1.65	1.63	1.62	1.63	1.63	1.65			1.69		1.68		1.62	1.69
q_{O2}	−0.68	−0.75	−0.73	−0.77	−0.76	−0.67	−0.68	−0.71			−0.70		−0.70		−0.68	−0.68
q_{Cat}	0.05	0.29	0.62	1.07	1.36	1.06	0.30	0.49			0.49		0.57		0.34	0.41
p_{SiO2}	0.71	0.76	0.81	0.69	0.76	0.54	0.62	0.60			0.61		0.62		0.61	0.59
$p_{Cat\text{-}O2}$	0.36	0.15	0.04	0.29	0.10	0.61	0.52	0.57			0.51		0.47		0.51	0.55
$-E_{HOMO}$[b]	12.18	12.18	11.81			12.21	6.61	8.30	6.40	8.30	12.99	12.24	12.91	13.33	8.16	12.91
$-E_{LUMO}$[b]	1.71	1.86	1.59	3.09	2.30	5.57	2.17	2.81	1.50	1.50	1.60	3.00	1.90	5.50	1.75	1.93

[a] Taken from Refs. 62–69.
[b] For neutral models.

So far, only small cluster models of cations coordinated to the zeolite skeleton have been studied[39,53,70–72] nonempirically [$H_3Si\overset{Cat}{O}AlH_3$, $(HO)_3Si\overset{Cat}{O}Al(OH)_3$, or $T_4O_4H_8Cat$], mostly in an attempt to describe the effect of the cation on the geometry and arrangement of the fragments in the zeolite skeleton. Although the cation coordination is poorly described in these models, the results describing the effect of the cation on the characteristics of the zeolite skeleton are in agreement with the conclusions of semiempirical studies of better models of the cationic sites.[57,58,61–69] Important differences were found only for the charges on transition metal cations, for which the semiempirical methods yield lower values than nonempirical methods.[72]

The HOMOs of the zeolite skeleton are identical with the lone electron pairs on the skeletal O atoms. On the other hand, the valence AOs of the metal cations are completely or partly unoccupied. Consequently, if a metal cation is coordinated by the lone electron pairs of these O atoms, then electron donor–acceptor bonds are formed, leading to transfer of the electron density from the lone electron pairs of the O atoms to the cation. The magnitude of this transfer depends on the type of cation (on its electronegativity) and its valence. The strength of the bonds between the cation and the zeolite skeleton is then proportional to the magnitude of this transfer (the bond length is inversely proportional). On the other hand, the strength of the TO bonds of the O atoms coordinating the cation decreases (the bond length increases) with increasing strength of the donor–acceptor bond. Both STO-3G calculations[39,71] for simple models of the $H_3Si\overset{Cat}{O}AlH_3$ or $(HO)_3Si\overset{Cat}{O}Al(OH)_3$ type and CNDO/2 calculations[57,58,60–69] for large cyclic models of the $T_6O_6(OH)_{12}Cat$ type lead to the same results (cf. Table 5.3). The CNDO/2 calculations for the $T_6O_6(OH)_{12}Cat$ clusters with various Si : Al ratios have also demonstrated that the O atoms coordinating cations that are bonded to both Si and Al (and are more basic than O atoms bonded only to Si) donate more electron density to the cation and thus form a stronger bond. Consequently, the positive charge on the cation increases with increasing Si : Al ratio.[60–69]

As mentioned previously, the LUMOs and often the HOMOs of cluster models of cationic sites in zeolites are mostly localized on the cations and their energies depend particularly on the type of cation and its valence. Thus it is apparent that the cations act as electron donor–acceptor sites. The abilities of these sites to accept or donate electron density can be roughly estimated on the basis of the frontier orbitals. It follows from the energies of the LUMOs calculated by the CNDO/2 method for neutral $T_6O_6(OH)_{12}Cat$ clusters (see Table 5.3) that, for example, the cationic sites formed by cations of the alkali metals exhibit lower ability to accept electron density (i.e., have a lower Lewis acidity) than zeolites with cations of the alkaline earth metals (Mg and Ca) and especially Al cations.[58,62–64] These conclusions correspond to the experimental observations of the electron acceptor abilities of individual cations.[1,3,73] It then follows (see Table 5.3) from the HOMO and LUMO energies for zeolitic windows containing transition metal cations that, while the Cu^{+}, Cr^{2+}, Cr^{3+}, and Ni^{+} cations exhibit a tendency to donate electrons (i.e., be oxidized), Fe^{3+} and especially Co^{3+} cations will prefer to accept electrons. This tendency is so strong for Co^{3+} that it has not yet been possible to oxidize Co^{2+} cations coordinated in zeolites[74] and Fe^{3+} is very readily reduced.[75]

Cationic OH Groups

In addition to bridging and terminal OH groups, zeolites containing polyvalent cations coordinated in the cation positions also contain OH groups bonded to these cations. For catalysis, it is desirable to know the properties of these groups, both because of their possible role in catalytic reactions and also because these groups can affect the properties of the cations to which they are bonded. In addition, these groups, which constitute another ligand to the cation (in addition to its coordination to the skeleton), could produce steric hindrance in the interactions of the cation with the reacting molecules.

CNDO/2 calculations of cluster models of the $T_6O_6(OH)_{12}Cat{-}OH$ type have shown[62–64,66,69] that coordination of the OH group to the cation causes a decrease in its positive charge (and thus its Lewis acidity) and understandably leads to a weakening of the bonds between the cation and the zeolite skeleton. The stability of the OH groups or the tendency of the individual cations to form Cat–OH groups (characterized by the Cat–OH bond orders given in brackets) exhibit the following trend for the studied cations: $Al^{3+}(0.82) > Co^{2+}(0.63) \cong Fe^{2+}(0.62) = Ni^{2+}(0.62) > Mg^{2+}(0.54) > Ca^{2+}(0.16)$.[62–64,66,69] The calculations thus indicate that the Al cation has the greatest tendency to be hydroxylated, the affinity of the transition metals for the OH group is somewhat lower, and this group is bonded to the Ca cation very weakly. This conclusion is in agreement with the experimental results.[73]

The acidity of the cationic OH group, characterized by the charge on its H atom (given in brackets), is demonstrably weaker than the acidity of the skeletal OH groups and exhibits the following order for the individual cations: skeletal OH group (0.16) > Al–OH (0.11) > Ni–OH (0.03) > Fe–OH (0.02) = Co–OH (0.02) > Mg–OH (−0.05) > Ca–OH (−0.16).[62,64,66,69]

It is interesting to note that the experimental concept of the basic character of the OH group on Mg and Ca is thus confirmed.[1,3,73]

Bridging Hydroxyl Groups

The bridging (skeletal) OH groups of zeolites represent important proton-donor sites, responsible for the strong Brönsted acidity of zeolitic catalysts.[1,3,76] As these sites play an important role in most catalytic processes occurring on zeolites, it is important to determine the manner in which the individual factors affect the properties of these sites. The bridging OH groups in zeolites can be affected by the structural characteristics of the zeolite (length of the TO bonds, TOT angles), their chemical composition (Si : Al ratio and presence of various metal cations), and also various defects in the crystal structure of the zeolite resulting, for example, from dehydroxylation or some other thermochemical treatment. We will now consider the type of information on the effect of these factors that can be obtained using quantum-chemical calculations on model systems.

Acidity. Because of the difficulties and ambiguities connected with the experimental determination of the acidity of Brönsted sites in zeolites, attempts were made to use quantum-chemical calculations to predict the position of bridging OH groups on the absolute acidity scale. It followed from comparison of the calculated heats of deprotonation ΔH_0° for bridging OH groups (obtained by 3-21G calculation of the dissociation energy of the OH group in the $H_3Si\overset{H}{O}AlH_3$ model) and various acidic molecules that the acidity of these bridging groups is greater than that of acetic acid.[40] STO-3G calculations of the dissociation energy, charge on the H atoms, and ionicity of the OH bonds for bridging OH groups [modeled by the $(HO)_3Si\overset{H}{O}Al(OH)_3$ cluster] and sulfuric acid indicate that the acidity of these two systems is comparable.[77] However, it should be noted that these conclusions should be accepted with caution because of the relatively small model used in the first case and the quality of the method (minimal base) in the second case. Nonetheless, it appears clear that bridging OH groups exhibit very high acidity.

The Effect of Structural Characteristics. The CNDO/2 calculations[46,47] of the charge distribution, OH bond dissociation energy, and stretching vibrational frequencies for $(HO)_3Si\underset{H}{O}Al(OH)_3$ and $Si_3AlO_4(OH)_8H$ cluster models of various OH groups in faujasites with different geometric characteristics have already demonstrated that the properties of bridging OH groups in zeolites depend on the geometry of their environment. It followed from later, more detailed CNDO/2 and STO-3G studies[37,79] of $(HO)_3Si\overset{H}{O}Al(OH)_3$ cluster models with various TO bond lengths and SiOAl angles that the dissociation energy of the bridging OH group, attaining the largest value roughly for the equilibrium SiOAl angle, decreases with increasing SiOAl angle (see Fig. 5.6). The calculations have also shown that shortening of the lengths of the SiO and AlO bonds in the $\equiv Si\overset{H}{O}Al\equiv$ fragment leads to a decrease in the dissociation

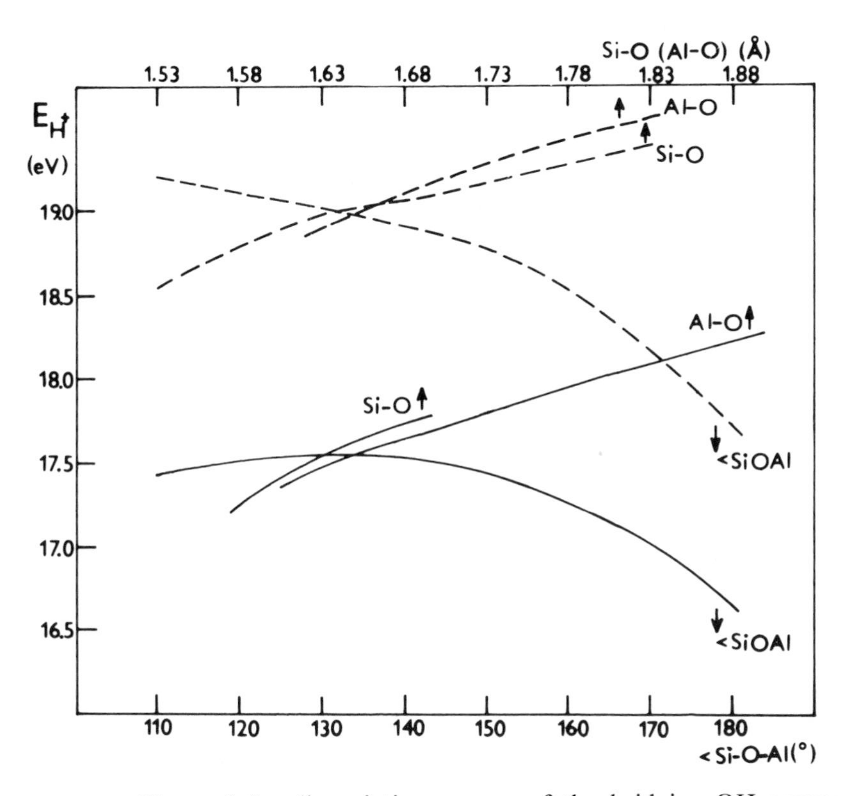

Figure 5.6. Plots of the dissociation energy of the bridging OH groups against the SiOAl angle or the lengths of SiO and AlO bonds calculated by *ab initio* (solid line) and CNDO/2 (dashed line) methods. [From I.N. Senchenya et al., Quantum chemical study of the effect of the structural characteristics of zeolites on the properties of their OH groups. 2, *J. Phys. Chem.* **90:** 4857–4859 (1989). Courtesy of Amer. Chem. Soc., Washington, DC.]

energy. However, the acidity of structurally different types of bridging OH groups will apparently not be as strongly affected by the structural characteristics. As mentioned previously, model STO-3G calculations[39] indicate that the length of the TO bonds and values of the SiOAl angles of structurally different types of OH groups should not differ from the equilibrium characteristics by more than ±0.02 Å and ±15°. Such changes in the geometry of the $\equiv Si\overset{H}{O}Al\equiv$ fragments then lead to a decrease in the dissociation energy of the OH bonds (and thus their acidity) by a maximum of 40 kJ mol.$^{-1}$

Until recently, the different values of the vibrational frequencies of bridging OH groups in zeolites were attributed to the different chemical compositions of the zeolites or differences in the electrostatic field of the zeolite at various sites in its lattice.[80] Simultaneously, it was assumed that the effect of structural factors on the frequency values is negligible.[40,81] As the IR data represent one of the most important characteristics of the Brönsted sites in zeolites, the effect of the structural characteristics of the bridging OH groups of zeolites on their vibrational characteristics was studied in detail using STO-3G calculations of $H_3Si\overset{H}{O}AlH_3$ clusters.[38] It followed from these calculations that a variation in the optimized structural characteristics of the TO bonds and SiOAl angles by ±0.02 Å and 10° (i.e., by values corresponding to differences between the structurally different OH groups) leads to changes in the stretching and bending vibrations of the bridging OH groups by 40 and 80 cm^{-1}, respectively, i.e., differences of the same order as that observed for structurally different types of these groups in zeolites.

The Effect of the Si : Al Ratio. The simplest model that can qualitatively describe the effect of the chemical composition around the bridging OH group on its properties is the $(XO)_3Si\overset{H}{O}Al(OX)_3$ cluster, where X is pseudoatom whose electron donor-acceptor properties model either Si (X_{Si}) or Al (X_{Al}). As the Loewenstein rule states that only pseudoatoms representing Si can be on the right-hand side of the model, this cluster can be used to model for chemically different environments around the OH group—$(X_{Al}O)_n(X_{Si}O)_{3-n}Si\overset{H}{O}Al(OX_{Si})_3$. CNDO/BW calculations[82] for this cluster have demonstrated that, for $n = 0, 1, 2$, and 3 (i.e., for Si : Al ratios of 7, 3, 1.7, and 1) the dissociation energy of the OH bond (and charge on H) attains values of 1443, 1467, 1505, and 1542 kJ mol^{-1} (0.404, 0.392, 0.384, and 0.376). Substitution of Al for Si (i.e., a decrease in the Si : Al ratio) thus results in a decrease in the acidity of the bridging OH groups. CNDO/2 calculations[47] for cyclic $Si_3AlO_4(OH)_8H$ and $SiAlSiAlO_4(OH)_8H_2$

clusters modeling the zeolite window with a Si : Al ratio of 3 or 1 lead to the same conclusions. Nonempirical calculations with a 3-21G base were used to estimate the effect of the chemical environment of the bridging OH groups on their properties using the very simple model system $H_3Si\overset{H}{O}AlH_3$.[83] Substitution of Si for Al was modeled by substitution of F for the terminal H atoms; i.e., the F : H ratio modeled the Si : Al ratio. These calculations confirmed that the acidity of the bridging OH groups, represented by the charge on H, increases with increasing F : H ratio, while their stretching vibrational frequency decreases. For example, symmetrical substitution of two H atoms for F leads to a decrease in the frequency from 3920 to 3909 cm^{-1}. However, it should be emphasized that for two reasons this model strongly overevaluates the effect of changes in the chemical environment of the OH group on its properties. In the zeolite, substitution occurs in the third coordination sphere of the OH group, while it occurs in the second sphere in the model. In addition, the difference in the electronegativities of H and F is greater than that between Si and Al.

STO-3G calculations[38] for the $H_2AlOSi(OH)_3$ and $H_3SiOSi(OH)_3$ systems modeling the effect of substitution of Si for Al in the third coordination sphere of the OH group indicate that the stretching vibrational frequency of the OH groups of these systems (with values of 4165 and 4164 cm^{-1}) practically does not differ. Calculations have thus demonstrated that the effect of chemical factors on the vibrational frequencies of the OH groups in zeolites is negligibly small.

The Effect of Cations. Another factor that can affect the properties of bridging OH groups of partly decationized forms of zeolites is the presence of various metal cations that, together with the protons, compensate the negative charge on the zeolite skeleton. The effect of the presence of various cations on the properties of bridging OH groups has been studied using CNDO/2 calculations[84] on the $Si_3Al_3O_7(OH)_{10}H_2M$ cluster (see Fig. 5.7) modeling two fourfold faujasite windows with two O_1H groups. M is either a proton forming an O_3H group or the Li, Na, or K cation localized in the S_I' cation position and coordinated to the O_3 atom. The calculated dissociation energy of the bridging OH groups, with values of 1832, 1859, 1880, and 1889 kJ mol^{-1} for the H, Li, Na, and K ions, indicates that the acidity of these groups decreases with decreasing electronegativity of the ion: H > Li > Na > K. Thus, it follows from the calculation that pure H-forms of the zeolite are more acidic than partly cationic forms and that the acidity of the OH groups in zeolites increases with the degree of decationization.

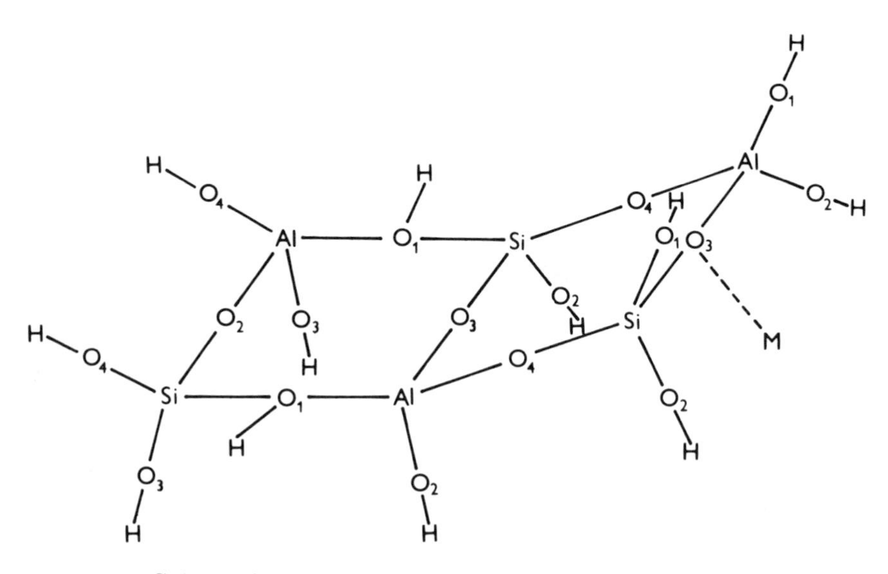

Figure 5.7. Schematic depiction of the model cluster used to estimate the effect of cations on the acid strengths of the bridging OH groups. M stands either for a proton forming an O_3H group or for Li, Na, and K cations located in the S_I cationic position.

The Effect of Dehydroxylation. The structure of dehydroxylated zeolites has not yet been completely clarified, complicating the modeling of the effect of dehydroxylation on the properties of OH groups in the zeolite. So far, it is clear only that the first step in the dehydroxylation leads to the formation of tricoordinated Si and Al. It is not, however, clear whether this structure is finite or is subject to further reorganization leading to the formation of extra-lattice aluminum.[1,3]

It follows from CNDO/2 calculations[85] on $Si_4Al_2O_6(OH)_{12}H_2$ and $Si_4Al_2O_6(OH)_{11}H$ clusters modeling nondehydroxylated and partly dehydroxylated (containing tricoordinated Si or Al) forms of zeolites that most of the bridging O atoms in models with tricoordinated Si or Al exhibit greater affinity for protons (i.e., higher dissociation energy of the OH bonds) than identical O atoms in models of nondehydroxylated zeolites. Thus, the acidity of the bridging OH groups in partly dehydroxylated forms of zeolites should be lower than that in the original H-forms.

Terminal OH Groups

Terminal OH groups are among the quantum chemically most intensively studied aluminosilicate systems, especially because they represent important Brönsted sites in amorphous silicates and aluminosilicates. Various models have been used[5,10,11,86–89] to describe their properties [from the simplest type, H_3SiOH or $(OH)_3SiOH$, to very large clusters with several TO_4 tetrahedra] using quantum-chemical methods from EHT to nonempirical with large bases. From the point of view of zeolites, it is important that these quantum-chemical calculations confirm the experimental assumption that the acidity of the terminal ≡SiOH groups is much lower than that of the bridging OH groups, reflecting their small importance for the catalytic activity of zeolites. Nonempirical 3-21G and 4-31G calculations[40] of the deprotonation enthalpy ΔH_0° of the terminal SiOH groups [in the $(OH)_3SiOH$ model] indicate that the acidity of these groups is comparable with that of phenol. It has also been demonstrated[86] that the acidity of the ≡SiOH groups, in contrast to their vibrational frequency, depends on the chemical composition of their environment. For example, a change in the electronegativity of pseudoatom X in the third coordination sphere of the ≡SiOH group [the $(XO)_3SiOH$ cluster and the CNDO/BW method] modeled by a change in its VOIP from 10 to 16 eV leads to a decrease in the dissociation energy of the OH bond (increase in the charge on H) from 1641 to 1428 kJ mol^{-1} (0.313 to 0.356), while the vibrational frequency remains unchanged.[86] Consequently, ≡SiOH groups on the surface of aluminosilicates with various chemical compositions and thus also various acidities exhibit identical vibrational frequencies.

Lewis Sites

Lewis sites in zeolites are understood to be any site containing an unoccupied orbital exhibiting the ability to accept electron density during interactions with molecules. It follows from experimental studies that, in addition to the cationic sites in zeolites, undefined sites formed in the dehydroxylation of H-forms of zeolites exhibit these properties. However, the structure of these sites has not yet been determined, leading to considerable problems in attempts to describe them quantum chemically. It is assumed[1,3] that the first step in the dehydroxylation leads to the formation of tricoordinated Si and Al according to the scheme:

$$2\ \equiv\overset{\text{H}}{\text{SiOAl}}\equiv\ \rightarrow\ \equiv\text{SiOAl}\equiv\ +\ \equiv\text{Si}\ +\ \text{Al}\equiv\ +\ H_2O$$

The electron acceptor abilities of various possible structures formed by dehydroxylation have been studied by CNDO/2, CNDO/BW, and nonempirical STO-3G calculations.[90–92] It follows from CNDO/2 calculations[91,92] on $Si_2Al_2O_3(OH)_8$, $Si_2Al_2O_4(OH)_7$, and $Si_5AlO_6(OH)_{11}$ clusters (modeling fourfold zeolite windows containing tricoordinated Si and Al simultaneously, fourfold windows with tricoordinated Si or Al, and sixfold window with either tricoordinated Si or Al) that, in systems with tricoordinated Si, the LUMO is mainly localized on this Si atom and has an energy of −4.5 to −2 eV. Therefore, systems with tricoordinated Si exhibit quite strong electron acceptor ability (cf. LUMOs in Table 5.3). On the other hand, the electron acceptor abilities of systems with tricoordinated Al, whose LUMO has higher energy than −1 eV, are relatively small. CNDO/BW and STO-3G calculations on the $(HO)_3SiOAl(OH)_2$ and $(HO)_2SiOAl(OH)_3$ models lead to identical conclusions.[90] As the electron acceptor ability of Al coordinated to three skeletal O atoms in the cationic positions is much larger (cf. Table 5.3) than that of tricoordinated Al in the zeolite skeleton, it is obvious that the ability of these sites to accept electron density depends both on the geometric arrangement and on the nature of the O atoms to which the Al atom is bonded, i.e., on the chemical environment.

The effect of the chemical environment on the electron acceptor ability of tricoordinated Al has been studied using the CNDO/BW method with the $Al(OX)_3$ cluster model.[93] Calculations have indicated that a variation in VOIP of the pseudoatom X from 7 to 16 eV, modeling a chemical change in the environment, causes a decrease in the energy of the LUMO localized on Al by 3.5 eV; i.e., a marked change in its electron acceptor ability.

Stability. In addition to suitable physical properties of its active sites, structural stability of the zeolite is essential for its use as a catalyst or sorbent. This fact has led to intensive experimental and theoretical study of zeolites to find and understand the relationships among their thermal and chemical stabilities and the chemical composition and structure.[1,3,94]

Quantum-chemical study of the structural stability of such systems and factors affecting this stability can be carried out in two ways. The first is based on calculations of changes in the overall bonding energy of the system with changes in its composition or structure. The second method involves evaluation of the local stability, especially that of the weakest points in the structure.[95]

CNDO/2 calculations[60] of isomorphous substitution of Al for Si in cluster models of the $T_6O_6(OH)_{12}$ type with various Si : Al ratios are examples of the first approach. It follows from these calculations that such a substitution always results in a decrease in the overall bonding energy of the cluster and thus in destabilization of the system. For example, substitution of Al for Si in the cyclic $Si_6O_6(OH)_{12}$ cluster leading to formation of the $Si_5AlO_6(OH)_{12}$ and $Si_3Al_3O_6(OH)_{12}^{3-}$ clusters is connected with a decrease in the overall bonding energy by 1403 and 5091 kJ mol^{-1}, respectively. On the other hand, subsequent compensation of the negative charge of the zeolite skeleton by coordination of a cation leads to an increase in the bonding energy of the cluster, i.e., stabilization. The degree of stabilization depends on the cation type. For example, coordination of the Na cation (to three O atoms) in the cation position of the $Si_5AlO_6(OH)_{12}^-$ cluster leads to its stabilization (an increase in its bonding energy) by 493 kJ mol^{-1}, while coordination of an Al^{3+} cation in the cation position of the $Si_3Al_3O_6(OH)_{12}^{3-}$ cluster results in stabilization by 6109 kJ mol^{-1}. However, these energy characteristics yield information only on the overall stability of the studied system, but provide no information on the stabilities of the individual bonds.

The strengths of the individual bonds in the zeolite skeleton and changes in these values can be found from the orders of the SiO and AlO bonds calculated for zeolite models with protons or cations. These calculations have shown that the SiO bonds are stronger than the AlO bonds (see Table 5.3), in agreement with the experimentally observed decrease in the stability of the zeolite skeleton with decreasing Si : Al ratio. The coordination of cations in the cation positions leads to a decrease in the strength of the skeletal TO bonds to the O atoms coordinating the cation, dependent on the type and valence of the cation (see Table 5.3). This weakening is relatively small for alkali metal cations and alkaline earth metal cations, while it can be large for polyvalent transition metal cations. The marked decrease in the strength of the TO bonds (the bond order decreases to almost one-half the original value) is connected with coordination of protons leading to the formation of bridging OH groups, explaining the lower structural stability of the H-forms of zeolites compared with the cation forms.[60]

The lower stability of the AlO bonds in the H-forms of zeolites compared to the cation forms can also be understood on the basis of comparison of model STO-3G calculations[55,95,96] of coordination of AlO_3 and BO_3 crystal fragments [modeled by $Al(OH)_3$ and $B(OH)_3$ molecules] with variously basic O atoms modeled by OH_2, ONaH, and OH^-. The transition of the AlO_3 and BO_3 systems from an energetically more favorable planar arrangement to a pyramidal state is connected with an energy gain of 123 and 236 kJ mol^{-1}, respectively. This gain can be compensated by interaction of $Al(OH)_3$ [$B(OH)_3$] with variously basic oxygens, OH_2, ONaH, and OH^-, leading to an energy loss of 185, 505, and 711 kJ mol^{-1} (94, 412, and 532 kJ mol^{-1}, respectively). It thus follows that the stability of tetrahedrally coordinated Al and B increases with increasing basicity of the O atoms that coordinate them, i.e., with decreasing electronegativity of the atom to which this O is coordinated, as is confirmed by the AlO bond orders calculated for zeolite models with, e.g., Li, Na, and K cations (cf. Table 5.3). On the other hand, these calculations indicate that, in the H-forms of boralites, B will be tricoordinated and the skeletal OH groups will have rather the character of terminal $\equiv$SiOH groups.

INTERACTION OF ZEOLITE SITES WITH MOLECULES

So far, most quantum-chemical studies of the interactions of molecules with zeolitic sites have been carried out to determine the structure of interaction complexes and interaction energies connected with the formation of these complexes, to evaluate the energetic suitability of the interaction of a given molecule with various types of zeolite sites or a given site with various molecules, and especially to determine the manner and degree to which the zeolite active site affects the properties of the interacting molecules. However, compared with the procedures used to study the interactions of molecules in the gas phase, the procedures for calculation of these characteristics are less adequate. For example, only the relative position of the molecules and zeolite sites is usually optimized, while the geometry of the two interacting systems is maintained rigid, with only a few exceptions. This approach (while acceptable for weak interactions of the molecules with zeolites) is a rather rough approximation for strong molecular interactions (e.g., with Lewis sites). For this reason and also because of the approximate nature of the quantum-chemical methods used, neglecting of the effect of the correlation energy and entropy and, on the other hand, the imperfections in the models of the zeolite sites, the energy characteristics obtained are usually only qualitative in character. Similarly, the description of the properties of the interacting molecules, usually only on the basis of changes in their electron distribution (i.e., from transfer of electron density between the molecule and the zeolite site, change in charge distribution or in the bond orders in the molecule forming the interaction complex, etc.) is also only qualitative. Nonetheless, in spite of this simplification, necessary because of the size of the systems studied, the quantum-chemical study of the interactions of molecules with zeolite sites has yielded a great deal of valuable information on the properties of the interacting molecules and contributed to understanding of the nature of the individual zeolite sites and their effects on the interacting molecules.

Water

Zeolites contain a number of sites with which water can form interaction complexes. The first group of sites, found both on zeolites and on other silicates or aluminosilicates, consists of bridging O atoms of the ≡SiOSi≡ type and terminal ≡SiOH groups. The water molecule can interact with these OH groups either as a proton donor (the H atom of water interacts with the O atom of the OH group) or a proton acceptor (H atom of the OH group with the O atom of water). The second group of sites that are typical for both zeolites and aluminosilicates consists of bridging O atoms of the ≡SiOAl≡ type, bridging OH groups, cationic and Lewis sites.

Interaction of water with the terminal ≡SiOH groups or bridging ≡SiOSi≡ sites has been studied by semiempirical[5,13,97,98] and nonempirical[9,70,99,100] methods (with various bases) using the cluster models of the $(HO)_3SiOH$ and $H_3SiOSiH_3$ types. All these calculations lead to qualitative agreement of the results from the point of view of the energetic probability of formation of the individual types of complexes. Interaction of the water molecule with the ≡SiOSi≡ site has been found to be least favorable (the 4-31G interaction energy equals -22.5 kJ mol^{-1}), while the interaction of water (as a proton acceptor) with the ≡SiOH groups is energetically most suitable (4-31G interaction energy of -38.2 kJ mol^{-1}).[9,100] However, because of the fixation of water on the zeolite surface, the formation of these complexes is connected with an entropy loss greater than the gain in enthalpy.[9] The positive value of ΔG of these processes, indicating their improbability, then explains the relative hydrophobicity of silicates or zeolites with high Si : Al ratio.[9,100]

Quantum-chemical calculations[9,70,97–102] of the interaction of water molecules with sites typical for zeolites have shown that these interactions are connected with much greater energy gain, indicating the formation of these complexes in spite of the expected entropy losses. STO-3G calculations[70] of the interaction of water (as a proton donor) with basic ≡SiOAl≡ sites modeled by the $H_3SiOAl_3^-$ cluster predict an interaction energy of -68.6 kJ mol^{-1} and values of -112, -73.3, and -259 kJ mol^{-1} for the formation of the interaction complexes of water with Li, Na, and Mg cationic sites (modeled by the $H_3Si\overset{\text{Cat}}{O}AlH_3$ clusters), respectively. Similarly, 3-21G calculations have indicated that the interaction of water (as a proton acceptor) with the $H_3Si\overset{\text{H}}{O}AlH_3$ model of the bridging OH group is connected with the energy gain of 91.2 kJ mol^{-1} (cf. Table 5.4).[101]

Table 5.4. Interaction energies, E (kJ mol^{-1}), optimized intermolecular distances, R (Å), and transfer of electron densities, Δq, for complexes of CO, H_2O, and NH_3 with models of terminal or bridging OH groups and Lewis sites.

	H_3SiOH[a]			$H_3Si\overset{\text{H}}{O}AlH_3$[a]			$Al(OH)_3$[b]		
Molecule	E	R	Δq	E	R	Δq	E	R	Δq
CO	15.5	2.208	0.0197	27.1	2.040	0.0334	165.3	2.080	0.357
H_2O	55.7	1.744	0.0728	91.2	1.744	0.0968	256.6	1.840	0.212
NH_3	56.2	1.853	0.0734	101.2	1.691	0.1145	321.0	1.940	0.231

[a] 3-21G, taken from Ref. 101.
[b] STO-3G, taken from Refs. 93 and 103.

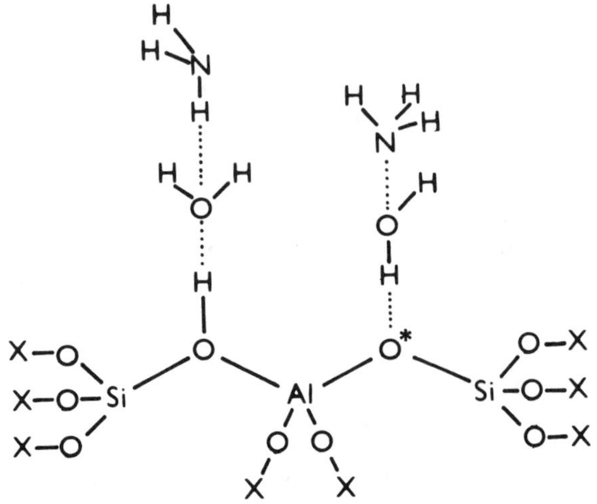

Figure 5.8. Schematic depiction of interactions of water and ammonia with bridging OH group or with the skeletal O atom and the cluster model.

The dependence of the interaction energy for the water molecules with bridging OH groups or centers of the ≡SiOAl≡ type on their chemical environment can be illustrated by the calculations of the energy for the interaction of water with these sites modeled by the $(XO)_3Si\overset{H}{O}Al(OX)_2O^xSi(OX)_3$ cluster (see Fig. 5.8), where X are pseudoatoms of the X_{Si} or X_{Al} type. CNDO/BW calculations[97] have indicated that the energy of the interaction of water molecules (as proton acceptors) with bridging OH groups decreases with increasing number of X_{Al} pseudoatoms in the left-hand part of the cluster (i.e., with decreasing acidity of the OH group), while the energy of interaction of water (as a proton donor) with the skeletal O^x atom increases with increasing number of X_{Al} in the right-hand part of the cluster.

Model 3-21G calculations[83] of interactions of a water molecule with the bridging OH group represented by the $H_3Si\overset{H}{O}AlH_3$ (for which the change in the chemical environment of the OH group is modeled by substitution of F for H) lead to the same conclusions. In addition, these calculations demonstrate that, as the acidity of the OH group increases (with increasing number of H substituted by F), the stronger interaction with the water molecule results in a lengthening and weakening of the OH bond, a decrease in the stretching vibrational frequency of the OH group, and an increase in the charge transfer from the molecule to the zeolite model.

From the point of view of the activation of water molecules interacting with zeolite active sites, it is important to know the manner and degree of the perturbation of this molecule by the individual sites. Consequently, the CNDO/2 method was used to study[104] the interaction complexes of water with Li, Na, Mg, and Ca cationic sites in zeolites (see Fig. 5.9). It has been shown that the electron acceptor ability of these cationic sites leads to transfer of electron density from the water molecule to the cation and the molecule is polarized by the cation charge. The electron density transfer is largest for the Mg cationic site (corresponding to its greatest electron acceptor ability characterized by the LUMO energy, cf. Table 5.3) and decreases through Li to the Ca and Na sites. The polarization of the interacting water molecules by the electrostatic field of the cation sites then exhibits the following trend: Mg > Ca > Na > Li.[104,105]

Because of these two effects, the acidity of the H atoms of the interacting water (characterized by the charge on H and given in brackets) and thus also the dissociability of the OH bonds in the water molecules has the following trend for the individual cation sites: isolated water (0.144) < Li–OH_2 (0.175) < Na–OH_2 (0.178) < Ca–OH_2 (0.181) < Mg–OH_2 (0.210).[104] This order explains the experimentally observed ability of poly-

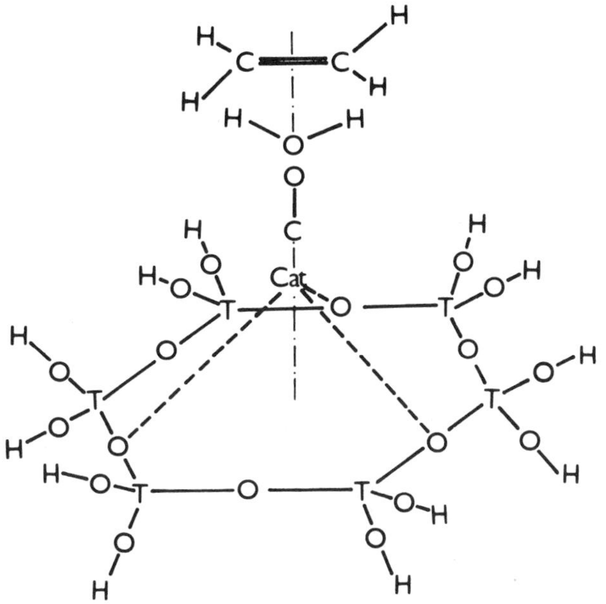

Figure 5.9. Schematic depiction of interactions of CO, water, and ethylene with the model of cationic sites in zeolites.

valent cations (Mg and Ca) to split a water molecule to form cationic and bridging OH groups. The change in the stretching vibrational frequency of the water molecule forming the interaction complex with the cationic sites can be estimated from the values of its OH bond orders calculated for the individual complexes. These bond orders then indicate that the stretching vibrational frequency should decrease in the order: isolated water > Li–OH_2 > Na–OH_2 > Ca–OH_2 > Mg–OH_2. The correlation between the calculated bond orders and the observed asymmetrical stretching vibrational frequencies of water interacting with these cation sites is apparent from Figure 5.10.

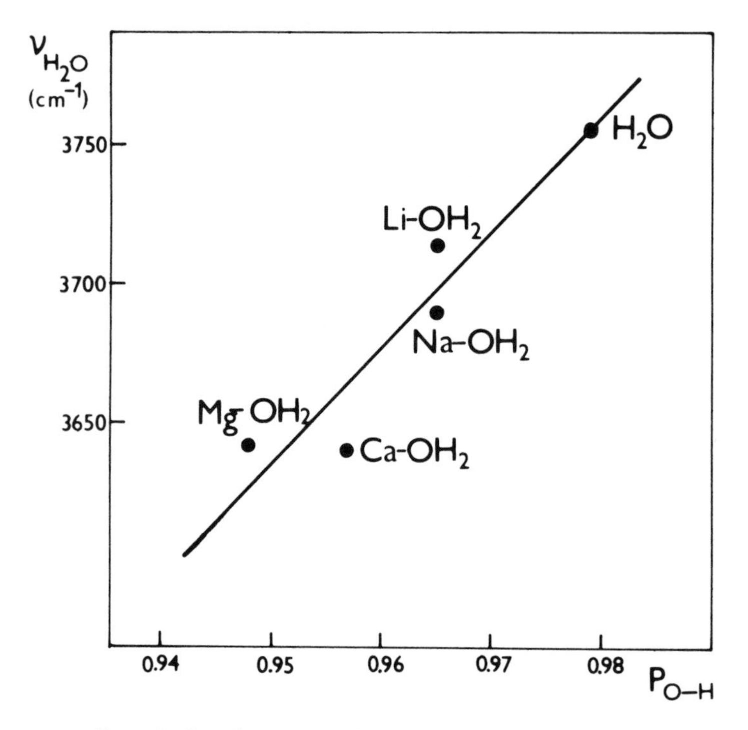

Figure 5.10. Correlation between the asymmetric stretching vibrational frequency, ν_{H_2O}, of water interacting with the cationic centers in zeolites and the Wiberg bond orders calculated for corresponding interaction complexes of water p_{OH}. [From Beran, S., Quantum chemical study of the characteristics of molecules interacting with zeolites, in *Structure and Reactivity of Modified Zeolites* (P.A. Jacobs et al., eds.). Amsterdam: Elsevier, 1984, pp. 99–106.]

Carbon Monoxide

The properties and number of catalytically interesting active sites in the zeolite can be estimated on the basis of adsorption of small molecules in the zeolite through study of the properties of the interaction complexes formed using various experimental techniques (especially spectroscopic) to determine the manner and degree of perturbation of the interacting molecules by the zeolite active sites. The CO molecule is often used in these types of tests. IR measurements of the frequencies of CO interacting with sites in various forms of zeolites has revealed the presence of at least two types of interaction complexes.[73,106,107] The vibration bands of interacting CO at lower frequencies than those for the isolated CO molecule are designated as nonspecific interactions, while vibrational bands at higher frequencies are assigned to the interactions of CO with the cation sites or bridging OH groups. One of the means of obtaining information on the type and nature of interaction complexes of CO involves the use of quantum-chemical calculations of models of these complexes and comparison of the results of these calculations with experimental data.

Consequently, the CNDO/2 method has been used to study[104,108] the interactions of CO with model Brönsted sites represented by the $Si_3AlO_4(OH)_8H$ cyclic cluster (interactions of the OH–CO and OH–OC type, see Fig. 5.11), with model Na, Ca, Mg, and Al cation sites, represented by the $T_6O_6(OH)_{12}Cat$ cyclic cluster (see Fig. 5.9) or a model of nonspecific interaction of CO with skeletal O atoms. Calculations have demonstrated that the interaction of CO with these sites is connected with transfer of electron density from the lone electron pair of the CO molecule to the zeolite active site, with a magnitude roughly corresponding to the electron acceptor ability of the site characterized by its LUMO energy. Similar transfer of electron density from CO to the zeolite is predicted by nonempirical calculations[101] of the interaction complexes of CO with the model of bridging OH groups and Lewis sites (cf. Table 5.4). Thus, transfer of electron density is the first way in which the zeolite active site affects the interacting molecules. As the removal of an electron from the CO molecule, leading to the formation of CO^+ results[107] in an increase in its vibrational frequency from 2143 to 2184 cm^{-1}, obviously transfer of part of the electron density from CO causes an increase in its vibrational frequency. On the other hand, this effect cannot explain the observed change (up to 100 cm^{-1}) in the frequency of the interacting CO.

However, calculations have also revealed that the interacting CO molecule is polarized by the electrostatic field of the zeolite site. Therefore, a second way in which the zeolite site acts on the interacting molecule is through its electrostatic field. Accordingly, a modified version of the CNDO/2 method has been used to study[105,108] the effect of the electro-

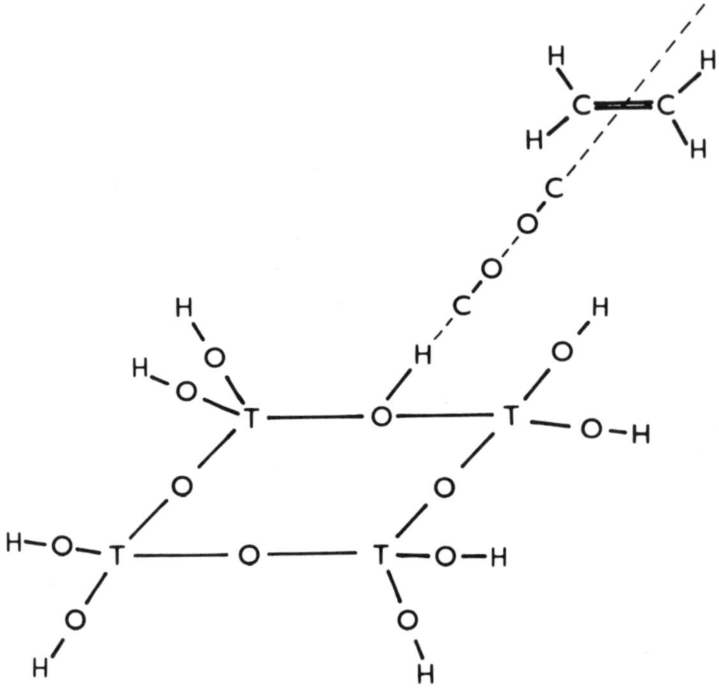

Figure 5.11. Schematic depiction of the types of CO and ethylene interaction complexes with models of the bridging OH groups in zeolites.

static fields of the individual sites (the site was represented by the corresponding point charge) on the properties of the CO molecule. These calculations have indicated that polarization of the CO molecule leading to the C^{-}–O^{+} dipole moment results in strengthening of the CO bond, while the opposite polarization causes bond weakening. Simultaneously, it has been found[104,108] that a great majority of the perturbation of CO molecules interacting with any site results from polarization. Thus, it follows that polarization of the molecule is the decisive factor for the way in which the site affects the strength of the CO bond. Model STO-3G calculations[101] of the CO force constants for the Mg^{2+}–CO and Mg^{2+}–OC complexes yielding values of 2657 and 1949 nm^{-1}, respectively, compared with 2458 nm^{-1} for the isolated molecule, have led to the same conclusions.

The following qualitative rule can thus be drawn for the change in the strength of the bond in the interacting CO molecule and thus for its vibrational frequency: interaction of CO through the C atom with a positively charged site or through the O atom with a negatively charged site results in strengthening of the CO bond and a shift in the vibrational frequency to higher values; the opposite types of interactions lead to the opposite effect.

The quantitative change in the strength of the CO bond (vibrational frequency) of CO interacting with individual types of zeolite sites can be estimated from the values of the CO bond orders. It follows from these bond orders that the vibrational frequencies of CO molecules interacting with the individual types of sites should exhibit the following trend: Al–CO > Mg–CO > Ca–CO > Na–CO ≅ OH–CO > isol. CO > OH–OC. Comparison of the bond orders for CO calculated[104,108] for the individual complexes with the vibrational frequency observed for CO interacting with zeolites with various sites reveals quite good correlation (see Fig. 5.12).

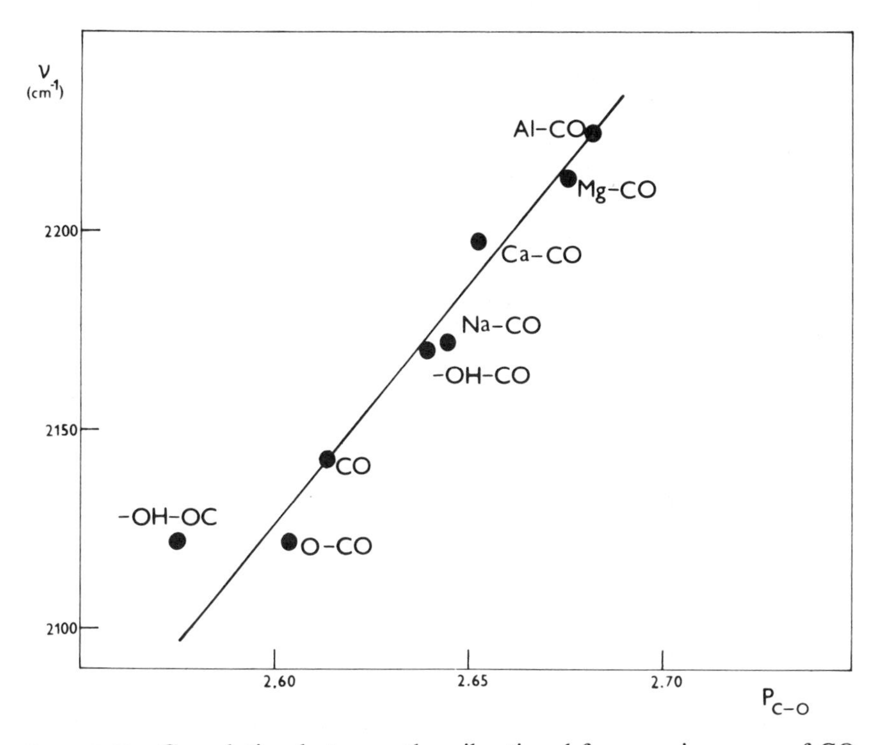

Figure 5.12. Correlation between the vibrational frequencies, ν_{CO}, of CO interacting with the cationic sites and OH groups in zeolites and the Wiberg bond orders calculated for corresponding interaction complexes of CO, p_{CO}. Data from Refs. 104 and 108.

Ammonia

The calorimetric measurement of the adsorption of ammonia at the acid sites in zeolites and the temperature-programmed desorption are important methods for determining the number and acid strength of these sites in zeolites.[1,3] Interpretation of the results of these measurements requires knowledge of the interaction energy of ammonia with the individual zeolite sites or at least their order. However, so far, quantum-chemical calculations have been used only to study the interaction complexes of ammonia with terminal and bridging OH groups or with model Lewis sites.

CNDO/BW calculations on $(XO)_3SiOH$ clusters have demonstrated that the interaction of the N atom of ammonia with the H atom of terminal OH groups ($E = -110.9$ kJ mol^{-1}) is energetically more favorable than the interaction of H atoms of ammonia with the O atom of these groups ($E = -47.3$ kJ mol^{-1}).[109] Similarly, the formation of the complex of ammonia with the bridging OH groups [cluster $(XO)_3Si\overset{H}{O}Al(OX)_2O^xSi(OX)_3$, see Fig. 5.8] is energetically more favorable ($E = -158.6$ kJ mol^{-1}) than the interaction of the H atoms of ammonia with skeletal O^x atoms ($E = -83.3$ kJ mol^{-1}). Calculations for an identical model illustrate that a change in the acidity of the bridging OH group (modeled by the variation in VOIP of the pseudoatom X from 7 to 17 eV) leads to a change in its interaction energy from -146 to -163 kJ mol^{-1}.[109] Interaction of ammonia with the model Lewis site $Al(OX)_3$ exhibits[93] the greatest energy gain, 275 kJ mol^{-1}. Moreover, the possible dissociation of ammonia at this site, leading to formation of a Brönsted site and NH_2 bonded to Al, has also been studied for this site, yielding a reaction heat of -162.2 kJ mol^{-1}.[93]

Nonempirical calculations[93,101–103] (cf. Table 5.4) have confirmed that the energy gain for the individual sites interacting with ammonia has the following order: $\equiv SiOH < \equiv Si\overset{H}{O}Al\equiv\ < \equiv Al$ and that, because of its greater basicity, the interaction energy of ammonia with the individual sites is greater than that for water and especially for CO.

Hydrocarbons

Knowledge of the character and energetics for the formation of the adsorption complexes of hydrocarbons with various zeolite sites is important for understanding and rationally employing their catalytic and adsorption (separation) ability. Consequently, the properties of the interaction complexes of hydrocarbons with Brönsted, cation, and Lewis sites of zeolites have been intensively studied using various experimental methods.[1,3]

An example of the interaction of hydrocarbons with Brönsted sites of zeolites is the formation of the H-complex of ethylene (or other unsaturated hydrocarbons), assumed to be the intermediate in various conversions of hydrocarbons or their derivatives.[110] The CNDO/2 method has been used in the study of the formation of the interaction H-complex of ethylene with two bridging OH groups, with geometry corresponding to the faujasite[111] [the $Si_3AlO_4(OH)_8H$ cluster, see Fig. 5.11] and ZSM-5 zeolite[112] [$Si_5AlO_6(OH)_{12}H$ cluster]. Because of the different geometry, the bridging OH group modeling the ZSM-5 zeolite (dissociation energy of the OH bond, E_H = 1870 kJ mol^{-1}, Wiberg bond order, p_{OH} = 0.92) is somewhat more acidic than the OH group representing faujasite (E_H = 1887 kJ mol^{-1}, p_{OH} = 0.93). Of the types of the studied interactions of ethylene with the OH group, the interaction leading to the formation of π-complex between ethylene and the OH group is energetically most favorable (the axis of the OH group is perpendicular to the ethylene plane and passes through the center of the C=C bond). The formation of this complex is connected with transfer of electron density from ethylene to the zeolite model, which is understandably greater for the more acidic OH group, similarly to the interaction energy (cf. Table 5.5). Transfer of electron density between ethylene and the cluster leads to a weakening of the OH bond (as can be seen from p_{OH} in Table 5.5), which should be reflected in the shift in the stretching vibrational frequency of the OH group interacting with ethylene. This shift has actually been observed[110] and equals 370 cm^{-1}. The formation of the H complex also causes a weakening of the C=C bond and the polarization (or a change in the

Table 5.5. Interaction energies, E (kJ mol^{-1}), transfer of electron densities, Δq, and a decrease in the C–C and O–H bond orders, Δp, for complexes of ethylene with clusters $Si_3AlO_4(OH)_8H$, $Si_5AlO_6(OH)_{12}H$, and $Si_3Al_3O_6(OH)_{12}Al$.

Cluster	E	q_{C_1}	q_{C_2}	Δq	Δp_{C-C}	Δp_{O-H}
$Si_3AlO_4(OH)_8H$[a]	60	−0.025	−0.045	0.108	0.102	0.118
$Si_5AlO_6(OH)_{12}H$[b]	80	−0.027	−0.053	0.118	0.120	0.135
$Si_3Al_3O_6(OH)_{12}Al$[c]						
π-complex	265	−0.002	−0.001	0.274	0.229	
σ-complex	270	−0.107	0.105	0.275	0.335	

[a] Taken from Ref. 111.
[b] Taken from Ref. 112.
[c] Taken from Ref. 113.

polarization) of bonds in the interacting ethylene molecule as a result of the electrostatic field of the cluster.

A second type of site that is important for catalysis is the Lewis electron acceptor site, represented, for example, by an Al cation localized in the zeolite cation position, i.e., by a cluster model of the $Si_3Al_3O_6(OH)_{12}Al$ type (see Fig. 5.9). Investigation of the interaction complex of ethylene with this site reveals[113] that, as for Brönsted[111,112] or other types of cation sites,[70,114] the formation of a π-complex between the ethylene molecule and the Al cation is energetically most favorable. It follows from CNDO/2 calculations[113] that the formation of this complex leads to much greater energy gain than the formation of the H-complex of ethylene (cf. Table 5.5) or the π-complex with Na cation site ($E = -94$ kJ mol^{-1}).[114] The greater strength of the electron donor–acceptor bond between ethylene and the Lewis site then results in greater transfer of electron density compared to the Brönsted site and, consequently, in greater weakening of the bonds in the interacting ethylene molecule (cf. Table 5.5). In addition, it has been found[113] that this π-complex of ethylene can be quite readily converted (with an energy barrier of less than 20 kJ mol^{-1}) to a σ-complex of the $-Al-CH_2-CH_2$ type, leading to greater polarization of the ethylene molecule.

STO-3G calculations on the energetics of the interactions of ethylene with models of Li, Na, and Mg cationic sites (the $H_3Si\overset{Cat}{O}AlH_3$ cluster) have demonstrated that (in agreement with semiempirical calculations) interaction leading to the formation of a π-complex between the cation and ethylene is energetically most favorable.[70,99] In agreement with experimental data, it was found that the calculated adsorption energy of ethylene (given in brackets) for the individual cationic sites exhibits the order: Mg (-98.4 kJ mol^{-1}) > Li (-60.0 kJ mol^{-1}) > Na (-10.3 kJ mol^{-1}).[70,99]

It follows from CNDO/2 calculations[114] of the interaction energies of ethylene, 2-methylpropene, and benzene with Na cation site of zeolite modeled by the $Si_5AlO_6(OH)_{12}Na$ cluster that this energy (in brackets) increases from ethylene (-94 kJ mol^{-1}) through 2-methylpropene (-113 kJ mol^{-1}) to benzene (-130 kJ mol^{-1}). This trend in the values of the interaction energy is understandable considering that the Na cation site corresponds to a weak Lewis site and that the basicity of the studied molecules decreases from benzene to ethylene. In this connection, it should be pointed out that CNDO/2 (and also nonempirical) calculations[115] of the interaction energies of these molecules with isolated Na cations yield an identical trend: benzene (-202 kJ mol^{-1} > 2-methylpropene (-176 kJ mol^{-1}) > ethylene (-140 kJ mol^{-1}), but the values are much larger, reflecting the unsuitability of using these models for cation sites in zeolites.

CONCLUSIONS

The preceding examples of quantum-chemical calculations of the properties of zeolites and their interactions with molecules demonstrate that, at present, this source of information on zeolites and the catalytic reactions occurring on them cannot compete with or even completely replace information obtained from experimental studies. The quantum-chemical description is still rather approximate because of the quality of the zeolite models used as well as of the quantum-chemical methods. On the other hand, it is obvious that even the quantum-chemical calculations of zeolite models carried out so far have made a great contribution to understanding, explaining, and rationalizing the experimental results; have yielded a considerable amount of information on the properties of the active sites at a molecular level; or have provided new ideas and impetus in this field of chemistry. Quantum-chemical descriptions of cluster models of the solid phase, which now constitute an independent field of research with a well-developed and constantly broadening methodological basis, have not yet been exploited to their full potential. It can be expected that the ever more effective computation procedures and the development of computer techniques as such, connected with the ability to use better models and quantum-chemical methods, will lead to improvement in the quality and amount of information yielded by quantum-chemical procedures on the properties of aluminosilicate systems, on their interaction complexes with molecules, and eventually, on the reactions occurring on the surfaces of these systems.

REFERENCES

1. Jacobs, P.A., *Carboniogenic Activity of Zeolites*. New York: Elsevier, 1977.
2. Rabo, J.A. (ed.), *Zeolite Chemistry and Catalysis*. Washington, DC: ACS Monograph 171, 1976.
3. Haynes, H.W., Jr., Chemical, physical and catalytical properties of large pore acidic zeolites, *Catal. Rev.-Sci. Eng.* **17:** 273–335 (1981).
4. Breck, D.W., *Zeolite Molecular Sieves*. New York: Wiley, 1974.
5. Dunken, H.H., Lygin, V.I., *Quantenchemie der Adsorption an Festkörperoberflächen*. Venheim, New York: Verlag Chemie, 1978.
6. Gibbs, G.V., Meagher, E.P., Newton, M.D., Swanson, D.K., A comparison of experimental and theoretical bond lengths and angle variation for minerals, in *Structure and Bonding in Crystals* (M. O'Keefe and A. Nawrotsky, eds.). New York: Academic Press, 1981, pp. 195–225.
7. Gibbs, G.V., Molecules as models for bonding in silicates, *Am. Mineral.* **67:** 421–451 (1982).
8. Zhidomirov, G.M., Chuvylkin, N.D., Quantum chemical methods in catalysis, *Usp. Khim.* **55:** 353–370 (1986).
9. Sauer, J., Zahradník, R., Quantum chemical studies on zeolites and silica, *Int. J. Quant. Chem.* **26:** 793–822 (1984).

10. Dunken, H.H., Hoffmann, R., Quantenchemische berechnungen von SiO_2-oberflächenclustern und ihren adsorptions- und oberflächen-reaktionen, *Z. Phys. Chem. N.F.* **125:** 207–238 (1981).
11. Zhidomirov, G.M., Kazansky, V.B., Quantum chemical cluster models of acid-base sites of oxide catalysts, *Adv. Catal.* **34:** 131–202 (1986).
12. Moore, E.B., Carlson, Ch.M., Extended Hückel theory calculations of atom migration energies into dimensional graphite, *Solid State Commun.* **4:** 47–49 (1965).
13. Mikheikin, I.D., Abronin, I.A., Zhidomirov, G.M., Kazansky, V.B., The cluster model calculations of chemisorption and elementary acts of catalytic reactions. I, Isotopic substitution of hydrogen atoms of surface OH groups of silica gel, *J. Mol. Catal.* **3:** 435–442 (1978).
14. Pelmenshchikov, A.G., Mikheikin, I.D., Zhidomirov, G.M., Cluster scheme of quantum chemical calculation of surface structures by MINDO/3 method, *Kinet. Katal.* **22:** 1427–1430 (1981).
15. Nishida, M., Cluster model approach for electronic structure of Si and Ge (111) and GeAs (110) surfaces, *Surface Sci.* **72:** 589–616 (1978).
16. Beran, S., Cluster modelling of ionic crystals by the CNDO/2 method, *Zh. Fiz. Khim.* **57:** 1178–1180 (1983).
17. Korsunov, V.A., Chuvylkin, N.D., Zhidomirov, G.M., Kazansky, V.B., Comparison of ionic and covalent schemes accounting for crystal chemical environment of active centres of oxides, *Kinet. Katal.* **21:** 402–407 (1980).
18. Chuvylkin, N.D., Cluster models in quantum chemical analysis of spectroscopic and activation characteristics of surface complexes on oxide catalysts, *Zh. Fiz. Khim.* **59:** 1085–1098 (1985).
19. Mikheikin, I.D., Lumpov, A.I., Zhidomirov, G.M., Kazansky, V.B., The cluster model calculations of chemisorption and elementary acts of catalytic reactions. VIII. Formation of proton-containing superacid centres in H-forms of zeolites during dehydroxylation, *Kinet. Katal.* **20:** 499–501 (1979).
20. Dempsey, E., The calculation on Madelung potentials for faujasite type zeolites I, *J. Phys. Chem.* **73:** 3660–3668 (1969).
21. Preuss, E., Linden, G., Peuckert, M., Model calculations of electrostatic field and potentials in faujasite type zeolites, *J. Phys. Chem.* **89:** 2955–2961 (1985).
22. Beran, S., CNDO/2 calculations on cluster models of zeolite crystals including their electrostatic field and its effect on the interacting molecules, *Chem. Phys. Lett.* **91:** 86–90 (1982).
23. Beran, S., Jirů, P., A quantum chemical study of the influence of an electrostatic field on the reactivity of methanol over zeolites, *React. Kinet. Catal. Lett.* **9:** 401–405 (1978).
24. Balhausen, C.J., Gray, H.B. *Molecular Orbital Theory*. New York: W.A. Benjamin, 1964.
25. Pople, J.A., Beveridge, D.L. *Approximate Molecular Orbital Theory*. New York: McGraw-Hill, 1970.
26. Boyd, J., Whithead, M.A., An SCF-MO-CNDO study of equilibrium geometries, force constants, and bonding energies. CNDO/BW. Part I. Parametrization, *J. Chem. Soc. Dalton Trans.* 73–77 (1972).
27. Bingham, R.C., Dewar, M.J., Lo, D.H., Ground states of molecules. XXV. An improved version of the MINDO semiempirical SCF-MO method, *J. Am. Chem. Soc.* **97:** 1285–1293 (1975).
28. Čársky, P., Urban, M., *Lecture Notes in Chemistry*. New York: Springer-Verlag, 1980, Vol. 16.
29. Beatzold, R.C., Application of molecular orbital theory to catalysis, *Adv. Catal.* **25:** 1–55 (1976).

30. Simonetta, M., Gavezzotti, A., The cluster approach in theoretical study of chemisorption, *Adv. Quant. Chem.* **12:** 103–187 (1980).
31. Pauling, L., *The Nature of the Chemical Bond,* 3rd ed. Ithaca, NY: Cornell U.P., 1960.
32. Brown, I.D., The bond-valence method: An empirical approach to chemical structure and bonding, in *Structure and Bonding in Crystals* (M. O'Keefe and A. Nawrotsky, eds.). New York: Academic Press, 1981. pp. 1–30.
33. Gutmann, V., *The Donor-Acceptor Approach to Molecular Interactions.* New York: Plenum Press, 1978.
34. Sanderson, R.T., *Chemical Bonds and Bond Energy,* 2nd ed. New York: Academic Press, 1976.
35. Newton, M.D., Gibbs, G.V., Ab initio calculated geometries and charges distributions for H_4SiO_4 and $H_6Si_2O_7$ compared with experimental values for silicates and siloxanes, *Phys. Chem. Miner.* **6:** 221–246 (1980).
36. Newton, M.D., O'Keefe, M., Gibbs, G.V., Ab initio calculation of interatomic force constants in $H_6Si_2O_7$ and the bulk modulus of α quartz and α cristobalite, *Phys. Chem. Miner.* **6:** 305–312 (1980).
37. Senchenya, I.N., Kazansky, V.B., Beran, S., Quantum chemical study of the effect of the structural characteristics of zeolites on the properties of their OH groups. 2, *J. Phys. Chem.* **90:** 4857–4859 (1986).
38. Pelmenshchikov, A.G., Pavlov, V.I., Zhidomirov, G.M., Beran, S., Effects of structural and chemical characteristics of zeolites on the properties of their bridging hydroxyl groups, *J. Phys. Chem.* **91:** 3325–3327 (1987).
39. Beran, S., The effect of coordination of protons and cations on the geometric characteristics of zeolites, *J. Phys. Chem.* **92,** 766–768 (1988).
40. Mortier, W.J., Sauer, J., Lercher, J.A., Noller, H., Bridging and terminal hydroxyls. A structural, chemical and quantum chemical discussion, *J. Phys. Chem.* **88:** 905–912 (1984).
41. Olson, D.H., Dempsey, E., The crystal structure of the zeolite hydrogen faujasite, *J. Catal.* **13:** 221–231 (1969).
42. Mortier, W.J., Pluth, J.J., Smith, J.V., Reevaluation of proton positions in hydrogen faujasite, *J. Catal.* **45:** 367–369 (1976).
43. Gibbs, G.V., Meagher, E.P., Smith, J.V., Pluth, J.J., Molecular orbital calculations for atoms in tetrahedral framework of zeolites, in *Molecular Sieves II* (J.R. Katzer, ed.). Washington, DC: ACS Symp. Ser. No. 40, 1977, pp. 19–21.
44. Bosáček, V., Beran, S., Jirák, Z., Distribution of protons and cations in Na,H-Y, zeolites, *J. Phys. Chem.* **85:** 3856–3859 (1981).
45. Olson, D.H., Kokotailo, G.T., Lawton, S.L., Maier, W.H., Crystal structure and structure-related properties of ZSM-5, *J. Phys. Chem.* **85:** 2238–2243 (1981).
46. Dubský, J., Beran, S., Bosáček, V., Quantum chemical study of the physical characteristics of the hydroxyl groups of zeolites, *J. Mol. Catal.* **6:** 321–326 (1979).
47. Beran, S., Quantum chemical study of the physical characteristics of the hydroxyl groups of zeolites. IV, *J. Mol. Catal.* **10:** 177–185 (1981).
48. Fripiat, J.G., Berger-André, F., André, J.M., Derouane, E.G., Non-empirical quantum mechanical calculations on pentasiltype zeolites, *Zeolites* **3:** 306–310 (1983).
49. Beran, S., Quantum chemical study of the physical characteristics of HZSM-5 zeolites and their comparison with faujasites, *Z. Phys. Chem. N.F.* **137:** 89–97 (1983).
50. Derouane, E.G., Fripiat, J.G., A non-empirical molecular orbital study of the siting and pairing of aluminium in mordenite, in *Proceedings of the 6th International Zeolite Conference* (D.H. Olson and A. Bisio, eds.). Guilford, UK: Butterworths, 1984, pp. 717–726.
51. Derouane, E.G., Fripiat, J.G., Non-empirical quantum chemical study of the siting and pairing of aluminium in the MFI framework, *Zeolites* **5:** 165–172 (1985).

52. Fripiat, J.G., Galet, P., Delhalle, J., André, J.M., Nagy, J.B., Derouane, E.G., A nonempirical molecular orbital study of the siting and pairing of aluminium in ferrierite, *J. Phys. Chem.* **89:** 1932–1937 (1985).
53. Hass, E.G., Mezey, P.G., Plath, P.J., A non-empirical molecular orbital study on Loewenstein's rule and zeolite composition, *J. Mol. Struct. THEOCHEM* **76:** 389–399 (1981).
54. Hass, E.G., Mezey, P.G., Plath, P.J., Non-empirical SCF molecular orbital studies on simple zeolite model systems, *J. Mol. Struct. THEOCHEM* **87:** 261–272 (1982).
55. Pelmenshchikov, A.G., Zhidomirov, G.M., Beran, S., Tino, J., Molecular approach in the study of the process of formation of the lattice of oxides. Characteristic features of coordination of AlO_3 elements, *Phys. Status Sol. (a)* **99:** 57–64 (1987).
56. Mikheikin, I.D., Abronin, I.A., Lumpov, A.I., Zhidomirov, G.M., The cluster model calculations of chemisorption and elementary acts of catalytic reactions. III. Effect of the cluster size on the results obtained. Stability of solution, *Kinet. Katal.* **19:** 1050–1057 (1978).
57. Mortier, W.J., Geerlings, P., Van Alsenoy, C., Figeys, H.P., A CNDO study of the electronic structure of faujasite type six-rings as influenced by the placement of magnesium and by the isomorphous substitution of aluminium for silicon, *J. Phys. Chem.* **83:** 855–861 (1979).
58. Beran, S. Quantum chemical study of Li-, Na- and K-faujasites, *J. Phys. Chem. Solids* **43:** 221–225 (1982).
59. Mombourquette, M.J., Weil, J.A., Mezey, P.G., Ab initio SCF-MO calculations on AlO_4 centres in alpha-quartz. I, *Canad. J. Phys.* **62:** 21–34 (1984).
60. Beran, S., Quantum chemical study of the stability of Faujasite type zeolites, *Z. Phys. Chem. N.F.* **123:** 129–139 (1980).
61. Beran, S., Dubský, J., Quantum chemical study of Na–X and Na–Y zeolites I and II, *J. Phys. Chem.* **83:** 2538–2544 (1979); *Chem. Phys. Lett.* **71:** 300–302 (1980).
62. Beran, S., Jírů, P., Wichterlová, B., Quantum chemical study of the physical characteristics of Al^{3+}, $AlOH^{2+}$ and $Al(OH)_2^+$ zeolites, *J. Phys. Chem.* **85:** 1951–1956 (1981).
63. Beran, S., Quantum chemical study of the physical characteristics of calcium-faujasites and their interactions with water, *J. Phys. Chem.* **86:** 111–116 (1982).
64. Beran, S., Quantum chemical study of the physical characteristics of Mg faujasites and their interactions with water, *Z. Phys. Chem. N.F.* **130:** 81–89 (1982).
65. Beran, S., Jírů, P., Quantum chemical study of the properties of Fe, Co, Ni and Cr ion-exchanged zeolites, in *Metal Microstructures in Zeolites* (P.A. Jacobs et al., eds.). Amsterdam: Elsevier, 1982, pp. 53–59.
66. Beran, S., Jírů, P., Wichterlová, B., Fe ions in the cationic sites and in the skeleton of faujasites. A quantum chemical study, *Zeolites* **2:** 252–256 (1982).
67. Beran, S., Quantum chemical study of the physical characteristics of Cu zeolites, *Chem. Phys. Lett.* **84:** 111–113 (1981).
68. Beran, S., Jírů, P., Wichterlová, B., Chromium ions within zeolites. Part 2. A quantum chemical study of the properties of chromium ions in faujasites, *J. Chem. Soc., Faraday Trans. 1* **79:** 1585–1590 (1983).
69. Beran, S., Quantum chemical study of physical characteristics of Co- and Ni-faujasites, *Coll. Czech. Chem. Commun.* **47:** 1282–1289 (1982).
70. Sauer, J., Hobza, P., Zahradnik, R., Quantum chemical investigation of interaction sites in zeolites and silica, *J. Phys. Chem.* **84:** 3318–3326 (1980).
71. Derouane, E.G., Fripiat, J.G., Quantum mechanical calculations on molecular sieves. 1. Properties of the Si–O–T (T = Si, Al, B) bridge in zeolites, *J. Phys. Chem.* **91:** 145–148 (1987).
72. Sauer, J., Haberland, H., Schirmer, W., Local structure and bonding in zeolites by

means of quantum chemical ab initio calculations: Metal cations, metal atoms and framework modification, in *Structure and Reactivity of Modified Zeolites* (P.A. Jacobs et al., eds.). Amsterdam: Elsevier, 1984, pp. 313–320.

73. Ward, J.W., Infrared studies of zeolite surfaces and surface reactions, in Ref. 2, pp. 118–284.
74. Naccache, C., Ben Taarit, Y., Spectroscopy applied to zeolite catalysis, *Acta Phys. Chem.* **24:** 23–38 (1978).
75. Wichterlová, B., Kubelková, L., Nováková, J., Jírů, P., Behaviour of Fe species in zeolite structure, in *Metal Microstructures in Zeolites* (P.A. Jacobs et al., eds.). Amsterdam: Elsevier, 1982, pp. 143–150.
76. Barthomeuf, D., General hypothesis on zeolites physicochemical properties. Application to adsorption, acidity, catalysis, and electrochemistry, *J. Phys. Chem.* **83:** 249–256 (1979).
77. Senchenya, I.N., Chuvylkin, N.D., Kazansky, V.B., Comparative quantum chemical analysis of electronic structure of ethyl fragment in ethoxy sulfuric acid and in ethoxy structure on the surface of high silicon zeolite, *Kinet. Katal.* **26:** 1073–1077 (1985).
78. Beran, S., Slanina, Z., Theoretical study of some thermodynamic and kinetic aspects of migration of hydrogen in zeolites of the faujasite type, *Zh. Fiz. Khim.* **57:** 1168–1171 (1983).
79. Beran, S., Quantum chemical study of the effect of the structural characteristics of zeolites on the properties of their bridging OH groups, *J. Mol. Catal.* **26:** 31–36 (1984).
80. Jacobs, P.A., Acid zeolites: An attempt to develop unifying concepts, *Catal. Rev.-Sci. Eng.* **24:** 415–440 (1982).
81. Mortier, W.J., Geerlings, P., CNDO study of the site II and the site III in faujasite-type zeolites, *J. Phys. Chem.* **84:** 1982–1986 (1980).
82. Mikheikin, I.D., Lumpov, A.I., Zhidomirov, G.M., Kazansky, V.B., The cluster model calculations of chemisorption and elementary acts of catalytic reactions. IV. Properties of the surface bridging OH groups in aluminosilicates and zeolites. Effect of the Si : Al ratio, *Kinet. Katal.* **19:** 1053–1057 (1978).
83. Datka, J., Geerlings, P., Mortier, W.J., Jacobs, P.A., Influence of the overall composition on zeolite properties. 2. Framework hydroxyls: A quantum chemical study, *J. Phys. Chem.* **89:** 3488–3493 (1985).
84. Beran, S., Quantum chemical study of the influence of cations on the acidity of the skeletal hydroxyl groups of zeolites, *J. Phys. Chem.* **89:** 5586–5588 (1985).
85. Beran, S., Quantum chemical study of the physical characteristics of hydroxyl groups in dehydroxylated faujasites, *Z. Phys. Chem. N.F.* **133:** 37–44 (1982).
86. Mikheikin, I.D., Abronin, I.A., Zhidomirov, G.M., Kazansky, V.B., The cluster model calculations of chemisorption and elementary acts of catalytic reactions. II. Properties of surface OH groups of oxides, *Kinet. Katal.* **18:** 1580–1583 (1977).
87. Lygin, V.I., Lygina, I.A., Investigation of surface structures and adsorption complexes by quantum chemical methods and spectroscopy, *Zh. Fiz. Khim.* **59:** 1180–1192 (1985).
88. Sauer, J., Molecular structure of orthosilic acid and importance of $(p\text{-}d)\ \pi$ bonding. An ab initio molecular orbital study, *Chem. Phys. Lett.* **97:** 275–278 (1983).
89. O'Keefe, M., Domengés, B., Gibbs, G.V., Ab initio molecular calculations on phosphates: Comparison with silicates, *J. Phys. Chem.* **89:** 2304–2309 (1985).
90. Senchenya, I.N., Chuvylkin, N.D., Kazansky, V.B., Quantum chemical study of the mechanism of dehydroxylation of crystalline and amorphous aluminosilicates, *Kinet. Katal.* **27:** 87–92 (1986).
91. Beran, S., Quantum chemical study of the Lewis sites in dehydroxylated faujasite zeolites, *J. Phys. Chem.* **85:** 1956–1958 (1983).

92. Dombrovsky, D., Lygin, V.I., Khlonova, E.G., Quantum chemical calculations on models of dehydroxylation of H-forms of X zeolites, *Zh. Fiz. Khim.* **57:** 1807–1809 (1983).
93. Senchenya, I.N., Mikheikin, I.D., Zhidomirov, G.M., Trokhimetz, A.I., The cluster model calculations of chemisorption and elementary acts of catalytic reactions. XIV. Lewis type acid centres, *Kinet. Katal.* **24:** 35–44 (1983).
94. McDaniel, C.V., Maher, P.K., Zeolite stability and ultrastable zeolites, in Ref. 2, pp. 285–331.
95. Pelmenshchikov, A.G., Zhidomirov, G.M., Khuroshvili, D.V., Tsitsishvili, G.V., Quantum chemical study of zeolite structure stability. Comparative discussion of zeolites and boralites, in *Structure and Reactivity of Modified Zeolites* (P.A. Jacobs et al., eds.). Amsterdam: Elsevier, 1984, pp. 85–90.
96. Pelmenshchikov, A.G., Pavlov, V.I., Zhidomirov, G.M., Quantum-chemical comparison of the coordination abilities of B and Al in oxides within the framework of the molecular model, *Phys. Status Sol. (b)* **130:** K1–K4 (1985).
97. Senchenya, I.N., Mikheikin, I.D., Zhidomirov, G.M., Kazansky, V.B., The cluster model calculations of chemisorption and elementary acts of catalytic reactions. XIII. Interaction of H_2O with silica gel and H-forms of zeolites, *Kinet. Katal.* **23:** 591 (1982).
98. Lygin, V.I., Magomedbekov, Kh.G., Lygina, I.A., Quantum chemical calculations on models of adsorption of water molecules on silicates, *Zh. Strukt. Khim.* **22:** 156–157 (1981).
99. Sauer, J., Fiedler, K., Schirmer, W., Zahradník, R., What can be expected from quantum chemistry in investigation of adsorption in zeolites, in *Proceedings of the Fifth International Conference on Zeolites* (L.V.C. Rees, ed.). London: Heyden, 1980, pp. 501–509.
100. Hobza, P., Sauer, J., Morgeneyer, Ch., Hurych, J., Zahradnik, R., Bonding ability of surface sites on silica and their effect on hydrogen bonds. A quantum chemical and statistical thermodynamic treatment, *J. Phys. Chem.* **85:** 4061–4067 (1981).
101. Geerlings, P., Tariel, N., Botrel, A., Lissillour, R., Mortier, W.J., Interaction of surface hydroxyls with adsorbed molecules. A quantum chemical study, *J. Phys. Chem.* **88:** 5752–5759 (1984).
102. Sauer, J., Nature and properties of acidic sites in zeolites revealed by quantum chemical ab initio calculations, *Acta Phys. Chem.* **31:** 19–24 (1985).
103. Senchenya, I.N., Chuvylkin, N.D., Kazansky, V.B., Quantum chemical study of interaction of nitrogen and carbon monoxide molecules with Lewis acid centres of alumina, *Kinet. Katal.* **27:** 608–613 (1986).
104. Beran, S., Quantum chemical study of the characteristics of molecules interacting with zeolites, in *Structure and Reactivity of Modified Zeolites* (P.A. Jacobs et al., eds.). Amsterdam: Elsevier, 1984, pp. 99–106.
105. Beran, S., Quantum chemical study of the interactions of CO and H_2O with cations in zeolites, *Z. Phys. Chem. N.F.* **134:** 93–97 (1983).
106. Kustov, L.M., Kazansky, V.B., Beran, S., Kubelková, L., Jírů, P., Adsorption of carbon monoxide on ZSM-5. Infrared spectroscopic study and quantum chemical calculations, *J. Phys. Chem.* **91:** 5247–5251 (1987).
107. Angell, C.L., Schaffer, P.C., Infrared spectroscopic investigations of zeolites and adsorbed molecules. II. Adsorbed carbon monoxide, *J. Phys. Chem.* **70:** 1413–1418 (1966).
108. Beran, S., Quantum chemical study of the interactions of CO molecules with Al, Na and H faujasites, *J. Phys. Chem.* **87:** 55–58 (1983).
109. Senchenya, I.N., Mikheikin, I.D., Zhidomirov, G.M., Kazansky, V.B., The cluster

model calculations of chemisorption and elementary acts of catalytic reactions. XII. Interaction of NH_3 with silica gel and H-forms of zeolites, *Kinet. Katal.* **22:** 1174–1179 (1981).
110. Poustma, M.L., Mechanistic consideration of hydrocarbon transformations catalyzed by zeolites, In Ref. 2, pp. 437–528.
111. Beran, S., Jírů, P., Kubelková, L., Quantum chemical study of the interactions of ethylene and propylene with the hydroxyl groups of zeolites, *J. Mol. Catal.* **12:** 341–349 (1981).
112. Beran, S., Quantum chemical study of the interaction of ethylene with hydroxyl groups in HZSM-5 zeolites, *J. Mol. Catal.* **30:** 95–99 (1985).
113. Beran, S., Jírů, P., Kubelková, L., Quantum chemical study of the interaction of ethylene with the Lewis sites of zeolites, *J. Mol. Catal.* **16:** 299–304 (1982).
114. Sauer, J., Deininger, D., Interaction of ethene, 2-methylpropene, and benzene with Na^+ ion. II. Quantum chemical study of sorption complexes in faujasites, *Zeolites* **2:** 114–120 (1982).
115. Sauer, J., Deininger, D., Interaction of ethene, 2-methylpropene, and benzene with Na^+ ion. 1. Quantum chemical study of gas-phase complexes, *J. Phys. Chem.* **86:** 1327–1332 (1982).

6
Theoretical Studies of Transition Metal Sulfide Hydrodesulfurization Catalysts

SUZANNE HARRIS AND R.R. CHIANELLI

INTRODUCTION

Hydroprocessing catalysts based on the transition metal sulfides (TMS) have been used widely for over 60 years, and catalysts such as Co/Mo/Al_2O_3 remain the industry "workhorses" in hydroprocessing of petroleum-based feedstocks. Applications include sulfur removal [hydrodesulfurization (HDS)], nitrogen removal [hydrodenitrogenation (HDN)], and product quality improvement (hydrotreating and hydroconversion). Original interest (prior to World War II) in these catalysts centered on their activity in the hydrogenation of coal liquids containing considerable amounts of sulfur, thus maintaining the transition metal in the sulfided state. It was quickly discovered that Co, Ni, Mo, and W sulfides and their mixtures were the most active and least expensive of the TMS. After World War II their major uses shifted to hydroprocessing of sulfur- and nitrogen-containing petroleum-based feedstocks.[1]

As petroleum feedstock supplies dwindle, we are now required to process larger quantities of "dirtier" feeds containing larger amounts of sulfur, nitrogen, and metals. In order to meet these requirements, a new generation of transition-metal-sulfide-based catalysts is needed that have higher activities, greater resistance to poisons, and greater selectivity to desired products. In addition to their inherent stability in sulfur-containing hydrotreating reactions, the TMS have also shown remarkable activity and selectivity in reactions that traditionally have been the domain of other catalyst types. For example, in recent work on CO/H_2 synthesis Co and K promoted MoS_2 catalysts show excellent selectivity to secondary alcohols.[2] In spite of the past and future importance of these TMS catalysts, an understanding of the general fundamental basis for and origin of their activity has only recently begun to emerge. Theory is playing an increasingly important role in developing this understanding, and in this chapter we discuss a number of theoretical studies that have concentrated on TMS catalyst systems.

Catalysis has traditionally been studied from the point of view of a specific reaction having commercial value. In the case of catalysis by TMS, for example, the reactions studied are related to removing S or N from commercial feedstocks or adding hydrogen to them. Although this approach, which focuses on a catalytic reaction, has led to scientific and commercial progress, another more general approach that focuses on the catalytic solid is also possible. This second approach promises us a broader understanding of catalysis by solids and a greater ability to control the selectivity and activity of catalysts over a wider range of reactions and conditions. It combines both experimental and theoretical approaches aimed at answering basic questions about solids of catalytic interest. For the TMS as HDS catalysts, some of these questions are:

1. What are the stable states of the transition metals in a sulfur-containing environment? (The answer to this question, as subsequently discussed, is the TMS.)
2. Which TMS are most active for HDS of dibenzothiophene (DBT)?
3. What relations exist between the properties (thermodynamic, electronic, or geometric) of the TMS and their ability to catalyze the preceding reaction?
4. What is the relation between the well-known promotional effect discussed subsequently and the properties of the simple TMS?
5. What is the role of crystal structure in stabilizing active surfaces in the HDS environment?
6. What is the relation between the states on the active surface and the properties of the bulk TMS?
7. What is the specific sterochemistry of an active site, and how does a reactant molecule interact with this site?

In this chapter we describe how theory and experiment are attempting to answer these questions. We begin with the experimental results that lay a groundwork for much of the theory. These results define which of the simple TMS are most active for HDS. Classical correlations can be made between the relative catalytic activity of the TMS and thermodynamic properties such as heats of formation, and these correlations are discussed. This is followed by a discussion of the promoted TMS catalyst systems and the relation between the thermodynamic properties of these promoted systems and the simple TMS. The next section focuses on the electronic structure of the simple TMS and promoted MoS_2 systems. Correlations can be made between the electronic structure of the simple TMS and the observed trends in relative catalytic activity. Moreover, the properties that are important for high catalytic activity in the simple TMS are found to be influenced by the presence of an effective promoter. Up to this point, the discussion focuses on the bulk properties of the simple and promoted TMS. The nature of the active surface sites and the interaction of a reactant molecule with these sites is of course crucial to the HDS catalytic reactions, and the next sections focus on the TMS surfaces. Most experimental and theoretical studies have concentrated on defining the active surfaces and surface–reactant interactions for MoS_2 and WS_2, and these sections discuss the information gained from these studies.

PERIODIC TRENDS IN THE HDS ACTIVITY OF SIMPLE TMS

A systematic series of experiments has shown that the HDS activity of the TMS is related to the position of the transition metal in the periodic table.[3] Using HDS of dibenzothiophene (DBT) as a model reaction, these experiments measured the C–S hydrogenolysis activity of a large group of unsupported TMS. The stable states of the TMS both prior to reaction (pretreated) and after reaction are indicated in Table 6.1. It is notable that while the TMS stable states exhibit considerable variation in both structure and stoichiometry, the activity trends that are discussed here are quite smooth.

Initially, the rates of desulfurization of DBT were measured at 400°C. A

Table 6.1. Stable binary sulfides.

H_2/15% H_2S (400°C)	HDS Reactor (400°C)
TiS_2	TiS_2
VS_x	VS_x
Cr_2S_3	Cr_2S_3
MnS	MnS
FeS_x	FeS_x
Co_9S_8	Co_9S_8
NiS_x	NiS_x
ZrS_2	ZrS_2
NbS_2	NbS_2
MoS_2	MoS_2
RuS_2	RuS_{2-x}
Rh_2S_3	Rh_2S_{3-x}
PdS	PdS
SnS_2	—
HfS_2	HfS_2
TaS_2	TaS_2
WS_2	WS_2
ReS_2	ReS_2
OsS_2	OsS_x
IrS_x	IrS_x
PtS	PtS
Au°	Au°

comparison of the activity of the TMS indicates that carbon–sulfur hydrogenolysis activity varies with the position that the metal occupies in the Periodic Table (Fig. 6.1). The hydrodesulfurization activity varies by about three orders of magnitude across a given period and down a given group. The maximum activity occurs in the second and third transition series with a peak occurring near RuS_2 in the second row and Os in the third row. The first row transition metal sulfides are relatively inactive compared to the second and third row transition series.

When normalized to surface area (Fig. 6.2), only slight changes occur in the curves. For example, the most active catalysts at the peak of the curves change position; Rh becomes slightly more active than Ru, Os

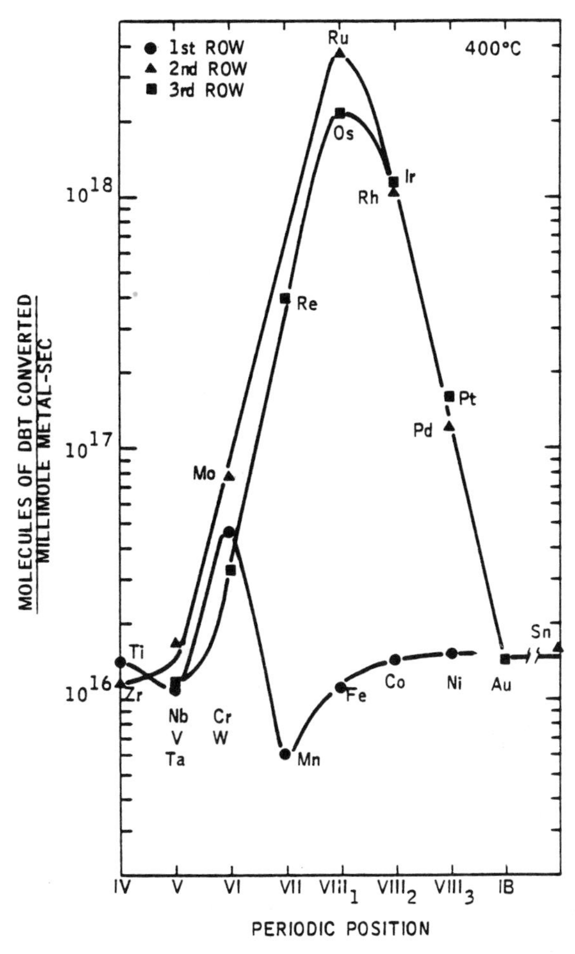

Figure 6.1. Periodic trends in the activities of the TMS for HDS of DBT at 400°C. Activities are normalized per millimole of catalyst (Ref. 3).

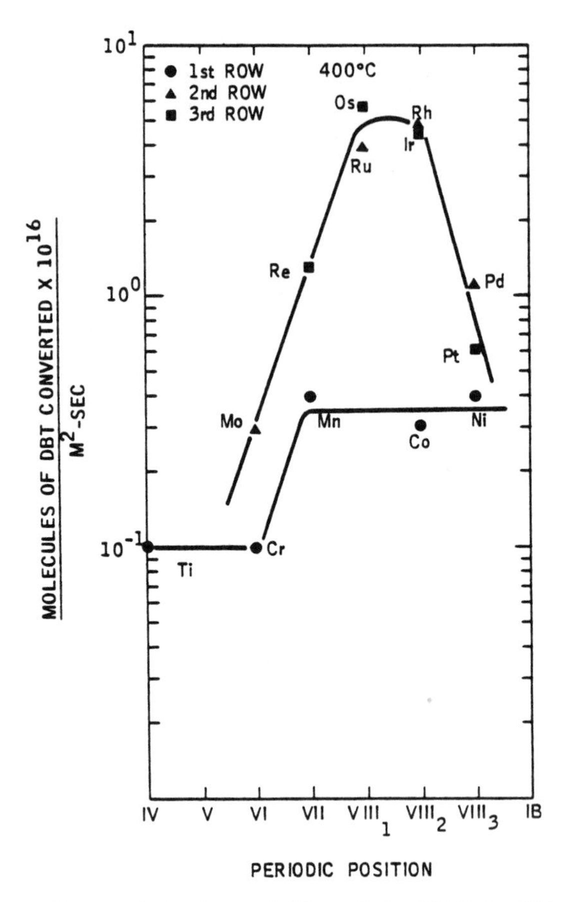

Figure 6.2. Periodic trends in the activities of the TMS for HDS of DBT at 400°C when activities are normalized per square meter of catalyst (Ref. 3).

becomes more active than Ir, etc. Also, the trends in the first row activities become smoother. These changes are not considered significant, however, because it is known that the HDS activities of the sulfides *do not* in general correlate to BET surface areas. This results from specific morphological effects of structural and geometric origin.[4] For example, catalysts with layered structures (MoS_2, ReS_2) have active sites at "edge" planes. These edge planes contribute most of the catalytic activity but add little to the total surface area.[5] Thus, the normalization of the activity to a per metal basis (Fig. 6.1) best reflects the intrinsic activity of the transition metal sulfides. The shape of the curves at 350°C remains essentially the same.

These HDS activity trends have been confirmed on carbon and SiO_2 supported catalysts with only slight variation in Group VIIIB maxima.[6,7]

The trends do not, however, appear on Al_2O_3 supported catalysts.[8] This is now known to result from the strong interactions of some TMS with Al_2O_3. These interactions mask the basic trends.[9,10] These periodic trends in activity indicate that the nature of the transition metal in the sulfide is the primary factor determining the intrinsic activity of the TMS. Although structure and geometry become important when considering activity optimization for a particular TMS, crystal structure is of secondary importance in determining the activity of one TMS relative to another. In other words, electronic rather than geometric factors play a dominant role in determining the HDS activity of the TMS.

The catalysis literature contains numerous examples of model reactions that display periodic maxima or "volcano" relationships. Sinfelt[11] has reviewed broad relationships between the catalytic activity of various metals in hydrogenation, hydrogenolysis, isomerization, hydrocarbon oxidation, and ammonia synthesis–decomposition reactions and the positions of these metals in the Periodic Table. The Group VIII transition metals display maximum activity when compared to Groups IV–VII and Groups I and IIB. Within Group VIII, the position of the maxima fluctuates depending on the reaction or on the particular transition series under study. In general, the catalytic activity of the metal in these studies can be correlated with the electronic configuration of the *d*-orbitals as "percentage *d* character" of the metallic bond (based on Pauling's valence bond theory) or with the strength of the metal adsorbate bond. It is perhaps surprising to find a similar "volcano" shape for the periodic trends in HDS activity, since the catalysts are now compounds rather than metals. It is equally surprising to find that Pauling percentage *d* character for the transition *metals* correlates with the HDS activity of the TMS (Fig. 6.3). Both the trends in activity and the correlations with *d* character, however, make it clear that the presence of 4*d* or 5*d* electrons are important in determining HDS activity.

A number of correlations can be made between thermodynamic quantities and the HDS activity of the TMS. For example, a correlation exists between the heats of formation of the TMS, the metal–sulfur bond strengths, and the catalytic activity of the sulfides.[3] The metal–sulfur bond strengths decrease continuously across the Periodic Table. As Figure 6.4 illustrates, the most active catalysts from the second and third transition series have intermediate values of the heat of formation (30–50 kcal mol^{-1}) and thus intermediate metal–sulfur bond strengths. Recent work shows that heats of formation of the sulfides also correlate linearly with heats of adsorption of sulfur on transition metal surfaces.[12] Taken together, these correlations indicate that maximum desulfurization activity is related to an optimum metal–sulfur bond strength on the surface of the catalyst. This appears to be consistent with the commonly accepted idea that sulfur vacancies on the surface of the catalyst are the active

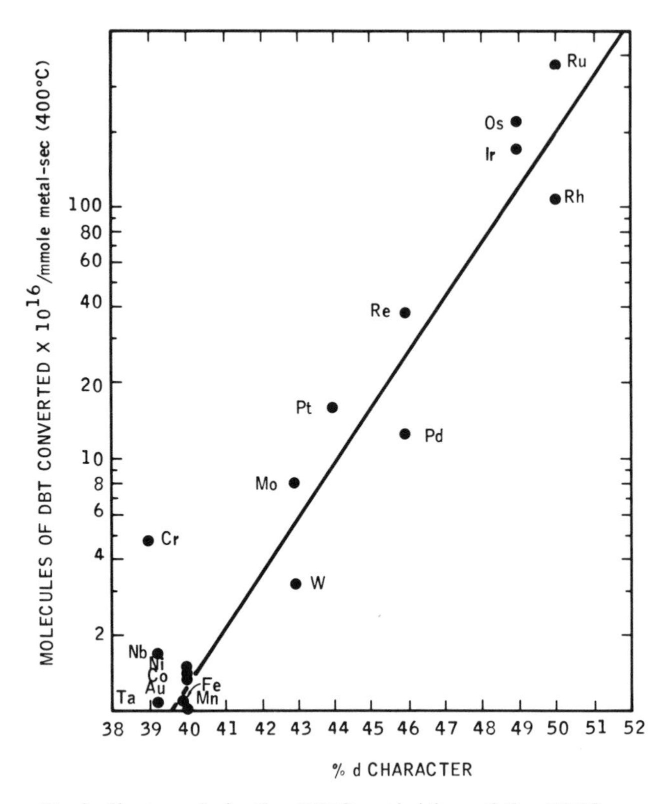

Figure 6.3. Periodic trends in the HDS activities of the TMS versus Pauling % *d* character for the transition metals.

HDS sites.[1] Under catalytic conditions, however, the metal–sulfur bond strength can be related not only to the ease of formation of sulfur vacancies but also to the binding strength of sulfur-bearing reactant molecules. The well-known principle of Sabatier[13] states that, for catalysts exhibiting maximum activity in a given reaction, the surface complexes formed between the catalyst and the reacting molecule will have intermediate heats of formation. Following this line of reasoning, the intermediate heats of formation of the most active TMS presumably reflect intermediate stabilities of the surface complexes formed between these catalysts and the sulfur-bearing molecule. Thus, sulfides of elements on the left-hand side of the Periodic Table have high heats of formation, bind sulfur or sulfur-bearing molecules too strongly, and are in a sense poisoned by sulfur. Sulfides of elements on the right-hand side of the Periodic Table have low heats of formation and probably bind sulfur-bearing molecules too weakly for reaction to occur. The sulfides that have intermediate heats of formations bind sulfur-bearing molecules neither too strongly nor too weakly and are effective catalysts. It should be noted that while these correla-

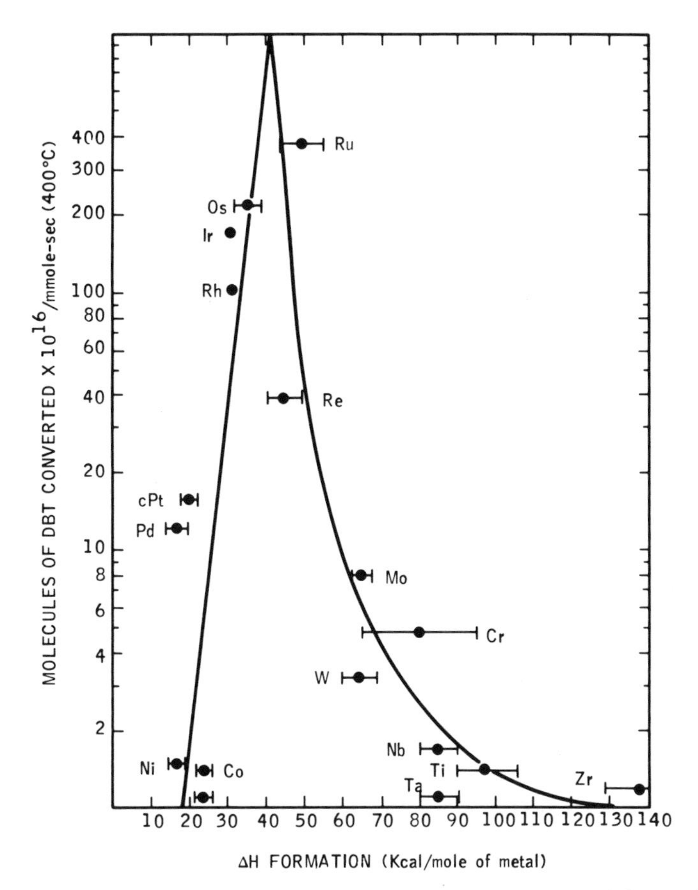

Figure 6.4. HDS activity of the simple TMS versus heats of formation.

tions are strictly obeyed for the 4*d* and 5*d* TMS, there is an inconsistency for the 3*d* TMS. Although the heat of formation of MnS (51 kcal) falls near the required range for high activity, MnS actually has low activity. Once again, this calls attention to the differences between the 3*d* and the 4*d* or 5*d* TMS. We will return to this point in the section discussing electronic structure of the TMS.

In summary, these experimental results and thermodynamic correlations indicate that:

1. Periodic or electronic factors play the primary role in determining the HDS activity of the TMS. Structural or geometric factors play only a secondary role.
2. The presence of 4*d* or 5*d* metals are necessary for high catalytic activity.
3. Metal–sulfur bond strength is important in determining the periodic trends in activity.

PROMOTED TMS CATALYSTS

It is well known that the presence of a second transition metal can lead to "promoted" TMS catalysts having HDS activities greater than the sum of the activities of the two simple sulfides. For example, Co_9S_8 or Ni_3S_2 act as "promoters" of MoS_2 or WS_2. Separately, Co_9S_8 or Ni_3S_2 has an activity that is of the same order as that of MoS_2.[3,14] Together, the "promoted" systems have activities that are far greater than can be accounted for by the simultaneous presence of two noninteracting phases. Such promotion of MoS_2 or WS_2 by Co or Ni occurs in either supported or unsupported catalysts. Although the subject of promotion has been well studied, no consensus really exists as to the origin of this effect. Studies of promotion have led to the idea of "synergy by contact" between the bulk sulfide phases Ni_3S_2 or Co_9S_8 and MoS_2 or WS_2.[15] Although this specific idea appears to be inaccurate on a microscopic scale, Ni/Mo, Co/Mo, Co/W, and Ni/W can still be viewed as "synergic pairs" whose members work together or cooperate. Recent work by Topsoe et al.[16] has shown the presence in both supported and unsupported CoMo catalysts of a unique form of sulfided Co (the "Co/MoS" phase). The presence of this phase correlates with activity, although the continuing presence of this phase after a catalyst has "worked" under real conditions has not been demonstrated. This finding suggests that for the synergic pairs a unique phase containing both of the metals actually serves as the active catalyst. This idea is further substantiated by thermodynamic correlations of the previously described type for the binary sulfide catalysts. These correlations, which involve the *average* of the heats of formation of the individual sulfides making up the synergic pairs, make it possible to relate the activity of the synergic pairs to the periodic trends in activity observed for the binary sulfides.[17]

As previously discussed, the heats of formation of the most active binary sulfides fall into an "optimum" range (Fig. 6.4). The maximum in activity for the second transition series occurs at RuS_2 (where $\Delta H_f = 49.2$ kcal mol^{-1}) and in the third transition series at Os (where $\Delta H_f = 35.3$ kcal mol^{-1}). As discussed previously, there is some uncertainty in the position of the exact maxima owing to incomplete correlation of HDS activity to BET surface area. The most active catalysts have heats of formation in the range of 30–50 kcal mol^{-1}. If the heats of formation of the individual sulfide components of the synergic pairs are averaged, these average values fall into the optimum range for the heats of formation of the binary sulfides. The average values, as well as the heats of formation of MoS_2, WS_2, and the first row TMS, are shown in Table 6.2. The averages for the known synergic pairs all lie near the center of the range ($\sim$40 kcal mol^{-1}) for the optimum value of the heat of formation. This suggests that the

Table 6.2. Average heats of formation of pairs of transition metal sulfides (kcal mol^{-1} of metal).

	Average Heat of Formation	
	MoS_2 (65.8)	WS_2 (62)
MnS (51.1)[a]	58.5	56.6
FeS_2 (42.6)	54.2	52.3
FeS (24)	44.9	43
Co_9S_8 (19.8)	42.8	40.9
Ni_3S_2 (17.2)	41.5	39.6
CuS (12.7)	39.3	37.4
ZnS (46)	57.3	54

[a] Heats of formation are in parentheses.

activity of the synergic pairs can be related to the periodic trends for the binary sulfides. This relation is illustrated in the plot of activity versus ΔH_f shown in Figure 6.5. Based on the trends for the binary sulfides, the average heats of formation of the various sulfide pairs can be used to predict qualitative trends in relative activity which may occur for the pairs. This is indicated by the line in Figure 6.5. This relation suggests that the synergic pairs may behave catalytically like second or third row *pseudobinary* sulfides. The sulfides of the synergic pairs may have surface properties that reflect the average surface properties and therefore the average bulk properties of the individual components. Thus, at the surface, the sulfided Co/Mo and Ni/Mo catalysts behave as if they were sulfides of hypothetical elements whose periodic position lies between the positions of the two members of the pair, hence the term "pseudobinary sulfide."

When the activities of the Co/Mo and Ni/Mo systems were measured for the HDS of DBT, this activity was indeed found to approximate that of hypothetical pseudobinaries having the average properties of Co_9S_8 and MoS_2 or Ni_3S_2 and MoS_2.[17] This is illustrated in Figure 6.6, where the activities of the Co/Mo and Ni/Mo systems are plotted along with the activities of a number of the binary sulfides. The fact that the activities of some of the synergic catalysts fall below the trend curve can be attributed to differences in the physical properties of the catalysts. The previously noted difficulty in correlating BET surface area to HDS activity is particularly troublesome in the case of the synergic pairs, where catalysts of the same nominal composition and surface area can have substantially different activities. It should be noted, however, that in general the surface areas of the synergic pairs are lower than those of the pure phases. Sur-

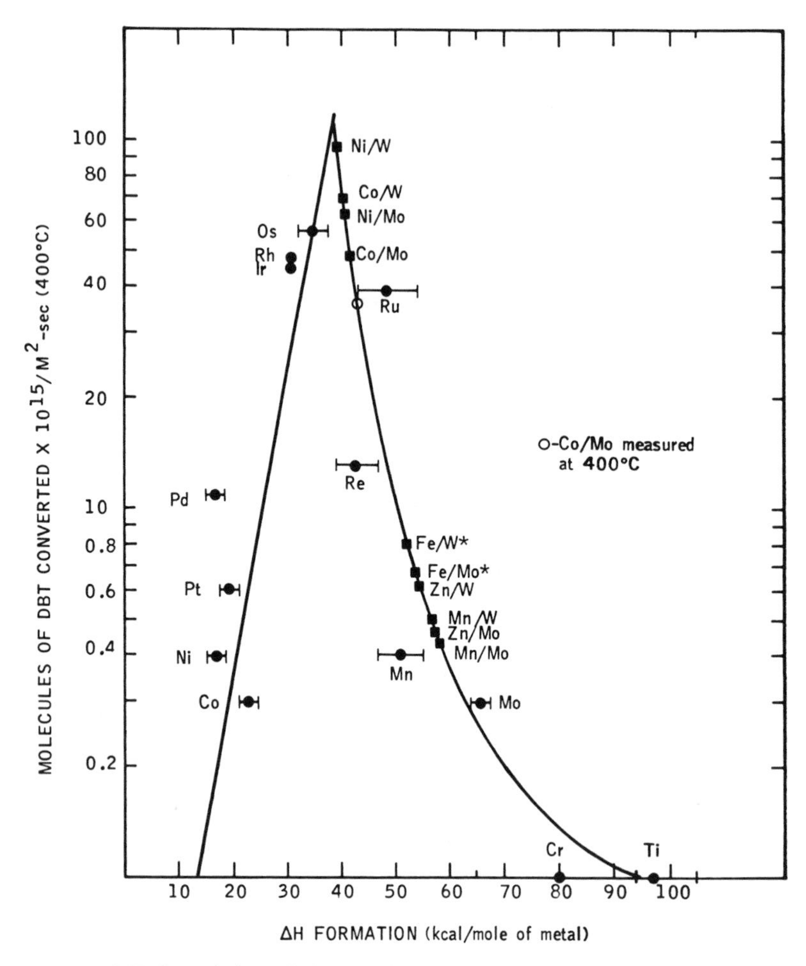

Figure 6.5. HDS activity of the TMS versus heats of formation (circles are the binary sulfides; squares are the average heats of formation of the pairs of simple sulfides taken from Table 6.2). Open circle is a measured activity for Co/Mo catalyst (Ref. 17).

face areas for the synergic pairs usually lie in the range of 15–25 m²/g, whereas MoS_2 prepared in a similar manner has surface areas in the range of 50 m²/g.[3,4] This suggests that in the synergic pairs the number of active sites does not increase but rather that the quality of the active site is enhanced. In fact, a datum point for the activity of Co/Mo at 400°C normalized to surface area falls quite close to the point predicted from the average value of the heat of formation of the synergic pair Co/Mo (Fig. 6.5).

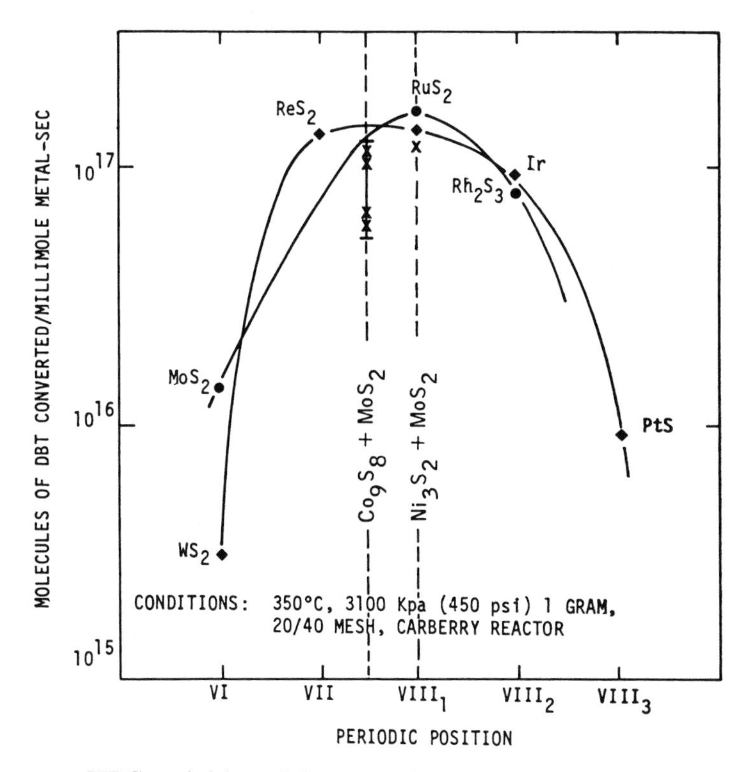

Figure 6.6. HDS activities of the second and third row TMS and the relation of the "synergic pairs" Co_9S_8/MoS_2 and Ni_3S_2/MoS_2 to the simple sulfides (Ref. 17).

The idea that the quality of the site might be changed is also supported by recent work on the role of Co in unsupported MoS_2 catalysts. This study showed that the surface area for a series of catalysts where $0 \leq M/(M + Mo) \leq 1$ is equal to or lower than that of unpromoted MoS_2.[18] Furthermore, the activation energy for the same series of catalysts was measured for dibenzothiophene conversion and found to be constant (20.7 ± 0.7 kcal mol^{-1}) for the entire series.

The relation between the average heats of formation and the activities of the synergic systems thus provides a link between the promoted systems and the trends in activity for the simple TMS systems. It should be noted, however, that just as for the simple TMS, not all of the average heats of formation correlate exactly with activity. There is some ambiguity, for example, in the Fe/Mo system. Because of the complexity of the Fe/S phase diagram, the Fe/S phase that is present under reaction conditions is not unequivocably known. If the heat of formation of FeS_2 is used

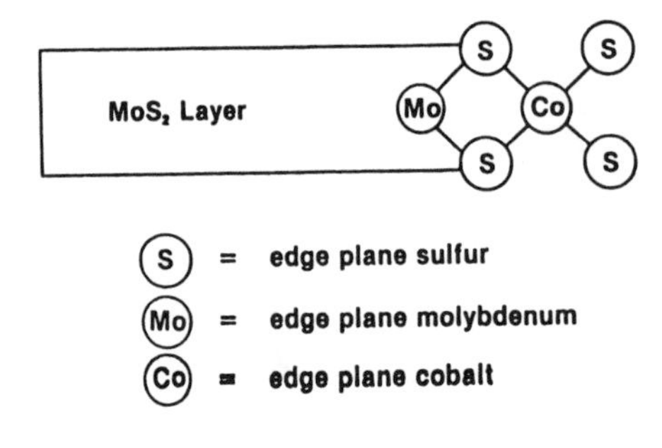

Figure 6.7. Schematic representation of sulfur shared between Co and Mo at MoS_2 edge. This shows one hypothetical position for Co at MoS_2 edge plane (Ref. 17).

for the averaging, the average heat of formation falls outside of the optimum range. Since iron sulfide does not promote MoS_2 catalysts,[1,19] this choice would be consistent with the known behavior of the Fe/Mo system. The Cu/Mo system is somewhat more bothersome, because while the average heat of formation for this system falls near the optimum value, Cu has actually been found to poison MoS_2 catalysts.[19] While this apparent discrepancy may also be due to a problem of choice of the appropriate Cu/S phase, it also illustrates that correlations of this type are not strictly obeyed for all possible pairs.

Several microscopic theories for promotion have been presented in the literature that are consistent with the suggested pseudobinary relation. This relationship calls for a Co/MoS unit to be present at the surface. (One such possible unit is shown in Figure 6.7.) For Co/Mo catalysts, this picture is consistent with the Co being located somewhere at the edge of the MoS_2 crystallites. The Co could be located at the edge of a single MoS_2 layer as suggested by Ratnasamy and Sivasanker,[20] or it could be located in octahedral or tetrahedral holes (Fig. 6.8) between the layers near the edge (pseudointercalation) as suggested by Voorhoeve and Stuiver.[21] Co located at the edge of MoS_2 is also consistent with the "surface enrichment" model of Phillips and Fote.[22] The essential point is that somewhere at the edge sulfur atoms (which are removed to create vacancies) are shared by Co and Mo and behave in some average electronic fashion. These special sulfur atoms may be similar to sulfur atoms attached to the surface of a noble metal sulfide such as RuS_2. A more elaborate model based on this idea has recently been presented by LeDoux.[23] Presumably, if such electronic interactions occur at the surface of promoted catalysts, metal sulfur bond strengths for these shared sulfurs are adjusted to the intermediate values that the binary periodic trends suggest are necessary for high activity. Thus, a picture of the Co/Mo and Ni/Mo systems emerges in which sulfurs are shared between the promoter and Mo. These shared sulfurs have the optimum metal–sulfur bond strength for vacancy formation.

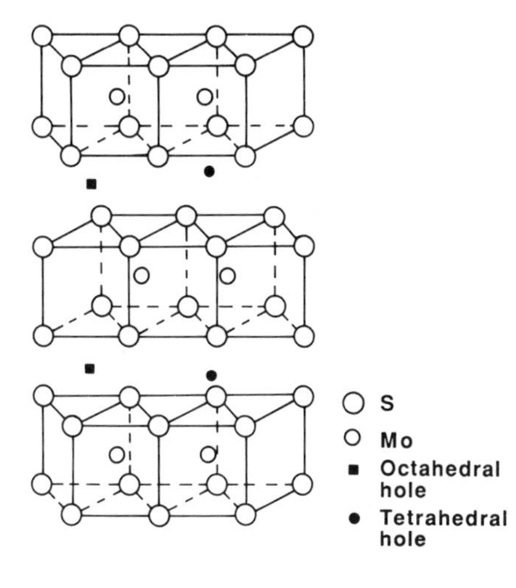

Figure 6.8. The stacking of layers in MoS_2 or WS_2. The octahedral and tetrahedral holes between the layers are indicated.

In summary:

1. Promoted or synergic systems can be related to the binary (simple) TMS through heats of formation averaging.
2. This relation suggests that surface sites on the promoted catalysts mimic the most active binary sulfides. These special sites may be created by the removal of sulfurs that are shared between the synergic metals, thus forming "average" vacancies.
3. This model is independent of particular details of structure.

ELECTRONIC STRUCTURE OF THE SIMPLE AND PROMOTED TMS

In the previous sections, the periodic trends for HDS by TMS and the relation between these trends and the promoted or "synergic systems" were discussed. The periodic trends show classical "heat of formation" and Pauling percent *d* character correlations, which indicate the importance of both the 4*d* and 5*d* electrons and the metal–sulfur bond strengths in determining HDS activity. Pseudobinary behavior of the promoted systems suggests that the same factors are important in the mechanism of promotion when sulfur vacancies are shared between appropriate metals. In this section, studies are reviewed that relate the electronic structure of both the simple TMS and the promoted MoS_2 systems to trends in catalytic activity.

Calculated Electronic Trends in the Simple TMS

The transition metal sulfides have been the subject of numerous experimental and theoretical studies aimed at elucidating their electronic structure. Both band structure[24–28] and cluster[29–35] calculations have been carried out for a number of the sulfides. In general, the metals tend to be far enough apart crystallographically that the metal–metal interactions are weak. Therefore, although the electronic structure changes as both the transition metal and structure are varied, the electronic structure of the sulfides is dominated by the covalent metal–sulfur bonds. Although a general picture of the bonding in the sulfides emerges from the various calculations on particular sulfides, it is not really possible to make detailed comparisons between the electronic structure of the different TMS because of the various techniques and approximations used for these calculations. In light of the periodic variations in HDS activity described previously, a more detailed systematic study and comparison of the electronic structure of the sulfides was desirable. With this end in mind, a consistent set of model calculations was carried out that would make it possible to compare some of the features of the electronic structure of the large group of transition metal sulfides. These calculations were aimed at determining how the electronic structure of the transition metal sulfides varies as a function of the periodic position of the metal and at identifying electronic factors that are related to the catalytic activity of the sulfides.[36–38]

Scattered wave $X\alpha$ calculations were employed to calculate energy levels and charge distributions for a series of octahedral MS_6^{n-} clusters. The central metal in these clusters was varied across the first transition series from Ti to Ni and across the second transition series from Zr through Pd. Enough electrons were included in each cluster that each sulfur was formally S^{2-} and each transition metal M had the formal oxidation state found in the corresponding metal sulfide. This resulted in each cluster having a negative charge ($n-$). A number of interesting trends in the electronic structure of the sulfides were observed. More important, considering the results of the cluster calculations in detail made it possible to identify electronic factors that correlate with the catalytic activity of the metal sulfides. A schematic diagram of the valence energy levels calculated for the clusters is shown in Figure 6.9. In each cluster a group of levels arising from the sulfur $3s$ orbitals lies at the lowest energy. Next highest in energy is a group of levels arising primarily from combinations of the sulfur $3p$ orbitals. Lying at the top of this group of sulfur levels is the $1t_{1g}$ level, a nonbonding combination of sulfur $3p$ orbitals. The sulfur-based orbitals are fully occupied in all of the clusters. Highest in energy (in most of the clusters) are the levels that correspond to the metal $3d$ or

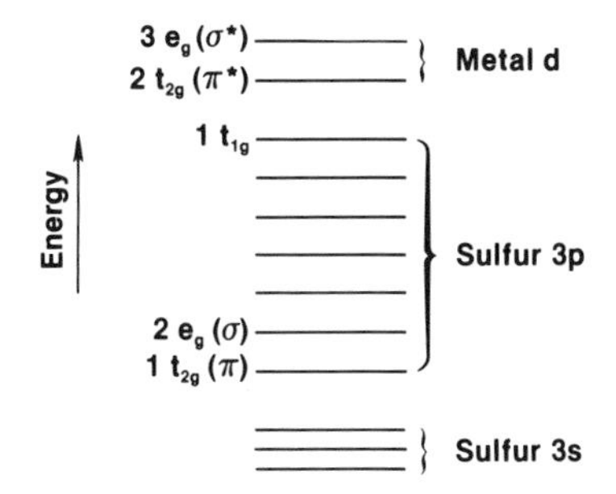

Figure 6.9. Schematic valence energy level diagram for an octahedral MS_6^{n-} cluster (Ref. 38).

4*d* orbitals. The occupations of these orbitals vary as the transition metal varies. The orbitals that are of interest are those having e_g and t_{2g} symmetry, because it is through these orbitals that the metal–sulfur *d–p* covalent interactions occur. The lower energy $2e_g$ and $1t_{2g}$ orbitals are mostly sulfur 3*p* in character but also have some metal *d* character. These two sets of orbitals are sigma and pi bonding, respectively, between the metal *d* and sulfur 3*p* orbitals. The higher energy $3e_g$ and $2t_{2g}$ orbitals are primarily metal *d* in character but also have some sulfur character. These orbitals are the antibonding counterparts of the lower energy bonding $2e_g$ and $1t_{2g}$ orbitals.

As the central transition metal is varied across both the 3*d* and 4*d* series, several systematic trends in the electronic structure of the clusters are observed. The first of these is simply the number of electrons that are formally associated with the transitional metal. For both the 3*d* and 4*d* metals, this number gradually increases upon moving from left to right in the Periodic Table. Another effect, proceeding from the left to right in each transition series, is the decrease in energy of the metal *d* orbitals relative to the energy of the sulfur 3*p* levels. Although observed for both sets of transition metals, the effect is more pronounced for the 4*d* metals. This systematic shift in the energy of the metal *d* orbitals brings them closer in energy to the sulfur 3*p* orbitals and results in a corresponding systematic increase in covalent mixing between the metal *d* and sulfur *p* orbitals. This is illustrated in Figure 6.10 where the metal contributions to the $2e_g$ (sigma) and $1t_{2g}$ (pi) metal–sulfur bonding orbitals (labeled D_σ and D_π respectively) are plotted versus the central metal of the cluster. A larger metal contribution to these orbitals, which are primarily sulfur 3*p* in character, is an indication of greater covalent mixing of the metal *d* and sulfur *p* orbitals and thus of a more covalent bond. For the sigma bonding $2e_g$ orbital, it can be seen from Figure 6.10a that although the metal orbitals in the 4*d* series generally make a larger contribution to the sigma bonding orbital, both the 3*d* and 4*d* orbital contributions to this sigma

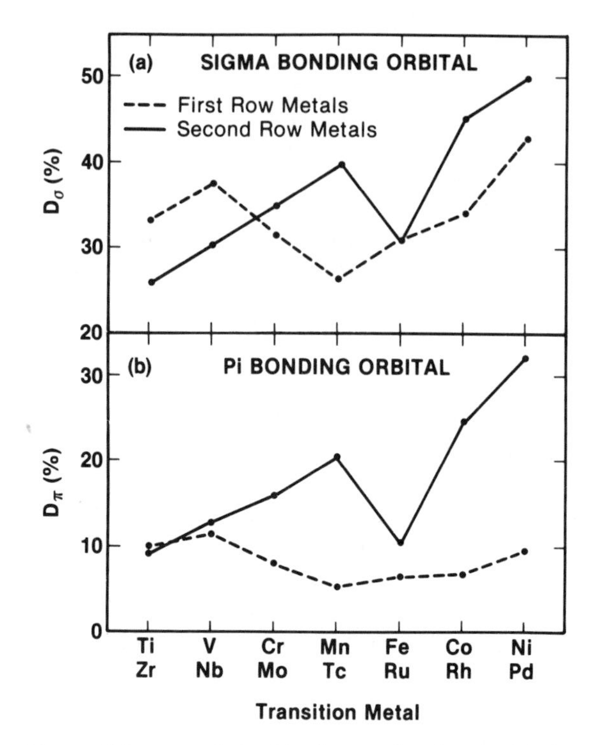

Figure 6.10. Plots of the metal d orbital contribution to (a) the sigma bonding $2e_g$ and (b) the pi bonding $1t_{2g}$ orbitals in the octahedral MS_6^{n-} clusters (Ref. 38).

bonding orbital increase across the Periodic Table. On the other hand, it can be seen from Figure 6.10b that, while the $4d$ metal contribution to the pi bonding $1t_{2g}$ orbital also increases across the Periodic Table, the $3d$ metal contribution to the pi-type bonding orbital is small across the entire $3d$ series. It is only for the larger $4d$ orbitals that a pi component to the metal–sulfur d–p bonding becomes important. The $4d$ metal sulfides are thus seen to be more covalent than their $3d$ counterparts, and this covalency is enhanced in the $4d$ metal sulfides by the pi contribution to the orbital mixing.

Although none of these trends correlates directly with the trends in catalytic activity, other factors that are related to these metal–sulfur bonding and antibonding orbitals were identified that do correlate with catalytic activity. For example, as was previously mentioned, the total number of d electrons alone does not correlate with the trends in activity, since the total d electron count simply increases across the Periodic Table and is very similar for $3d$ and $4d$ metals having very different activities. When the number of d electrons in the highest occupied molecular orbital (HOMO) of each cluster is considered, however, a different pattern emerges. This pattern can be seen in Figure 6.11, where the d electron count in the HOMO (labeled n) is plotted for each of the first and second row sulfides. This plot exhibits the same general features as the experimental activity plot for the sulfides. That is, the number of electrons n for the first row sulfides peaks with Cr, while for the second row sulfides this number peaks at Ru and Rh. The maximum value of n for the second row is twice the maximum value for the first row. Thus, the maximum in catalytic activity for the sulfides occurs in those sulfides with the maximum number of d electrons in the highest occupied energy level. It is informative to note why the number of electrons in the HOMO varies so between corresponding $3d$ and $4d$ metals, while the total number of d electrons does not. A major reason for this difference is the weaker metal–sulfur interaction for the $3d$ metals. This results in weaker d orbital splittings in the $3d$ series so that for Mn and Fe both the $2t_{2g}$ and $3e_g$ levels are partially occupied in high-spin d^5 and d^6 configurations, respectively. For the $4d$ metals Ru and Rh, where the d orbital splittings are larger, only the $2t_{2g}$ level is occupied in a low-spin d^6 configuration. For the other metals on the far right of both transition series (Co, Ni, and Pd), occupation of the $3e_g$ level is forced by the total d electron count. Thus, in the less active $3d$ metal sulfides (Mn, Fe, Co, and Ni), the HOMO is the $3e_g$ (sigma antibonding) level. For the $4d$ sulfides the HOMO for all the metal sulfides except Pd is the $2t_{2g}$ level, and the activity increases as the number of electrons in the $2t_{2g}$ level increases. Therefore, not only do the

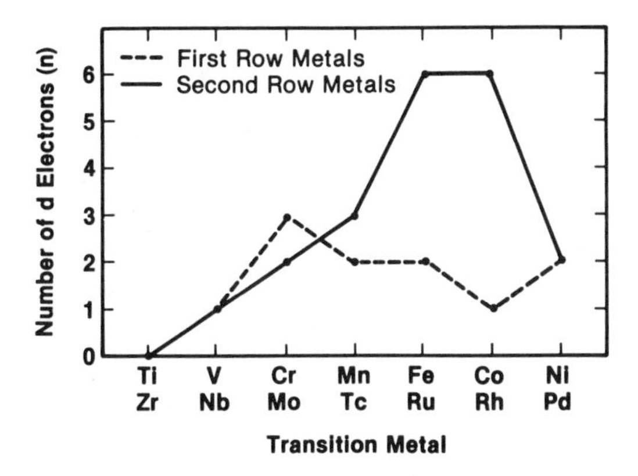

Figure 6.11. Formal d electron occupation of the HOMO in each MS_6^{n-} cluster (Ref. 38).

more active catalysts have more electrons in the HOMO, but also the HOMO in these active catalysts is the $2t_{2g}$ level.

As previously discussed, the covalency of the metal–sulfur bond increases from right to left in the Periodic Table and is greater for the $4d$ metals than for the $3d$ metals. The covalency, as measured by the metal contribution to the sigma and pi bonding orbitals (D_σ and D_π in Fig. 6.10), does not correlate directly with activity. It was possible, however, to define another calculated parameter whose behavior more closely resembles the observed trends in catalytic activity. This parameter, which provides a measure not only of the covalency of the metal–sulfur bond but also of the covalent contribution to the metal–sulfur bond *strength,* takes into account both the net number of bonding electrons and the covalency of the metal sulfur bonds in each sulfide. This parameter B is defined by

$$B = n_\sigma D_\sigma + n_\pi D_\pi$$

where D_σ and D_π have been defined previously as the metal d orbital contribution to the sigma and pi bonding orbitals (Fig. 6.10) and n_σ and n_π are the net number of sigma and pi bonding electrons, respectively. That is,

$$n_\sigma = (4 - n[3e_g])$$

where 4 is the number of electrons in the $2e_g$ (sigma bonding) orbitals and $n[3e_g]$ is the number of electrons in the $3e_g$ (sigma antibonding) orbitals. Likewise,

$$n_\pi = (6 - n[2t_{2g}])$$

where 6 is the number of electrons in the $1t_{2g}$ (pi bonding) orbitals and $n[2t_{2g}]$ is the number of electrons in the $2t_{2g}$ (pi antibonding) orbitals. Since D_σ and D_π provide a measure of the sigma and pi covalency and n_σ and n_π measure the net number of electrons that take part in the d–p sigma and pi bonds, the products $n_\sigma D_\sigma$ and $n_\pi D_\pi$ provide a relative measure of the sigma and pi contributions to metal–sulfur d–p bond strengths. The sum of these quantities B gives a relative measure of the overall covalent contribution to the metal–sulfur d–p bond strength. This parameter takes into account the fact that the covalent bond strength depends on both the magnitude of the metal–sulfur interaction and the net number of bonding electrons. A plot of B versus transition metal is shown in Figure 6.12. The variations in B reflect not only the greater covalency of the $4d$ metal sulfides but also the fact that, even though the metal–sulfur orbital mixing is stronger on the right-hand side of the transition series,

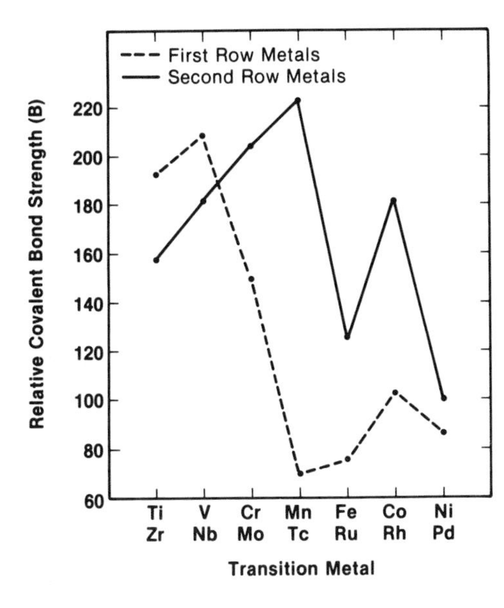

Figure 6.12. Calculated relative covalent contribution to the metal–sulfur bond strength B for each transition metal sulfide (Ref. 38).

the number of antibonding electrons is also larger, so that the actual covalent bond strength decreases. The parameter B differentiates quite effectively between the bonding in the metal sulfides, and a comparison of the shape of the plot of B with the shape of the experimental activity plot (Fig. 6.1) suggests a correlation between the covalent contribution to the metal–sulfur bond and HDS activity. That is, a larger value of B can be associated with higher HDS activity. This result suggests one reason why heat of formation and metal–sulfur bond strength data alone cannot predict catalytic activity. These quantities do not give any indication of the relative sizes of the ionic and covalent components of the metal–sulfur bond, while the relation between B and catalytic activity suggests that it is the covalent component that is related to activity. As discussed earlier, although it was found for the second and third row sulfides that intermediate heats of formation and intermediate metal–sulfur bond strengths correlate with high activity, it was also found that such correlations were not as successful for the first row sulfides. This is consistent with the greater covalency of the second row sulfides. In these sulfides where the covalent component to the metal–sulfur bond is higher, we might expect better correlations between heats of formation, bond strengths, and activity, while for the more ionic first row sulfides we might expect poorer correlations.

Comparison of the two factors n and B with the catalytic activity of the metal sulfides suggests that the better catalysts have larger values of n or

B or both. By assuming that both n and B are directly related to activity, a correlation between the calculated electronic structure and the experimental catalytic activity of the sulfides was obtained. This was done by defining an activity parameter A_2, which is the product of n and B:

$$A_2 = nB$$

A plot of A_2 calculated for each cluster, versus the transition metal in the cluster is shown in Figure 6.13. Also shown on this figure, plotted against the right-hand scale are the experimentally measured HDS activities of the corresponding sulfides. Although the agreement is not exact, the overall behavior of the activity parameter A_2 follows very closely the trends in measured activities. Particularly noticeable is the strong differentiation between the 3d and 4d transition metals. The general behavior of the parameter A_2 suggests that certain electronic factors are indeed related to the catalytic activity of the metal sulfides. These factors include the number of electrons in the highest occupied orbital, the degree of covalency in the metal–sulfur bond, and the covalent contribution to the metal–sulfur covalent bond strength. Since the parameter A_2 takes into account all three of these factors, it provides a good criterion for catalytic activity.

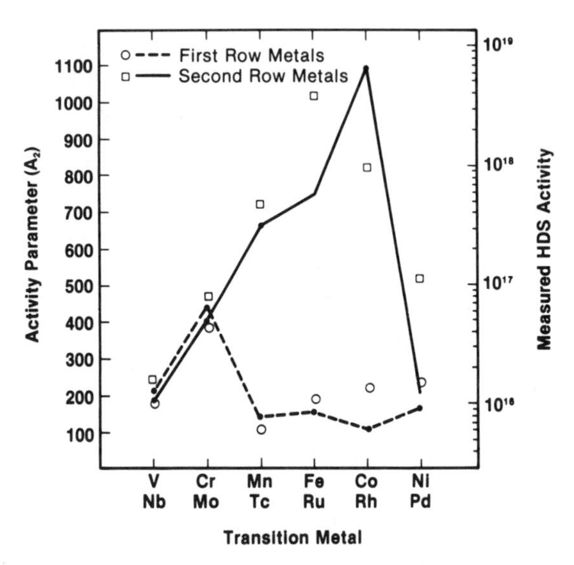

Figure 6.13. Calculated activity parameter A_2 for each TMS. Shown for comparison, using the right-hand scale, are the measured HDS activities of the TMS (Ref. 38).

That is, a sulfide with a large calculated value of A_2 would be expected to have high activity.

In summary, the correlations obtained for the simple TMS indicate that

1. The more active $4d$ TMS catalysts exhibit a larger covalent contribution to the metal–sulfur bond strength. This covalency differentiates the $4d$ TMS from the less active $3d$ TMS.
2. An optimum $4d$ or $5d$ TMS catalyst has a high d electron density available on the transition metal.

Calculated Electronic Trends in the Promoted MoS_2 Systems

If the electronic factors discussed previously are related to catalytic activity, then it is likely that they are also related to the promotion effect observed in the MoS_2-based systems. A systematic study of the ability of $3d$ transition metals to increase or decrease the HDS activity of MoS_2 showed that only three first row metals have a significant effect on HDS activity.[19] Both Co and Ni serve as promoters, while Cu acts as poison. The other first row metals have little effect on activity. In order to first establish whether there is an electronic basis for these trends in promotion and to then determine whether the electronic factors which had been related to catalytic activity for the binary sulfides are also important for promotion in the MoS_2 systems, another series of model cluster calculations were carried out.[19,39]

Scattered wave Xα calculations were carried out for a group of model clusters $MoM'S_9^{n-}$, where the first row transition metal M′ was varied systematically across the series from V to Zn. Enough electrons were included in each cluster so that each sulfur was formally S^{2-}, the molybdenum was formally Mo^{4+}, and the first row metal M′ had the formal oxidation state appropriate for the corresponding metal sulfide stable under reactor conditions. This resulted in each cluster having a negative charge ($n-$). The geometry of the model cluster, which consists of two face-sharing octahedra, is illustrated in Figure 6.14. Both metals are octahedrally coordinated by six sulfurs, three of which are shared between the

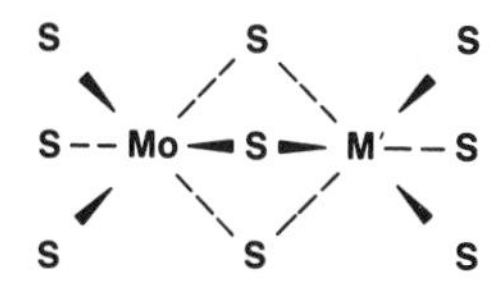

M′ = V, Cr, Mn, Fe, Co, Ni, Cu, Zn

Figure 6.14. Geometry of the cluster $MoM'S_9^{n-}$ used to model the promoted MoS_2 catalyst system. It consists of two face sharing MS_6 octahedra (Ref. 19).

two metals. The choice of octahedral coordination around Mo facilitated comparisons between the Mo–S bonding and Mo electronic configuration in these model clusters and in the simple binary cluster used previously. The coordination about the first row transition metal might have been chosen as either octahedral or tetrahedral. The choice of octahedral coordination for M′ once again allowed ready comparisons with the earlier calculations. Selected calculations using tetrahedrally coordinated M′ verified that a change in geometry did not change the major trends resulting from the calculations.

A schematic diagram of the valence energy levels calculated for the $MoM'S_9^{n-}$ clusters is shown in Figure 6.15, and it shares many of the features of the simpler diagram for the octahedral MS_6^{n-} clusters (Fig. 6.9). Once again, a group of sulfur 3*s* levels lie below a larger group of levels resulting from the sulfur 3*p* levels. Because of the large number of sulfur levels, these groups are represented as blocks. Although not shown specifically there are a group of levels lying toward the bottom of the 3*p* group that contain some Mo or M′ character or both and are the Mo–S and M′–S bonding orbitals. The top of the sulfur 3*p* group of levels is once again delineated by a nonbonding combination of sulfur 3*p* orbitals, in this case the $2a_2$ level as shown on the diagram. The notable feature of the diagram is the presence, above the sulfur 3*p* levels, of levels resulting from both the Mo 4*d* and M′ 3*d* metals. The levels for each metal fall into groups which are labeled "t_{2g}" and "e_g." These labels are not strictly correct, because even though each metal is octahedrally coordinated, the

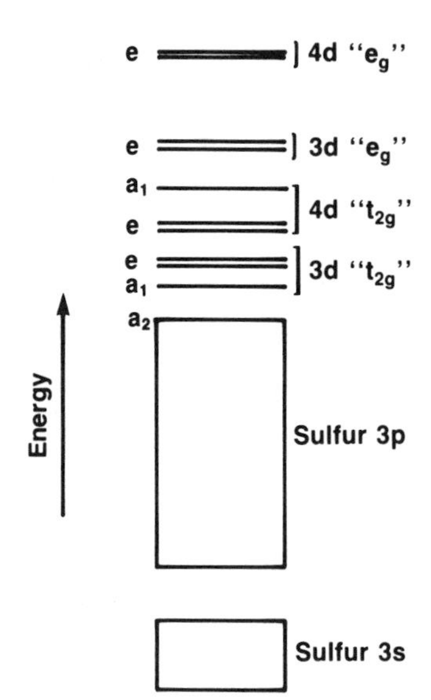

Figure 6.15. Schematic valence energy level diagram for an $MoM'S_9^{n-}$ cluster (Ref. 19).

overall symmetry of the cluster is C_{3v}. In this lower symmetry the five d orbitals actually transform as $a_1 + e + e$, and these labels are also shown on the diagram. Since the remnants of the octahedral splitting between the t_{2g} (now a_1 plus e) and e_g (now e) orbitals are still apparent, however, it is useful to continue to refer to the two groups of levels as "t_{2g}" and "e_g." The schematic diagram in Figure 6.15 corresponds to the case where the M' levels lie low enough in energy that the M' 3d "t_{2g}" and "e_g" sets of levels each lie below the corresponding Mo 4d sets of orbitals but not so low in energy that the entire 3d manifold of levels lies below the 4d levels. This relative energy separation applies to most of the clusters. The relative energies of these levels, the number of electrons and their distributions within these levels, and the mixing of the 3d and 4d orbitals all depend on M′. These quantities change as M′ is varied, and it was found that these changes can be related to the promoter ability of the 3d metals.

As was found for the simple sulfide clusters, the metal 3d orbitals become more stable upon proceeding from the left- to the right-hand side of the transition series, i.e., dropping in energy relative to the sulfur 3p orbitals. This means that the metal 3d orbitals also drop in energy relative to the Mo 4d orbitals. Proceeding across the 3d series from V to Zn this drop in energy is quite large. Figure 6.16 illustrates the relative energies of the sulfur 3p, the M′ 3d "t_{2g}" and "e_g," and the Mo "t_{2g}" sets of orbitals for all the $MoM'S_9^{n-}$ clusters. (The Mo 4d "e_g" orbitals, which are not important to the discussion, were omitted from the diagram.) For ease of comparison, all the energies shown on the diagram are measured relative to the nonbonding $2a_2$ combination of sulfur 3p orbitals that always lies at the top of the sulfur 3p group of orbitals. It is obvious from this diagram that the relative energies of the Mo 4d and M′ 3d orbitals change dramatically in going from the left- to the right-hand side of the 3d transition series. For V, on the left, the 3d levels are slightly higher in energy than the Mo 4d levels. For Cr, Mn, and Fe, the 3d levels shift downward somewhat. Moving to the right this shift becomes larger, so that finally for Cu and Zn the 3d levels lie well below the Mo 4d levels.

At the same time that the 3d levels decrease in energy, the number of electrons occupying these levels increases. While Mo contributes two "d" electrons to each cluster, the number of "d" electrons contributed by the first row metal depends on the metal. Thus, the total number of "d" electrons in the clusters varies from four in the V–Mo cluster all the way up to 12 in the Zn–Mo cluster. The distribution of electrons among the metal orbitals depends on both the relative energies of the orbitals and on spin effects. (It should be noted that although the "up" and "down" spin orbitals are not shown in Figure 6.16, the actual occupations of the orbitals shown in the figure were determined by a series of spin unrestricted calculations.) An examination of these metal orbital occupations

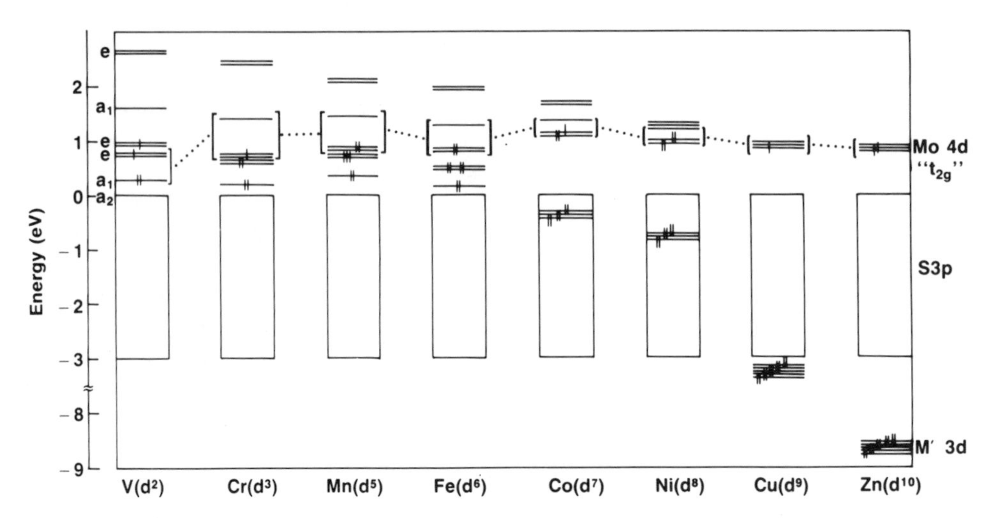

Figure 6.16. Calculated valence energy levels for the $MoM'S_9^{n-}$ clusters, where M′ is the 3*d* metal shown below each diagram. The sulfur 3*s* and the Mo 4*d* "e_g" levels have been omitted. The block labeled S 3*p* represents 20 levels. The energies of all the levels are shown relative to the energy of the nonbonding a_2 level lying at the top of the sulfur 3*p* group of orbitals. The three Mo 4*d* "t_{2g}" levels are bracketed so as to distinguish them from the M′ 3*d* levels (Ref. 19).

reveals important differences between the clusters containing those 3*d* metals that are known to promote or poison the activity of MoS_2 and the clusters containing those 3*d* metals that have no effect on the activity. The clusters can be divided into three groups.

In the first group of clusters (those containing Co, Ni, or Cu), Mo is formally reduced or oxidized, relative to the Mo in MoS_2. Co and Ni donate electrons to Mo, while Cu withdraws electrons from Mo. These electron transfers are the direct consequence of both the number of "*d*" electrons contributed to the cluster by the 3*d* metal and the relative energies of the 3*d* and 4*d* orbitals. Co and Ni contribute seven and eight "*d*" electrons, respectively, to the cluster, and in a pure Co–S or Ni–S cluster, the 3*d* "e_g" orbitals would be occupied. In the Mo–Co and Mo–Ni clusters, however, the Mo "t_{2g}" levels lie at an energy between the 3*d* "t_{2g}" and "e_g" orbitals. Thus, electrons are transferred from the 3*d* "e_g" to the lower energy Mo 4*d* "t_{2g}" orbitals, and Mo is formally reduced. Cu, on the other hand, contributes nine "*d*" electrons to the cluster, but all five of the Cu 3*d* orbitals lie well below the Mo 4*d* levels. An electron is transferred from a Mo "t_{2g}" level to a lower energy Cu 3*d* level and Mo is formally oxidized.

In the second set of clusters (those containing Fe and Zn), no such transfer of electrons occurs between Mo and the 3*d* metal. In the Zn–Mo cluster, the Zn 3*d* orbitals are so low in energy that they are not really involved in M–M or M–S bonding. Zn formally contributes 10 "*d*" electrons to the cluster, and thus the Zn 3*d* orbitals are fully occupied and have no effect on the Mo in the cluster. This Mo is very similar electronically to the Mo in MoS_2. In the Fe–Mo cluster, Fe formally contributes six "*d*" electrons to the cluster, but these electrons occupy the Fe "t_{2g}" set of orbitals, which lie lower in energy than the Mo orbitals. Thus, the Mo in the Fe–Mo cluster is also very similar electronically to the Mo in MoS_2.

In the third set of clusters (those containing V, Cr, and Mn), the results are not so clear cut. The "t_{2g}" sets of 3*d* and 4*d* orbitals lie very close in energy, and particularly in the a_1 set there is considerable mixing of the 3*d* and 4*d* orbitals. This mixing pushes one a_1 orbital down and one a_1 orbital up in energy, relative to the *e* orbitals. The *e* orbitals, on the other hand, are quite close in energy. In these clusters the number of electrons varies from four to seven, and because of the near degeneracy of the two sets of *e* levels, numerous occupations and spin states are possible. Other than observing that there is no clearcut transfer of electrons between Mo and the 3*d* metal, it is very difficult to assign with any certainty an actual configuration of electrons in the metal *d* orbitals.

In summary, when the first row transition metal M′ is varied in these Mo–M′ clusters, the changes in electronic structure depend on changes in both the relative energies of the Mo 4*d* and M′ 3*d* orbitals and the number of "*d*" electrons that M′ contributes to the cluster. As the 3*d* levels drop in energy across the first transition series, the number of 3*d* electrons occupying these levels increases. It is only when M′ is Co, Ni, or Cu, however, that these effects combine in such a way that the electronic state of Mo is formally affected. In the presence of Co or Ni, Mo is reduced relative to the Mo present in MoS_2. That is, the formal Mo 4*d* orbital occupation increases from 2 in MoS_2 to 3 and 4 in the presence of Co and Ni, respectively. In the presence of Cu, on the other hand, Mo is formally oxidized so that the formal 4*d* occupation decreases to 1. These effective reductions and oxidation of Mo correlate with an increase and decrease in catalytic activity. These results suggest that the promotion effect is indeed electronic in origin and that it arises from the ability of the effective promoters to donate electrons to Mo and thus increase the number of "4*d*" electrons. Likewise, a poison such as Cu has the ability to remove electrons from Mo and thus decrease the number of "4*d*" electrons. None of the other 3*d* metals appears to have the ability unambiguously to formally donate to or withdraw electrons from Mo.

As well as suggesting that there is an electronic origin for the promotion effect, these results are also consistent with the earlier theory relating catalytic activity in the binary sulfides to several electronic factors. These calculations indicate that in the promoted Co/Mo and Ni/Mo systems one of these electronic factors for Mo, n, the number of electrons in the HOMO, is increased by the presence of the promoter. Since these extra electrons occupy an antibonding orbital the Mo–S bonds are somewhat weakened. Thus, the factor B that enters the activity parameter and measures the relative Mo–S covalent bond strength decreases in the presence of the promoter. On the other hand, the presence of a poison such as Cu has just the opposite effect on both these factors. To a first approximation, for all the other systems considered here, the two electronic factors n and B for Mo are affected very little by the presence of the $3d$ transition metal.

A simple activity parameter was also constructed for the model clusters treated here. This was done by taking n as the occupation number of the highest occupied Mo based set of orbitals. When n is the same as in MoS_2, it is assumed that B is unchanged from the value for MoS_2 and the activity parameter A_2 is thus the same as in MoS_2. If n is different from MoS_2, the previous results for MoS_2 were used to recalculate the value of B, taking into account the increase or decrease in the number of antibonding electrons. This approach was used, rather than attempting to redefine and calculate B for these mixed metal clusters, because the sulfur atoms that are shared between the two metals make such a redefinition far from straightforward. Moreover, since any change in B is considerably smaller than the corresponding change in n (in the Ni/Mo cluster, for example, B decreases by only 15% while n increases by 100%), the variations in A_2 are dominated by changes in n. Thus, a new definition of B would not have a large effect on these variations. The activity parameter for the model clusters, calculated by the method just described, is plotted in Figure 6.17, along with the experimental HDS activities of the real catalyst systems. The trends in activity are clearly reproduced by the trends in the value of the activity parameter. The correlations obtained here suggest that promotion does have an electronic basis and that the electronic factors identified earlier in the study of the binary sulfides are influenced by the presence of a promoter.

The following picture of the electronic structure of an active HDS catalyst thus emerges from the combined results of these cluster calculations. In an active binary sulfide catalyst (one containing a $4d$ transition metal) the transition metal is able to covalently bond very effectively (in both a sigma and pi manner) to sulfur. In the very best catalysts this ability is combined with a high electron density in the metal d orbitals. The dominant electronic factor related to the promotion of MoS_2 is the increase in this number of "d" electrons associated with Mo. This effect

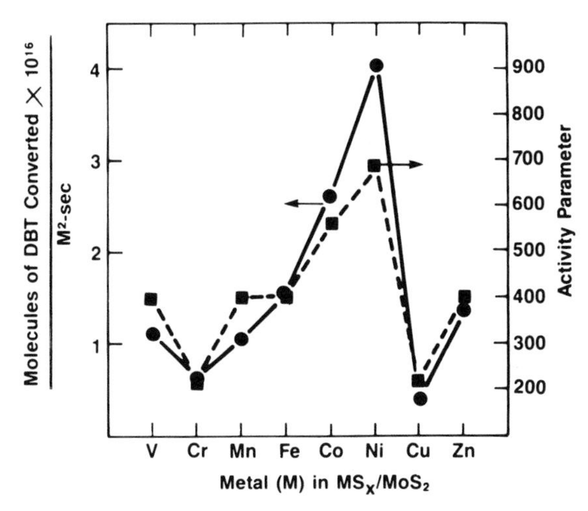

Figure 6.17. Calculated activity parameter for each mixed metal sulfide system (right-hand scale). Shown for comparison are the measured HDS activities (Ref. 19).

appears to be more important than the smaller changes in the strength of the Mo–S bond. Of the $3d$ transition metals, only Co and Ni clearly act as promoters, because it is only for these two metals that the number of $3d$ electrons and the relative energies of the $3d$ and $4d$ orbitals combine in such a way that electrons are donated to Mo. The relation described earlier between catalytic activity and average heats of formation for the synergic pairs takes into account changes in metal–sulfur bond strength but does not take into account any type of electronic interaction between the metals. Although these calculations suggest that the electronic interaction dominates, changes in metal–sulfur bond strengths do accompany these electronic interactions. An estimate of the magnitude of these changes is provided by changes in the quantity B, but a more quantitative measure would be useful. The electronic interactions and the variations in metal–sulfur bond strengths are interdependent, and both probably play a role in promotion. For example, a weaker metal–sulfur bond would make it easier to remove a sulfur and thus create an active site. At the same time, electronic interactions between the promoter and Mo at the active site would alter the electronic configuration of the active site.

These results provide (1) a correlation between the electronic structure of the bulk transition metal sulfides and their catalytic activity and (2) an electronic explanation for the promotion effect. Since these results apply to models for the bulk sulfides, however, they only provide indirect information about the surface interactions that immediately affect the catalytic process. The active sites in these catalysts are most likely sulfur vacancies, i.e., exposed metal centers, and in most proposed HDS mechanisms a sulfur-containing molecule such as thiophene, benzothiophene, or dibenzothiophene binds to the catalyst surface through such a metal center. The fact that a dominant factor in the correlations is n, the number of electrons in the HOMO, suggests that the number of electrons available at this exposed metal center (in solid-state terminology, the metal d electron density of states at the Fermi level) is particularly important. That is, greater available d electron density on the metal center leads to higher activity. This in turn suggests that donor rather than acceptor properties of the metal center are important to catalytic activity. Direct information about these surface interactions requires a better understanding of the nature of the metal sulfide surfaces, the active sites, and the interactions between these sites and the reacting molecules. Most of the work aimed at these problems, both experimental and theoretical, has concentrated on MoS_2 and WS_2, and in the next sections some of this work is described.

STRUCTURAL EFFECTS IN LAYERED TMS

The previous sections describe the dominant role played by electronic structure in determining total HDS activity of the TMS. In this section the effect of crystal structure on catalysis by MoS_2 and WS_2 will be described. MoS_2 and WS_2 exhibit the same highly anisotropic layered structure (Fig. 6.8). The structural anisotropy is a consequence of the chemical bonding. Within one layer, the structure can be viewed as a two-dimensional macromolecule. Each metal atom is bound to six sulfur atoms and each sulfur atom is bound to three metal atoms. Because the sulfur is so tightly bound, its interaction with the next layer of sulfur above it is extremely weak. This creates the "van der Waals" gap, which is the main feature of interest in regard to intercalation and lubrication properties.[40] At the edge of the layer, however, incompletely coordinated and thus highly reactive atoms can occur. Thus, although the basal planes (002) have been the general focus of studies in the vast intercalation literature, the "edge" planes (100) of the layered TMS become the focus of catalytic studies. Because of this structural anisotropy, MoS_2 crystallites tend to grow in thin plates so that many basal plane sites and few edge plane sites occur.

Recent examinations of the edge plane properties of MoS_2 consist of chemical and physical studies of MoS_2 single crystals, microcrystals, and powders aimed at understanding the relation between edge plane properties and catalytic activity.[41,42] The chemical reactivity of the edge planes both in oxidation and in segregation of metallic impurities, such as promoters, has been studied.[43] The electronic structure of these edge planes has also been measured by photoemission and optical absorption.[5] This study showed that optical absorption below 1.2 eV is due to exposed edge planes. Surface defects such as "dangling bonds" or surface vacancies would be expected to have midgap states and thus increase the optical absorption in this region. In a recent set of calculations that modeled different types of sulfur vacancies (or edge defects), Xα calculations were used to calculate optical transitions for the various defect sites.[44] The allowed transitions were calculated to fall into the energy range below 1.2 eV. These results thus suggest that sulfur vacancies are responsible for the optical absorptions measured for the edge planes.

The potential importance of MoS_2 edge planes in hydrotreating catalysts has long been recognized.[20,45] Evidence for the reactivity of edge planes in MoS_2 can be found in the linear correlation between O_2 chemisorption, which measures edge plane surface, and the HDS of DBT.[4] In general, HDS activity does not correlate to N_2 BET surface area measurements. This is because the basal plane area contributes to the total surface area but not to the catalytic activity. Therefore, MoS_2 catalysts made by a variety of preparative methods will have widely different edge to basal plane ratios and only the O_2 chemisorption will give a good correlation to activity. If the preparative method is constant, however, the basal plane area can be proportional to the edge and a good correlation between total surface area and activity can be obtained.[46]

The potential importance of edge planes in the promotion of MoS_2 and WS_2 has also been long recognized. The concept of pseudointercalation of Ni atoms at the edge of WS_2 was introduced to account for the increase in benzene hydrogenation activity[21] and was extended to HDS activity in the Co/MoS system.[47] The term pseudointercalation is used to make the point that unlike other layered sulfides such as NbS_2 and TaS_2, MoS_2 and WS_2 do not normally intercalate most metals. To pseudointercalate means to insert in octahedral or tetrahedral holes near the edges (Fig. 6.8). Other promotion models have also been presented that emphasize the role of Co or Ni at the MoS_2 or WS_2 edge planes.[16,20]

To further verify the importance of the edge planes, a geometrical model was developed for unpromoted and promoted MoS_2 and WS_2 catalysts. Using this model, it was possible to correlate the number of edge sites with catalytic activity and thus verify the importance of these edge sites in HDS and HYD catalysis.[48,49] Although the experimental data used for the correlations was for supported catalysts, the authors assumed, based on experimental evidence, that the catalyst consisted of small patches of single slabs of MoS_2 (WS_2). These slabs were viewed as fragments of a single layer of MoS_2 and thus modeled by slabs of different geometrical shapes and sizes. The shapes [chain (C), triangle (T), rhomb (R), and hexagon (H)] are illustrated in Figure 6.18. For each shape, the number of corner, edge, and basal sites can be calculated as a function of

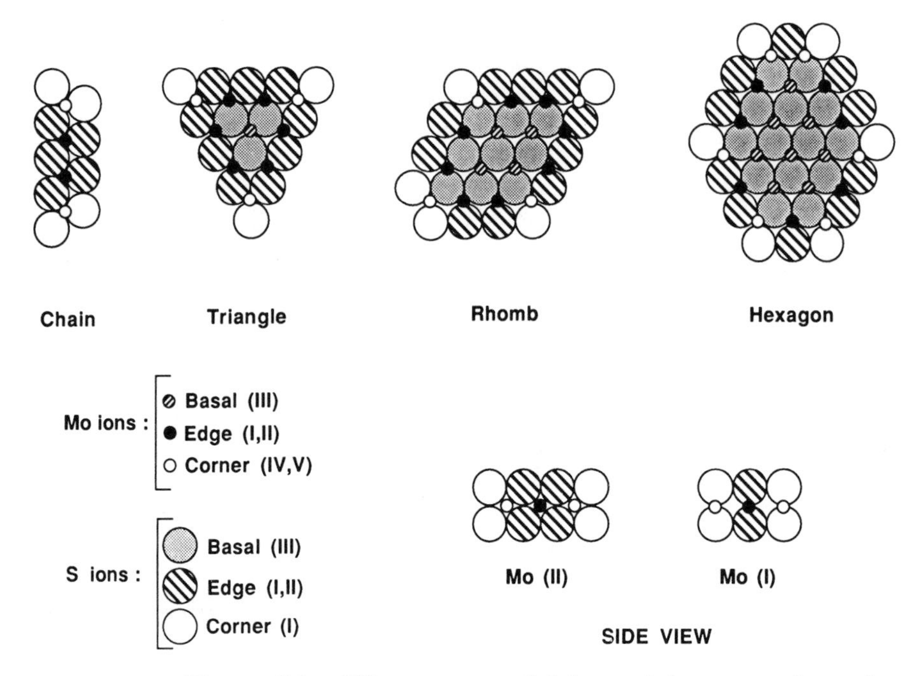

Figure 6.18. Views of the different types of slabs used for geometric modeling by Kasztelan et al.[49] The top half of the diagram shows top views of the different symmetrical slabs. The side views illustrate the two types (I and II) of local environment of Mo at the edges of a slab.

the slab size. Based on this model, it was possible to predict for each slab shape the specific activity for a given reaction, the amount of chemisorbed oxygen, and the promoter effect as a function of Mo (W) content. These predictions could then be compared with published experimental results.

Different types of Mo and S ions can be identified for the various slab shapes, and these types are indicated in Figure 6.18. The largest differentiation of the Mo and S ions depends on whether they occupy basal (b), edge (e), or corner (c) positions. Depending on their local environment, further differentiation of the ions can also be made. Each basal S is coordinated to three Mo's, and all basal Mo's and S's are thus labeled III. Along the edge, two types of Mo and S ions can be identified. They are distinguished as types I and II and are determined by whether the edge S is coordinated to one or two edge Mo's. Finally, each corner S is of type I, but the corner Mo's are further differentiated by their S coordination. Besides being coordinated to the type I corner S's, a corner Mo can be coordinated either to four other type II edge S's or to two type II edge S's and two type III basal S's. These corner Mo's are differentiated as type IV and V, respectively. It is informative to note these distinguishable edge and corner sites, because they result in different types of sites when edge sulfurs are removed and thus might be distinguishable catalytically. For example, a type I edge Mo can at most accommodate two sulfur vacancies, while up to four sulfur vacancies can occur for a type II edge Mo or a corner Mo. We will return to this point subsequently.

The number of each type of ion, M_b, M_e, M_c, S_b, S_e, S_c, or the total number of metal and sulfur ions, M and S, can be calculated as a function of the geometry and slab size. Then if each type of site has an activity a_x (x = b,e,c) for a given reaction, the activity resulting from the x sites in a single slab is

$$A_x = a_x \mathrm{M}_x$$

For N slabs per gram of support, the total activity per gram of support is therefore

$$A_x = N a_x \mathrm{M}_x$$

The metal loading in grams per gram of carrier is $m_\mathrm{M} N \mathrm{M}$, where m_M is the atomic weight of the metal, so that the specific activity of the catalyst as a function of metal loading is expressed as

$$A_x^s = (a_x/m_\mathrm{M})(\mathrm{M}_x/\mathrm{M})$$

It is important to note that this expression depends only on the geometrical characteristics (through M_x/M) of a single slab. For a given shape with a total of M metal atoms, the relative numbers of metal atoms in an edge position M_e/M, a corner position M_c/M, or the basal plane M_b/M vary differently as the size of the slab and thus M increase. These variations are illustrated in Figure 6.19 for a rhomb. If the number of slabs N remains constant as the metal loading on the carrier increases, then the slab size "M" increases and the M_x/M ratios vary. If a given catalytic reaction specifically involves one type of Mo ion, a curve similar to that for the appropriate ratio should also result for the variation in specific activity versus metal loading. The curve in Figure 6.19 for the variation of the M_e/M ratio was found to be particularly interesting because it shows a maximum. The variation of M_e/M for all four of the geometries is plotted in Figure 6.20, and it can be seen that except for the chain all of the curves show a maximum. The maxima occur for different slab sizes for the three

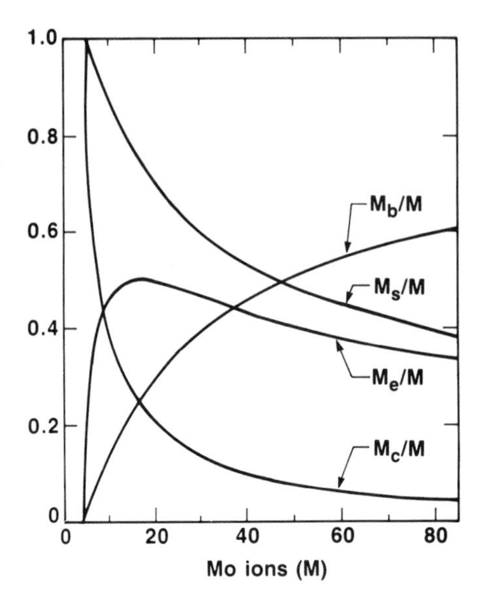

Figure 6.19. Variation of the specific number of each type of Mo ion (M_c, M_e, M_b, and M_s) versus the slab size M (number of Mo ions) for the rhomb model of MoS_2 (Ref. 49).

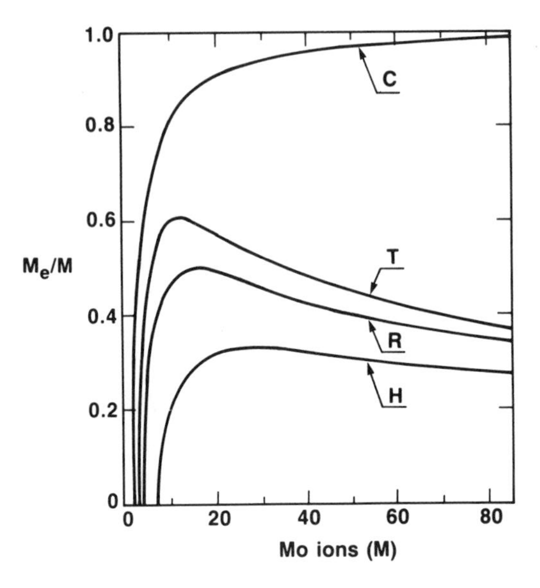

Figure 6.20. Variation of the specific number of edge Mo ions [M_e/M versus the slab size M for the different symmetrical MoS_2 slab shapes (chain, triangle, rhomb, and hexagon] (Ref. 49).

shapes. These plots apply to symmetric slabs, and it should be noted that in general this maximum does not occur for an unsymmetric slab.

The curves for M_e/M are emphasized because comparisons with experimental data for HDS, HYD, and O_2 chemisorption show that curves for activity and chemisorption as a function of metal loading all resemble the curves in Figure 6.20. (That is, they resemble the curves for the triangle, rhomb, and hexagon, not the chain. This eliminates the chain as a possible structure.) This suggests that increased metal loading results in growth of the slabs and not an increase in the number of slabs. More important, it indicates that only edge sites are involved in the rate-limiting step for the HDS and HYD reactions studied and in O_2 chemisorption. These results cannot distinguish, however, between either the different types of edge sites or the three different symmetric slab types. Further comparisons between calculated and experimental S/M ratios, however, suggest that the symmetric rhomb or hexagon geometry is more likely than the triangular geometry.

These models were also extended successfully to promoted MoS_2 systems. It was assumed that effective promoter ions are located at edge positions of the slab. Different positions are possible, but no assumptions were made about these positions. It was assumed that decoration of the edge of the slab can occur up to a maximum of one promoter ion for every edge Mo and that beyond this, excess promoter ions begin to form a stable bulk sulfide (e.g., Ni_3S_2 when the promoter is Ni). Finally, an increase in activity by a factor $q = a_p/a_e$, where a_p is the activity of a promoted site, is expected for each promoted site, so that the promoter is assumed to improve the "quality" of each site. The change in activity that occurs with decoration can then be calculated for the various model slabs under different conditions, and the influence of parameters such as the number of Mo's, M, the total number of promoter atoms, P_T, the total number of metal ions, T (where $T = M + P_T$), the ratio of promoter to total metal ions r [where $r = P_T/(M + P_T)$], and the promotion ratio q can be calculated. The types of curves which result from this analysis are illustrated for a hexagon in Figure 6.21. Figure 6.21a shows the calculated variation in activity for a slab of constant size as the number of promoter ions increases, while Figure 6.21b shows the calculated variation in activity for a slab of constant total metal content T as the ratio r increases. For the case where the slab size is constant as the number of promoter ions increases, Figure 6.21a shows that stepwise decoration first results in an increase in activity. When the number of promoters exceeds M_e, however, the curve levels off, because a bulk promoter sulfide begins to form. The activity may remain constant (α) if the bulk sulfide formed has no activity, it may increase slightly (β) if the bulk sulfide has low activity, or it may decrease slightly (γ) if the bulk sulfide obstructs the edge sites or

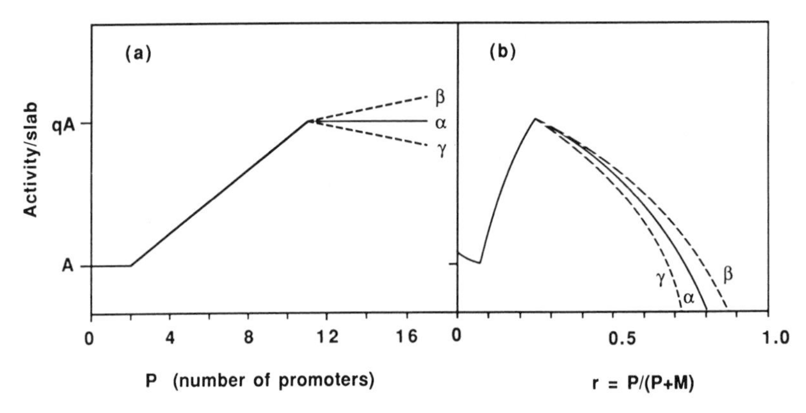

Figure 6.21. (a) Variation of activity of a hexagonal slab containing 27 M atoms (M_e = 9) versus the number of promoters at edge positions. (b) Variation in activity versus r (r = promoter/promoter + Mo) for samples at constant metal (promoter + Mo) content (Ref. 48).

depletes the edges of promoter. For the case where the total number of metal atoms remains constant while the ratio of promoter to total metal increases, Figure 6.21b shows that the activity increases to a maximum and then decreases. The steepness of the decrease will vary for the three cases, α, β, or γ described previously. The volcano shape is a consequence of the geometry of the MoS_2 slab. The optimum r (0.25 for the hexagon) varies with geometry and slab size, but generally falls in the range between 0.25 and 0.375. Both of these plots are particularly interesting because the results of a number of experimental studies at these conditions were available for comparison. It was found that experimental activity curves for constant Mo loading closely resemble the curve in Figure 6.21a for both HDS and HYD reactions. Likewise the experimental curves for constant metal (Mo + promoter) loading, probably the most numerous in the literature, show the same volcano shape found in Figure 6.21b. More important, not only is the shape similar, but also the optimum ratio r usually falls between 0.2 and 0.4. These and other correlations between the experimental activity and the predicted activity based on this geometric model provide evidence that the active sites do occur at edges. Once again, however, although these results clearly indicate the importance of edge sites and the association of promoter ions with these edge sites, this purely geometric model does not provide information about the nature of the sites.

Several conclusions were reached based on these modeling results. For the cases considered, the hydrotreating catalysts can be described as small rhombohedral or hexagonal patches of MoS_2 having optimum sizes in the range of 10–20 Å. Since for all of the reactions considered the variation of the specific activity for unpromoted MoS_2 catalysts is similar to the variation of the relative number of edge sites M_e/M on a single slab of MoS_2, the rate-limiting step for all these reactions occurs only at edge sites. The nature of these edge sites cannot be defined. Two different types of edge Mo's (Mo_I and Mo_{II}) exist. These may accommodate different levels of S unsaturation, but this geometrical model cannot differentiate between all the possible types of sites. Both the effect of reconstruction at the edges as well as electronic factors associated with the different types of sites become important if one wishes to try to describe the edge sites in detail. Finally, the fact that the variation in catalytic activity for the promoted MoS_2 catalysts resembles the variation predicted assuming edge decoration by promoter atoms indicates that the promoters do lie at the edge planes. Once again, however, the model does not provide detailed information as to the nature or exact position of the promoter. Just as in the CoMoS phase described by the Topsøe group,[16] this model suggests a close association of Co and Mo around the edges of a MoS_2 slab, but it does not provide information about the nature of the active site. The role of the promoter may be to enhance the activity of the Mo edge site or it may provide a new site, either alone or in concert with an edge Mo. Once again, a better understanding of both the structure of the edges of the promoted MoS_2 catalysts and the electronic properties associated with the different possible sites is required for a detailed description of these sites. This study clearly indicates, however, the importance of geometric factors in MoS_2- and WS_2-based hydrotreating catalysts.

ACTIVE SITE THEORY

Although the actual nature of active sites in hydrotreating catalysts remains an uncertainty, a number of researchers have modeled the adsorption of sulfur-containing molecules on active sites. For the most part, this work has involved very simple models. Semiempirical molecular orbital techniques have been used to study the electronic structure of these model systems, and these studies have led to diverse and often conflicting conclusions regarding adsorption and adsorption sites. All of the work has used thiophene, SC_4H_4, as the sulfur-containing molecule and has attempted to study the interaction of thiophene with some model of the active site on an unpromoted or promoted MoS_2 catalyst. Assuming that an incompletely coordinated edge Mo forms the active site, there are numerous possible types of sites. Using the terminology of Kasztelan et al., Figure 6.18, an active site at Mo_I could have either one or two sulfur vacancies, while an active site at Mo_{II} could have anywhere from one to four sulfur vacancies. Complicated combinations of these two could also be envisioned. A thiophene molecule could interact with such an active site in several ways. The two interaction modes that are most often discussed are a vertical adsorption mode, where the thiophene binds to an edge Mo through the ring S and thus stands up perpendicular to the edge, and a horizontal mode, where the ring lies flat along the edge. In this mode the thiophene could bind to an edge Mo in several ways, but the two ways most often considered are through all five ring atoms (an η^5 mode) or through the ring S only. The criterion for the most favorable site is usually chosen to be the amount of weakening of the C–S bonds of the adsorbed species relative to free thiophene.

The first and most approximate model calculations were carried out by Zdrazil and Sedlacek.[50] Assuming that the active site acts as an electron acceptor, they modeled the active site with an H^+ ion and used both Complete-Neglect-of-Differential-Overlap (CNDO/2) and Hückel-molecular-orbital (HMO) calculations to study different adsorption modes of thiophene on this model site. They concluded that a horizontal binding mode with the ring bound to the active site through the ring sulfur atom was most favorable. The extreme simplification of the approximation of the active site limits the validity of these conclusions.

The next model calculations were carried out by Duben.[51,52] Modeling the active site with a Mo^{3+} ion, HMO calculations were used to study both the vertical adsorption of thiophene and the mechanism of HDS. Once again the use of such a simplified model limits the usefulness of the results, but it is important to note that this work recognized the character of the molecular orbitals of free thiophene and related this character to interactions between the thiophene and the Mo center. The three highest-energy occupied orbitals and the lowest-energy unoccupied orbital (LUMO) of thiophene are illustrated in Figure 6.22. Assuming that bonding occurs in a vertical mode through the ring sulfur, the highest-energy occupied orbital (HOMO) in thiophene ($1a_2$) has no S character and thus would not be involved in binding to the Mo. Instead, binding could occur through electron donation from the filled $6a_1$ (σ) or $2b_1$ (π) orbitals or back donation from the metal into the empty $3b_1$ (π^*) orbital or both. Donation into this empty π^* orbital would result in a weakening of the C–S bonds and could thus serve to aid the C–S bond breaking process. Although Duben was considering only a vertical adsorption model when he noted the possible importance of the occupation of this π^* orbital, occupation of this orbital could also occur in horizontal adsorption. Duben later extended this work to a model for the adsorption of pyridine.[52] (This is apparently the only attempt to model the electronic interaction between a nitrogen containing heterocycle and an HDN catalyst.) Based on these

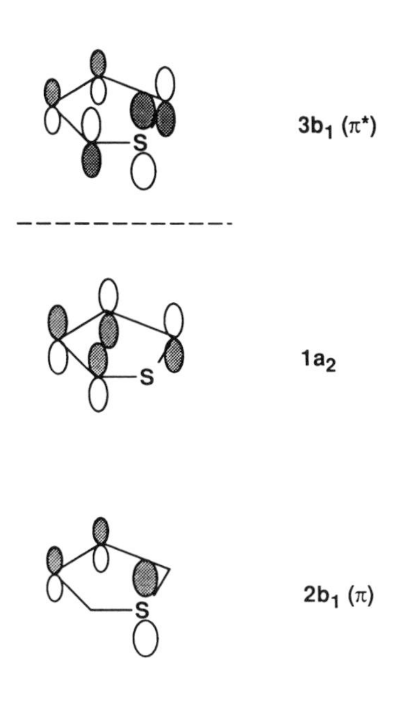

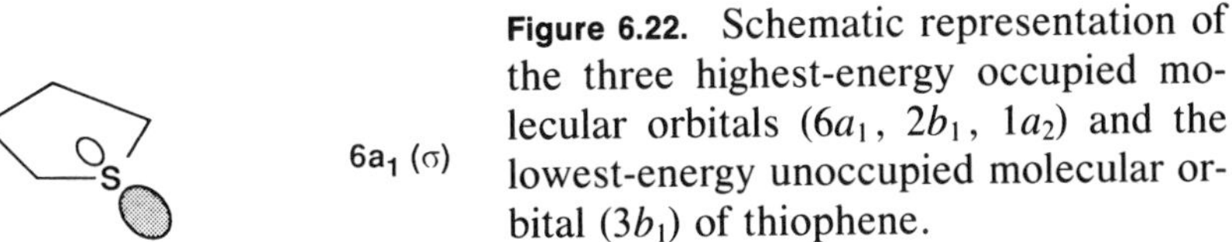

Figure 6.22. Schematic representation of the three highest-energy occupied molecular orbitals ($6a_1$, $2b_1$, $1a_2$) and the lowest-energy unoccupied molecular orbital ($3b_1$) of thiophene.

results, it was concluded that the differences in HDS and HDN mechanisms are related to the different electronic properties of five- and six-membered rings rather than to the different heteroatoms. Again the validity of these conclusions is somewhat doubtful.

A more complicated model for both adsorption and HDS involves a molybdenum oxysulfide model for the HDS catalyst. A brief report of this type of model was given by Vladov et al.,[53] while a much more complete study was described by Ruette and Ludena.[54] They assumed that an active HDS catalyst is based on an oxidic precursor consisting of Mo–O chains in which each Mo is tetrahedrally coordinated by four oxygens. They modeled this chain with an Mo_3O_{10} cluster (Fig. 6.23a) and assumed that O_3, O_7, and O_{11} in Fig. 6.23a are the surface O's. The energetics of several reaction pathways were calculated, and the following pathway was found to be energetically feasible. The first steps involve hydrogena-

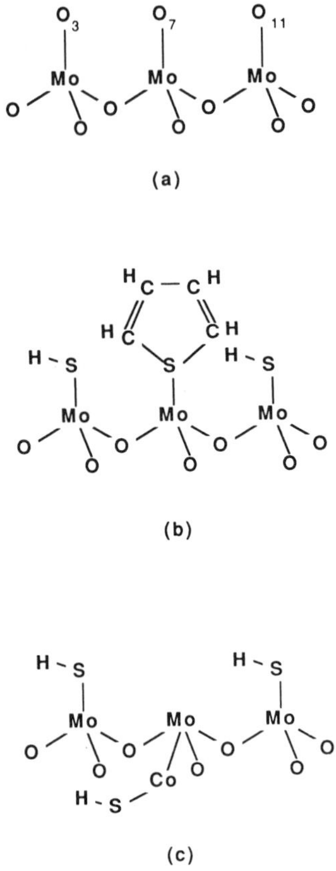

Figure 6.23. Model clusters used by Ruette and Ludena[54] for molecular orbital calculations: (a) the Mo_3O_{10} cluster, (b) the cluster used for studying interaction between thiophene and the catalyst, (c) the cluster used as a model for the promoted catalyst system.

tion of the three surface O's and the removal of an –OH group to form a vacancy on this surface:

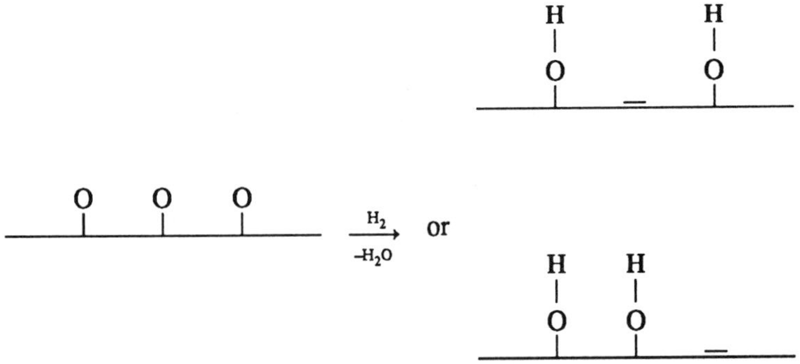

(6.1)

A number of steps then leads to sulfidation of the surface and finally the formation of a vacancy on this sulfided surface:

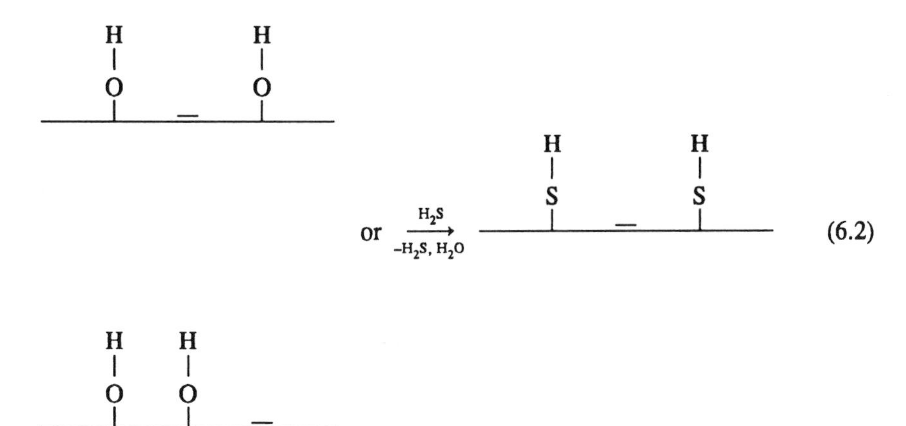

The site formed in Eq. (6.2) was found to be energetically favorable and thus considered as the active site for HDS. The interaction of the model site with thiophene was then considered. A vertical binding mode was chosen (Fig. 6.23b) and an attempt was made to elucidate the mechanism for the C–S bond breaking. One of the adjacent H's was allowed to interact with the thiophene ring—in particular with either the S or the adjacent C individually, or with both the S and adjacent C simultaneously. The energetically most favorable process was found to be one resulting in an intermediate structure where the H is shared by the S and C atoms. Subsequent hydrogenation would occur elsewhere on the ring. The authors concluded that simple C–S bond cleavage may not occur and that instead the HDS reaction may go through complicated transition states.

An attempt to understand the effect of adding a Co promoter was also included in this study. Two possible positions were considered for the Co. In the first configuration, where the Co forms an oxygen bridge with the Mo at which the vacancy occurs, the presence of the Co was found to have little effect on the vacancy site. In the second configuration, Figure 6.23c, the Co replaces an oxygen in an adjacent (but missing in this model) molybdenum oxygen chain and is assumed to be coordinated to an –SH group. This configuration, which results in a direct interaction between Mo and Co, decreases the charge on the Mo and also changes the character of the LUMO at the vacancy site. The authors concluded that an explanation for the promoter effect might lie in the electronic changes in the LUMO that are induced by the presence of the Co when it interacts directly with Mo.

This work illustrates how difficult it often is, in model calculations of this sort, to separate the results from the model. In fact, it is very difficult to assess the results of this study because of the model. The authors have attempted to choose a more realistic model than the simple ones described previously, but in doing this they have introduced the added complication of an oxymolybdenum precursor and have had to make assumptions about its structure. In addition, assumptions have had to be made about the degree of sulfidation and hydrogenation. Although the work presents a careful set of systematic calculations, it is difficult to determine the significance of the conclusions of the study because it is simply not possible to determine how model dependent these conclusions are.

In another series of cluster calculations, Nikishenko et al.[33,55–57] used Extended-Hückel-Molecular-Orbital (EHMO) calculations to model MoS_2, Co promoted MoS_2, thiophene adsorption, H_2 activation, and mechanisms for HDS of thiophene. They used a cluster fragment taken from the MoS_2 structure, which eliminates some of the problems described above for the molybdenum oxysulfide cluster but also introduces some new ones. A fragment removed from the 2H–MoS_2 crystal lattice is illustrated in Figure 6.24a. In their early calculations, Nikishenko et al. used part of this fragment as their basic cluster. The cluster, having the formula Mo_7S_{24}, is indicated in Figure 6.24a by the heavy lines and circles. Most of their work was carried out, however, for the smaller cluster illustrated in Figure 6.24b. This basic cluster, which has the formula MoS_{12}, was used to study thiophene adsorption, Co promotion, and H_2 activation. Sites I, II, and III where Co can be substituted into the lattice are also indicated in Figure 6.24b. Calculations for clusters having a large range of Co/Mo ratios were carried out. These are listed in Table 6.3. (The clusters with 10 S's have two S vacancies.) Very extensive calculations were carried out to consider the effect of Co promoters, the optimum sites for thiophene and H_2 adsorption, and the possible HDS mechanism. Conclusions were reached based on the relative energy of the HOMO and on the relative stabilization energies of various configurations. These conclusions are of limited usefulness, however, because of very serious drawbacks in the model clusters. First, the Co's and Mo's in these clusters are assumed to occupy adjacent trigonal prismatic sites. This is a configuration that brings the metals into much closer contact than is probably reasonable. It is more likely that the Co's either occupy sites between the MoS_2 layers or substitute for Mo's in the MoS_2 lattice. Second, these clusters were all assumed to be neutral. Given the stoichiometry of the clusters in Table 6.3, this means that the electronic structure of these clusters in no way models the actual electronic structure of the species that the clusters are attempting to model. Once again, then, an

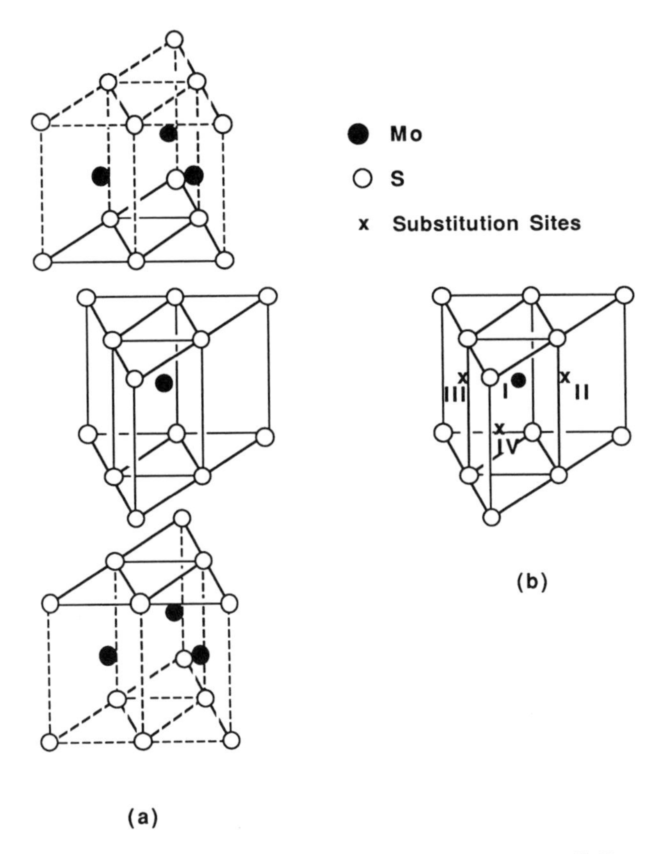

Figure 6.24. Fragments of MoS_2 used by Nikeshenko et al.[33,55] for molecular orbital calculations: (a) The Mo_7S_{36} fragment. Calculations were carried out on the Mo_7S_{24} cluster indicated by the solid lines. (b) The MoS_{12} cluster used for most of the calculations.

attempt to improve the model actually limits the validity of the conclusions reached in the study.

Another research group also employed clusters to model MoS_2 and to study the possible modes of thiophene adsorption on MoS_2. Using EHMO theory, Joffre, Lerner, and Geneste[58,59] carried out a series of calculations aimed at identifying both an optimum adsorption site and an optimum adsorption mode. All of the clusters were based on an MoS_6^{8-} trigonal prismatic fragment (Fig. 6.25a) having the geometry found within a sheet of MoS_2. Using this as the base cluster, HDS site models were obtained by removing one, two, and three sulfurs. These type of vacancy models are also illustrated in Figures 6.25b–d. (These are the same type of

Table 6.3. Calculated cluster structures for model cluster in Figure 6.24b.

	Sites	
Cluster Type	Co	Mo
$MoS_{10,12}$	—	I
$MoS_{12,12}Co$	III	I
$MoS_{10,12}Co_2$	II,IV	I
$MoS_{10,12}Co_3$	II,III,IV	I
$CoS_{10,12}$	I	—
$CoS_{10,12}Mo$	I	III
$CoS_{10,12}Mo_2$	I	II,IV
$CoS_{10,12}Mo_3$	I	II,III,IV

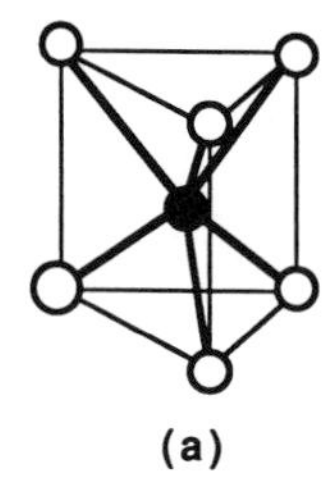

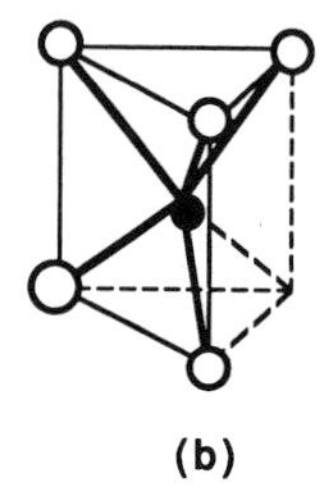

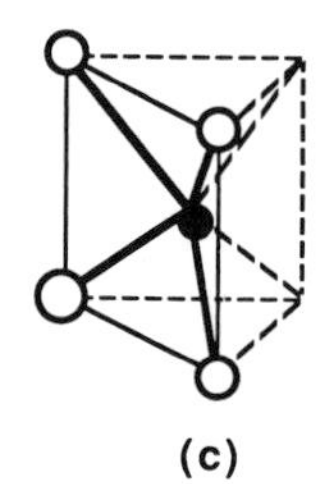

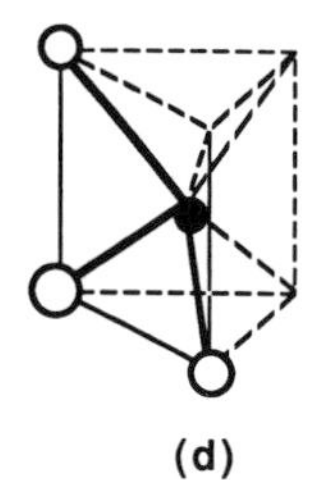

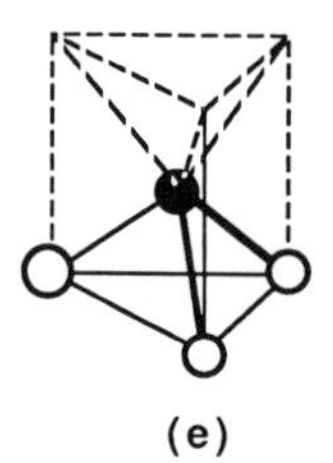

Figure 6.25. Structures of the model clusters used by Joffre et al.[58,59] for molecular orbital calculations: (a) trigonal prismatic MoS_6 cluster; (b) one vacancy site model; (c) two vacancy site model; (d) and (e) two different three vacancy site models.

clusters used by Horsley[44] to calculate optical transitions for defect sites.) There are two distinct ways to create three sulfur vacancies, and both of these models were employed. The actual clusters employed had formulas such as $MoS_5H_5^-$, MoS_4H_4, and $MoS_3H_3^-$. The protonation of the clusters provided moderately charged clusters that still allow molybdenum to be formally Mo^{4+} and each sulfur to be formally S^{2-}. Vertical as well as several horizontal modes of thiophene adsorption were considered. Energies of interaction were calculated as the difference between the energy of the site model–thiophene complex and the sum of the energies of the isolated site model cluster and thiophene molecule. Energy curves were calculated as a function of the thiophene–molybdenum distance. For all the adsorption modes it was found that the interaction energy is unfavorable for the one vacancy model site but then becomes more favorable as the number of vacancies increases. Vertical adsorption, however, is preferred over horizontal adsorption for all of the site models, although the difference between vertical and horizontal adsorption energy is smaller for the three-vacancy-site model than for the other sites. These calculations therefore suggest that vertical adsorption at either a three-vacancy or two-vacancy site is a likely mode of adsorption at the edge of a MoS_2 sheet. Although both the model clusters and the computational technique used for these calculations are very simple, this study provided the first systematic comparison of the different types of vacancies which might occur on a MoS_2 catalyst.

A series of calculations that utilized probably the most realistic model for MoS_2 was recently carried out by Zonnevylle, Hoffmann, and Harris.[60] Using the extended-Hückel-tight-binding method, they calculated the electronic structure of an extended edge section of the MoS_2 lattice and then studied the interaction of thiophene with the edge surface. A section of the model for MoS_2 is shown in Figure 6.26. This can be considered as an infinite three-layer ribbon that lies along the edge of an MoS_2 sheet. In this representation, the view of MoS_2 has been turned 90° from the more familiar view of an MoS_2 sheet. Here the basal plane lies in the yz plane and the edge lies in the xy plane. The z axis is perpendicular to the edge. This structure is translated in the y direction to give an infinite model that is neutral and has the stoichiometry MoS_2. It provides a representation of two particular MoS_2 edges. The first, which includes Mo(3) and Mo(6), corresponds to an edge where two sulfurs have been removed from every Mo. In Kasztelan's terminology (Fig. 6.18) each Mo along this edge is an Mo_I with two sulfur vacancies. The second edge, which includes S(7), S(8), S(10), etc., corresponds to a fully sulfided edge. In Kasztelan's terminology, each Mo along this edge is a Mo_{II} with no sulfur vacancies. The calculations concentrated on the sites with two sulfur vacancies, because not surprisingly for a fully sulfided edge, interactions

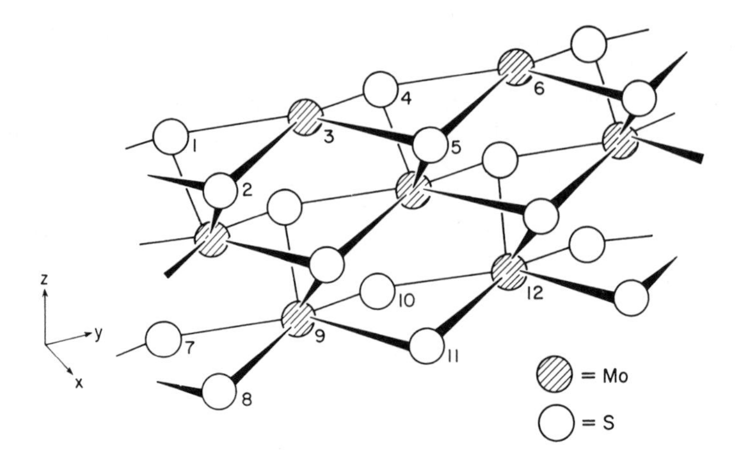

Figure 6.26. A portion of the infinite MoS_2 ribbon model used by Zonnevylle et al.[60] for band structure calculations. The top and bottom *xy* planes represent two kinds of MoS_2 edge planes. The basal plane lies in the *yz* plane.

between thiophene and the sulfided edge were found to be negligible. A number of possible adsorption geometries on the Mo_I-type atoms protruding from the other edge surface were considered. These included vertical adsorption directly above a Mo or bridging two Mo's, adsorption with the ring sulfur directly above a Mo or bridging two Mo's but with the ring tilted 45° from the vertical, and finally horizontal adsorption with the ring S directly over a Mo or with the ring shifted to give an η^5 adsorption mode. The criteria used to judge the effectiveness of the binding site were the occupation of the $3b_1$ π^* orbital of thiophene (Fig. 6.22) and the weakening of the C–S bond in the thiophene ring (as measured by the C–S overlap population). These criteria led to the conclusion that the most effective binding mode is the η^5 horizontal one. This resulted in the largest π^* occupation and the smallest C–S overlap populations. Interestingly enough, the calculated binding energies show the same trends as the cluster calculations of Joffre et al.,[58,59] and an analysis based on binding energy would suggest that vertical adsorption is favored. For the type of sites studied here, however, the η^5 mode of adsorption clearly is more effective in weakening the C–S bonds in the thiophene ring. This suggests that initial adsorption might occur through an η^1 vertical mode, although subsequent reaction would require the ring to flip over into the η^5 horizontal binding mode. The results of this study also show that removal of surrounding sulfurs increases the C–S bond weakening resulting from η^5 coordination. Thus, increasing the coordinative unsaturation of an edge

Mo appears to increase its effectiveness as an active site. Even with the much more realistic model used for these calculations, it is important to recognize that the nature of the model itself may still influence the conclusions. There are numerous other possible choices of active sites, and how dependent the conclusions are on the particular choice of surface and site cannot be determined without a number of further calculations.

This study also addresses the question of promotion and how the addition of known promoters or poisons affects the adsorption of thiophene. Since Co and Ni are known promoters of the HDS reaction and Cu has been shown to be a poison,[19] the effect of adding these three 3*d* metals was studied. The effect of adding other 3*d* metals was not considered. Both substitution for Mo in the MoS_2 lattice and pseudointercalation into octahedral or tetrahedral holes between the MoS_2 layers were studied. The model for substitution was constructed from the model shown in Figure 6.26. Half of the protruding Mo's were simply replaced by Co, Ni, or Cu. The model for pseudointercalation was constructed by adding a second MoS_2 ribbon to that shown in Figure 6.26 and then placing the 3*d* metal atoms in octahedral or tetrahedral holes. In all cases, the 3*d* metals were added as neutral atoms. Substitution of any of the 3*d* metals for Mo was found to oxidize the adjacent surface Mo's. Placement of any 3*d* metal in octahedral holes was found to reduce the surface Mo's. Placing Cu in the tetrahedral holes also reduces Mo, while placing Co or Ni in the tetrahedral holes oxidizes the surface Mo's. These results suggested that if charge transfer effects are important for HDS activity the promoters and poisons may occupy different types of positions in the MoS_2 structure. The η^5 adsorption mode of thiophene was also studied in these bimetallic systems. Both adsorption over one of the protruding Mo's in the pseudointercalate model systems and adsorption over either the Mo or 3*d* metal in the substitution model were considered. Adsorption over the 3*d* metal was found to be unfavorable since the interaction between the metal and ring is weak. The effects on the adsorption over the Mo were much smaller and were not conclusive. It is interesting to observe, however, that even in the cases where the surface Mo is reduced relative to the surface Mo in MoS_2 (octahedral pseudointercalation of the promoter), this surface Mo does not become a better donor. In the intercalated systems, sulfurs interact with both Mo's and the 3*d* metal intercalates rather than just with Mo's. The Mo–S bonds are weakened relative to the Mo–S bonds in MoS_2 and the Mo 4*d* orbital, which serves as the donor orbital in the Mo–thiophene interaction, is stabilized. Thus, the energy difference between the Mo donor orbital and thiophene acceptor orbitals widens and the interaction between the two orbitals weakens.

The results of these calculations that involve charge transfer must be viewed with some caution, not because of the model, but because of the calculational technique. Since EH calculations are parametrized with valence orbital ionization potentials, the relative orbital energies of the two metals in systems like this are predetermined by the input values. Furthermore, since the calculations are not self-consistent, the relative energies of the metal orbitals are never allowed to adjust to the redistributions of charge that occur upon interaction. Thus, the change transfer predicted by the calculations is often biased by the original choice of oxidation state and orbital energies.

This brings up a very important problem that often arises in attempts to study electronic effects in catalyst systems in general, not just in HDS systems. Using a model that realistically represents a very complex catalyst system often requires the use of very approximate calculational procedures. The picture of the electronic structure that emerges may then depend on the choice of calculational methods. Using a simpler model allows the use of more exact calculational procedures, but now the resulting picture of the electronic structure probably depends on the choice of model. The results of the studies described in this section as well as in earlier sections illustrate how both these approaches have provided us with important information about the electronic structure of the metal sulfides themselves and about possible active sites and adsorption modes. Still, neither approach can provide us with definitive answers to the many questions that can be asked regarding HDS. Attempting to model accurately the very complex catalyst systems involved in HDS, while at the same time attempting to treat the electronic structure of the systems with less approximate nonempirical self-consistent calculations is a formidable task. Better experimental characterization of active sites and adsorption modes on MoS_2 and other sulfide HDS catalysts is really the input required by the theoretician. This might allow the development of fairly simple realistic models that could be studied with less approximate calculational techniques.

CONCLUSIONS

Various theoretical studies, in conjunction with experimental results, are helping to answer a number of important questions about the TMS catalyst systems. The approaches are diverse, but by combining the results of the various studies we gain a better understanding of the catalytic materials and the relation between electronic and structural properties of these materials and their catalytic activity. Thermodynamic correlations suggest that HDS activity is related to the strength of the metal–sulfur bond on the surface of the TMS catalyst. Further thermodynamic correlations for the promoted HDS catalysts suggest that surface sulfurs are shared between the "synergic" metals. The surface properties of the resulting "pseudobinary" sulfide then mimic the surface properties of a very active noble metal sulfide. The calculated electronic structure of the TMS suggests that it is the covalent contribution to the metal sulfur bond strength that is related to catalytic activity. The 4*d* and 5*d* TMS, which are more active than the 3*d* TMS, exhibit a larger covalent contribution to the metal–sulfur bond. Moreover, the most active 4*d* or 5*d* TMS catalysts also have a high *d* electron density available on the metal. Electronic structure calculations for models of the promoted HDS catalysts indicate that an effective promoter can serve to reduce Mo and thus increase the *d* electron density available on Mo. This suggests that Mo acts as an electron donor rather than an acceptor when it interacts with a sulfur-bearing molecule. Geometric models for MoS_2 verify that the edge planes of MoS_2 provide the active HDS sites. Electronic structure calculations for models of these active sites suggest that C–S bond weakening in thiophene (and thus activation of the thiophene ring) occurs through donation of electronic charge from a coordinatively unsaturated Mo into a π^* thiophene orbital. These calculations indicate that η^5 coordination of thiophene to Mo results in the greatest C–S weakening. They also suggest that the effect of a promoter on thiophene adsorptions may be quite complex.

Although these studies provide us with a better understanding of the basis for catalytic activity, a number of questions clearly remain. The complexity of the TMS based catalytic systems, however, makes any theoretical study of their electronic properties (particularly their surface properties) a formidable task. A more detailed description of the catalyst surface, active sites, and adsorption modes is sorely needed. As increasingly sophisticated experiments provide this information, we can expect that theory will provide us with a better understanding of the role played by the electronic structure of the TMS surfaces and active sites.

REFERENCES

1. Weisser, O., Landa, S., *Sulfide Catalysts: Their Properties and Applications*. Oxford: Pergamon, 1973.
2. Dianis, W.P., *Applied Cat.* **301:** 99 (1987).
3. Pecoraro, T.A., Chianelli, R.R., *J. Catal.* **67:** 430 (1981).
4. Tauster, S.J., Pecoraro, T.A., Chianelli, R.R., *J. Catal.* **63:** 515 (1980).
5. Roxlo, C.B., Deckman, H.W., Gland, J., Cameron, S.D., Chianelli, R.R., *Science* **235:** 1629 (1987).
6. Vissers, J.P.R., Grost, C.K., Van Oers, E.M., deBeer, V.H.J., Prins, R., *Bull. Soc. Chem. Belg.* **93:** 813 (1984).
7. LeDoux, M.J., Michaux, O., Agostini, G., Pannisod, P., *J. Catal.* **102:** 275 (1986).
8. Wakabagashi, K., Abe, H., Orito, Y., *Kogio Kagaku Zasshi* **74** (7): 1317 (1971).
9. Vissers, J.P.R., Scheffer, B., de Beer, V.H.J., Moulin, J.A., Prins, R., *J. Catal.* **105:** 277 (1987).

10. Hayden, T.F., Dumesic, J.A., Sherwood, R.D., Baker, R.T.K., *J. Catal.* **105:** 299 (1987).
11. Sinfelt, J., *Prog. Solid State Chem.* **10** (2): 55 (1975).
12. Bernard, J., Oudar, J., BarBouth, N., Margot, E., Berthier, Y., *Surf. Sci.* **88:** L35–L41 (1979).
13. Sabatier, P., *Ber. Deutsch. Chem. Ges.* **44:** 2001 (1911).
14. De Beer, V.H.J., Duchet, J.C., Prins, R., *J. Catal.* **72:** 369 (1981).
15. Delmon, B., in *Proceeding of the Third International Conference on the Chemistry and Uses of Molybdenum,* Ann Arbor, Michigan, August 19–23, 1979, pp. 73–85.
16. Topsoe, H., Clausen, B.S., Candia, R., Wivel, C., Morey, S., *J. Catal.* **68:** 433 (1981).
17. Chianelli, R.R., Pecorara, T.A., Halbert, T.R., Pan, W.-H., Stiefel, E.I., *J. Catal.* **86:** 226–238 (1984).
18. Vrinat, M.L., DeMourgues, L., *Appl. Catal.* **5:** 43 (1983).
19. Harris, S., Chianelli, R.R., *J. Catal.* **98:** 17 (1986).
20. Ratnasamy, P., Sivasanker, S., *Catal. Rev.-Sci. Eng.* **22:** 401 (1980).
21. Voorhoeve, R.S.H., Stuiver, J.C.M., *J. Catal.* **23:** 228 (1971).
22. Phillips, R.W., Fote, A.A., *J. Catal.* **41:** 168 (1976).
23. LeDoux, M.J., Michaux, O., Agostini, G., Pannisod, P., *J. Catal.* **96:** 189 (1985).
24. Mattheiss, L.F., *Phys. Rev. B* **8:** 3719 (1973).
25. Zunger, A., Freeman, A.J. *Phys. Rev. B* **16:** 906 (1977).
26. Bullett, D.W., in *Surface Properties and Catalysis by Non Metals* (J.P. Bonnelle et al., eds.). Dordrecht: D. Reidel Publishing Co., 1983 p. 47, and references therein.
27. Holzwarth, N.A.W., Harris, S., Liang, K.S., *Phys. Rev. B* **32:** 3745 (1985).
28. Guo, F.Y., Liang, W.Y., *J. Phys. Chem.* **19:** 995 (1986).
29. Li, E.K., Johnson, K.H., Freeouf, D.E., *Phys. Rev. Lett.* **32:** 470 (1974).
30. deGroot, R.A., Haas, C., *Solid State Commun.* **17:** 887 (1975).
31. Tossell, J.A., *J. Chem. Phys.* **66:** 5712 (1977).
32. Gagarin, S.G., Kovtun, A.P., Krichko, A.A., Sachenko, V.P., *Kinet. Katal.* **20:** 935 (1979).
33. Nikishenko, S.B., Slinkin, A.A., Antoshin, G.V., Minachev, Kh. M., *Kinet. Katal.* **20:** 1103 (1979).
34. Freidman, S.P., Zhukov, V.P., Gubanov, V.A., *Solid State Commun.* **36:** 559 (1980).
35. Friedman, S.P., Gubanov, V.A., *J. Phys. Chem. Solids* **44:** 187 (1983).
36. Harris, S., *Chem. Phys.* **67:** 229 (1982).
37. Harris, S., Chianelli, R.R., *Chem. Phys. Lett.* **101:** 603–605 (1983).
38. Harris, S., Chianelli, R.R., *J. Catal.* **86:** 400–412 (1984).
39. Harris, S., *Polyhedron* **5:** 151 (1986).
40. Whittingham, M.S., Jacobson, A.J., eds., in *Intercalation Chemistry*. New York: Academic Press, 1982.
41. Roxlo, C.B., Daage, M., Leta, D.P., Liang, K.S., Rice, S., Ruppert, A.F., Chianelli, R.R., *Solid State Ionics* **22:** 97 (1986).
42. Chianelli, R.R., Ruppert, A.F., Behal, S.K., Kear, B.H., Wold, A., Kershaw, R., *J. Catal.* **92:** 56 (1985).
43. Roxlo, C.B., Daage, M., Ruppert, A.F., Chianelli, R.R., *J. Catal.* **100:** 176 (1986).
44. J. Horsley (personal communication).
45. Chianelli, R.R., in *Surface Properties and Catalysis by Non-Metals* (J.P. Bonnelle et al., eds.). Dordrecht: D. Reidel Publishing Co., 1983, p. 361.
46. Fretz, R., Breysse, M., Lacroix, M., Vrinat, M., in Second Workshop on Hydrotreating Catalysts, Louvain laNeiove, Oct. 1984.
47. Farragher, A.L., Cossee, P., in *Proceedings of the 5th International Congress on Catalysis* (Palm Beach, 1972), (J.W. Hightower, ed.). Amsterdam: North Holland, 1973, p. 1301.

48. Kasztelan, S., Toulhoat, H., Grimblot, J., Bonnelle, J.P., *Bull. Soc. Chem. Belg.* **89:** 807 (1984).
49. Kasztelan, S., Toulhoat, H., Grimblot, J., Bonnelle, J.P., *Applied Catal.* **13:** 127 (1984).
50. Zdrazil, M., Sedlacek, J., *Collection Czechoslav. Chem. Commun.* **42:** 3133 (1977).
51. Duben, A.J., *J. Phys. Chem.* **82:** 348 (1978).
52. Duben, A.J., *J. Phys. Chem.* **85:** 245 (1981).
53. Vladov, Ch., Neshev, M., Petrov, L., Shopov, D. in *Proceedings Vth International Symposium on Heterogeneous Catalysis, Part II.* Varna, 1983, p. 479.
54. Ruette, F., Ludena, E.V., *J. Catal.* **67:** 266 (1981).
55. Nikishenko, S.B., Slinkin, A.A., Antoshin, G.V., Minachev, Kh. M., Nefedov, B.K., *Kinet. Katal.* **23:** 283 (1982).
56. Nikishenko, S.B., Slinkin, A.A., Antoshin, G.V., Minachev, Kh. M., Nefedov, B.K. *Kinet. Katal.* **23:** 695 (1982).
57. Nikishenko, S.B., Slinkin, A.A., Antoshin, G.V., Minachev, Kh. M., Nefedov, B.K. in *Proceedings of the Climax Fourth International Conference on the Chemistry and Uses of Molybdenum* (H.F. Barry and P.C.H. Mitchell, eds.). Ann Arbor, MI: Climax Molybdenum Company, 1982, p. 51.
58. Joffre, J., Lerner, D.A., Geneste, P., *Bull. Soc. Chim. Belg.* **93:** 831 (1984).
59. Joffre, J., Geneste, P., Lerner, D.A., *J. Catal.* **97:** 543 (1986).
60. Zonnevylle, M.C., Hoffmann, R., Harris, S., *Surf. Sci.* **199:** 320 (1988).

7 Factors Affecting the Reactivity of Organic Model Compounds in Hydrotreating Reactions

CLAUDE MOREAU AND PATRICK GENESTE

INTRODUCTION

This chapter deals with the study of the molecular factors that affect the reactivity of organic model compounds over sulfided hydrotreating catalysts. For the foreseeable future, hydrotreating will remain a major processing for any modern refinery. However, it is difficult to determine how the use of hydrotreating will grow. The major reason for this uncertainty arises from the many alternate processes being considered for converting residual fractions and heavy fuels to clean liquid fuels. Another major factor affecting the growth of hydrotreating is the future of the synthetic fuels industry, particularly the upgrading of liquids from coal, shale, and tar sands, where the levels of oxygen and nitrogen are much higher than those of conventional crude oils; consequently, relatively more hydrotreating will be required.[1]

Numerous data are available in the literature concerning hydroprocessing of model compounds, but these data are often given for catalysts that differ in their composition or their mode of sulfidation, and conflicting results are sometimes obtained from different laboratories on allegedly similar catalysts.[2] On the other hand, the operating conditions can also differ, for example, batch reactor or flow reactor. All the data reported in this chapter will be given for the same catalyst and under the same operating conditions.

Up to now, most of the studies have been concerned with the catalysts that are generally a mixture of molybdenum or tungsten oxide plus nickel or cobalt oxide or both supported on γ-Al_2O_3. Either molybdenum or tungsten appears necessary for good activity. Cobalt and nickel do not provide significant activity when present alone, but they increase activity when combined with molybdenum or tungsten (the promoting effect). Hydrotreating catalysts are generally manufactured in an oxide state and must be converted to the proper sulfided state in order to achieve the desired activity and selectivity. The different studies on catalysts have dealt with the structure of the active phases, the role of sulfur, the influence of sulfidation, the nature and the localization of active sites, the role of metal promoter, etc. The economic aspect is also worth mentioning. The level of activity of hydrotreating increases with the refining of heavier, dirtier, more sour crudes, and recent studies indicate that hydrotreating catalyst use will grow at 6% yearly from 1985 to 1990. This is the fastest growth in volume among refinery catalysts.[3]

A better knowledge of the catalysts can be expected to give information leading to the improvement of a particular catalytic system or to define new concepts for better catalysts. A careful examination of the reactivity of representative organic model compounds in hydrotreating reactions is also expected to give the same kind of information. The main objective of this chapter is to study this latter aspect; the results we have obtained will also be developed herein. A short reexamination of some general features will be presented first.

GENERAL FEATURES

Representative Organic Models

The most representative organic molecules present in petroleum feedstocks, or in other potential sources of hydrocarbons, are aliphatic compounds (thiols, thioethers, alcohols, ethers, primary or substituted amines), but, for the most part, aromatics (thiophenols, phenols, anilines), heteroaromatics (thiophenes, furanes, pyrroles), benzo-fused heteroaromatics (benzofuranes, benzothiophenes, indoles, quinolines), and dibenzo-fused heteroaromatics (dibenzothiophenes, dibenzofuranes, carbazoles, acridines, phenanthridines).[4,5] This list must be completed by the addition of aromatic hydrocarbons such as benzene, naphthalene, anthracene, and phenanthrene (Fig. 7.1). Most of the molecules shown in Figure 7.1 thus belong to the chemistry of aromatic compounds. These aromatic compounds are known to be particularly more stable and more resistant to hydrotreating than aliphatic derivatives.

Main Chemical Reactions

The principal catalytic reactions involved in hydrotreating processes are referred to as hydrodesulfurization (HDS), hydrodenitrogenation (HDN), and hydrodeoxygenation (HDO), in which the removal of heteroatoms is in some cases accompanied by hydrogenation of aromatic rings.[1]

Hydrodesulfurization. The chemical reactions most frequently of interest in hydrotreating are those involving the removal of sulfur from

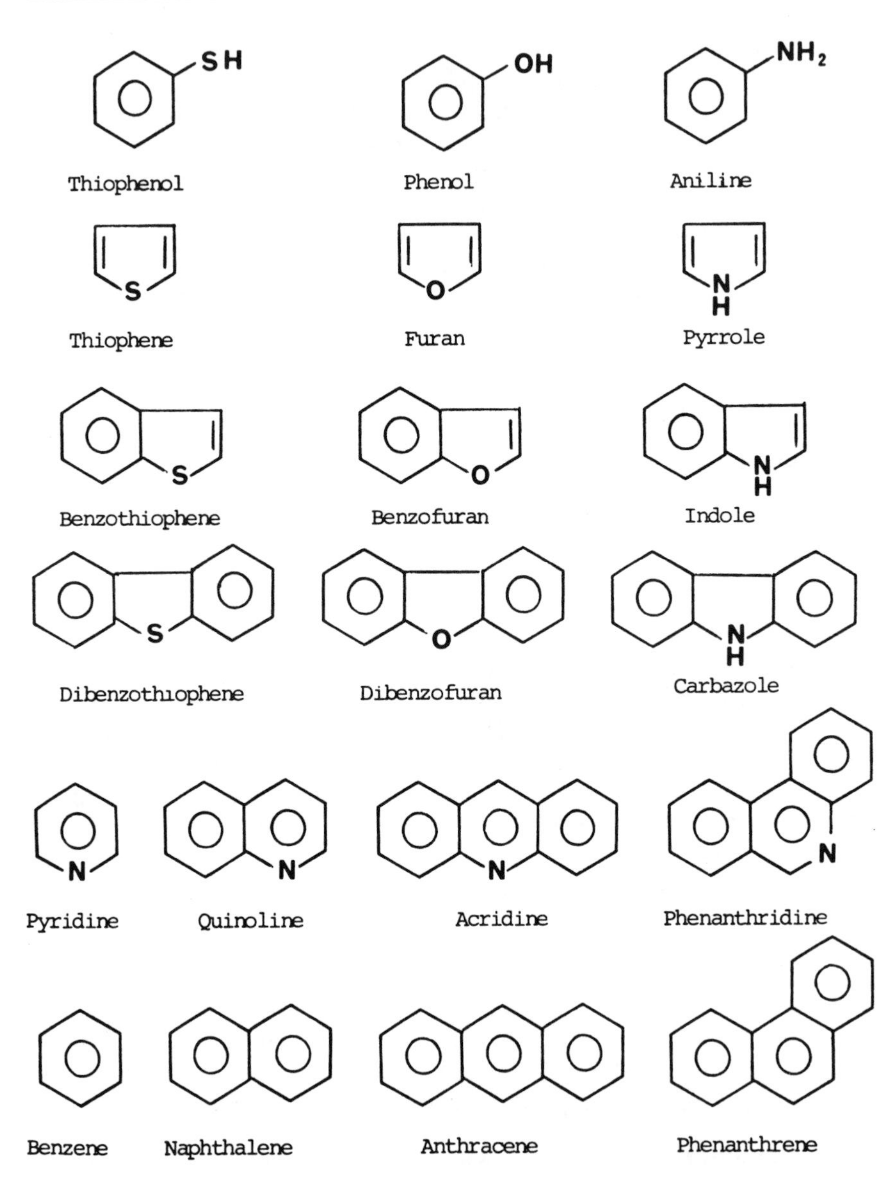

Figure 7.1. Basic structure of organic compounds in liquid fuels.

hydrocarbon molecules. Hydrogen is required not only to transform sulfur into hydrogen sulfide but also to saturate, or partially saturate, the desulfurized hydrocarbon. In fact, when cyclic molecules are involved, ring saturation normally takes place before the sulfur or nitrogen or oxygen atoms are removed.

Hydrodenitrogenation. Removal of nitrogen is important for many petroleum streams such as those being pretreated for feed to catalytic reformers, fluid catalytic crackers, and hydrocrackers, especially when feed is derived from high-nitrogen crude oils. Nitrogen removal will become more critical in the near future when significant amounts of high-nitrogen synthetic crudes from coal and shale oils must be upgraded by hydrotreating.

Hydrodeoxygenation. Most petroleum crudes do not contain large amounts of oxygen. The pretreater required to remove sulfur and nitrogen generally also removes oxygen adequately under the same operating conditions. High-oxygen synthetic crudes from coal, oil shale, and tar sands will make oxygen removal more critical in the future.

Aromatics Saturation. Saturation of monoaromatic rings is sometimes desirable for improvement of the smoke point, diesel index, etc., for certain middle-distillate products. A significant reduction in monoaromatic rings requires fairly severe hydrotreating conditions because the single ring is quite stable. It should also be noted that monoaromatic saturation of any petroleum stream that eventually yields a gasoline fraction is generally undesirable because of the resultant decrease in octane.

Reactivity of Organic Models Compounds

In order to allow the reader more concerned with catalysts to better understand the next discussion on the reactivity of organic molecules in hydrotreating reactions, some concepts concerning the main elementary chemical steps that occur in hydrotreating reactions and the molecular factors involved in these reactions will be presented.

Elementary Chemical Steps. Hydrodesulfurization, hydrodenitrogenation, and hydrodeoxygenation are generally associated with the removal of the corresponding heteroatom. In some cases heteroatoms are removed only after partial or total saturation of the aromatic rings. In order to gain a better understanding of the reactivity of the molecules, it is necessary to divide the reaction processes into their elementary chemical steps.

Two main types of chemical reactions are involved in catalytic hydrotreating over sulfided catalysts: (i) hydrogenation of aromatic and heteroaromatic rings and (ii) cleavage of C_{sp^2}-heteroatom bonds in unsaturated molecules and, more generally, cleavage of C_{sp^3}-heteroatom bonds. The necessity of dividing the chemical reactions into several elementary steps is illustrated in Figure 7.2 for HDN or quinoline. As shown in Figure 7.2, there are four elementary chemical steps before nitrogen removal: (i) hydrogenation of the heteroaromatic ring, (ii) hydrogenation of the aromatic ring, (iii) cleavage of C_{sp^3}-N bonds, and (iv) cleavage of C_{sp^2}-N bonds. It should be noted that there are some slight differences in the cleavage of C_{sp^3}-N bonds. In 2-propylcyclohexylamine (PCHA), the C–N bond is exocyclic; in 1,2,3,4-tetrahydroquinoline (1,2,3,4-THQ) and decahydroquinoline (DHQ), the C–N bonds are endocyclic. Finally, it should be noted that the nitrogen atoms are different in the two latter compounds. The nitrogen atom is capable of being conjugated with the aromatic ring in 1,2,3,4-tetrahydroquinoline but not with the saturated decahydroquinoline. All these elementary steps must be taken into account for an exhaustive discussion on the reaction mechanisms that are affected by the molecular properties of the model compounds.

Molecular Factors. As already mentioned, most of the molecules present in feedstocks are aromatic, and it is quite obvious that their reactivity will not only be largely influenced by the general aromaticity of

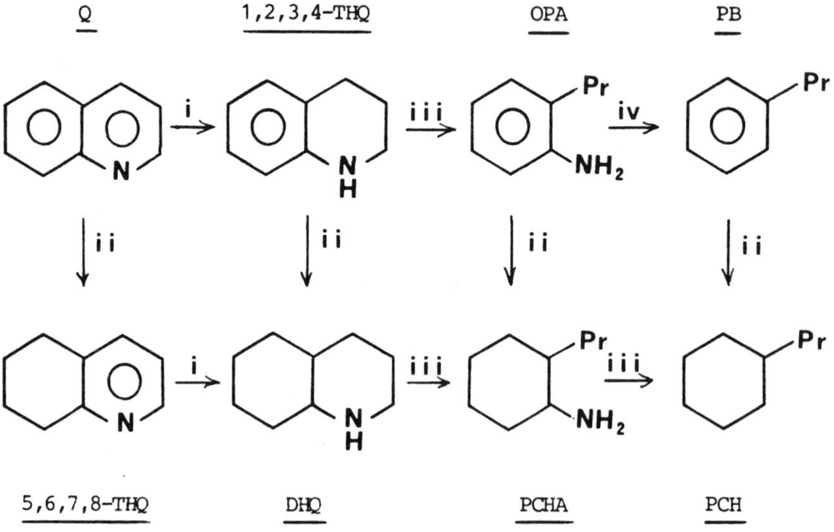

Figure 7.2. Reaction network for hydrodenitrogenation of quinoline.

these molecules but also by modifications resulting from the presence of heteroatoms or from structural effects.

Aromaticity. The aromatic character of a molecule gives information on its degree of unsaturation and on its thermodynamic stability. Aromatic compounds are generally characterized by a special stability and undergo substitution reactions more easily than addition reactions. The total aromaticity is generally given by resonance energy (RE), which is defined as the value obtained by subtracting the actual energy of the molecule from that of the most stable contributing structure.[6–10] Resonance energy always has the effect of increasing the stability of a molecule and is generally calculated from heats of combustion or hydrogenation of molecules. Resonance energy increases with the number of aromatic rings independently from the presence or the absence of heteroatoms in the rings. For example, the resonance energies for single aromatic six-membered rings (benzene, toluene, phenol, aniline, pyridine) are on the order of 35–40 kcal mol^{-1}. For two aromatic rings (biphenyl, diphenylmethane, fluorene, carbazole, diphenylamine, 9,10-dihydroanthracene, naphthalene, quinoline), the resonance energies vary from 70 to 80 kcal mol^{-1}, and for three aromatic rings (anthracene, phenanthrene, acridine), they vary from 80 to 100 kcal mol^{-1}.[6,11]

Electronegativity of Heteroatoms. The presence of heteroatoms in aromatic or saturated molecules results in differences in their acid–base properties. Acid and basic properties are generally accounted for in terms of the pK_as of the acid for acid compounds and of the conjugate acid for basic ones.[12] Most of the available pK_a data are reported in water and the data in liquid phase generally correlate with those in gas phase.[13–15]

The literature is relatively scarce with respect to the pK_as of molecules involved in hydrotreating reactions. Nevertheless, the following tendencies can be noted concerning aromatic compounds: (i) basicity increases in the following order for the series thiophenol, phenol, aniline: $S < O < N$ (Table 7.1, entry 1); (ii) for nitrogen-containing compounds, basicity increases with decreased conjugation: benzylamine is a stronger base than aniline and diphenylamine (Table 7.1, entry 2); (iii) for six-membered heteroaromatics such as pyridine, quinoline, and acridine, basicities are similar, whatever the number of aromatic rings (Table 7.1, entry 3); (iv) for five-membered heteroaromatics such as pyrrole and indole, basicities are also similar (Table 7.1, entry 4). At this point the difference in basicity between five- and six-membered heterorings is worth mentioning: for pyrrole, the unshared electron pair of the nitrogen atom is used to complete the aromatic sextet of the electrons needed for the aromaticity of the system.[10] For pyridine the lone electron pair of the nitrogen atom does not

Table 7.1. pK_as values (in water) for aromatic heteroatom-containing compounds (data from Ref. 12).

Entry	Compounds		
1	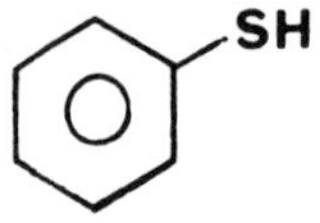 Thiophenol (6.5)[a]	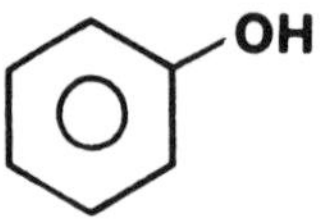 Phenol (10)[a]	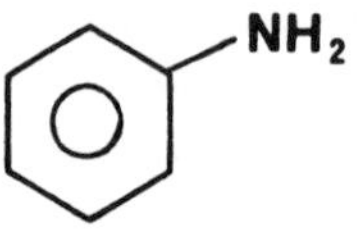 Aniline (27)[a]
2	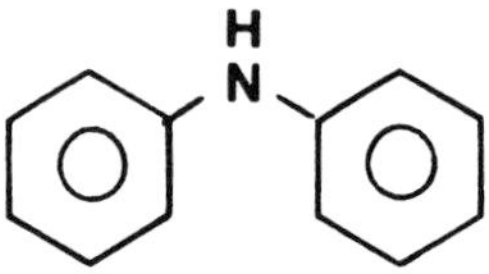 Diphenylamine (0.8)[b]	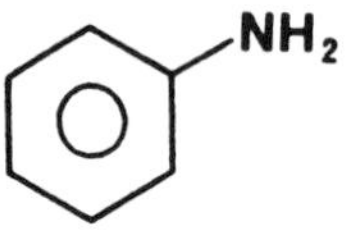 Aniline (4.6)[b]	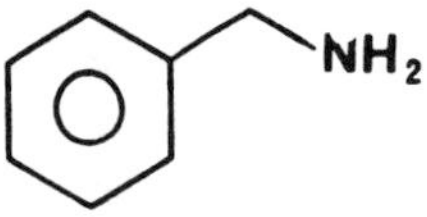 Benzylamine (9.3)[b]
3	 Pyridine (5.2)[b]	 Quinoline (4.9)[b]	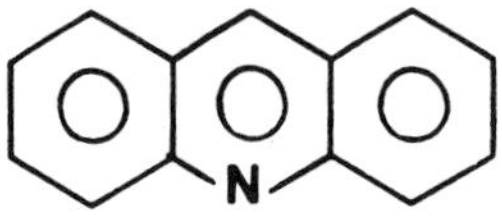 Acridine (5.6)[b]
4	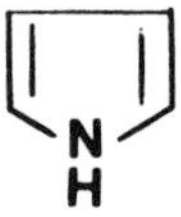 Pyrrole (−3.8)[b]	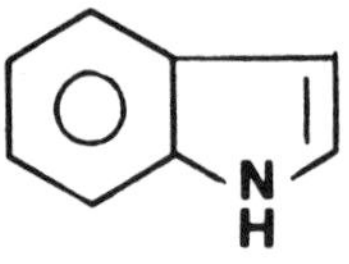 Indole (−3.5)[b]	

[a] pK_a of the acid.
[b] pK_a of the conjugate acid.

participate in the aromaticity, and this is the reason why pyrrole is a much weaker base than pyridine.

Concerning saturated compounds, parallel tendencies are observed: (i) amines are stronger bases than alcohols and mercaptans (Table 7.2, entry 1); (ii) for five- and six-membered N-compounds, heterocyclic or not, basicities are nearly identical (Table 7.2, entries 2 and 3).

Structural Effects. Structural or geometrical effects can appear in both saturated and aromatic molecules.

Table 7.2. pK_as values (in water) for saturated heteroatom-containing compounds (data from Ref. 12).

Entry	Compounds		
1	C_2H_5SH Ethanethiol (10.5)[a]	C_2H_5OH Ethanol (15.5)[a]	$C_2H_5NH_2$ Ethylamine (10.8)[b]
2	Cyclohexylamine (10.9)[b]	Dicyclohexylamine (10.9)[b]	
3	Piperidine (11.2)[b]	Pyrrolidine (11.3)[b]	

[a] pK_a of the acid.
[b] pK_a of the conjugate acid.

For saturated molecules, we have mentioned that some differences can exist in the cleavage of C_{sp^3}-N bonds depending on whether the nitrogen atom is endocyclic or exocyclic. It has been shown experimentally that the C–N bond is more easily cleaved in cyclohexylamine than in decahydroquinoline or 1,2,3,4-tetrahydroquinoline (Table 7.3).[16] For the latter two compounds, the C–N bond rupture requires opening of the ring. This factor should also be taken into account in the removal of heteroatoms that often occurs only after saturation of aromatic rings.

For planar aromatic molecules, the resonance energy increases with what is called the angularity of the system.[17] For example, anthracene and phenanthrene are isomeric hydrocarbons containing the same total number of aromatic rings (Table 7.4, entry 1). Nevertheless, the resonance

Table 7.3. Influence of structural effects on the rates of C_{sp^3}-N bond cleavage.[16]

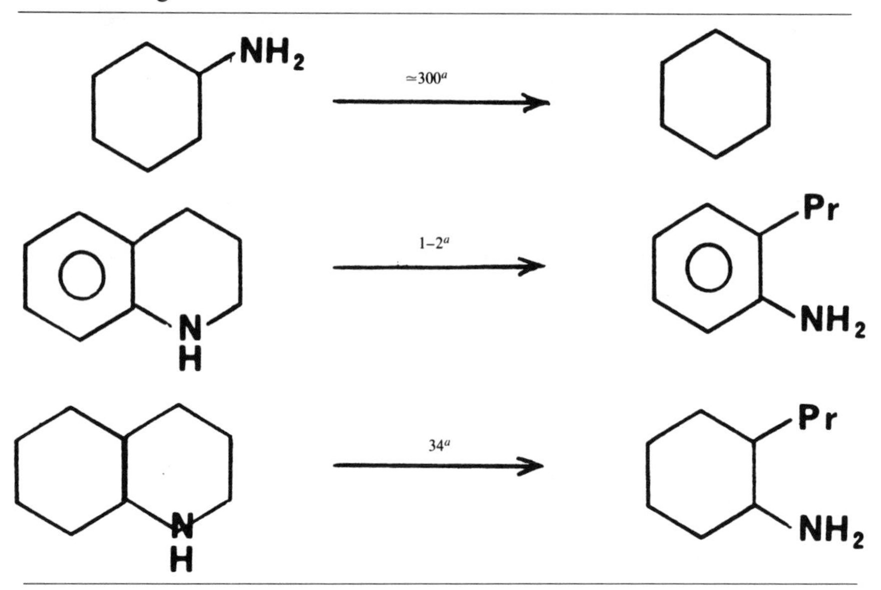

[a] Rate constants × 10^3 min^{-1} (g cat)$^{-1}$.

Table 7.4. Influence of structural effects on the resonance energies (RE) of aromatic compounds.

Entry	Compounds	
1	Anthracene RE = 84 kcal mol^{-1} [a] RE = 71 kcal mol^{-1} [b]	Phenanthrene RE = 92 kcal mol^{-1} [a] RE = 85 kcal mol^{-1} [b]
2	Fluorene RE = 76 kcal mol^{-1} [a]	9,10-Dihydroanthracene RE = 72 kcal mol^{-1} [a]

[a] Reference 11.
[b] Reference 18.

energy depends on the arrangement of the ring. Thus, phenanthrene is more stable than anthracene by about 10 kcal mol^{-1}. On the other hand, resonance energies can be similar for compounds with different geometries. For example, fluorene and 9,10-dihydroanthracene have similar resonance energies (76 and 72 kcal mol^{-1}, respectively) (Table 7.4, entry 2). However, the fluorene molecule is planar[19] whereas 9,10-dihydroanthracene is folded along the C_9–C_{10} axis.

These possible structural and geometrical effects must particularly be taken into account when discussing the reactivity of organic molecules over heterogeneous catalysts where the interaction between the molecule and the catalyst surface is an important parameter to be considered.

REACTIVITY OF AROMATIC MOLECULES

In this section the reactivity of organic molecules generally used as models to evaluate the activity of hydrotreating catalysts will be discussed. We will consider the different molecular factors capable of being involved in the three main classes of chemical reactions: (i) hydrogenation of aromatic and heteroaromatic compounds, (ii) hydrogenolysis of C_{sp^2}-heteroatom bonds, and (iii) cleavage of C_{sp^3}-heteroatom bonds.

Most of the reaction networks are well established on the whole and will not be detailed again in this section. Only the rate constants will be considered, making it possible to focus on one particular factor, the others being roughly constant. In most cases we will present results from our own laboratory, which have been obtained under typical hydrotreating operating conditions, i.e., batch reactor, high temperature (340°C), and high hydrogen pressure (70 bar). Under these conditions, hydrogenation equilibria are shifted toward the saturated products. The catalysts used were a sulfided NiO–MoO_3/γ-Al_2O_3 catalyst (HR 346 from Procatalyse) as used in our previous studies[20] and, in some cases, a sulfided NiO–WO_3/γ-Al_2O_3 catalyst (named GS8).[21] The activity and the selectivity of this latter catalyst have been shown to be close to those of the NiMo catalyst.[16] Comparison with literature data will also be considered when possible.

Hydrogenation Reaction

It is experimentally well known that aromatic and heteroaromatic compounds mainly react through hydrogenation reactions. We have thus considered several families of aromatic and heteroaromatic molecules that only differ in their aromatic properties. For each family, basicity (expressed in terms of the pK_as of the conjugate acid) and structural or geometrical effects are approximately constant.

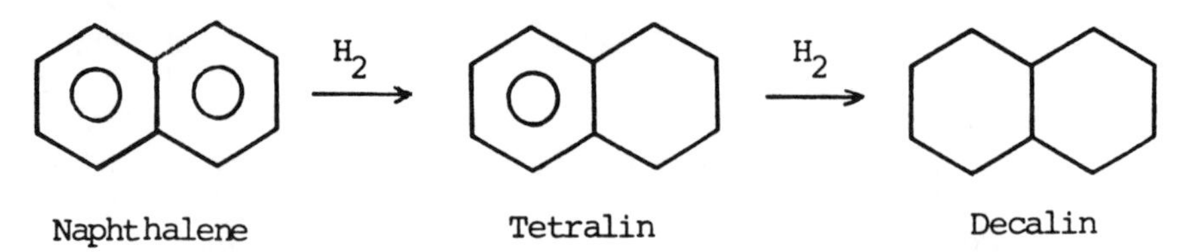

Figure 7.3. Reaction network for hydrogenation of naphthalene.

Hydrogenation of Aromatic Hydrocarbons. Hydrogenation of aromatic hydrocarbons is a key step in catalytic hydrotreating reactions. Up to now, benzene and naphthalene are the two models which have been extensively studied on several sulfided catalysts: unsupported MoS_2, WS_2[22]; γ-alumina-supported $CoO{-}MoO_3$[23,24]; $NiO{-}WO_3$[25].

On these catalysts, it has been clearly shown that hydrogenation of naphthalene into tetralin occurs more readily than hydrogenation of benzene alone (Fig. 7.3). The reason invoked for this higher reactivity is that the resonance energy of the second ring of naphthalene is less than that of benzene.[26]

For hydrogenation of higher fused systems such as anthracene and phenanthrene, the literature provides only partial information.[26] However, these two compounds, like naphthalene, undergo preferential hydrogenation of one ring (Figs. 7.4 and 7.5).

The relative rate constants for hydrogenation of aromatic hydrocarbons are reported in Table 7.5. We have also reported in this table several sources of resonance energies to illustrate the differences that can result from the mode of calculation of the resonance energies and from the

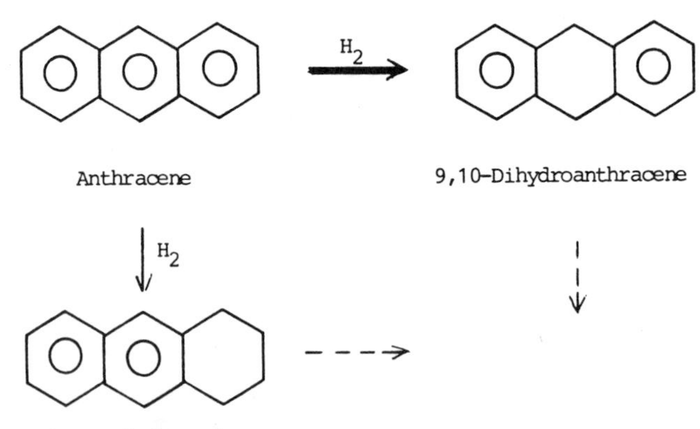

Figure 7.4. Reaction network for hydrogenation of anthracene.

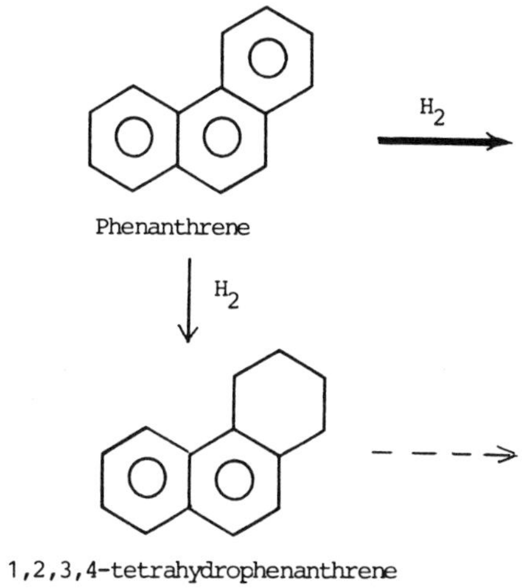

Phenanthrene
H_2
9,10-Dihydrophenanthrene
1,2,3,4-tetrahydrophenanthrene

Figure 7.5. Reaction network for hydrogenation of phenanthrene.

reaction that serves as support for these calculations, i.e., combustion and hydrogenation.[8] All the scales lead to the same sequence in the aromaticity order: benzene < naphthalene < anthracene < phenanthrene.

For these planar aromatic molecules, the only variable factor of reactivity is the aromatic property accounted for by total or partial resonance energies.

As already mentioned, aromatic compounds are characterized by a particular stability and undergo substitution reactions more easily than addition reactions. The low aromatic character of one of the naphthalene rings is experimentally illustrated by its ability to undergo addition reactions across the 1,2 positions[31] (Fig. 7.6) and by the recent calculations of resonance energies per ring.[27]

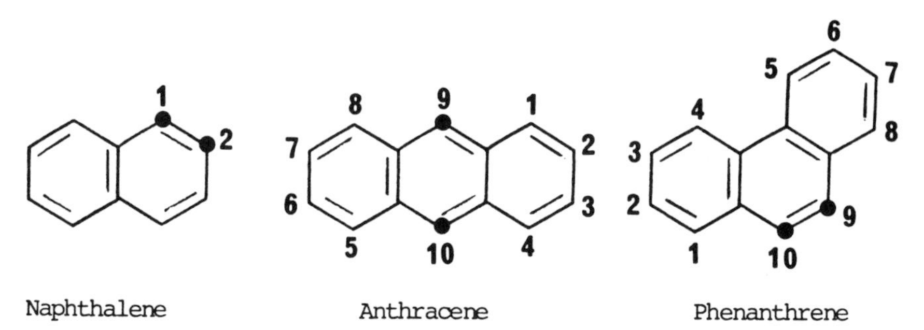

Figure 7.6. Favorable positions for addition reactions to naphthalene, anthracene, and phenanthrene.

Table 7.5. Resonance energies and relative rate constants for hydrogenation of aromatic hydrocarbons over sulfided catalysts.

Compounds	Total RE[11] (kcal mol^{-1})	Total RE[18] (kcal mol^{-1})	Total RE[6] (kcal mol^{-1})	RE/ring[27] (kcal mol^{-1})	k_{rel}^{22} (MoS_2)	k_{rel}^{22} (WS_2)	k_{rel}^{23} ($CoMo/Al_2O_3$)	k_{rel}^{28} (NiW/Al_2O_3)	$k_{rel}^{29,30}$ $NiMo/Al_2O_3$
	36	38	39	40	1	1	1	1	1
	61	59	75	28	14	23	21	18	10
	84	71	105	—	—	62	—	40	36
	92	85	—	—	—	—	—	—	4

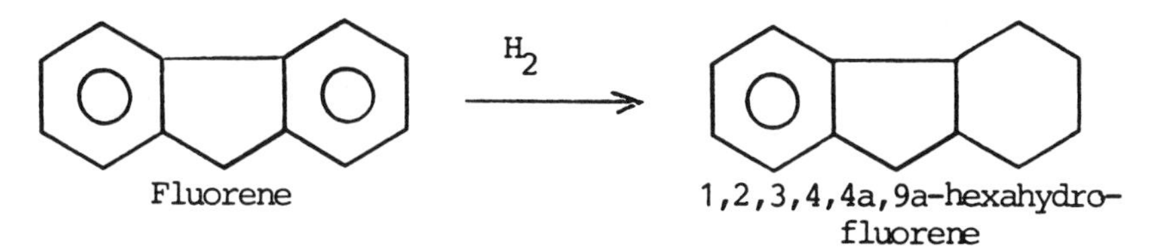

Figure 7.7. Reaction network for hydrogenation of fluorene.

Anthracene and phenanthrene also undergo many addition reactions across the 9,10 positions[32] (Fig. 7.6). The explanations given for hydrogenation of naphthalene into tetralin are therefore valid for hydrogenation of the two former products.

The greater stability of phenanthrene versus anthracene by about 10 kcal mol^{-1} and, consequently, the increased stability of the ring to be hydrogenated clearly explains the decrease of reactivity in the case of phenanthrene.

Everything else being equal, the lower the aromaticity of the ring to be hydrogenated, the easier it is to hydrogenate the aromatic hydrocarbons.

Remark: It is both interesting and important to note a parallelism in the hydrogenation rate constants whatever the catalytic system is, unsupported MoS_2 or γ-Al_2O_3-supported CoMo, NiW, and NiMo. Our own results are in good agreement with the hypotheses of Sapre and Gates[23] concerning the effect of the support in CoMo catalysts: supporting the sulfided catalyst on alumina results primarily in an increase in the number of catalytic sites but not in a significant change in the nature of the catalytic activity for aromatic hydrogenation reactions.

Another feature is the difference in geometry for similar aromatic hydrocarbons. A good illustration of this is given by comparison of the hydrogenation of fluorene (Fig. 7.7) and 9,10-dihydroanthracene (Fig. 7.8). The resonance energies are nearly the same (76 and 73 kcal mol^{-1},

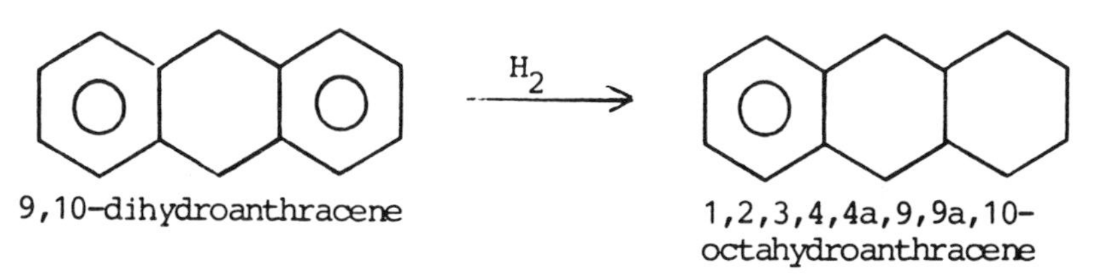

Figure 7.8. Reaction network for hydrogenation of 9,10-dihydroanthracene.

respectively), and the rate constants do not differ in a significant manner (Table 7.6, entries 2 and 3). The fluorene molecule is planar, whereas 9,10-dihydroanthracene is folded along the C_9–C_{10} axis with a dihedral angle of about 150–160°. The reactivity thus seems to be less affected by the geometry of the molecule than by its aromaticity.

This assumption is easily confirmed by examination of the hydrogenation of diphenylmethane (Fig. 7.9), a nonrigid system with aromaticity

Table 7.6. Resonance energies and rate constants relative to benzene for hydrogenation of aromatic hydrocarbons over sulfided catalysts.

Compounds	Entry	Total RE[11] (kcal mol^{-1})	RE/ring (kcal mol^{-1})	k_{rel}[29]
	**1**	36	36	1
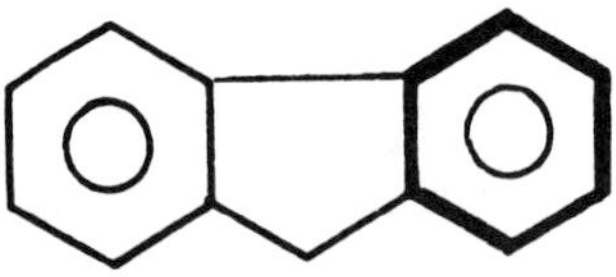	**2**	73	36.5	2
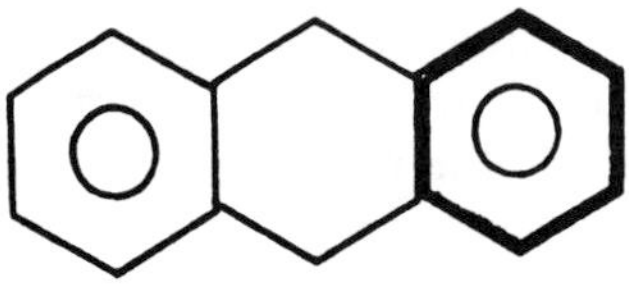	**3**	76	38	4
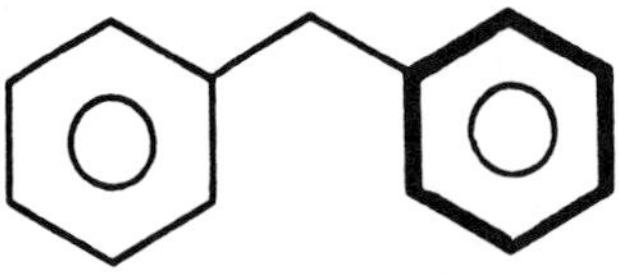	**4**	68	34	1.5
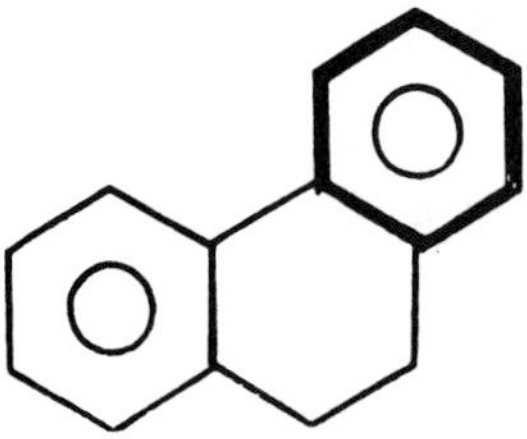	**5**	—	—	2

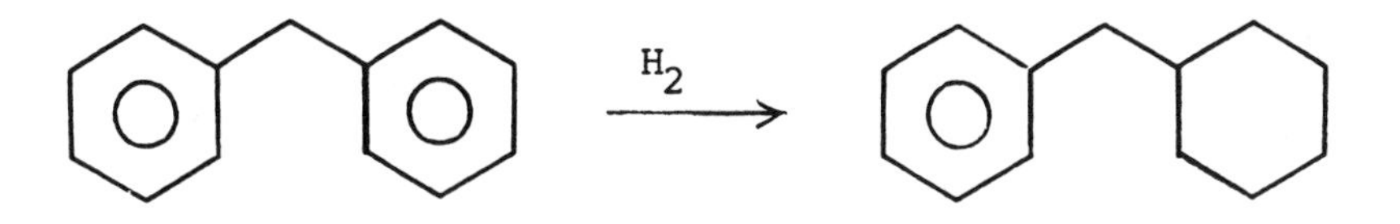

Figure 7.9. Reaction network of hydrogenation of diphenylmethane.

close to that of the preceding hydrocarbons (68 kcal mol^{-1}). The rate of hydrogenation is almost equal to that of fluorene, although the geometry is different (Table 7.6, entry 4).

The slight influence of geometrical effects is still confirmed by examination of the hydrogenation of 9,10-dihydrophenanthrene (Fig. 7.10), a rigid and twisted system. The rate constant of hydrogenation is similar to that of fluorene and diphenylmethane (Table 7.6, entry 5). All the rate constants summarized in Table 7.6 are relative to benzene taken as the reference aromatic model. The aromaticity, as given by resonance energies per ring, is nearly constant.

> Everything else being equal, the rates of hydrogenation do not depend in a significant manner on the geometrical modifications of the molecules: folding or free rotation.

Hydrogenation of Six-Membered Aromatic Heterocyclic N-Rings. In this series, we will only consider the N-containing heterorings in which a nitrogen atom replaces a carbon atom. When nitrogen is the heteroatom, there is little difference in the aromatic sextet and the unshared electron

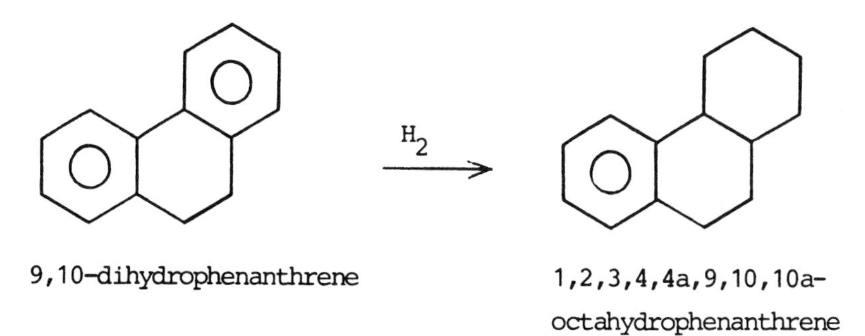

Figure 7.10. Reaction network of hydrogenation of 9,10-dihydrophenanthrene.

pair of the nitrogen atom does not participate in the aromaticity. With respect to benzene, pyridine has the same aromatic properties and, consequently, very similar resonance energies, whatever the method of determination of these energies. Moreover, the presence of the lone electron pair gives these molecules basic properties in addition to aromatic ones.

As for aromatic hydrocarbons, hydrogenation of aromatic N-hetero-rings is also a key step, since hydroprocessing of N-containing molecules generally requires saturation of aromatic rings prior to any C–N bond cleavage.[28,33,34]

The well-known similarity in the aromatic properties of pyridine, quinoline, and acridine as compared to their respective parent hydrocarbons (benzene, naphthalene, and anthracene) has led to the investigation of the hydrogenation of pyridine, quinoline, acridine, and phenanthridine under the same operating conditions as those reported in Table 7.5 for their parent hydrocarbons. The relative rate constants for hydrogenation of N-rings over a sulfided $NiO–WO_3/\gamma\text{-}Al_2O_3$ and the relative rate constants for hydrogenation of the corresponding hydrocarbon rings over the same catalyst are reported in Table 7.7. We have also reported in this table the resonance energies, total or per ring. It can be seen that hydrocarbons and N-heterorings are similar in their aromatic properties.

For these planar N-heteroaromatics, the only variable factor of reactivity is the aromaticity of the ring to be hydrogenated. Basicity, expressed as pK_as of the conjugate acids, is nearly constant for these N-compounds.

It can be seen from Table 7.7 that the rates of hydrogenation of the pyridine nucleus closely parallel those of the benzene nucleus and, as a consequence, the factors invoked in one series are readily transposable to the other. The weaker aromaticity found for hydrogenation of naphthalene into tetralin (see Fig. 7.3) and quinoline into 1,2,3,4-tetrahydroquinoline (Fig. 7.11) can thus be attributed to a loss of resonance of the ring to be hydrogenated.

A similar decrease in the aromaticity of the N-ring of acridine is also invoked to account for its higher reactivity. The relative rate constants of phenanthridine and acridine, 2 and 24, respectively, are comparable to those of phenanthrene and anthracene, 4 and 36, respectively (Figs. 7.12 and 7.13; see also Figs. 7.4 and 7.5), although the resonance energies of phenanthridine and acridine cannot be determined by the same mode of calculation as their hydrocarbon counterparts. We can assume that this decrease in reactivity is directly related to an increase in total aromaticity of phenanthridine and, consequently, in the aromaticity of the N-ring.

As in the case of hydrocarbon rings, the hydrogenation of six-membered N-rings is essentially related to the aromaticity of the ring to be hydrogenated. Moreover, the similarity in the absolute rate constants

Table 7.7. Resonance energies, pK_as, and relative rate constants for hydrogenation of six-membered N-heteroaromatics; comparison with the parent hydrocarbons.

Compounds	Total RE[6] (kcal mol^{-1})	RE/ring[27] (kcal mol^{-1})	$k_{rel}^{29,30}$ (NiW/Al_2O_3)	pK_a^{12}	Compounds	Total RE[6] (kcal mol^{-1})	RE/ring[27] (kcal mol^{-1})	k_{rel}^{28} (NiW/Al_2O_3)	$k_{rel}^{29,30}$ (NiMo/Al_2O_3)
	43	42	1	5.2		39	40	1	1
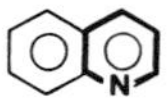	69	30	22	4.9	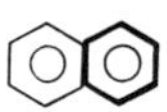	75	28	18	18
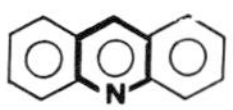	106	—	24	5.6	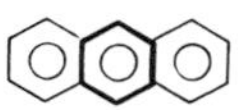	105	—	40	36
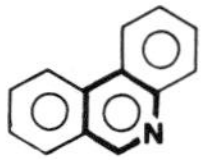	—	—	2	—	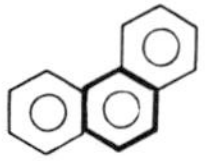	—	—	—	4

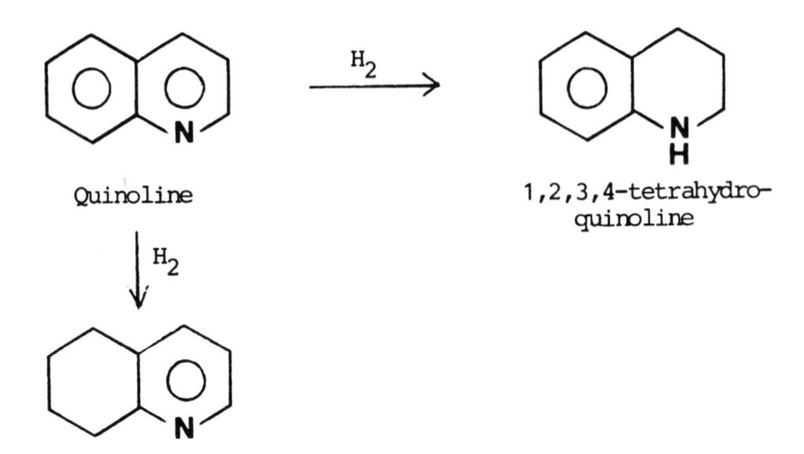

Figure 7.11. Reaction network for hydrogenation of quinoline.

(N-ring/hydrocarbon ≅ 2) for the two series indicates that basicity is not an important factor in hydrogenation of N-heterorings.

Remark: In the hydrocarbon series, it was possible to compare the relative hydrogenation rates for different catalytic systems. In the nitrogen series, this comparison is less evident.[34] Using sulfided NiMo and NiW catalysts, we could not detect any difference in the relative and

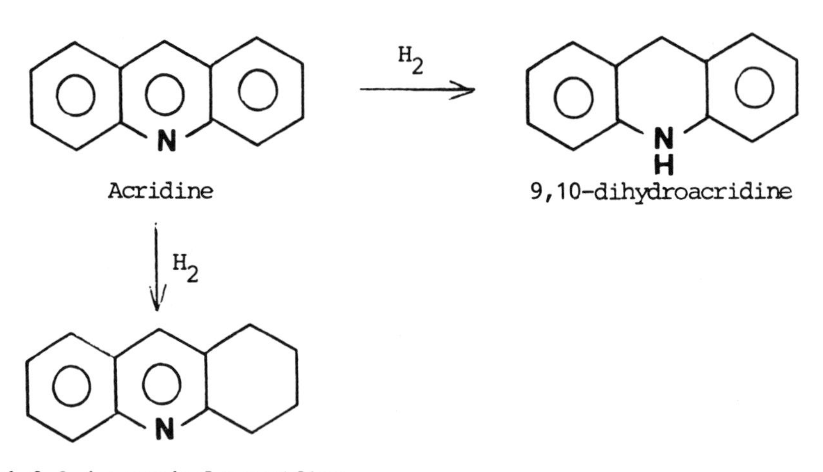

Figure 7.12. Reaction network for hydrogenation of acridine.

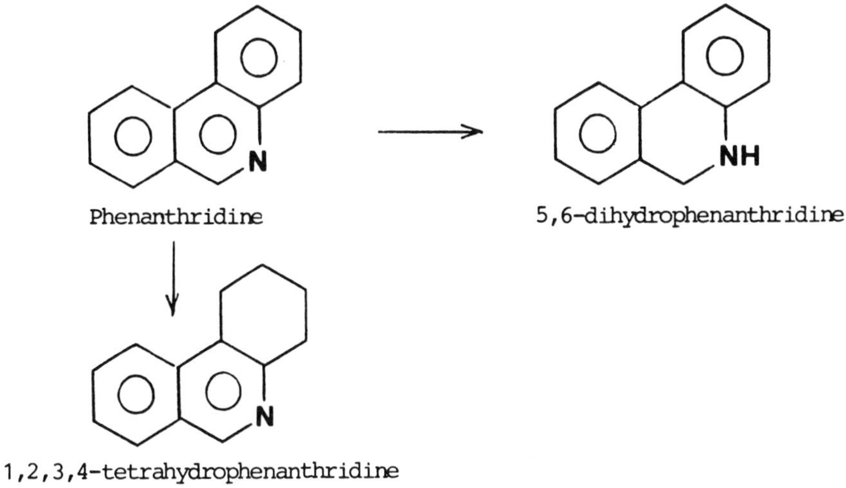

Figure 7.13. Reaction network for hydrogenation of phenanthridine.

absolute rate constants of quinoline and acridine. Our results confirm those cited in the literature for similar catalysts.[35] It has also been shown that a CoMo catalyst is slightly less hydrogenating than NiMo or NiW catalysts (factor $\simeq 2$).[35] This behavior was not confirmed by the results we obtained on hydrogenation of pyridine and quinoline over promoted CoMo and NiMo catalysts.[36] The observations concerning the catalytic sites in hydrogenation of aromatic hydrocarbons are thus the same for hydrogenation of six-membered N-heteroaromatics.

Hydrogenation of Aromatic Five-Membered Heterorings. As mentioned previously, the aromatic properties of the ring in six-membered heteroaromatics are due to the presence of the nitrogen atom. On the contrary, five-membered analogs containing sulfur, oxygen, and nitrogen are all known to have aromatic properties. We can then compare the reactivity of systems for a given heteroatom and also study the influence of the heteroatom on the reactivity for a given system.

Resonance energies of six-membered aromatic rings are well defined. This is not the case for five-membered rings, as illustrated in Table 7.8. Pyrrole is known to have the most important aromatic character, whereas furan has the least. For thiophene, this character depends on the mode of calculation of the resonance energy. Thiophene is aromatic as pyrrole from heats of combustion[11] or as furan from heats of hydrogenation.[27]

In order to determine the importance of aromaticity in hydrogenation of five-membered heteroaromatics, we have studied the reactivity of various

Table 7.8. Resonance energies (in kcal mol^{-1}) for aromatic five-membered heterorings.

Furan	Thiophene	Pyrrole	Reference
22	28	25	11
16	29	22	11
23	31	31	37
22	22	31	27

N-, O-, and S-containing models: pyrrole, indole, carbazole, furan, benzofuran, dibenzofuran and thiophene, benzothiophene, dibenzothiophene over a NiMo catalyst.

Five-Membered N-rings. The relative hydrogenation rate constants of the five-membered N-rings are reported in Table 7.9. We have also reported in this table the resonance energies, total or per ring, and the pK_a values when available. The rates of hydrogenation of the N-rings are nearly the same for pyrrole and indole, which have a similar structure and basicity, and for tetrahydrocarbazole. These results cannot be related directly to the resonance energies per ring. This points out the complexity of the determination of resonance energies in five-membered rings. Without any complementary information, the experimental results seem to indicate a conservation of aromaticity in the pyrrole ring for the three systems considered. This assumption is not completely unexpected. Car-

Table 7.9. Resonance energies, pK_as, and relative rate constants for hydrogenation of five-membered N-heteroaromatics.

Compounds	Total RE[11] (kcal mol^{-1})	RE/ring[27] (kcal mol^{-1})	pK_a[12]	k_{rel}[29,30,38] ($NiMo/Al_2O_3$)
 N H	25	30	−3.8	1
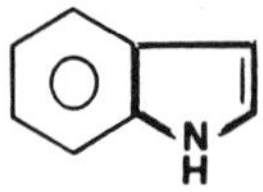 N H	49	17	−3.5	2
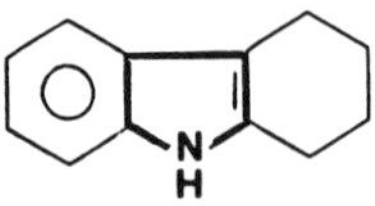 N H	—	—	—	1

bazole and tetrahydrocarbazole are hydrogenated at similar rates (Table 7.10), indicating a similar stability. We will see subsequently that the stability of pyrrole is equivalent to that of purely aromatic pyridine or benzene.

Remark: As already observed in the preceding series of aromatic or heteroaromatic models, NiMo and NiW catalysts behave in the same manner, confirming the observations in the literature.[34] Likewise, the CoMo catalyst does not differ to a significant extent from that of the NiMo catalyst as far as its hydrogenating properties are concerned. The rate constant ratios k_{CoMo}/k_{NiMo} for pyrrole, indole, and carbazole have been found to be equal to 1.70, 0.92, and 2.17, respectively.[39]

Comparison between Six-Membered and Five-Membered N-rings. Such a comparison is extremely important due to the large difference in the basicities between the two systems. In pyrrole, the lone electron pair of the nitrogen atom is required to complete the aromatic sextet of electrons needed for the aromaticity of the system, whereas in pyridine the unshared electron pair of the nitrogen atom does not participate in the aromaticity. Consequently, six-membered N-rings are more basic than their five-membered analogs.

The hydrogenation rate constants are very similar for pyrrole and pyridine [15×10^{-3} and 10×10^{-3} min^{-1} $(g\ cat)^{-1}$, respectively, over sulfided NiW/Al_2O_3 catalyst[16]]. Depending on the mode of calculation of the resonance energies, the aromatic properties are identical[8,11] or slightly higher

Table 7.10. Relative rate constants for hydrogenation of carbazole and tetrahydrocarbazole.

Reaction	k_{rel}^{29} ($NiMo/Al_2O_3$)
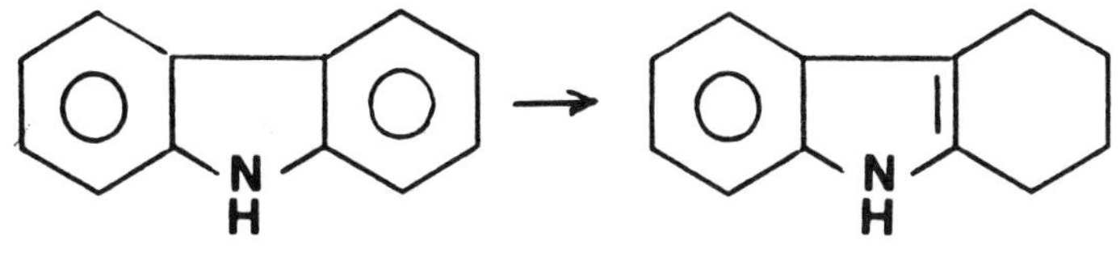	1
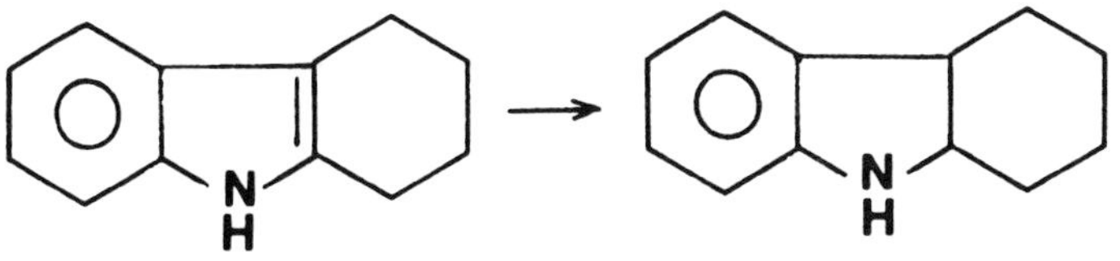	2

for pyridine than for pyrrole.[27] This demonstrates again the difficulty in the determination of resonance energies. These slight differences in the aromaticity between the two models cannot account for their large difference in basicity. This would confirm that hydrogenation reactions depend mainly on the aromaticity of the ring to be hydrogenated rather than on the basic properties of the heteroatoms present in the ring.

On the other hand, this could also indicate that hydrogenation sites are not affected by the basic character of the molecules. This situation would be possible for π-adsorbed species as already assumed from our results[40] and those in the literature.[23]

Five-Membered O- and S-Rings. Data concerning the resonance energies and the basicities for O- and S-analogs of five-membered N-rings are not readily available. We will assume that geometrical effects and basicity do not vary in each series as is the case for the N-compounds.

The relative rate constants for hydrogenation of five-membered O- and S-rings are reported in Table 7.11 along with those of N-rings. As seen in this table, the rates of hydrogenation of O- and S-containing models closely parallel those of N-compounds. The hydrogenation rate constants of the tetrahydro derivatives of dibenzofuran and dibenzothiophene could not be measured because they disappeared more rapidly than they were formed.

As for N-compounds, the experimental results obtained for the O- and S-models suggest that the aromatic properties in O-rings and S-rings are conserved.

Remark: The rates of hydrogenation of five-membered N-rings are not markedly affected by the catalytic system: CoMo, NiMo, or NiW. Oxygen compounds cannot really be compared but five-membered S-rings were also shown to be only slightly or not at all dependent on the catalyst.[43] The rate constant ratios for hydrogenation of benzothiophene with respect to thiophene vary from 1 to 2 for CoMo, CoW, NiMo, and NiW catalysts.

Influence of the Electronegativity of the Heteroatoms. We have shown that five- and six-membered N-rings are hydrogenated at similar rates although the difference in basicity is important in these two series. Likewise, six-membered N-rings are more basic than their corresponding hydrocarbons, but the hydrogenation rates are still similar. We will now compare a series of compounds with similar geometrical and aromatic properties to determine the influence of the heteroatom on the hydrogenation of the adjacent benzenic rings.

The relative rate constants for the hydrogenation of fluorene, dibenzothiophene, dibenzofuran, and carbazole are given in Table 7.12. All these

Table 7.11. Relative rate constants for hydrogenation of five-membered O-, S-, and N-heteroaromatics.

O-Compounds	k_{rel}[41,42] (NiMo/Al_2O_3)	S-Compounds	k_{rel}[41,42] (NiMo/Al_2O_3)	N-Compounds	k_{rel}[29,30,38] (NiMo/Al_2O_3)
 O	1	 S	1	 N H	1
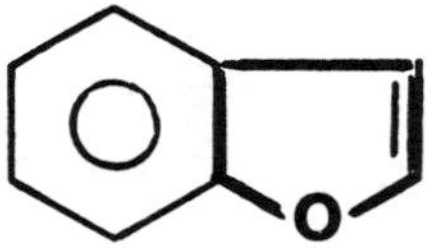 O	3	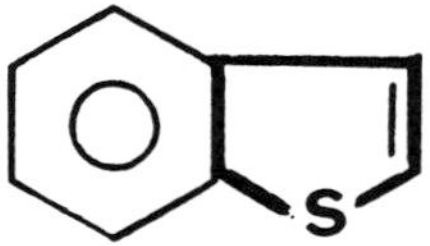 S	2	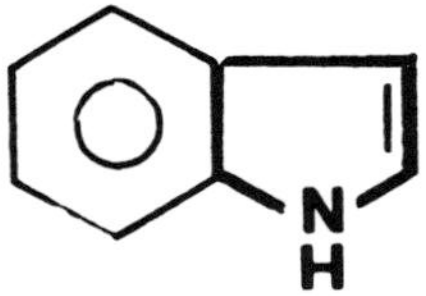 N H	2
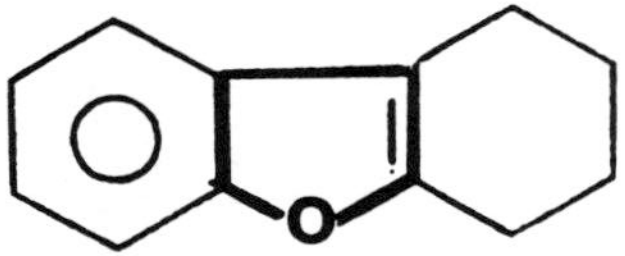 O	See text	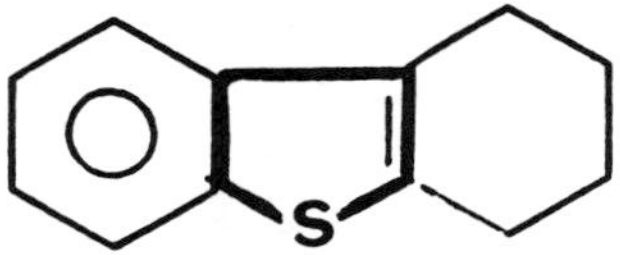 S	See text	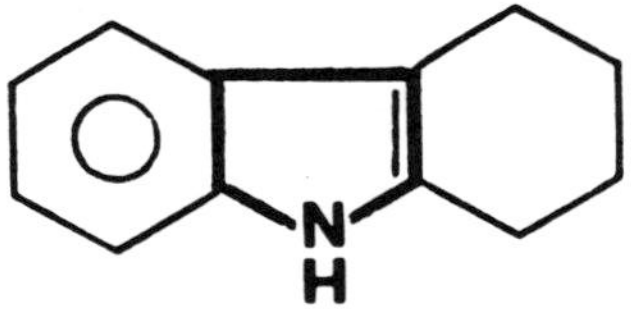 N H	1

Table 7.12. Influence of the basicity on hydrogenation of planar heteroaromatics.

Compounds	Total RE[11] (kcal mol^{-1})	k_{rel}^{29} ($NiMo/Al_2O_3$)	Basicity
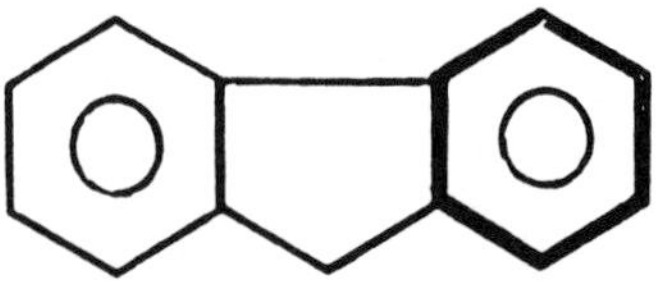 [fluorene structure]	76	1	Increasing basicity ↓
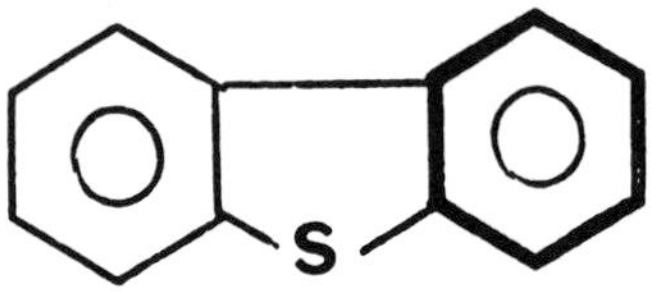 [dibenzothiophene structure, S]	—	0.5	
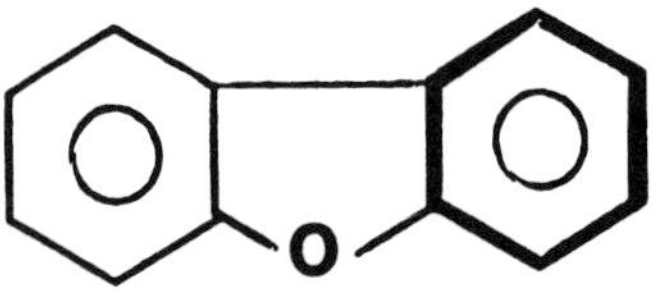 [dibenzofuran structure, O]	—	0.5	
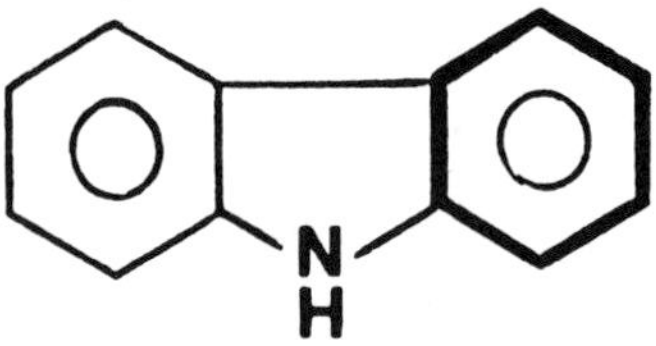 [carbazole structure, NH]	75	1	

compounds have planar structures[19,44,45] and similar aromatic properties.[11] As was mentioned, the basicity is expected to increase in the order $CH_2 < S < O < N$. From Table 7.12, it can be seen that the hydrogenation rates are nearly the same in the presence of a methylene group or heteroatoms.

For folded molecules such as 9,10-dihydroanthracene, thioxanthene,[46] xanthene,[47] and 9,10-dihydroacridine (Table 7.13), it can be seen that the hydrogenation rates do not vary in a significant manner by decreasing the basicity of the heteroatom.

The same reasons can also be invoked in the series diphenylmethane, diphenylsulfide, diphenylether, and diphenylamine (Table 7.14). For these molecules with free rotation around the C–X bond, the slight variation in the rates of hydrogenation of the aromatic ring cannot be related to differences in basicity of the heteroatoms.[40]

The data reported in the last three tables indicate that aromaticity is nearly constant for all models. The largest variation in the hydrogenation rate constants does not exceed a factor of 4, confirming that hydrogenation reactions do not depend on the basicity of heteroatoms or on possible structural effects.

In the series pyrrole, furan, thiophen and indole, benzofuran, benzothiophene (Table 7.15), the variation in the rates of hydrogenation is more important than in the series carbazole, dibenzofuran, dibenzothiophene (see Table 7.12). The role of the basicity of the heteroatom was not shown to be important for hydrogenation of most of the models considered up to now. The increase in rates observed in the series pyrrole, furan, and thiophene can thus be reasonably attributed to the lower aromatic properties observed in O- and S-containing five-membered molecules. The differences in reactivity are less important than for six-membered heterocycles and can account for the difficulties involved in determining the resonance energies of five-membered heteroaromatic rings.

Table 7.13. Influence of the basicity on hydrogenation of folded heteroaromatics.

Compounds	Total RE[11] (kcal mol^{-1})	k_{rel}^{29} ($NiMo/Al_2O_3$)	Basicity
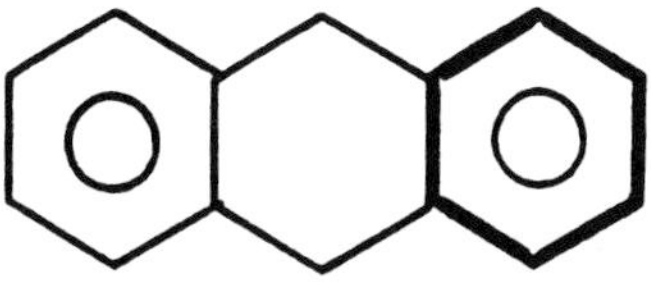	73	1	
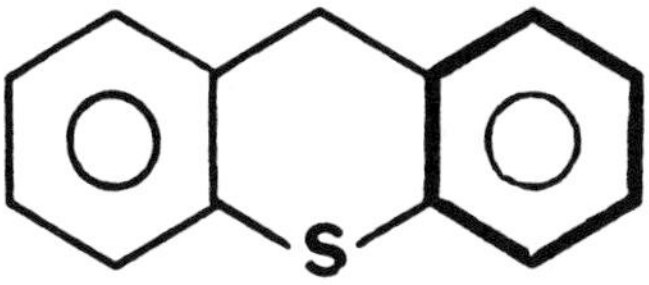 S	—	0.25	Increasing basicity
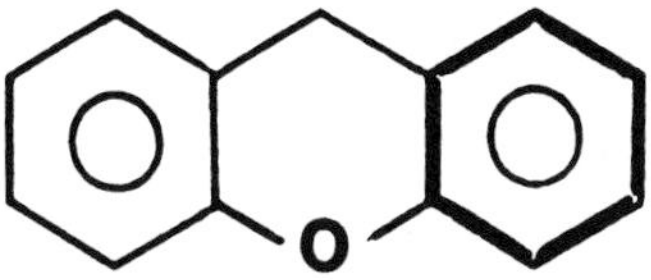 O	—	0.5	
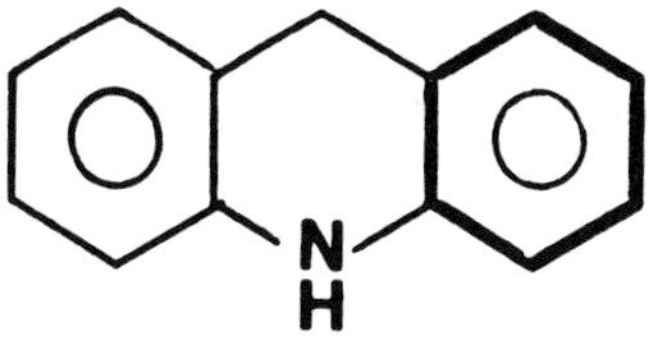 NH	—	1	

Table 7.14. Influence of the basicity on hydrogenation of free-rotation heteroaromatics.

Compounds	Total RE[11] (kcal mol^{-1})	k_{rel}^{40} ($NiMo/Al_2O_3$)	Basicity
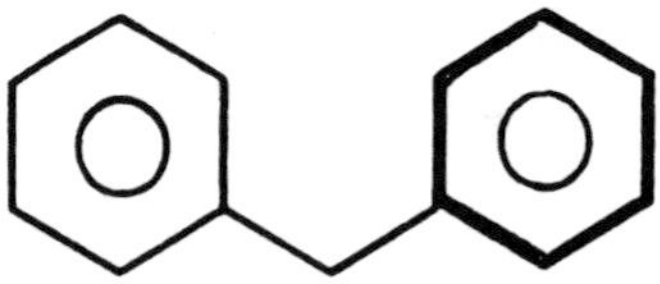	68	1	↓
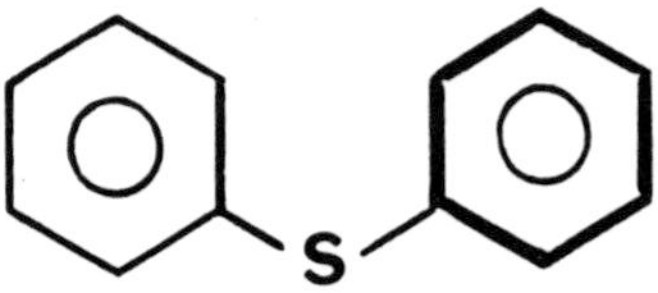 S	—	0.6	Increasing basicity
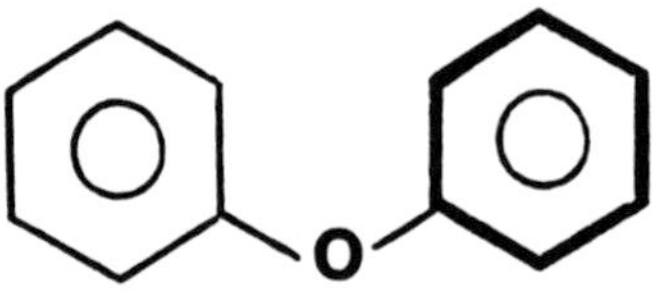 O	—	0.6	↓
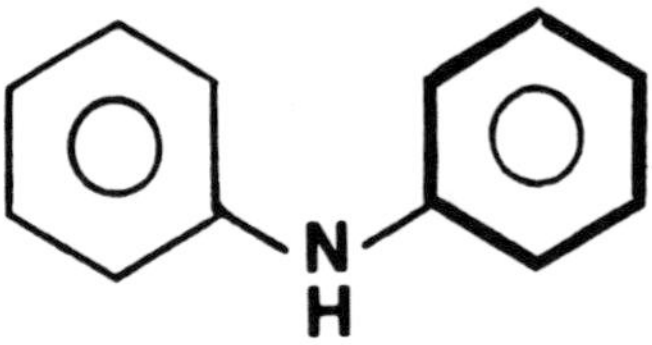 N H	77	2.5	↓

Remark: As already observed in other series, aromatic or heteroaromatic compounds, the rates of hydrogenation do not markedly depend on the catalytic system. For example, CoMo catalysts behave like NiMo catalysts: $k_{benzothiophene}/k_{benzofuran}$ = 2.5 (NiMo catalyst), 3.4 (CoMo catalyst)[48]; $k_{benzothiophene}/k_{indole}$ = 8.5 (NiMo catalyst), 3.5 (CoMo catalyst)[48]; $k_{benzothiophene}/k_{thiophene} \simeq 2$ (CoW, NiW, CoMo, NiMo catalysts).[43]

Summary on Hydrogenation Reactions. In this systematic study of the factors of reactivity involved in hydrogenation of aromatic model compounds such as hydrocarbons, five- and six-membered heterocycles over sulfided catalysts and under typical hydrotreating operating conditions, the following conclusions can be drawn:

1. Hydrogenation is mainly affected by the aromatic properties of the rings to be hydrogenated rather than by the electronegativity of the het-

Table 7.15. Influence of the aromaticity and basicity on hydrogenation of five-membered heteroaromatics.

Compounds	Total RE[27] (kcal mol^{-1})	k_{rel}^{42} ($NiMo/Al_2O_3$)	Basicity	k_{rel}^{41} ($NiMo/Al_2O_3$)	Compounds
NH	30	1	Decreasing basicity ↓	1	NH
O	22	3.3		3.4	O
S	22	7.7		8.5	S

eroatoms or the structural properties of the model considered. In the absence of important steric factors, hydrogenation reactions are favored for low aromatic character rings. There is no reason to make a particular distinction among S, O, or N atoms.

2. From the comparison of our data with those available in the literature for the same molecules using different catalytic systems, it appears that the rates of hydrogenation are only slightly sensitive to the catalytic system, whether supported or not and promoted or not. This would confirm the presence of MoS_2 or WS_2 as the active phase for hydrogenation reactions, in good agreement with previous assumptions.[23,49] The active phase should not be sensitive to structural or basic properties or both of the molecules. The predominant role of aromatic properties leads to the consideration that hydrogenation reactions occur through π-adsorbed species, as already suggested.[40]

3. The presence of alkyl or aryl groups does not, in general, affect the hydrogenation reactions. The slight differences observed in reactivity are easily accounted for in terms of electronic effects.[50] On the contrary, the presence of bulkier substituents and particularly of rigid cycles seems to affect the reactivity in a significant way (factor $\cong$ 10) due to steric effects. These effects are important to be considered when hydroprocessing heavy products. This is illustrated by hydrogenation of the heavy model compounds such as the 1,2,3,4,5,6,7,8-octahydro derivatives of anthracene, acridine, and phenanthridine (Fig. 7.14). Under classical operating conditions, these intermediates react about 10 times slower than their monocyclic reference models, benzene or pyridine.[16] A more difficult adsorption due to the rigid structure of these compounds is a possible explanation.

Hydrogenolysis of Intramolecular C_{sp^2}-Heteroatom Bonds

This section will be shorter than the section devoted to hydrogenation reactions, since most of the reactions generally proceed first through

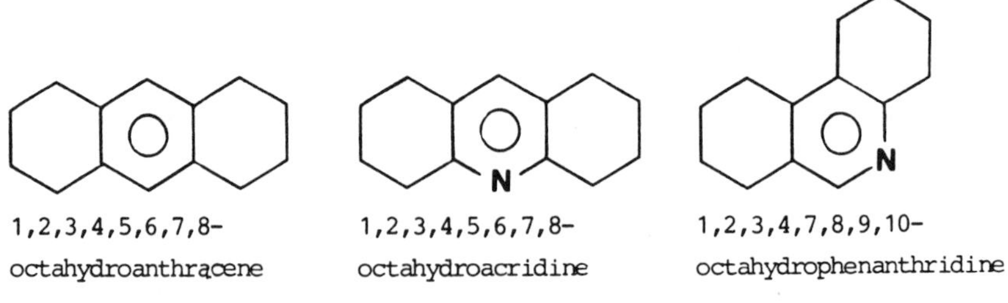

Figure 7.14. Sterically hindered aromatic and heteroaromatic compounds.

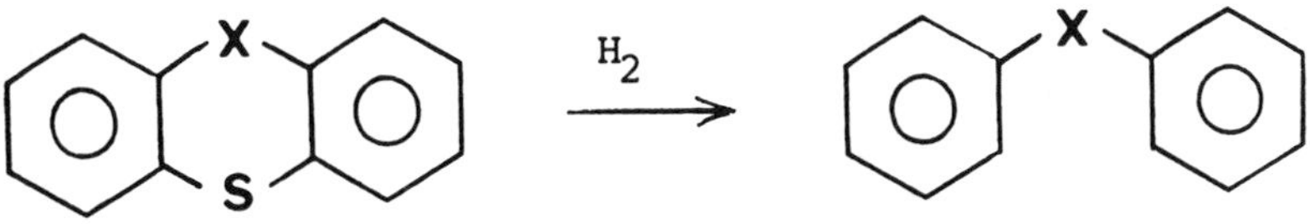

X = CH_2 Thioxanthene
X = S Thianthrene
X = O Phenoxathiin
X = NH Phenothiazine

Figure 7.15. Reaction network for desulfurization of thioxanthene, thianthrene, phenoxathiin, and phenothiazine.

hydrogenation steps under high-temperature and high-pressure operating conditions. Nevertheless, to have more information on this important competitive step, we will proceed in a similar manner to that developed for the hydrogenation reactions, i.e., systems with only one variable factor will be considered.

Numerous studies have dealt with intermolecular competitions. For example, it is well known that nitrogen bases such as pyridine or acridine inhibit the hydrodesulfurization of thiophene or dibenzothiophene.[51–53] Few studies have been reported concerning HDS reactions where a second heteroatom is present on the same molecule.[54] We carried out a systematic study on models that allowed us to determine this influence.[40,55] The following models were thus considered (Fig. 7.15): thioxanthene (X = CH_2), thianthrene (X = S), phenoxathiin (X = O), and phenothiazine (X = NH). These compounds have similar aromatic properties. The resonance energies vary between 70 and 80 kcal mol^{-1}. They also have comparable structures. The degree of folding along the X–S axis, as given by the dihedral angle θ, is about 130°–140°.[46,47,56,57]

The relative rate constants for hydrogenolysis of the C–S bond over a sulfided $NiMo/Al_2O_3$ catalyst are reported in Table 7.16. The basicity

Table 7.16. Relative rate constants for hydrogenolysis of C_{sp^2}-S bonds.

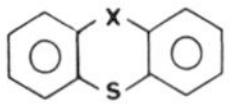		Dihedral angle	$k_{rel}^{40,55}$ ($NiMo/Al_2O_3$)
Thioxanthene	X=CH_2	135°	1
Thianthrene	X=S	128°	1.5
Phenoxathiin	X=O	138°	1.5
Phenothiazine	N=NH	140°	1.4

increases from X = CH_2 to X = NH and, as it can be seen in this table, the relative rate constants are nearly the same. For these model compounds, the hydrogenolysis rates of C–S bonds are not affected by the presence of a second heteroatom or a methylene group. Consequently, hydrogenolysis reactions are not related to the basicity of the molecule. This rules out the hypothesis of adsorption through heteroatoms. A π-adsorbed state involving both the aromatic ring and the heteroatom was preferred.[40,55] This point will be discussed later in detail.

Structural Effects. Structural effects have been shown to have no influence on hydrogenation reactions. On the contrary, they may significantly affect hydrogenolysis reactions as shown in Table 7.17 for planar and folded molecules. The aromatic properties are nearly the same for these molecules. From Table 7.17, it can be seen that, for a given heteroatom, the rates of hydrogenolysis are markedly increased for folded systems, thioxanthene and xanthene, compared to the planar reference compounds, dibenzothiophene and dibenzofuran. These results have been interpreted in terms of π-adsorption through aromatic rings, more important for planar than for folded molecules.[40,55]

Influence of the Heteroatom. The relative rate constants for hydrogenolysis of C–S bonds with respect to C–O bonds for a series of compounds with similar structure and aromaticity are reported in Table 7.18. It can be seen that S-containing molecules always react faster than their O-containing analogs. This behavior differs from that reported for hydrogenation reactions.

In summary, hydrogenolysis of intramolecular C_{sp^2}-heteroatom bonds does not depend on the total aromaticity of a molecule but rather on

Table 7.17. Influence of structural effects on hydrogenolysis of C_{sp^2}-S and C_{sp^2}-O bonds.

S-Compounds	k_{rel}^{55} ($NiMo/Al_2O_3$)	O-Compounds	$k_{rel}^{29,55}$ ($NiMo/Al_2O_3$)
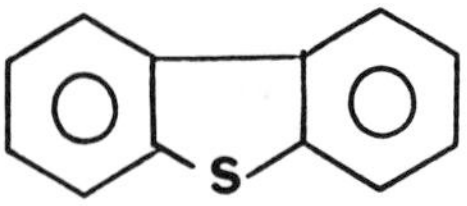 S	1	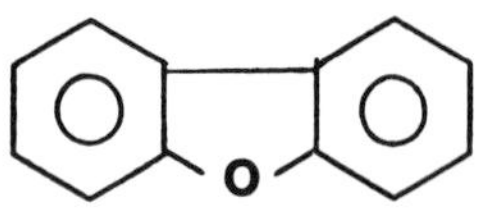 O	1
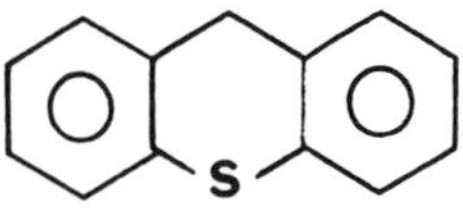 S	11	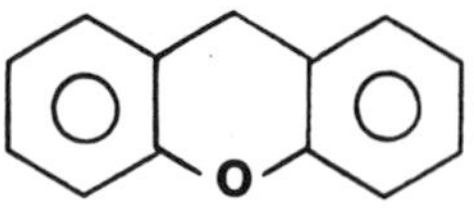 O	5

Table 7.18. Influence of the heteroatom on hydrogenolysis of C_{sp^2}-S and C_{sp^2}-O bonds.[29,40,55]

Compounds	k_{rel} ($NiMo/Al_2O_3$)	Compounds	k_{rel} ($NiMo/Al_2O_3$)
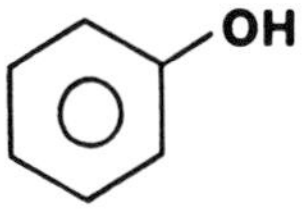	1	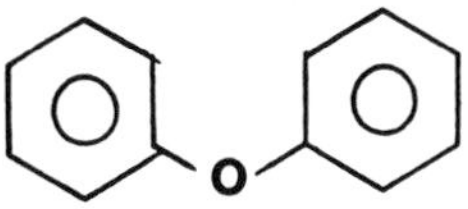	1
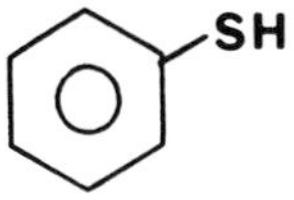	200	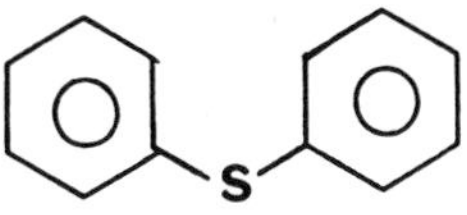	100
Compounds	k_{rel} ($NiMo/Al_2O_3$)	Compounds	k_{rel} ($NiMo/Al_2O_3$)
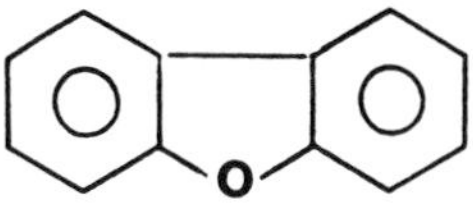	1	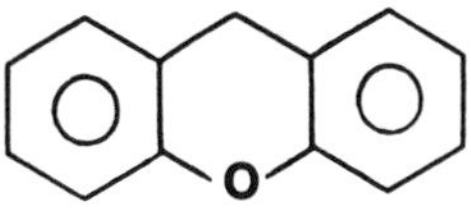	1
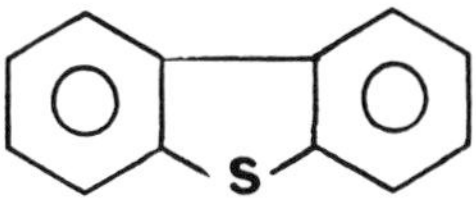	13	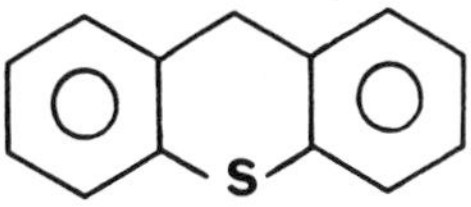	33

both structural effects and electronegative properties of the heteroatoms.

In order to rationalize the factors affecting the reactivity of organic model compounds in hydrogenation as well as in hydrogenolysis reactions, we have studied a simple reaction involving these two elementary reactions simultaneously.

Competition Between Hydrogenation and Hydrogenolysis

The particularly appropriate reaction for this study is the hydroprocessing of substituted benzenes Ph–X, where X = SH, SPh, OH, OPh, NH_2, NHPh, F, Cl, Br, which are for the most part postulated or identified intermediates during the course of HDS, HDO, or HDN reactions and which may react through hydrogenation of the aromatic ring or through hydrogenolysis of the C_{sp^2}-X bond, depending on the nature of the substituent X.[58] All these compounds are similar in their structure and in their

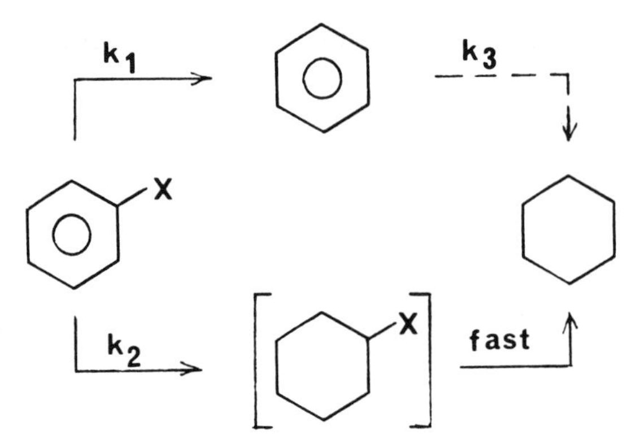

Figure 7.16. General reaction network for hydroprocessing of monosubstituted benzenes.

total aromaticity.[11] They only differ in the electronegative properties of the substituent X through inductive effects or resonance effects or both.

The reaction network that accounts for the hydroprocessing of substituted benzenes is given in Figure 7.16.

A similar reaction scheme has also been proposed recently for the hydrodeoxygenation of methyl-substituted phenols over a sulfided CoMo/Al_2O_3 catalyst.[59]

From the initial rates of formation of benzene and cyclohexane or from their relative concentrations, we have shown that it was possible to calculate the rate constant for the hydrogenation of the benzene ring (k_2) and the rate constant for the hydrogenolysis of the C_{sp^2}-X bond (k_1). These rate constants obtained on a sulfided NiMo/Al_2O_3 catalyst at 340°C, 70 bar H_2, are given in Table 7.19.

From this table, it can be seen that low hydrogenation rates correspond to high hydrogenolysis rates and vice versa. For S, Cl, and Br substrates, cyclohexane is barely detectable by GC–MS analysis, so that a kinetic consecutive reaction scheme (Fig. 7.17) can easily account for the experimental data when $k_1 \gg k_2$. Therefore, the limiting value given by the rate of hydrogenation of benzene, $k_3 = 2 \times 10^{-3}$ min^{-1} (g cat.)$^{-1}$, is a good approximation.[58]

Figure 7.17. Consecutive reaction network for hydroprocessing of monosubstituted benzenes.

Table 7.19. Hydrogenolysis (k_1) and hydrogenation (k_2) rate constants [$\times 10^3$ min^{-1} (g cat)$^{-1}$]; inductive (σ_I) and resonance (σ_R) substituent parameters.

Substrates	σ_I	σ_R	k_1	k_2
Thiophenol	0.27	−0.17	400	Lower limit[a]
Diphenylsulfide	0.31	−0.13	300	Lower limit[a]
Phenol	0.24	−0.42	2	33
Diphenylether	0.40	−0.34	3	3
Aniline	0.17	−0.47	1	10
Diphenylamine	0.30	−0.50	1	9
Chlorobenzene	0.47	−0.21	75	1[a]
Bromobenzene	0.47	−0.19	100	1[a]
Fluorobenzene	0.54	−0.33	8	2
Benzene	0.00	0.00	—	2

[a] The lower limit (see text) corresponds to $\simeq$0.5% of hydrogenation for thiophenol and diphenylsulfide and to $\simeq$1% of hydrogenation for chlorobenzene and bromobenzene.

The most convenient method to describe the effects of substituents on the chemical reactivity is to separate them into localized (inductive or field or both) and delocalized (resonance) effects.[60] The localized effects are represented by the substituent constant σ_I and the delocalized effects by σ_R. These constants are reported in Table 7.19 and are positive or negative depending on the electron-withdrawing or electron-donating ability of the substituent. The reactivity can thus be represented by Hammett-type equations $\log k/k_0 = \rho_I \sigma_I$ or $\log k/k_0 = \rho_R \sigma_R$, where k and k_0 are the reaction rate constants for substituted benzenes and unsubstituted benzene substrates, respectively. By plotting the logarithms of the rate constants of hydrogenation and hydrogenolysis against σ_I and σ_R it is possible to determine which effect will predominate.

We have shown there was no correlation between the logarithms of the rate constants k_1 and k_2 and the substituent constant σ_I, thus ruling out the direct influence of the basic properties of the heteroatom on the reactivity of aromatic molecules.[58] On the contrary, a good correlation is found when plotting the logarithms of the rate constants of hydrogenolysis of C_{sp^2}-X bonds against the resonance parameter σ_R (Fig. 7.18). A satisfactory correlation is also found for the hydrogenation reaction (Fig. 7.19) up to the lower limit (X = F, Cl, Br, SH, SPh) corresponding to the hydrogenation rate of benzene as illustrated in Figure 7.16. The results show that the reactions of hydrogenation and hydrogenolysis are mainly influenced by π-electron delocalization through resonance. Hydrogenation will be favored by highly electron-donating substituents (N or O), and, inversely, hydrogenolysis will be favored by slightly electron-donat-

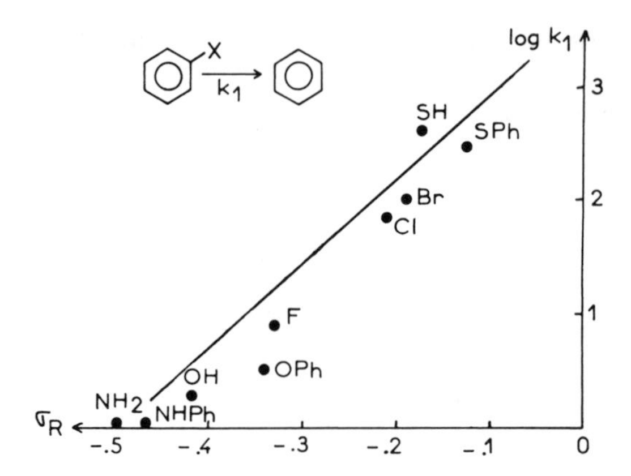

Figure 7.18. Plot of the hydrogenolysis rate constant k_1 vs the resonance substituent constant σ_R.

ing substituents such as Cl, Br, or SH. By analogy with homogeneous catalysis, a positive slope for the hydrogenolysis of C_{sp^2}-X bonds would imply an increase in the electron density during the course of the reaction, while a negative slope for the hydrogenation reaction would, in turn, imply a decrease in the electron density of the kind observed in electrophilic substitution reactions.[61] In other words, as compared to the initial state, the activated complex would be more electron-rich for hydrogenolysis and more electron-deficient for hydrogenation. The contrasting behavior observed for these two elementary steps shows unambiguously the

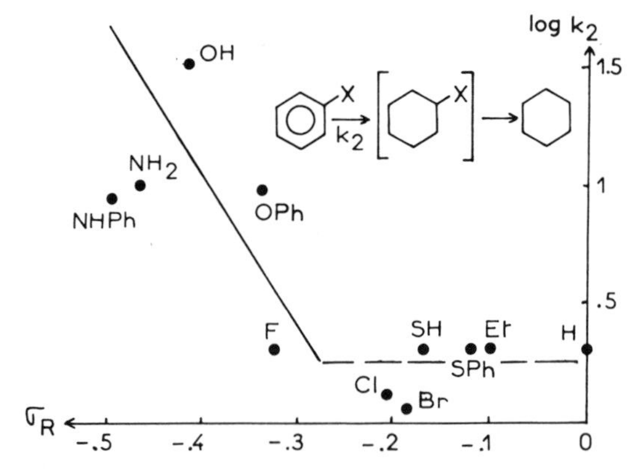

Figure 7.19. Plot of the hydrogenation rate constant k_2 vs the resonance substituent constant σ_R.

existence of two different catalytic sites, one responsible for hydrogenation associated with an electron-withdrawing character ($\rho < 0$) and the other one responsible for hydrogenolysis of C_{sp^2}-X bonds associated with an electron-donating character ($\rho > 0$). The interactions between the organic substrate and the catalyst can be depicted as illustrated in Figure 7.20, which shows a kind of electron transfer between the organic molecule and the catalytic sites. The existence of two catalytic sites has often been proposed in the literature to account for hydrotreating reactions.[59,62–64] The chemical nature and the mode of action of these sites will be discussed subsequently.

The correlations that have been found for hydrogenation and hydrogenolysis of substituted benzenes are very important in that these correlations also apply to the heavier models previously studied. For example, small differences in the rate constants are observed in hydrogenation of fused systems such as fluorene, dibenzothiophene, dibenzofuran, and carbazole. Only a factor of 2 was found between the hydrogenation rate constants of carbazole and dibenzothiophene (see Table 7.12). When the heteroatom is incorporated into a six-membered ring as in xanthene, thioxanthene, and 9,10-dihydroacridine, the rate constant ratio k_{NH}/k_S is equal to 4 (see Table 7.13).

On the other hand, greater differences in the rate constants are observed for hydrogenolysis reactions of the compounds mentioned previously. The rate constant ratio k_S/k_O is equal to 13 when the heteroatom is incorporated into a five-membered ring and 33 in the case of six-membered rings (see Table 7.18).

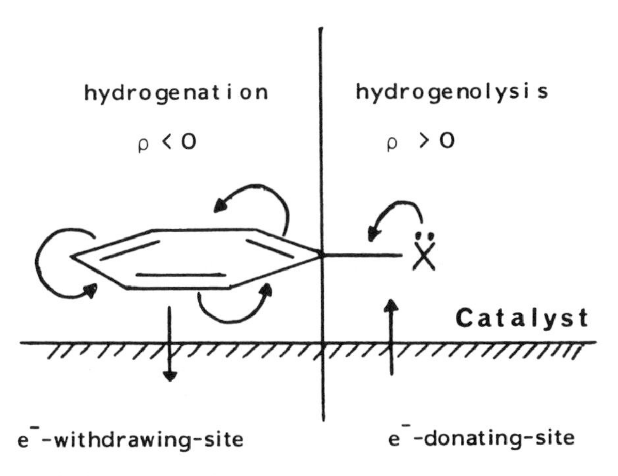

Figure 7.20. Dual-site mechanism for hydroprocessing of monosubstituted benzenes over sulfided $NiMo/Al_2O_3$ catalyst.

Table 7.20. Rate constants [$\times 10^4$ min^{-1} $(g\ cat)^{-1}$] for hydroprocessing of aniline, 1-naphthylamine, and 9-aminoanthracene over sulfided $NiMo/Al_2O_3$ catalyst.[70]

Compounds	k_{total}	Percentage of C–N Bond Cleavage	Percentage of Ring Hydrogenation	k_{C-N} Bond Cleavage	k_{ring} Hydrogenation
NH_2	110	10	90	10	100
NH_2	340	29	71	80	200
NH_2	1400	57	43	800	600

These results are in complete agreement with those reported for simple models and confirm the predominant influence of the aromatic factors in the two elementary chemical reactions.

The consequences of these correlations are also important to consider to improve a catalytic system or to define new concepts for new catalysts. Indeed, hydrogenolysis reactions are favored for the less electron-donating substituents (see Fig. 7.18).

One of the major problems in hydrodenitrogenation of heavy fuels is the conversion of alkylanilines,[33] and the question that arises from our results on substituted benzenes is how to favor the cleavage of C_{sp^2}-N bonds. A relatively simple way is to make the adjacent benzene ring less aromatic, rendering the nitrogen atom less conjugated with this ring. We have already shown that the rates of hydrogenation of aromatic rings increase with the decreased aromatic character of the ring to be hydrogenated (see Table 7.5). We can expect that amino groups present on less aromatic rings will thus be less conjugated with these rings. For this purpose, we have performed the hydroprocessing of 1-naphthylamine and 9-aminoanthracene compared to aniline over a sulfided $NiMo/Al_2O_3$ catalyst at 340°C and 70 bar H_2.[65] The experimental results are reported in Table 7.20. These results clearly show that reactivity increases when the conjugation of the N-atom with the ring bearing the amino group decreases. The differences in reactivity are important in both hydrogenolysis and hydrogenation reactions. The percentage of C–N bond cleavage increases up to 60% as expected from our assumptions on the diminution of the electron-donating character of the substituent. Conversely, the percentage of ring hydrogenation is reduced up to 40%, thus confirming the interdependence of the two chemical reactions and the duality of the mechanism proposed from the correlations observed on simple models. Finally, as already observed in other systems, the enhancing effect is larger in the hydrogenolysis reaction (by a factor of 80) than in the hydrogenation reaction (factor of 6) when going from an aromatic system like aniline to a lesser aromatic one like 9-aminoanthracene. The last point to be noted is the progressive change in the hybridization of the carbon bearing the amino function, which has a pure sp^2 character in aniline and which tends to a sp^3 character in 9-aminoanthracene. This change is easily shown by the rapid denitrogenation of cyclohexylamine under the same operating conditions.[16,66]

Summary

Hydrogenation reactions of aromatic and heteroaromatic rings are mostly affected by the aromatic properties of the rings to be hydrogenated rather than by the electronegativity of the heteroatom or the geometry of the molecule. Hydrogenation is easier when the ring to be hydrogenated is less aromatic.

Unlike hydrogenation reactions of heteroaromatics that are not influenced by the nature of the heteroatom, hydrogenolysis reactions are affected by the change of heteroatom in systems with similar aromatic and structural properties. This is the case for molecules where the heteroatom is capable of being conjugated with the adjacent aromatic ring. It has been clearly shown that the reactivity depends on the electron-donating ability through resonance of the S, O, or N substituent. The hydrogenolysis of a C_{sp^2}-heteroatom bond is favored by slightly electron-donating substituents. This rules out the possibility for the heteroatoms to act through their basic properties and still reinforces the preponderance of their aromatic properties through π-electron delocalization.

For simultaneous hydrogenation and hydrogenolysis reactions, an opposite correlation was found in hydrogenation reactions: strongly electron-donating substituents favor hydrogenation.

It should be noted finally that these two correlations lead unambiguously to the existence of two distinct catalytic sites, one responsible for hydrogenation with the higher oxidation level and the other one responsible for hydrogenolysis with the lower one.

After this detailed examination of the reactivity of aromatic molecules, the reactivity of saturated molecules will be briefly developed, since most of N-aromatic compounds are hydrogenated prior to any C–N bond cleavage.

REACTIVITY OF SATURATED MOLECULES

Cleavage of C_{sp^3}-N Bonds

This reaction is of fundamental interest since several mechanisms may operate as mentioned previously. The reactivity may be affected by ring strain, heteroatom basicity, conformational effects, steric effects, etc. There is also a particular interest in denitrogenation of saturated N-compounds, since information can be expected concerning the interactions between the organic molecules and the catalyst, the acid–base properties of the catalyst, the active phase, the role of the support, etc.

Primary amines generally are rapidly transformed into olefins (Fig. 7.21), even in the absence of hydrogen. However, under standard hydrotreating operating conditions, these intermediates are not often detected.

$$R-CH_2-CH_2-NH_2 \longrightarrow R-CH=CH_2$$

Figure 7.21. Olefin formation from aliphatic amines.

$$R-CH_2-CH_2-NH_2 \longrightarrow (R-CH_2-CH_2 \rightarrow_2 NH$$

Figure 7.22. Duplication reaction from aliphatic amines.

It should also be pointed out that condensation products may be formed depending on whether the catalyst is supported or not (Fig. 7.22).

According to general elimination mechanisms,[67] the acid properties of the catalyst might account for the formation of olefins and condensation products. This was clearly shown for alcohol dehydration over solid catalysts.[68] Olefin formation is favored by the presence of Brönsted acid sites, whereas condensation products are favored by Lewis acid sites. These two types of sites have recently been hypothesized in hydrotreating reactions.[63,64] The increase in the rates of C–N bond cleavage in the presence of H_2S or H_2O or both[69] would result from a shift of the equilibrium B-sites $\rightleftharpoons$ L-sites in favor of supplementary B-sites. In fact, this is not so simple, since the effect of added N-compounds also facilitates cleavage reactions as shown in recent results obtained in our laboratory.[70]

Olefin formation is usually interpreted in terms of β-elimination, through the unimolecular E1 mechanism involving the formation of a carbocation intermediate or through the concerted bimolecular E2 mechanism (Fig. 7.23).[71]

For rigid systems (Fig. 7.24), E2 eliminations are more difficult and require precise substituent orientation (trans). A reaction that cannot occur through E2 elimination will proceed more slowly but will always be observable.[72]

Remark: Intermediate olefins cannot be formed from some molecules because of the absence of β-hydrogens in triphenylcarbinol or for steric reasons in 1-adamantanol and 1-adamantanamine (Fig. 7.25). Two mechanisms are then possible: a true hydrogenolysis involving hydrogen partici-

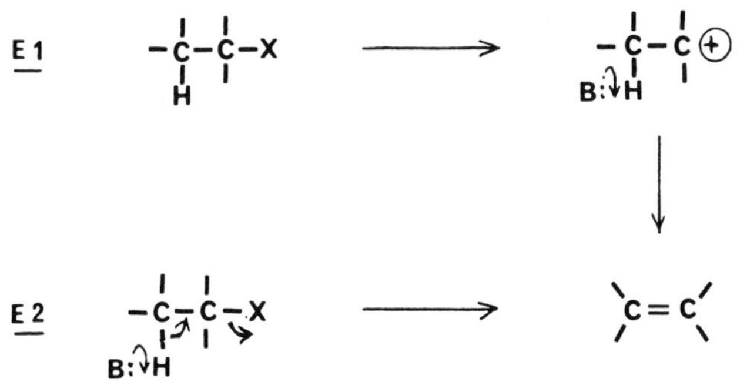

E1 –C(H)–C–X ⟶ B:→H –C–C⊕

E2 B:→H –C–C–X ⟶ >C=C<

Figure 7.23. E1 and E2 β-elimination mechanisms.

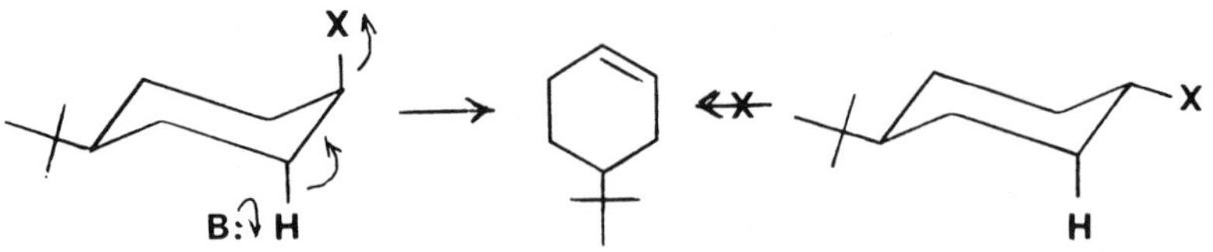

Figure 7.24. Stereocontrolled β-elimination.

pation or a mechanism involving the formation of carbocation intermediates, which are particularly stable for these three compounds.

Another feature to be noted is that the removal of nitrogen proceeds more slowly than that of oxygen by a factor of 10. It should also be mentioned that C-heteroatom bonds in benzylic systems are very easily broken as compared to alkyl systems. In the former case, the resulting carbocation is particularly stable (tropylium ion) as compared to the classical alkyl carbocation in the latter case (Fig. 7.26).

Influence of the Electronegativity of the Heteroatoms

It has been experimentally shown that C_{sp^3}-N bonds are more easily cleaved when nitrogen atoms are more basic (Fig. 7.27).[66] These results would support a β-elimination mechanism favored by the basicity of the leaving group. On the contrary, when β-elimination cannot take place, the less basic functional group will then be removed (1-adamantanol > 1-adamantanamine).[73]

For N-heterocyclic molecules resulting from prior saturation of heavier aromatic ones, the removal of nitrogen can also contribute to the overall reaction rate. Saturated N-heterorings are generally less reactive than their corresponding aliphatic analogs of similar basicities for entropic reasons due to the opening of rings.

1-adamantanol 1-adamantanamine Triphenylcarbinol

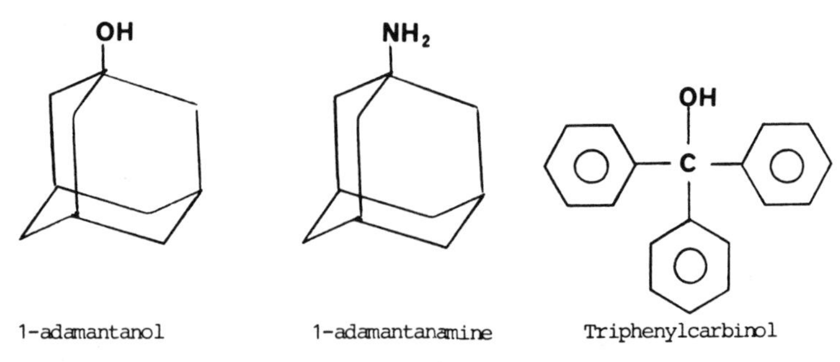

Figure 7.25. Models known for absence of β-elimination.

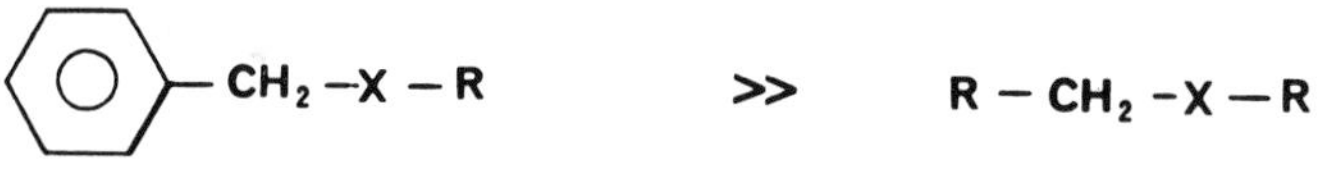

Figure 7.26. Influence of the leaving group on the cleavage of CH_2–X bonds.

In a similar manner, ring-opening of five-membered rings (pyrrolidine) is easier than that of six-membered rings (piperidine) for similar basicities as given by their pK_as, 11.3 and 11.1, respectively.[16] The effects of ring strain will thus favor the reactivity of five-membered heterocycles as compared to six-membered ones.

Another illustration of this difference in reactivity is given by comparison of the reactivity of indoline and 1,2,3,4-tetrahydroquinoline (Fig. 7.28). Under the same operating conditions (NiMo catalyst, 340°C, 70 bar H_2), indoline gives 2-ethylaniline, whereas 1,2,3,4-tetrahydroquinoline is hydrogenated to decahydroquinoline instead of yielding 2-propylaniline.[16] These observations are valid in liquid phase[16] as well as in gas phase.[34]

On the other hand, we have already mentioned that in some cases C–N bond cleavage could be rate-limiting. This was experimentally shown in hydrodenitrogenation of acridine and phenanthridine, where the perhydro derivatives (Fig. 7.29) accumulate in the medium prior to the cleavage of C–N bonds. The denitrogenation rates are very much lower than the rate of denitrogenation of the reference heterocycle piperidine, by a factor of about 10 and 50, respectively, for similar basicities. This decrease in rate constants can result from the presence of rigid cycles at the α and β

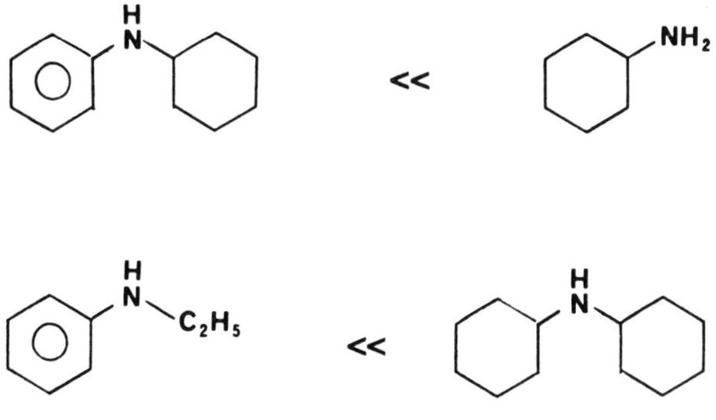

Figure 7.27. Influence of the basicity on the cleavage of C_{sp^3}-N bonds.

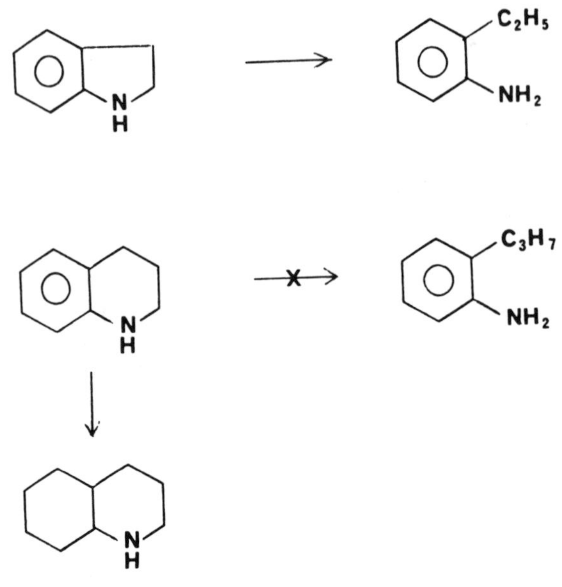

Figure 7.28. Influence of the ring-strain on the cleavage of C_{sp^3}-N bonds.

positions of the nitrogen atom. On the contrary, the presence of methyl groups at the α positions of nitrogen in 2,2,6,6-tetramethylpiperidine increases the rate of C–N bond cleavage.[73]

For the sterically hindered molecules, perhydroacridine and perhydrophenanthridine, it is possible to consider a difficult adsorption through N-atoms. On the other hand, the absence of well-oriented β–C–H bonds to give β-elimination would rule out this last possibility for the two compounds in their stable conformation (Fig. 7.30). Nevertheless, the required trans orientation of β–C–H bonds can be found for less stable conformers (Fig. 7.31).

In the absence of β-elimination, one can also consider the α-elimination mechanism involving a carbene intermediate (Fig. 7.32).[74] This mecha-

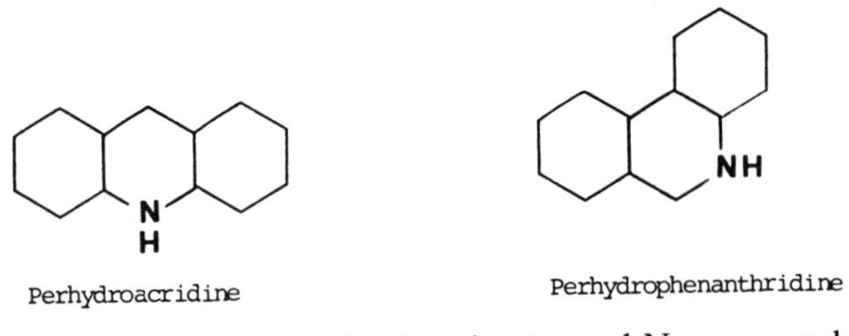

Figure 7.29. Sterically hindered saturated N-compounds.

(a)

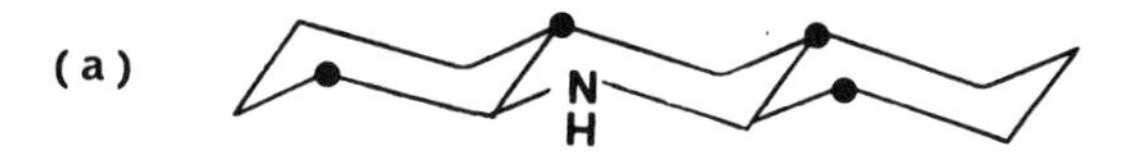

(b)

Figure 7.30. β-positions in perhydroacridine (a) and perhydrophenanthridine (b) under their most stable conformation.

(a)

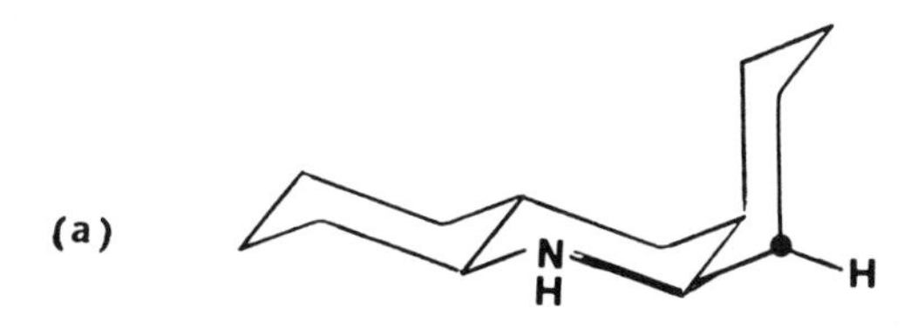

(b)

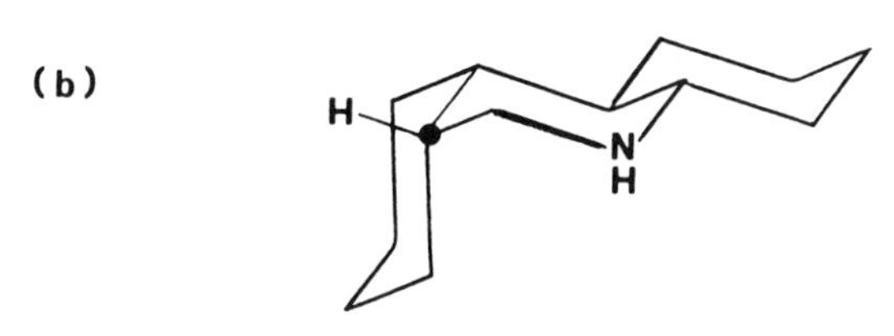

Figure 7.31. Conformers of perhydroacridine (a) and perhydrophenanthridine (b) allowing β-elimination.

$$R-CH_2-CH_2-X \longrightarrow R-CH_2-\ddot{C}H + HX$$

Figure 7.32. α-Elimination mechanism.

Figure 7.33. Metal-complexation mechanism according to Laine.

nism can be compared, to a certain extent, to that proposed by Laine, who described a metal complexation of the N-ring (Fig. 7.33).[75,76] As mentioned by Laine, "a true test for a proposed mechanism's viability is that it must explain the characteristic anomalies of the reaction." The metal-complex intermediates in question might account for the favorable effect of nucleophiles like H_2O and H_2S on the mechanism of cleavage of C–N bonds.[69] The reasoning would be the same for NH_3, and we have just shown that 2,6-diethylaniline favors the cleavage of 1,2,3,4-tetrahydroquinoline to give 2-propylaniline as does H_2S.[70] According to Satterfield, H_2S increases the number of hydrogenolysis sites, but, since basic

Table 7.21. Relative rate constants for cleavage of C_{sp^3}-heteroatom bonds.[42]

Compounds	k_{rel} ($NiMo/Al_2O_3$)
[piperidine, NH]	1
[tetrahydropyran, O]	12
[tetrahydrothiophene-ring, S]	45

N-compounds have the opposite effect on hydrogenation sites, the resulting effect is nearly the same. In the absence of more precise data, it is difficult to come to an unambiguous conclusion on this point.

On the other hand, the true mechanism of carbon–heteroatom bond cleavage should take into account the nature of the heteroatom. Indeed, in heterocyclic compounds, the C_{sp^3}-S bond is more easily cleaved than the C_{sp^3}-O and C_{sp^3}-N bonds as experimentally shown over a sulfided NiMo catalyst at 340°C and 70 bar H_2 (Table 7.21).[42] According to elimination mechanisms, the observed reactivity order would exclude the presence of strong Brönsted acid sites as recently shown by IR measurements on sulfided catalysts.[77] Protonation of the nitrogen atom would favor the cleavage of the C–N bond. Moreover, in terms of adsorption, the more basic N-compounds are more strongly adsorbed than their S- or O-analogs and would thus be less reactive.

THEORETICAL MODELING OF THE REACTIVITY

In order to better understand the reactivity of organic molecules in the presence of hydrotreating catalysts, a theoretical approach was carried out to attempt to model the hydrogenation and hydrogenolysis reactions. Prior to modeling the reactivity, it is necessary to model the catalytic sites first.

Modeling of the Catalytic Sites

The structure of hydrotreating catalysts has been well studied, and it is generally accepted that the active phase of these catalysts consists of MoS_2 slabs in which the promoter atoms are incorporated.[21,49,78] The catalytic sites are generally viewed as anionic vacancies, and recent results performed on molybdenum monocrystals supported this general view.[79,80]

Different models for catalytic sites have been used in calculations based on quantum-chemical methods.[81] These studies involve mainly the hydrodesulfurization of thiophene. The models vary from simple proton[82] to clusters containing several metallic atoms,[83] and the most often used quantum method is the Hoffmann EHT method.[84]

For the modeling studies performed in our laboratory, catalytic sites with one, two, or three anionic vacancies have been considered (Fig. 7.34). The active center is a molybdenum atom with five, four, or three first sulfur neighbors, respectively.

It was shown that vertical adsorption of thiophene on these sites calculated by the EHT method is favored over the horizontal adsorption whatever the degree of unsaturation of molybdenum and the size of the cluster simulating the catalyst.[81,85] Few studies have been reported concerning

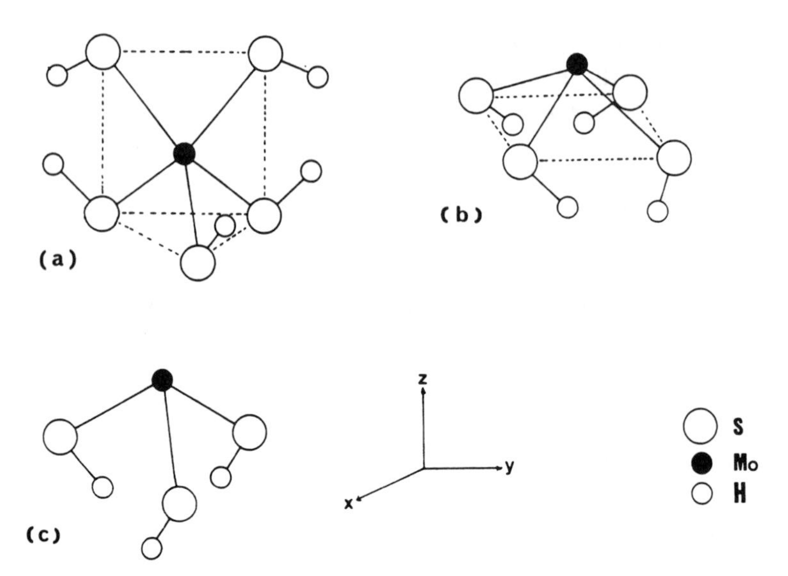

Figure 7.34. Models of catalytic sites with one (a), two (b), and three (c) anionic vacancies.

other molecules. A systematic study of S-models (thiophene, 2-methylthiophene, 3-methylthiophene), O-models (furan, 2-methylfuran), and N-models (pyrrole, pyridine) was performed in our laboratory.[42] It was confirmed that vertical adsorption through the heteroatom is more energetically favored than horizontal adsorption through the heteroatom or through the π-system. We can therefore reconsider the reaction mechanisms of aromatic molecules where the occurrence of hydrogenation and hydrogenolysis reactions depends on the operating conditions. This can be done by using our experimental results and Sabatier's principle, which states "that the most active catalyst is the one for which heats of formation are not too large and not too small."[86] In the section dealing with the factors of reactivity, we showed that hydrogenation reactions are mainly affected by the aromatic properties of the molecules and not by the nature of the heteroatoms; this implied a π-adsorbed state. By contrast, hydrogenolysis reactions depend on the nature of the heteroatoms and a σ-adsorbed state was recently proposed.[59]

From the calculations performed on thiophene for vertical and horizontal adsorptions, it is thus possible to imagine an energetic scheme such as that given in Figure 7.35, where the activation barriers $\Delta G_{I}^{\ddagger}$ and $\Delta G_{II}^{\ddagger}$ are characteristic of hydrogenation and hydrogenolysis reactions, respectively. Depending on the nature of the heteroatom and on the operating conditions, the differences in reactivity would be based only on the difference in energy between the two adsorbed states: (i) for the hydrogenolysis reaction, $\Delta G^{\ddagger}_{\rm II}$, through vertical adsorption by the heteroatom depending on the electronegativity of the heteroatom and (ii) for the hydrogenation reaction, $\Delta G^{\ddagger}_{\rm I}$, through horizontal π-adsorption depending on the aromatic properties of the molecules. We can then assume that, for weakly basic heteroatoms, the difference in energy between the two adsorbed states is such that the reaction could proceed through either hydrogenation or hydrogenolysis depending on both reaction temperature and H_2 pressure. On the contrary, for more basic compounds, the difference in energy would be too important for hydrogenolysis of carbon–heteroatom bonds to occur. These assumptions agree well with Furimsky's data obtained from hydrogenation equilibria.[27]

The preceding hypotheses are also important to consider in the case of intermolecular competitions, for example, in the inhibition of hydrodesulfurization reactions by nitrogen bases. Heteroaromatic molecules could be adsorbed either on the hydrogenation site through the π-system or on the hydrogenolysis site through the heteroatom. Nitrogen bases have been reported to inhibit the hydrodesulfurization of thiophene,[52] and we have further shown that this inhibiting effect takes place whatever the

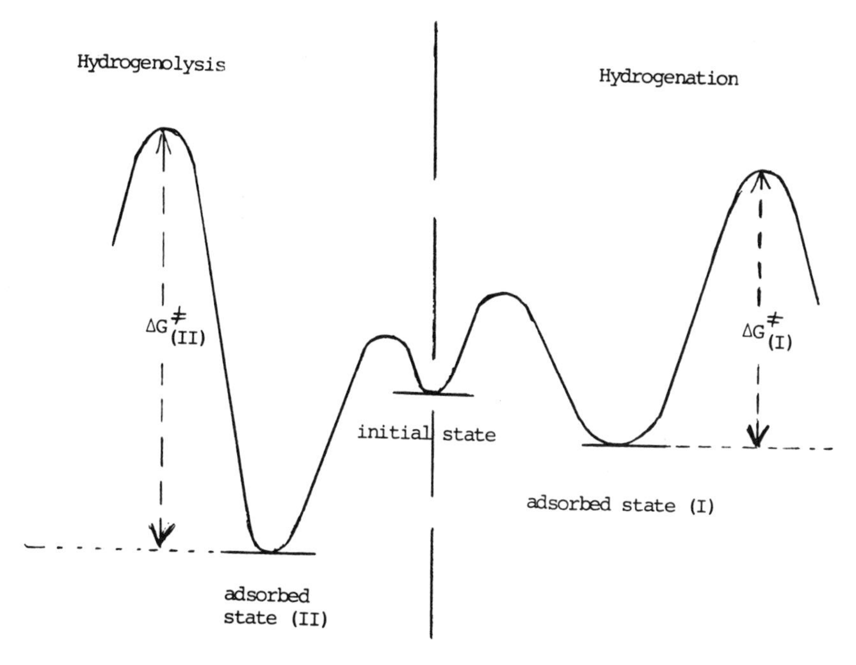

Figure 7.35. Energy profiles for hydrogenation and hydrogenolysis reactions.

Table 7.22. Inhibiting factor for thiophene HDS in the presence of nitrogen-bases (NiW/Al_2O_3 catalyst).[30]

Nitrogen Bases	pK_a	Inhibiting Factor
Pyrrole	−3.8	2
Aniline	4.6	2
Quinoline	4.9	2
Pyridine	5.3	4
Piperidine	11.1	5

basicity of molecules (Table 7.22).[30] It is worth mentioning that the presence of the heteroatom is necessary. Indeed, no inhibiting effect was observed on thiophene hydrodesulfurization in the presence of benzene, the aromatic analog of pyridine. This would be a new indication of the existence of two catalytic sites, one inhibited by nitrogen (hydrogenolysis site) but not the other site (hydrogenation site).[87] However, heavy nitrogen compounds are also known to inhibit the hydrodenitrogenation of alkylanilines[33,70,88] and such a conclusion on simultaneous reactions cannot be generalized.

In order to compare the results on adsorption of different compounds, another important parameter to be considered is the distance between the catalytic site and the molecule. Indeed, a small variation in the distance

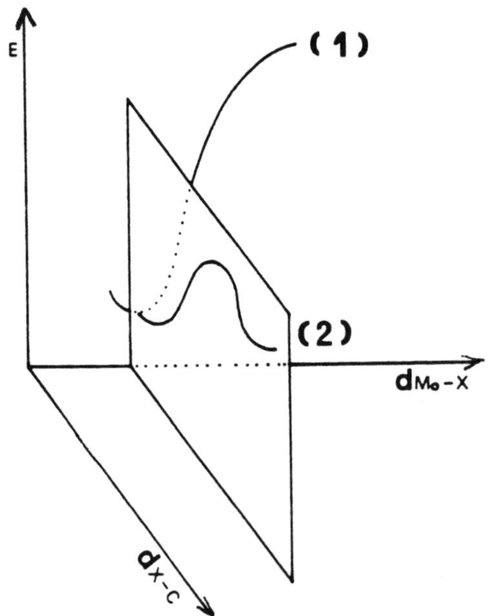

Figure 7.36. Adsorption (curve 1) and desulfurization (curve 2) of thiophene.

Mo-heteroatom can reduce the energetic gap between the two modes of adsorption, vertical and horizontal.[81] The influence of this parameter is not easy to evaluate, but the consequences are important both on the adsorption step and on the reaction modeling.

Modeling of the Reactivity

According to general concepts of heterogeneous catalysis, the reactant adsorbed on the surface of the catalyst is transformed into reaction products. By stretching progressively the carbon–heteroatom bond in the adsorbed reactant (Fig. 7.36), one determines in this way the energetic barriers representative of the cleavage of carbon–heteroatom bonds.[85]

Concerning desulfurization of thiophene, the energetic profiles obtained on two or three vacancy sites are very similar (Fig. 7.37). A slight

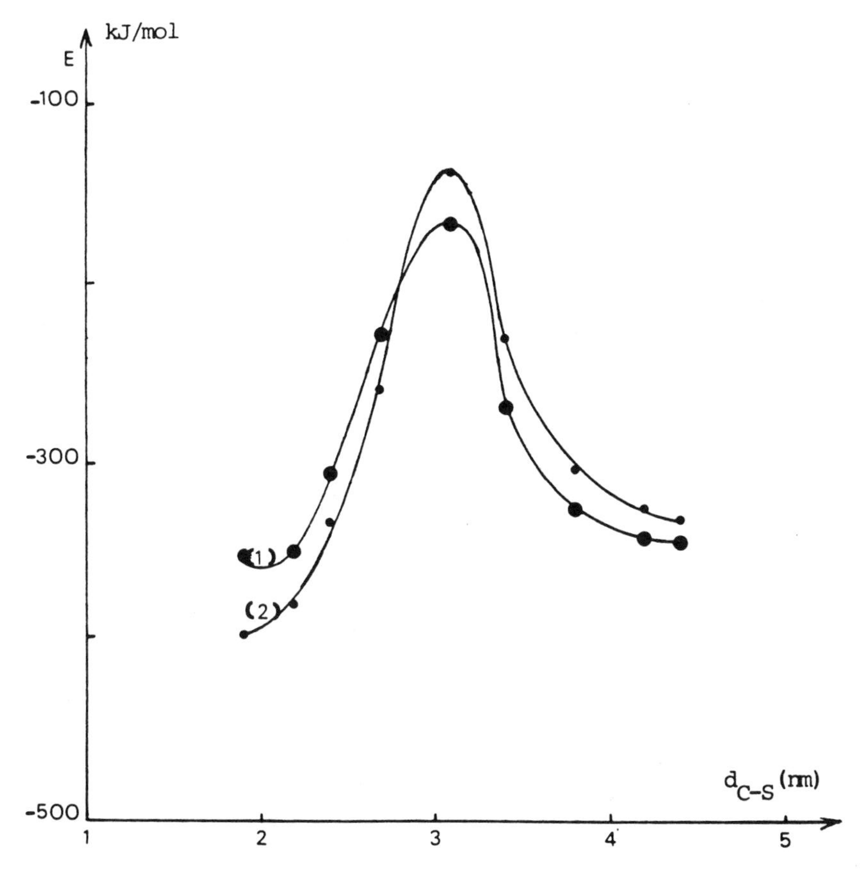

Figure 7.37. Desulfurization of thiophene over a two (curve 1) and three (curve 2) vacancies site model.

but meaningful rise of energy is initially observed showing that adsorption weakens the C–S bond. From the maximum of energy, an activation energy can be estimated. The weakening of the C–S bond is more pronounced in the case of two vacancies; this is in agreement with the lower activation energy obtained in this case. These energetic barriers are higher than the experimental barrier, but could be reduced by consideration of some other factors.[85] The activation energy depends to a large extent on the starting configuration.[81] As there is no experimental data on the actual Mo–S distance, the minimum Mo–S distance measured in Mo complexes was chosen for illustration. Even if the calculated activation energies are higher than the experimental ones, it is possible to verify that the cleavage of a C–S bond in thiophene is more difficult than that of a C–S bond in tetrahydrothiophene. Calculations are under way to model C–O and C–N bond cleavage.

Nature of the Catalytic Sites

From the preceding theoretical calculations, it is thus possible to model hydrotreating reactions over coordinatively unsaturated sites. The desulfurization site could sit at an edge,[89] whereas a molybdenum atom with three unsaturations could be situated at crystal corners.[90] Although the modeling needs some simplification, these observations agree with the results obtained from the reactivity of organic molecules, where it was unambiguously shown there are two catalytic sites with different oxidation levels. Recent results also support the fact that hydrodesulfurization reactions occur on edges and hydrogenation reactions occur on corners.[91] In a similar way, hydrogenation of *cis*-1,3-pentadiene has been shown to require three coordinatively unsaturated Mo ions.[92]

The quantum-chemical approach of hydrogenation and hydrogenolysis reactions brings some insight to the problem of hydrotreating. The results obtained could lead to a model for a heterogeneous hydrotreating catalyst containing several metal atoms and to the study of its interaction with different organic compounds.

CONCLUSIONS

From the results presented in this chapter and from comparison with available literature, the following conclusions can be drawn.

Hydrogenation of aromatic hydrocarbons as well as of that of five- and six-membered heteroaromatic compounds is essentially affected by the aromatic properties of the rings to be hydrogenated. Hydrogenation is facilitated by the decreased aromatic character of the ring in question. No significant influence was observed by replacing a methylene group by S, O, or N heteroatoms, thus ruling out the role of the heteroatom basicity on the hydrogenation reaction. No influence was found either by changing the structure of the starting materials: planar on folded, or with free rotation.

For reactions involving hydrogenolysis of C_{sp^2}-heteroatoms and hydrogenation of the adjacent aromatic ring, the reactivity depends on the aromaticity through π-electron delocalization. Hydrogenation is favored by strongly electron-donating substituents and, conversely, hydrogenolysis is favored by slightly electron-donating substituents. The slopes of the correlations found for both reactions, hydrogenation and hydrogenolysis, as a function of the electron-donating character are of opposite sign. The presence of such correlations constitutes chemical evidence for the existence of two distinct catalytic sites, one responsible for hydrogenation associated with an electron-withdrawing character and the other one responsible for the hydrogenolysis of C_{sp^2}-heteroatom bonds associated with an electron-donating character. A dual-site mechanism involving Mo or W atoms at different oxidation levels is proposed, the higher oxidation state for the hydrogenation site and the lower one for the hydrogenolysis site.

Theoretical calculations on adsorption and reactivity of S, O, and N aromatic molecules can be related, to a large extent, to the experimental results. Adsorption was shown to occur on catalytic sites consisting of two or three anionic vacancies, through vertical adsorption involving the heteroatom, and through horizontal adsorption involving the aromatic π-system. This would agree with our experimental data, which have led us to define two catalytic sites of different oxidation levels. This is also in agreement with the fact that hydrogenation reactions occur through π-adsorbed species, whereas hydrogenolysis reactions occur through adsorption of heteroatoms. All these assumptions might better explain the results obtained in simultaneous HDS/HDN reactions.

Although the quantum-chemical approach of reactions involved in hydrotreating requires improvements, there is good agreement between our own results and those available in the literature concerning sulfided catalysts. It can be expected in the near future that this approach could lead to a better model for a more effective hydrotreating catalyst.

ACKNOWLEDGMENTS

Most of the results reported in this chapter have been obtained in the framework of the Groupement Scientifique "Hydrotraitement Catalytique" of C.N.R.S., related to the industrial partners of the G.I.E. ASVAHL: ELF, TOTAL, and I.F.P.

The authors thank all the participants in this Groupement Scientifique for many fruitful discussions and particularly their co-workers: Dr. R. Durand, Dr. J. Joffre, and Dr. J.L. Olivé; and their research students: C. Aubert, L. Bekakra, S. Biyoko, T. Jei, J.B. Mensah, A. Messahli, C. Moulinas, and N. Zmimita.

REFERENCES

1. McCulloch, D.C., *Applied Industrial Catalysis* (B.E. Leach, ed.). New York: Academic Press, 1983, p. 70.
2. News Brief, *Appl. Catal.* **14:** 394 (1985).
3. Stinson, S.C., *Chem. Eng. News* **64**(7): 27 (1986).
4. Weisser, O., Landa, S., *Sulphide Catalysts, Their Properties and Applications.* Oxford: Pergamon Press, 1973.
5. Le Page, J.F., *Catalyse de Contact.* Paris: Technip, 1978.
6. Acheson, R.M., *The Chemistry of Heterocyclic Compounds—Acridines.* New York: Interscience, 1953, p. 53.
7. Wheland, G.W., *Resonance in Organic Chemistry.* New York: Wiley, 1955, p. 75.
8. Cook, M.J., Katritzky, A.R., Linda, P., *Adv. Heterocycl. Chem.* **17:** 255 (1974).
9. Gilchrist, T.L., *Heterocyclic Chemistry.* New York: Wiley, 1985, p. 5.
10. March, J., *Advanced Organic Chemistry.* New York: Wiley, 1985, p. 37.
11. See Ref. 7, Table 3.7, pp. 98–100.
12. Albert, A., Serjeant, E.P., *The Determination of Ionization Constants.* London: Chapman and Hall, 1971.
13. Taft, R.W., *Prog. Phys. Org. Chem.* **14:** 247 (1983).
14. Headley, A.D., *J. Am. Chem. Soc.* **109:** 2347 (1987).
15. Headley, A.D., *J. Org. Chem.* **53:** 312 (1988).
16. Moreau, C., Aubert, C., Durand, R., Zmimita, N., Geneste, P., *Catalysis Today* **4:** 117 (1988).
17. See Ref. 7, p. 104.
18. Moyano, A., Paniagua, J.C., *J. Org. Chem.* **51:** 2250 (1986).
19. *Structure Reports* **19:** 583 (1955).
20. Moreau, C., Durand, R., Zmimita, N., Geneste, P., *J. Catal.* **112:** 411 (1988).
21. Breysse, M., Bachelier, J., Bonnelle, J.P., Cattenot, M., Cornet, D., Décamp, T., Duchet, J.C., Durand, R., Englehard, P., Frety, R., Gachet, C., Geneste, P., Grimblot, J., Gueguen, C., Kasztelan, S., Lacroix, M., Lavalley, J.C., Leclercq, C., Moreau, C., De Mourgues, L., Olivé, J.L., Payen, E., Portefaix, J.L., Toulhoat, H., Vrinat, M., *Bull. Soc. Chim. Belg.* **96:** 829 (1987).
22. See Ref. 4, Table 27, pp. 126–127.
23. Sapre, A.V., Gates, B.C., *Ind. Eng. Chem., Process Des. Dev.* **20:** 68 (1981).
24. Broderick, D.H., Sapre, A.V., Gates, B.C., Katzer, H., Schuit, G.C.A., *J. Catal.* **73:** 45 (1982).
25. Wilson, M.F., Fisher, I.P., Kriz, J.F., *J. Catal.* **95:** 155 (1985).
26. See Ref. 4, pp. 132–133.
27. Furimsky, E. *Erdöl, Erdgas, Petrochem. Brennst. Chem.* **36:** 518 (1983).
28. Aubert, C., Durand, R., Geneste, P., Moreau, C., Zmimita, N., *Actas Simp. Iberoam. Catal.* (Soc. Iberoam. Catal., ed.). Venezuela: Merida, 1986, p. 1153.
29. Aubert, C., Doctorat Thesis, Montpellier, 1986.
30. Zmimita, N., Doctorat Thesis, Montpellier, 1987.
31. See Ref. 10, p. 39.

32. See Ref. 10, p. 40.
33. Toulhoat, H., Kessas, R., *Rev. Fr. I.F.P.* **41:** 511 (1986).
34. Schulz, H., Schon, M., Rahman, N.M., *Stud. Surf. Sci. Catal.* **27:** 201 (1986).
35. Bhinde, M.V., Shih, S., Zawadski, R., Katzer, J.R., Kwart, H. *Proceedings of the Third International Conference on Chemistry and Uses of Molybdenum* (H.F. Barry and P.C.H. Mitchell, eds.). Ann Arbor, MI: Climax Molybdenum Company, 1979, p. 184.
36. Jei, T., Doctorat Thesis, Montpellier, 1984.
37. Normant, H. *Chimie Organique,* Paris: Masson, 1963, p. 407.
38. Olivé, J.L., Biyoko, S., Moulinas, C., Geneste, P., *Appl. Catal.* **19:** 165 (1985).
39. Stern, E.W., *J. Catal.* **57:** 390 (1979).
40. Aubert, C., Durand, R., Geneste, P., Moreau, C., *J. Catal.* **97:** 169 (1986).
41. Biyoko, S., Doctorat Thesis, Montpellier, 1984.
42. Mensah, J.B. Doctorat Thesis, Montpellier, 1987.
43. De Beer, V.H.J., Dahlmans, J.G.J., Smeets, J.G.M., *J. Catal.* **42:** 467 (1976).
44. *Structure Reports* **35B:** 225 (1970).
45. *Structure Reports* **39B:** 230 (1973).
46. Structure Reports **39B:** 240 (1973).
47. Aroney, M.J., Hoskins, G.M., Le Fèvre, R.J.W., *J. Chem. Soc.* (*B*) 980 (1969).
48. Rollmann, L.D., *J. Catal.* **46:** 243 (1977).
49. Topsoe, H., Clausen, B.S., *Appl. Catal.* **25:** 273 (1986), and references therein.
50. Aubert, C., Durand, R., Geneste, P., Moreau, C., *J. Catal.* **112:** 12 (1988).
51. Cerny, M., *Collect. Czech. Chem. Commun.* **47:** 1465 (1982).
52. La Vopa, V., Satterfield, C.N., *J. Catal.* **110:** 375 (1986).
53. Nagai, M., Masunaga, T., Hana-Oka, N., *J. Catal.* **101:** 284 (1986).
54. Geneste, P., Olivé, J.L., Biyoko, S., *J. Catal.* **83:** 245 (1983).
55. Aubert, C., Durand, R., Geneste, P., Moreau, C., *Bull. Soc. Chim. Belg.* **93:** 653 (1984).
56. Hosoya, S., *Acta Crystallogr.* **16:** 310 (1963).
57. *Structure Reports* **20:** 628 (1956).
58. Moreau, C., Bachelier, J., Bonnelle, J.P., Breysse, M., Cattenot, M., Cornet, D., Décamp, T., Duchet, J.C., Durand, R., Englehard, P., Fréty, R., Gachet, C., Geneste, P., Grimblot, J., Gueguen, C., Kasztelan, S., Lacroix, M., Lavalley, J.C., Leclercq, C., De Mourgues, L., Olivé, J.L., Payen, E., Portefaix, J.L., Toulhoat, H., Vrinat, M., *Amer. Chem., Prepr.* **32:** 298 (1987).
59. Gevert, B.S., Otterstedt, J.E., Massoth, F.E., *Appl. Catal.* **31:** 119 (1987).
60. Charton, M., *Prog. Phys. Org. Chem.* **13:** 119 (1981).
61. See Ref. 10, p. 447.
62. Shabtai, J., Nag, N.K., Massoth, F.E., *J. Catal.* **104:** 413 (1987).
63. Yang, S.H., Satterfield, C.N., *J. Catal.* **81:** 168 (1983).
64. Kwart, H., Katzer, J., Horgan, J., *J. Phys. Chem.* **86:** 2641 (1982).
65. Moreau, C., Bekakra, L., Olivé, J.L., Geneste, P., *Proceedings of the Ninth International Congress on Catalysis* (M.J. Phillips and M. Ternan, eds.). Calgary, Canada: Chemical Institute of Canada, 1988, Vol. 1, p. 58.
66. Finiels, A., Geneste, P., Moulinas, C., Olivé, J.L., *Appl. Catal.* **22:** 257 (1986).
67. Saunders, W.H., Jr., Cockerill, A.F., *Mechanisms of Elimination Reactions,* New York: Wiley, 1973.
68. See Ref. 67, p. 221.
69. Satterfield, C.N., Smith, C.M., Ingalis, M., *Ind. Eng. Chem., Process. Des. Dev.* **24:** 1000 (1985).
70. Moreau, C., Bekakra, L., Messahli, A., Olivé, J.L., Geneste, P., *Stud. Surf. Sci. Catal.* **50:** 115 (1989).

71. See Ref. 67, p. 1.
72. Gould, E.S., *Mechanism and Structure in Organic Chemistry*. London: Holt, Rinehart and Winston, 1969, p. 472.
73. Moreau, C., Durand, R., Geneste, P. (unpublished results).
74. See Ref. 67, p. 29.
75. Laine, R.M., *Catal. Rev.-Sci. Eng.* **25:** 459 (1983).
76. Laine, R.M., *New. J. Chem.* **11:** 543 (1987).
77. Duchet, J.C., Lavalley, J.C., Ouafi, D., Bachelier, J., Cornet, D., Aubert, C., Moreau, C., Geneste, P., Houari, M., Grimblot, J., Bonnelle, J.P., *Catalysis Today* **4:** 97 (1988).
78. Kasztelan, S., Jalowiecki, L., Wambecke, A., Grimblot, J., Bonnelle, J.P., *Bull. Soc. Chim. Belg.* **96:** 1003 (1987).
79. Roberts, J.T., Friend, C.M., *J. Am. Chem. Soc.* **108:** 7204 (1986).
80. Bussell, M.E., Somorjai, G.A., *J. Catal.* **106:** 93 (1987).
81. Joffre, J., Geneste, P., Lerner, D.A., *J. Catal.* **97:** 543 (1986).
82. Zdrazil, M., Sedlacek, J., *Collect. Czech. Chem. Commun.* **42:** 3133 (1977).
83. Nikishenko, S.B., Slinkin, A.A., Antoshin, G.V., Minachev, Kh.M., Nefedofk, B.K., *Proceedings of the Fourth International Conference on the Chemistry and Uses of Molybdenum* (H.F. Barry and P.C.H. Mitchell, eds:,). Ann Arbor, MI: Climax Molybdenum Company, 1982, p. 51.
84. Hoffmann, R., *J. Chem. Phys.* **39:** 1937 (1963).
85. Joffre, J., Lerner, D.A., Geneste, P., *Bull. Soc. Chim. Belg.* **93:** 831 (1984).
86. Boudart, M., *Kinetics of Chemical Processes*. Englewood Cliffs, NJ: Prentice-Hall, 1968.
87. Massoth, F.E., Muralidhar, G., *Proceedings of the Fourth International Conference on the Chemistry and Uses of Molybdenum* (H.F. Barry and P.C.H. Mitchell, eds.). Ann Arbor, MI: Climax Molybdenum Company, 1982, p. 343.
88. Moreau, C., Bekakra, L., Durand, R., Zmimita, N., Geneste, P., *Stud. Surf. Sci. Catal.* **50:** 115 (1989).
89. Kasztelan, S., Toulhoat, H., Grimblot, J., Bonnelle, J.P., *Appl. Catal.* **13:** 127 (1984).
90. Massoth, F.E., Muralidhar, G., Shabtai, J., *J. Catal.* **85:** 53 (1984).
91. Blekkan, E.A., Mitchell, P.C.H., *Bull. Soc. Chim. Belg.* **96:** 961 (1987).
92. Wambeke, A., Jalowiecki, L., Kasztelan, S., Grimblot, J., Bonnelle, J.P., *J. Catal.* **109:** 320 (1988).

8
Theoretical Investigation of Metal–Support Interactions and Their Influence on Chemisorption

HELMUT HABERLANDT

INTRODUCTION

Phenomena of Metal–Support Interactions: General Remarks and Definitions

It is a well-known fact and a subject of continuous scientific and technological research[1–5] that activity and selectivity of heterogeneous catalytic reactions depend on the nature of the support. As a specific example the CO hydrogenation (Fischer–Tropsch synthesis)[6] has been investigated extensively since the oil crisis in the early 1970s. The activity of supported Pt was found to increase with variation of the support as 1:10:100 in the sequence SiO_2, Al_2O_3, TiO_2.[6,7] A similar situation is met with supported nickel catalysts.[8,9] There is also a pronounced support effect on selectivity: The methane yield is generally lower using TiO_2 and Al_2O_3 instead of SiO_2 as a support.[9,10]

But, unfortunately, the explanation of the support effect is not obvious. For example, Bartholomew et al.[9] showed in a paper on CO hydrogenation that one has to investigate simultaneously the influence of support, metal loading and dispersion, catalyst preparation and pretreatment, as well as the extent of reduction of the metal in order to understand what is going on.

Roughly speaking the metal catalyst preparation proceeds in three steps (see, e.g., Ref. 11): First, a so-called precursor is formed (impregnation, precipitation, ion exchange). Second, the precursor is subjected to a preactivation treatment (drying, calcination). Third, the system is activated (mostly by reduction in hydrogen) and a metallic (active) phase is formed. In the case of oxidic supports we end up with a solid-state system consisting of a reduced metal, an unreduced metal bound in the metal oxide (or in a metal–support compound, e.g., silicate, aluminate), and the supporting oxide. The extent of reduction of the metal and its dispersion depend on the nature of the support and on its pretreatment as well as on the kind of preparation and on the treatment conditions of the individual steps. The mutual interdependencies of these parameters may be illustrated, e.g., for nickel on alumina and silica: For coprecipitation catalysts it was deduced that the strength of the interaction of the precursor with the alumina support is stronger than that with the silica support.[12] Narayanan et al.[13] found a considerably decreased metal–support interaction in the Ni/Al_2O_3 precursor as revealed by increased reduction of oxidic nickel to the metallic state. This was attributed to the calcination of the support before loading the metal.[13] Using model catalysts obtained from vacuum deposition of first silica and alumina films and then of nickel on these films Arai et al.[14] concluded, based on the observation of Ni surface migration, that the magnitude of the metal–support interaction in Ni/SiO_2 is even larger than in Ni/Al_2O_3. From a comparison of the standard preparation techniques, impregnation and precipitation for nickel on silica, it is observed that the precursor–support interaction is smaller in the former case resulting in a larger amount of reduced nickel and in larger crystallite sizes.[15] This difference in metal–support interaction also has considerable consequences with respect to structural changes in calcination steps.[16]

These rather puzzling interdependencies of preparation conditions in formation and reduction of heterogeneous catalysts and in particular the role of metal–support interaction in this connection are of huge scientific interest and technological importance. However, nowadays they are as a whole not accessible to theoretical investigations. Rather, the "final" reduced heterogeneous catalysts are modeled. This approach will be followed in this chapter. This theoretical modeling procedure parallels in many respects investigations on "experimental model catalysts" (cf. the experimental section) in the literature.

Let us recall at this point the final aim of any catalytic research: It is the design of catalysts with high activity and selectivity toward desired products. Roughly, there may be four steps for a catalytic reaction: adsorption of reactants (Langmuir–Hinshelwood mechanism), diffusion to active sites, reaction via the formation of intermediates, and desorption of products. Each of these steps may be rate-determining and may depend on the geometric or electronic structure or both of the catalyst surface. Coming back to a given reduced metal catalyst a change of its surface structure can be brought about in the first line by variations of particle sizes and shapes[17] but also by support effects on the metallic particles. Particle size effects and support effects on the catalytic activity and selectivity are as a rule experimentally very difficult to discriminate.[18–20] Furthermore, another complication arises: In addition to the real support effects (vide infra) there are some apparent ones, e.g., incomplete reduction of the metal, bifunctional catalysis, and spillover phenomena,[21] which may play a role in each of the four steps of the catalytic reactions.

In order to make clear what theoreticians have in mind when they are going to study support effects, we return to the reduction step in preparing a heterogeneous metal catalyst: Small metallic particles of different sizes are stabilized on the support. Their sizes and morphologies (spherical, pill-box shape, etc.) are governed by both the geometric and the electronic structure of the support surface; even contamination of the metal by the support may occur.[21] This is what we call "geometric effect of the support." The geometric effect cannot be considered theoretically without a tremendous effort (cf. Section 3). For a given particle size and shape there may occur a more or less pronounced support effect on the electronic properties of the metal particle. It is this "electronic effect of the support" that is usually addressed by theoretical investigations. This electronic effect of the support has been termed "Schwab Effect of the Second Kind" by Boudart and is "the only metal–support interaction which may be called as such in a strict manner."[21] In part, the geometric effect is involved, however, in the special geometry of the model and cannot be completely separated from the electronic effect. Thus, properly chosen models are of crucial importance in order to obtain reliable results (vide infra). With respect to the four steps of a catalytic reaction, theoretical investigations as a rule are restricted to the first one, the adsorption step.

Bond[22] proposed a classification of support effects; he distinguished between weak (WMSI), medium (MMSI), and strong (SMSI) metal–support interactions. The WMSI occurs predominantly with transition metals supported on "nonreducible" oxides like SiO_2, Al_2O_3, and MgO. The MMSI is ascribed to transition metal species in and on zeolites. The SMSI is found mainly with group VIII transition metals supported on reducible (transition-metal) oxides like TiO_2 and Nb_2O_5. The so-called SMSI effect is usually defined as the strong diminution, indeed suppression, of H_2 and CO chemisorption on the latter transition metal/oxidic support systems after a high-temperature reduction (HTR) at about 773 K in hydrogen. This effect was observed for the first time by Tauster, Fung, and Garten[23] in the system Pt/TiO_2. The modifying effect of the support on catalytic reactions on Ni/TiO_2 had already been pointed out[24] and had been ascribed to the electr(on)ic properties of the support.[24] The SMSI effect has been found meanwhile also with transition metals supported on "nonreducible" oxides[25,26] after reduction in a hydrogen atmosphere at temperatures near 1100 K. Besides affecting chemisorptive properties of the catalysts, the SMSI also influences activity and selectivity. Alkane hydrogenolysis activity is drastically suppressed, but the activity for CO–H_2 reactions is frequently increased for the same systems, often accompanied by improved selectivity to higher hydrocarbons.[27] Recently it has been questioned, however, whether the creation of the SMSI state is really a prerequisite for obtaining high activity or unusual selectivity[10,28] (vide infra). The origin of the SMSI effect is still unclear and remains a subject of current interest (see, e.g., Ref. 27).

In the present review on theoretical contributions to the understanding of metal–support interaction phenomena we focus on the WMSI and SMSI and in particular on their interrelation; in addition, some general comments concerning support effects based on theoretical work are made.

The Impact of Related Subjects

It is appropriate here to give a short survey on the experimental and theoretical results from related subjects that could be of some significance for the understanding of the electronic support effects.

First, we mention the special properties of bimetallic catalysts and alloys.[29] In dependence on the portion of a second metallic component of the active phase (e.g., Cu in a Ni–Cu system), the activity of "structure-sensitive" reactions (e.g., ethane hydrogenolysis) may drop by orders of magnitude.[17] Both the "electronic and the geometric factor in catalysis" come into play.[17] Interestingly, support effects cause sometimes similar activity patterns (see, e.g., a comparison of ethane hydrogenolysis on Ni–Cu and Rh/TiO_2 catalysts[30]).

Second, the influence of additives such as alkaline or sulfur compounds acting as promoters or poisons of catalytic reactions should be recalled (see Ref. 31 for a review). Addition of alkali metals results in an increased selectivity toward higher hydrocarbons in the $CO–H_2$ reaction and in a decrease of the rate of hydrogenolysis.[32] Using the highly surface-sensitive metastable quenching spectroscopy Lee et al.[33] investigated the properties of CO and metallic K co-adsorbed on the Ni(111) single-crystal face. They found evidence (i) for increased CO binding energy, (ii) for local changes of the work function due to the presence of K, and (iii) for enhancement of the $2\pi^*$ CO-derived peak directly observed in the spectrum.[33] The latter is brought about by an increase of back-bonding due to the presence of K and should result in a decrease of activation energy of dissociation of CO. Sachtler et al.[34] found that highly oxophilic Mn, Ti, Zr, and Nb ions on Rh surfaces enhance CO dissociation (as the alkaline additives do) but through a direct interaction with the oxygen atom of tilted adsorbed CO. We conclude that additives have electronic effects, whether direct or indirect ones, on surface reactions. Again the "geometric factor" connected with the position and distribution of additives can hardly be discriminated.

Third, in recent years some work has been devoted to the investigation of catalytic properties of—at first glance—rather unusual intermetallic compounds and alloys (e.g., Refs. 35 and 36). The chemisorptive and surface properties of such compounds have been studied as well.[37–39] Intermetallics like $NiSi_2$ and Pt_3Ti exhibit catalytic and chemisorptive behavior similar to that of catalysts in the SMSI state. In this sense they serve as models for supported catalysts[40] and may be termed "model catalysts." Increased methanation rates over Ni_xSi_y systems are primarily attributed to a modified chemical state of the nickel.[36]

Last but not least we turn to a field that may seem to lie far away from heterogeneous catalysis for the moment: the physics and chemistry of metal–semiconductor interfaces (cf. Ref. 41 for a recent review). This subject is of primary importance in science and technology of electronic circuits and devices. The formation of a contact between, say, an n-type semiconductor and a metal produces an adjustment in their Fermi levels at the interface caused by a charge transfer between the metal and the semiconductor. A band bending occurs in the latter subsystem near the interface forming the Schottky barrier for charge transport. The Fermi level is "pinned" at a nearly constant value near the interface. The still open question is: What is the origin of this Fermi-level pinning? Are surface defects on the semiconductor side responsible, or perhaps a penetration of the semiconductor by the tails of the wavefunction of the metal ("metal-induced gap states"), or interfacial states connected with chemical bonding at the interface, or does even a mutual interdiffusion occur forming a new interfacial compound by chemical reactions?[42] Similar causes may serve as a "microscopic explanation" of support effects on the metallic catalyst. But considering the materials involved another striking link between the fields of metal–semiconductor interfaces and supported metal catalysts is obvious. For example, at nickel–silicon interfaces, nickel silicides are formed.[43] Interestingly, the previously mentioned $NiSi_2$, which serves as a model catalyst for the SMSI phenomenon, occurs at the temperature of 1023 K,[43] i.e., just at about that temperature at which Ni/SiO_2 catalysts must be reduced to show this phenomenon[25] (vide supra).

After all, we may expect a large impact of these four subjects related to the metal–support interaction. We will benefit from both experimental and theoretical results.

Demands on Theoretical Models: The Interplay Between Experiment and Theory

There are two key questions to be answered in this review: (1) What can theoretical research contribute to the understanding of support effects, and (2) how can this be done? The beginning of theoretical investigations may be characterized best by the headline "electronic theory of catalysis." This concept introduced earlier by Wolkenstein, Hauffe, and Schwab (e.g., Ref. 44) was extended to cover also the electronic support effect on the active metallic phase and on catalytic reactions.[24,45] In the metal–support system an electron flow leads to the adjustment of the Fermi levels of both subsystems. This is of course the same situation met with metal–semiconductor interfaces. By controlling the Fermi level by specifically doping the support different activation energies of donor or acceptor reactions could be obtained.[45]

The electronic theory of catalysis has been used successfully to explain support effects on catalytic reactions (e.g., Refs. 46 and 47). Unfortunately, it suffers from a severe limitation: Only the bulk electronic properties of both subsystems can be taken into account. However, the particular electronic structure and geometry of the support surface, the metal particle, and the metal–support interface are of importance for a deeper insight in support phenomena. Hence, a microscopic theory like quantum chemistry or solid-state electronic theory (Section 3) is needed. Thermodynamic and kinetic considerations, molecular dynamic simulations, Monte Carlo calculations, and empirical concepts are also appropriate. We will focus here primarily on quantum chemistry and to a lesser extent on solid-state theory. These theoretical approaches allow for the study even of geometric support effects when conducted on a highly sophisticated level (Section 3). For electronic support effects a knowledge as specific as possible of the geometry of systems under study is an essential prerequisite. In other words, carefuly chosen geometric models have to be used as an input for the calculations. A critical inspection of available experimental data on the structure is required (cf. Section 2).

But what results may we expect from the calculations? The kind and reliability of results depend on the level of sophistication of the method used (Section 3). In principle, however, insight can be obtained into the nature of the chemical bond at the interface; the direction of electron transfer between the subsystems; the modification of electronic, magnetic, and chemisorptive properties of metal species on the support; and the variation of bonding in adsorbates. Hence, theoretical results may be considered as an important supplement to spectroscopic investigations in that they confirm or refute one or the other model as suggested on the basis of experimental results. The proposal of new experiments should be possible as well. Using highly sophisticated methods the theory is still more in a position of a "theoretical spectroscopy" allowing even for the determination of distances at the interface. A brief summary of theoretical methods is given in Section 3.

EXPERIMENTAL INVESTIGATIONS OF GEOMETRIC STRUCTURES AND ELECTRONIC PROPERTIES OF METAL–SUPPORT SYSTEMS: SUPPORT INFLUENCE ON ADSORBATES

The rapid development and application of advanced highly surface-sensitive spectroscopic techniques such as X-ray photoelectron spectroscopy (XPS), ultraviolet photoelectron spectroscopy (UPS), Auger electron spectroscopy (AES), extended X-ray absorption fine structure (EXAFS), X-ray absorption spectroscopy (XAS), and transmission electron microscopy (TEM) have extended our knowledge of the structure of heterogeneous metal catalysts in recent years.[2–5,11,48–50] The findings of these studies are of importance, on the one hand, for the proposal of well-founded structural models as required for quantum-chemical and solid-state theoretical calculations (Section 2.1), while, on the other hand, they yield data in particular on the electronic structure of the system under study for comparison with the theoretical results (Section 2.2). A review of the huge amount of data now available on different metal–support systems is beyond the scope of this chapter. Rather, we will focus here on the systems Ni/SiO_2, Ni/TiO_2, and Ni/Al_2O_3, keeping in mind our intention to address mainly theoretical work on the WMSI and the SMSI in Section 4. To get a more complete picture, some essential results on other metals deposited on the same support materials will be included. We will consider "real" metal–support catalysts as well as specially prepared metal–support systems, designated as model catalysts, which often exhibit chemisorptive and catalytic properties very similar to the real systems. This section is completed with the inspection of some experimental results aimed at the elucidation of support effects on adsorbates (Section 2.3) and on the catalytic activity and selectivity (Section 2.4).

Geometric Structure and Bonding at Metal–Support Interfaces

The Metal Particles. In connection with investigations on the nature of the SMSI effect, most of the work is concerned with the determination of size distributions or mean particle sizes of metal crystallites (for Ni/TiO_2, see Refs. 28, 51–58) or with the question of the migration of reduced support particles onto the active metallic phase (for Ni/TiO_2, see Refs. 51, 53, 55, 59–64). Reducing "real" catalytic systems at about 700 to 800 K, mean particle diameters between 3 nm[58] and 13 nm[56] have been observed depending on catalyst preparation and metal loading. About 10% of the particles have a diameter of about 1 nm.[52–54] The particles are considered to be flat rather than hemispherical.[28,52–55,59,62,65] The TiO_x migration hypothesis is widely favored.[51,53,59–64] There is also evidence that Ni particles move simultaneously.[63,64] In our opinion, these findings confirm the idea of an intimate contact between the metal and the supporting oxide through a large interface.

With respect to Ni/SiO_2, the situation is different: While Ni/TiO_2 is structurally better characterized in the SMSI state, there are only a few papers dealing with the structure of Ni/SiO_2 in this state. Particle diameters as small as 0.5 nm[66] are found in the "normally" reduced [low-temperature reduction (LTR)] state, the mean diameters being in the range between 3.9 nm[25] and 10 nm.[52] In the SMSI state the mean diameter is enlarged to 12 nm.[25] Evidence has been presented by electron microscopy that platinum particles on silica become encapsulated when annealed at about 1250 K[67]; this reminds us of the TiO_x migration on Ni. Keck and Kasemo[68] found Pt–Si–O overlayers to occur on polycrystalline Pt in an ambient atmosphere of $H_2 + O_2$ even at temperatures as low as 320 K. The SMSI state was observed also in Pt/SiO_2.[69]

The system Ni/Al_2O_3 has been characterized by several authors.[9,12,13,52,54,56–58,70–74] Depending on nickel loading, preparation methods, and reduction temperatures particle diameters between 2 nm[73] and 9 nm[56] have been determined. For both Ni/SiO_2 and Ni/Al_2O_3, three-dimensional crystallites were observed in contrast to two-dimensional ones for Ni/TiO_2, indicating a stronger metal–support interaction in the latter case.[9] Interestingly, Turlier and Martin[70] were unable to detect the SMSI state in Ni/Al_2O_3, while they succeeded for Ni/SiO_2.[25] In both cases they used a precursor containing a relatively high metal loading of about 20 wt% Ni. In our opinion[75,76] a possible reason may be the considerably stronger interaction in the Ni/Al_2O_3 precursor state (cf. the Introduction) in comparison with the Ni/SiO_2 precursor. This interaction strongly influences the reducibility at the nickel–support interface. One can speculate that under appropriate preparation conditions an SMSI state of Ni/Al_2O_3 may be realized. In fact, Narayanan et al.[13] observed a considerably decreased metal–support interaction in the Ni/Al_2O_3 precursor due to a support calcination prior to metal loading (cf. the Introduction). Such a support calcination was not performed by Martin et al.[70]

Using very low metal loadings between 0.5 wt% and 3 wt% Ni on previously calcinated alumina supports Bartholomew et al.[9] found evidence for SMSI effects even after reducing the catalyst at only 725 K. The average crystallite diameters were determined to be rather small down to 1.6 nm.[9] This again reflects an intimate contact between metal and support, which seems to be due to another origin here compared to the Ni/TiO_2 system, since the percentage of metal reduction in this special catalyst amounts to only 29%.[9] Interestingly, recently support effects of "a new type of metal–support interaction," named "redox" metal–support interaction (MSI), have been ascribed "to the presence of unreduced surface–nickel species"[73] on less reducible supports including Al_2O_3.[73,74]

The geometry of the metal particles has frequently been investigated.[50,77–80] EXAFS measurements show that the metal–metal distances in small particles are as a rule slightly decreased (up to a maximum of about 5%,[50] for a review, see, e.g., Ref. 4) compared with their bulk metal values. This holds for nickel as well.[66,81]

The Interface. The results on the interfacial structure are not unequivocal.[50,57,77,78,82–95] From EXAFS spectra the existence of Ni–O–Si subunits in "normally" reduced Ni/SiO_2 is inferred[82,83] whereas in Co/TiO_2 no indication of Co–O–Ti bonds has been found.[84] The Ni–O bond length was determined to be 205 pm[83] and 200 pm[82] depending on the preparation. Based on ferromagnetic resonance (FMR) studies and electron spin resonance (ESR) investigations, Bonneviot et al.[87] assumed the presence of Ni^+ and Ni^{2+} ions at the metal–support interface acting as "anchoring" sites of small nickel particles. This model is consistent with the abovementioned EXAFS findings keeping in mind the Ni–O distance in bulk nickel oxide to be 209 pm (see, e.g., Ref. 96). Very small paramagnetic Ni^0 clusters anchored through Li^+ (or Ni^+) on SiO_2 have been also observed.[88] To our knowledge there is up to now no EXAFS study on Ni/SiO_2 in the SMSI state. Hence, we may at present only refer to the Ni–Si alloy hypothesis[25] to explain the interfacial structure.

Turning to Ni/Al_2O_3 an EXAFS investigation on the precursor state of this system provides evidence for a diffusion of Ni ions into tetrahedral and octahedral support sites.[89] The Ni–O distances are found to range between 203 pm and 208 pm depending on the metal loading.[89] Ni–O distances of 183 pm and 207 pm were assigned to Ni in tetrahedral and octahedral positions of a $NiAl_2O_4$ spinel phase, respectively.[95] Other authors determined the Ni–Al distance to be about 177 pm for this system.[94] Careful EXAFS investigations on Rh/Al_2O_3 and Pt/Al_2O_3 revealed the existence of two different metal–oxygen distances, the shorter being related to incomplete reduction of the metal oxide and the longer to the coordination of Rh to two or three surface oxygen atoms.[77,90,91] Alloy formation as the cause of the SMSI effect in Pt/Al_2O_3 is a widely accepted assumption.[26,97–99] Bond distance estimations from EXAFS results are not available.

Finally some remarks concerning titania-supported metals should be made. Unfortunately, there are no EXAFS studies on Ni/TiO_2, either in the "normally" reduced or in the SMSI state. Short et al.,[85] however, using EXAFS found no evidence of Pt–Ti or Pt–O bonds in Pt/TiO_2 in the SMSI sta e. On the other hand, using the same method evidence for direct metal–me al bonding in high-temperature reduced Rh/TiO_2 has been obtained recently.[92] The Rh–Ti lengths were determined to be about 255 pm, considerably smaller than in the intermetallic compound RhTi (268 pm) indicating "that the Ti (and perhaps the Rh) has cationic character."[92] The mean coordination number of Rh is between 2 and 3, while that of Ti is about 2.[92] In the oxidized catalysts probably Rh–O–Ti linkages are present.[92] Almost simultaneously a paper by Koningsberger et al.[93] appeared using the same spectroscopic method for the same system. Rh–O bond lengths of 271 pm and 267 pm were detected in the LTR and HTR state, respectively, whereas the Rh–Ti bond, found only in the latter state, had a length of 342 pm.[93] No alloy or suboxide formation was ascertained. Rather, Rh is assumed to sit on a reduced TiO_2 surface, which "closely resembles a shear plane in Ti_4O_7."[93] As in silica- and alumina-supported metals, the SMSI is often ascribed to alloy formation in titania supported metals. Only recently Ruckenstein et al.[100] attributed a shape alteration at elevated temperatures of nickel particles on titania partly to the formation of the intermetallic compound Ni_3Ti. Preparing titania-impregnated Pt model catalysts Beard and Ross[101] reported XPS and EXAFS evidence for the formation of the ordered alloy phase Pt_3Ti during HTR.

The Binding Energy. There is also an intimate connection between the bonding at the interface and the binding energy of metals on the support. This is evident from the onset of migration and aggregation of metals on different surfaces at different temperatures. A careful analysis has been performed, e.g., by Sushumna and Ruckenstein[102] for the system Fe/Al_2O_3. "It is suggested that strong chemical interactions may lead to a considerable decrease in the interfacial tension between crystallite and substrate and, therefore, to a tendency for the crystallites to spread out."[102] The authors stress that the SMSI, the sintering, and the redispersion phenomena as well as changes in the shape of the crystallites "are all interlinked parts of a complex surface phenomenon."[102] Based on similar considerations heats of adsorption have been estimated: From the work of adhesion of Ni on ceramics Geus[103] obtained 20 ± 2 kJ mol^{-1} for Ni on oxidic supports. For Pt and Pd on α-Al_2O_3 values in the range 40 to 80 kJ mol^{-1} are given[104] and for Pd on MgO 46 kJ mol^{-1} has been found.[105] Yermakov et al.[106] considering the high mobility of metal atoms even near room temperature conclude that relatively weak van-der-Waals-type forces are acting between the metal and the support. On the other hand, secondary ion mass spectroscopic (SIMS) studies[107] indicated that a stronger Rh-support bond should exist for SMSI systems compared with "normally" reduced (LTR) catalysts. This view is in accordance with the assumption of alloy formation (vide supra). Interaction energies have not been estimated for this case. Investigating the reversibility of the SMSI state under CO–H_2 reaction conditions Anderson et al.[108] stated "that a Pt–Ti bond is significantly stronger than a Rh–Ti bond" since the extent of recovery of Pt/TiO_2 is slower and less complete than that of Rh/TiO_2.

The Structural Model. From the above survey on experimental findings the following requirements for realistic structural models can be formulated:

1. Even assuming a highly dispersed active phase, metal species of at least 10 atoms interacting with the support surface should be considered; the use of bulk Ni–Ni distances is justified as a first approximation. Concerning the SMSI state an intimate contact between the metal atoms and the support is to be realized (e.g., layers of nickel atoms).
2. With respect to a simulation of the WMSI and of the SMSI situation, models should be chosen which exhibit different oxidation states of the metal cation (Si^{n+}, Ti^{n+}, Al^{n+}) in order to account for a partial support reduction.
3. The surface sites to be considered as interacting with metal atoms are both surface defects and "atomic groups" on clean surfaces (cf. Section 4). The former may lead to a chemical bonding at the interface, while through the latter primarily van-der-Waals bonding is established.
4. As to the SMSI modeling, direct interactions between metal atoms and support cations (e.g., Ni–Ti^{3+}) have to be dealt with.
5. The choice of bond distances at the metal–support interface is problematic. Values of EXAFS measurements are recommended if available. Otherwise, distances have to be estimated from similar inorganic compounds or atomic/ionic radii. Using more sophisticated theoretical methods a calculation of optimum distances is possible (Section 3).

As it will become clear (Section 4) our models do not meet, in particular, the first requirement. Rather, we confine ourselves for the moment to single-metal-atom models, since (1) the understanding of these "lower-limit" models is certainly a prerequisite for investigations of extended models and (2) realistic metal–support models with more than one metal atom are still very hard to treat with thorough theoretical methods so that only a few results are available.

The Electronic Structure of Metal–Support Interfaces

The electronic state of supported metal particles is the subject of extensive experimental study.[3] Mostly the XPS method was used (see, e.g., Refs. 59, 65, 101, 107, 109–114; for a brief review see Ref. 115), but also X-ray absorption edge,[79,85,116] and electric conductivity investigations,[24,47] as well as studies of magnetic properties[57,58,87,88,117] have been performed. Different metal–support systems including Ni/SiO_2,[57,58,87,88,114] Ni/Al_2O_3,[57,58] and Ni/TiO_2[24,47,58,65,117] have been considered. The general opinion is that an electron transfer occurs between the metal and the support, which is directed toward the metal in the case of SMSI systems or corresponding model systems, whereas the direction is reversed in "normally" reduced heterogeneous catalysts. There are, however, a number of papers reporting that there is no evidence of an electron transfer.[4,85,118] It should be recalled that XPS chemical shifts may be affected not only by a charge transfer but also by relaxation effects that depend on the particle size (see, e.g., Ref. 119). It is a very difficult task to separate initial- and final-state effects.[120] Furthermore, a determination of particle size changes during oxidation–reduction cycles of catalysts seems necessary.[109] Sexton et al.,[109] assuming that a flattening of metal clusters on the surface occurs (so-called wetting), estimated a lower bound to that part of the chemical shift that originates from the charge transfer in their SMSI systems. This confirms the conclusion of an electron transfer to the metal for such systems.

With respect to Ni-containing intermetallic compounds the results are not unequivocal. An electron transfer to nickel is predicted for Ni–Ti[121,122] and Ni–Al[123] alloys. A low Ni coverage on the Si(001) surface, however, was shown to yield an electron transfer to Si.[43] A careful theoretical analysis for bulk nickel silicides led Xu Jian-hua et al.[124] to propose a slight electron transfer to Ni. Yet, simultaneously the Ni_{3d} population decreases making a positive chemical shift conceivable.[124] Thus, the XPS shift seems to indicate a change in the nickel electronic configuration compared to that in bulk nickel rather than a charge transfer between the two constituents. The same conclusion was drawn from an experimental XPS investigation of Pt on SiO_2.[125] On the other hand, in γ-Ni_2SiO_4 Ni is positively charged.[126]

We focused the discussion on the direction of electron transfer not having given any numbers. This is a consequence of experimental uncertainties as indicated previously. Furthermore, it should be kept in mind that the net charge of an atom in a molecule or solid is not directly observable and that no unambiguous definition of this quantity exists. All experimentally or theoretically derived net charges of atoms are based on more or less justified models, e.g., the Mulliken population analysis, and give only a rough picture of the charge transfer in the systems under discussion.

For completeness, next we will give an impression of the support influence on magnetic properties of the active metal species. Gregory and Moody[127] found that upon alloying Ni with Ti, the mean atomic magnetic moment decreases from $0.616\mu_B$ (Bohr magnetons) for bulk Ni to $0.103\mu_B$

for an alloy containing 14 at% of Ti. A drastic decrease in saturation magnetization after a high-temperature reduction of Ni/SiO_2 has been ascribed to a Ni–Si alloy formation.[25] Different dependencies of the magnetization on the temperature led to the prediction of different SMSI mechanisms in Ni/TiO_2 and Ni/aluminum phosphate systems.[58] The slight decrease of Curie temperature T_C as against the bulk value, however, may be due to either a particle size effect or the support influence.[58] Recently, in a single magnetic measurement the saturation magnetization and the paramagnetic moment of Ni/TiO_2 samples were obtained.[117] The former indicates a "loss of bulk nickel" after HTR ("metastable Ni–Ti–O phase") while the latter points to the presence of localized, unpaired electrons ascribed to paramagnetic Magneli phases (presence of Ti^{3+} species).[117] There is also some evidence for the existence of Ti^{3+}-related defects in reduced transition-metal/TiO_2 systems from ESR (see, e.g., Refs. 128 and 129).

THE SUPPORT INFLUENCE ON CHEMISORPTION

Recalling the definition of the SMSI effect (see the Introduction) the question of its origin arises. But also in the WMSI state, support effects on chemisorption may occur. Mostly, the investigations have been focused on the H_2 and CO uptake, respectively.[4] Only recently several studies address changes of heats of adsorption, of activation energies of dissociation, or of spectroscopic properties of adsorbed molecules (vide infra). Generally, the diminution of H_2 and CO uptake is more pronounced for elements of higher transition metal rows (Ir, Os, Pt, Ru, Rh, Pd) compared to those of the first row (Ni, Co, Fe).[4] In principle, there are two completely different explanations of the SMSI effect, a geometric and an electronic one:

1. simple physical blockage of adsorption sites on the metal by migration of partially reduced oxidic particles from the support.[130–136]
2. modification of the electronic state of the metal by the influence of the support[137–140] resulting in a change of heats of adsorption or in an occurrence of activation barriers for dissociation.

Many authors claim that both causes are involved simultaneously, i.e., some of the sites are simply blocked whereas others are electronically influenced. This opinion is widely favored for Ni/TiO_2,[55,60,61,141,142] Pt/TiO_2,[143–149] and Rh/TiO_2,[30,129,148–150] whereas the SMSI effect in Pd/TiO_2 generally is ascribed to simple site blocking.[132–135] Site blocking was concluded to be responsible for the diminution of chemisorption capacity by some authors also in Pt/TiO_2[130,131] and Rh/TiO_2[136] while others prefer the

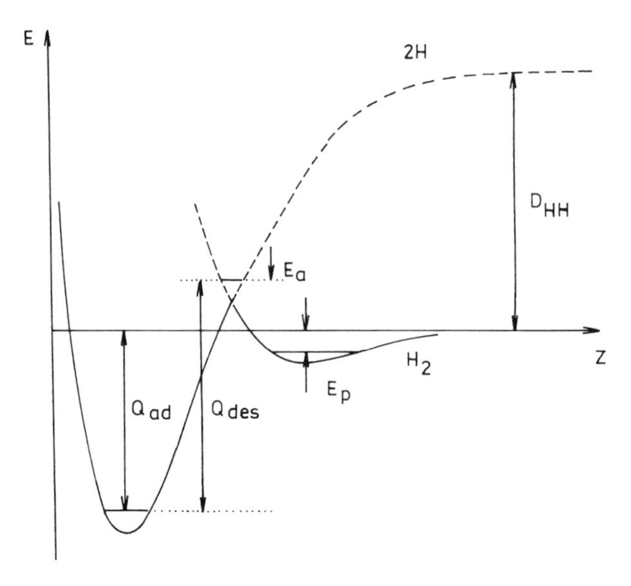

Figure 8.1. Sketch of the schematic one-dimensional potential diagram of activated dissociative chemisorption as proposed by Lennard-Jones (adapted in part from Ref. 151). D_{HH} and E_p denote the dissociation energy of the gas phase hydrogen molecule and the heat of adsorption in a molecular precursor state weakly bound on the surface, respectively. The distance above the surface is given by z. Solid lines in the potential wells symbolize the ground vibrational states. For further explanation see text. [Reprinted with permission from *The Nature of the Surface Chemical Bond* (T. N. Rhodin and G. Erel, eds.). Amsterdam: North-Holland, 1979, p. 316, Figure 5.1.]

concept of a modification of the electronic state by the support influence for the same systems (see, e.g., Refs. 138–140).

General Remarks Concerning the Chemisorption on Group VIII Transition Metals. H_2 and CO adsorb dissociatively and molecularly, respectively. The dissociative chemisorption may be roughly described by means of the schematic one-dimensional potential diagram as proposed by Lennard-Jones (Fig. 8.1, see, e.g., Ref. 151). Accordingly, the heat of (dissociative) adsorption Q_{ad} of, e.g., an H_2 molecule is related to the molecular desorption activation energy Q_{des} through

$$Q_{des} = Q_{ad} + E_a \tag{8.1}$$

where E_a designates the activation energy (including also the vibrational energy in the transition state, Fig. 8.1) of dissociative chemisorption.*

* Q_{ad}, Q_{des}, and E_a are given in kJ mol^{-1} throughout.

For dissociative H_2 chemisorption on low-index crystal faces E_a is, as a rule, nearly zero (see, e.g., Ref. 152, Table 11) but it amounts, e.g., on Cu(100), to about 21 kJ mol^{-1},[153] considering the high-index Ni(997) surface H_2 adsorption turns out to be an activated process[154] on the terraces as on the flat (111) surface[155] but to be nonactivated on the step sites similarly to that on the more open (110) surface.[155]

The effect of crystallographic orientations of single-crystal planes on the molecular heats of desorption Q_{des} is rather small,[156] in particular, values between 85 and 98 kJ mol^{-1}[157] are obtained for H_2 on Ni. On Ni(110) in addition a high-energy site (123 kJ mol^{-1}, β-state) is found.[157] However, temperature programmed desorption (TPD) indicates, e.g., a lowering of H_2 desorption energy for Ni(100) contaminated with electronegative modifiers[158] and an increase of this quantity (for D_2) in the presence of (electropositive) potassium.[159] Recently, potassium nitrate addition to Ni/SiO_2 has been shown to decrease the hydrogen heat of adsorption.[160] Hence, a support influence on the heats of adsorption/desorption [neglecting for the moment E_a in Eq. (8.1), i.e., considering nonactivated dissociation] is well imaginable. To be a little bit more specific it must be added that there exist different sites of adsorption exhibiting different heats of adsorption.[161] Furthermore, with increasing hydrogen coverage ($\Theta_H \gtrsim 0.5$) the heat of adsorption decreases drastically, probably due to a gradual increase of the mutual hydrogen–hydrogen repulsion.[161] To render coverage effects unlikely it is recommended to investigate "initial" heats of adsorption. The dependence of heats of adsorption on particle sizes has to be considered as well.[132,133] Thus, it is a difficult task to investigate the support influence on chemisorption.

Finally, we may ask for a connection between the molecular heat of (dissociative) adsorption, Q_{ad}, and the energy of the bond between the hydrogen atom and the surface, E_{HM}. This is given by[162]

$$Q_{ad}(H_2) = 2E_{HM} - D_{HH} - \eta E_{MM} \tag{8.2}$$

where D_{HH} is the dissociation energy of the gas-phase hydrogen molecule, η is the number of metal–metal bonds that have to be broken for the chemisorption of two adatoms H, and E_{MM} is the respective metal–metal bond energy.* The last term is usually neglected, but this neglect has to be justified by independent measurements, e.g., by vibrational spectroscopy of the M–H "surface molecule."[162] There is indeed experimental evi-

* It should be noted that E_{HM} and E_{MM} in Eq. (8.2) contain contributions from the vibrational energy as Q_{ad} does. The electronic part of E_{HM} is related to the well depth and is correlated in Ref. 162 with the stretch vibrational frequency of the H/surface complex ("surface molecule").

dence for the importance of the term ηE_{MM}, resting primarily on the structural dependence of small metal particles on adsorbate coverage. Renouprez et al.,[163] e.g., found a relaxation of nearest-neighbor distances in small Pt particles encaged in zeolites from smaller to larger (bulklike) values after hydrogen adsorption. CO adsorption on very small Rh crystallites leads to a disruption of a significant number of the metal–metal bonds.[164] All this is a reminder of the situation described by Sachtler with the terminus "demetallization effect,"[165] i.e., a weakening of metallic bonding in the substrate.

As mentioned previously, vibrational spectroscopy of the M–H "surface molecule" in principle permits an estimation of the bond strength (which is related to the electronic part of E_{HM}) of an adsorbed atom on a special surface site. Characteristic frequencies are assigned to surface sites of different ligancy exhibiting different metal–adsorbate bond strengths. The most favorable adsorption site is characterized by the highest population and, hence, by the largest band intensity. However, a careful study of the vibrational characteristics [e.g., H/Ni(110)[166] and CO/Ni(100)[151]] shows a sensitive dependence on the coverage. Another interesting feature of CO vibrational characteristics concerns the well-known bonding-back bonding picture of this molecule on metal species[167]: The back donation from metal $3d$ orbitals to the antibonding $2\pi^*$ CO orbital is higher at higher coordinated sites thus lowering the C–O stretch frequency. Simultaneously, the metal–C stretch frequency is decreased due to a decreased σ donation per bond.[167] Hence, it seems to be very difficult to elucidate the support influence on the vibrational properties. Concerning CO, some more qualitative attempts have been made (vide infra), though a further complication with metal catalysts is the presence of unreduced metal ions; the adsorption of CO on these yields quite different characteristic vibrational frequencies (for Ni/Al_2O_3 see Ref. 168).

The Support Effect on H_2 Dissociation. It is obvious from the selected experimental results (Table 8.1) that the occurrence and heights of activation barriers depend not only on the support but on the nature of the metal and the amount of its loading as well. Activation energies are small or absent for the case of SiO_2 supports and small or medium with Al_2O_3 indicating a stronger support effect in the latter case. The results on Ni/TiO_2 are not unequivocal (Table 8.1); the small observed barrier is not believed to be responsible for the diminution of hydrogen uptake.[52] Interestingly, Raupp et al.[61] found the adsorption at the most favorable of three adsorption sites occurring in their TiO_x/Ni model catalyst to be an activated process. No attempt has been made to our knowledge to determine the change of activation energy of dissociation comparing the same catalysts in the WMSI and SMSI states. With respect to Ni/SiO_2 in its WMSI

Table 8.1. Support effects on activation barriers of dissociative H_2 adsorption; selected experimental results.

System	T_R[a] (K)	R[b] (%)	Physical Characterization D[c] (%)	d[d] (nm)	Activation Energy[e] (kJ mol^{-1})	Method	Reference
Co (unsupported)	673	100	0.3		5.8	Shift of TPD peaks with temperature as a function of adsorption temperature	169
3% Co/SiO_2	673	75	11		43		
10% Co/SiO_2	673	92	10		18		
3% Co/Al_2O_3	648	22	10		Very large[f]		
10% Co/Al_2O_3	648	34	9.9		39		
10% Co/TiO_2	673	47	4.5		Yes[g]		
14% Ni/Al_2O_3	LTR	84[h]	17[h]	5.7[h]	10	Shift of TPD peaks with temperature as a function of adsorption temperature	157
10% Ni/Al_2O_3	LTR	Not given			No		
10% Ni/TiO_2	873	Not given			No		
7% Ni/TiO_2	773	89		8–12	Small[i]	Adsorption isobars	52
6.5% Ni/η-Al_2O_3	773	83		8	Small[j]		
6.8% Ni/SiO_2	773	100		10	No		

[a] Temperature of reduction.
[b] Extent of reduction.
[c] Dispersion.
[d] Mean particle diameter.
[e] This activation energy is according to its determination not necessarily equal to the quantity E_a (Fig. 8.1); rather it is an apparent one including also kinetic contributions.
[f] No detectable uptakes at any temperature.
[g] Overlapping of peaks inhibits a quantitative determination.
[h] From Ref. 9.
[i] A shallow maximum is found near 473 K.
[j] A shallow maximum is found near 373 K.

state, the absence of an activation barrier is seen from Table 8.1. Concerning the SMSI state of this system, there is evidence from experiments with the intermetallic compound $NiSi_2$ (also showing the SMSI effect, cf. the Introduction) that adsorption of predissociated hydrogen occurs while dissociative adsorption is inhibited, probably by a large activation barrier.[37,40] Although there is clearly a support effect on the activation barrier of H_2 dissociation for several systems, almost nothing is known on its "microscopic" origin. Weatherbee et al.,[170] e.g., speculate that the support effect might be due either to perturbations of electronic or geometric surface structure or both caused by metal–support interactions or to residual oxygen at the surface perhaps present as a metal oxide. Hence, this question is a challenge for theoreticians.

The Support Effect on Heats of Adsorption/Desorption. For several catalytic and model systems, an appreciable effect has been found (cf. Table 8.2 for hydrogen). In particular for Pt/TiO_2 the heat of adsorption is more or less drastically decreased due to HTR treatment depending on the metal loading. Although Q_{ad} depends strongly on the dispersion of the metal, the authors point out that there exists also a pronounced support effect.[143,145] Besides this, a site blocking with TiO_x species is another reason for the diminution of hydrogen uptake.[145] One may ask, however, whether in the SMSI state new sites holding hydrogen less strongly may be created due to support influence or treatment conditions. In this case we were concerned indirectly with a geometric effect causing a modified binding of adsorbed hydrogen. The latter seems to be the case in fact for Ni/TiO_2[61,157] and for Ni/Al_2O_3[142,157] at least with respect to the strongest binding site (not given in Table 8.2 for Ni/Al_2O_3). Again, site blocking plays an important role. Concerning TiO_x/Ni, Raupp et al.[142] claim an increase of the adsorption strength of hydrogen on two surface sites. This may be questioned, however, since the coverage Θ_H decreases appreciably with increasing Θ_{TiO_x}, and, indeed, the authors concede that "a portion of the observed upward shift in peak temperature . . . is due to initial hydrogen coverage differences."[142] Interestingly, the effects observed for TiO_x and Al_2O_3 adspecies on Ni are very similar.[142] However, the amount of TiO_x required to obtain nearly equivalent H_2 desorption spectra compared with Al_2O_3-contaminated Ni is much lower than the necessary amount of Al_2O_3.[142] This reminds us of the remarkably higher reduction temperatures required for the onset of the SMSI effect in Al_2O_3-supported systems.

A similar situation is met in CO adsorption. Sites exhibiting a large adsorption energy are mostly blocked by oxidic species, while the heat of adsorption of the remaining CO molecules in these sites is slightly decreased (from 135 to 128 kJ mol^{-1} for Al_2O_3/Ni[142]). Furthermore, sites holding CO more weakly become occupied.[142] In a later work minimizing diffusional processes of titania species onto the nickel, the same authors[55] found a decreased CO heat of adsorption and a lower saturation coverage with increased extent of titania reduction due to an electronic metal–support interaction between Ni and the underlying titania. This type of interaction is important only for Ni coverages not thicker than three atomic layers.[55] Using Ni/Al_2O_3 model catalysts Hedge et al.[172] have shown by means of UPS that at low nickel coverages complete dissociation occurs at about 300 K. Appreciably lowered heats of CO adsorption have been reported also for Pt/TiO_2 after HTR.[144,147–149] A preferential blocking of high-energy sites on the Pt surface cannot be excluded,[144] but "clean metal" sites and highly reactive TiO_x-metal-related sites have been proposed as well.[147,149] Dwyer et al.,[130] drawing on low-energy inelastic ion scattering (LEISS) and TPD results of H_2 and CO adsorbed on TiO_x-modified Pt foils, state that simple site blocking is the main cause for the diminution of chemisorption since practically no adsorbed species in new adsorption sites were observed. The same conclusion was drawn for CO and H_2 adsorption on Pd/TiO_2 after a high-temperature treatment.[132–134]

A support influence on H_2 adsorption even for low-temperature reduced catalysts is evident from Table 8.2. The heat of adsorption is decreased for Ni/SiO_2 compared to the chemisorption on unsupported Ni[171] while for Ni/Al_2O_3[157] two forms of adsorbed hydrogen have been found, the one less strongly and the other more strongly bound compared with clean Ni. The reason for this behavior is not clear up to now. Again one may speculate whether new adsorption sites are created or filled due to the support influence or whether there is an electronic effect (at least for the less strongly bound hydrogen form on Ni/Al_2O_3). In this connection the oxidation state of the metal—perhaps different from zero owing to incomplete reduction or electron transfer to the support—might be of importance.

Support Influence on CO Vibrational Characteristics. High-resolution electron energy loss spectroscopic (HREELS) results on TiO_x/Ni model catalysts provide evidence for a relative increase of on-top CO concentration compared with bridge-bonded CO with increasing reduction time.[141] It is claimed that the concentration of bridge sites drops down more rapidly than that of on-top sites owing to the presence of inactive TiO_x species.[141] Additionally, a peak at 1850 cm^{-1} is found, which is attributed to CO adsorbed "on a site near the surface titanium oxide."[141] Comparing IR spectra of a 19.7 wt% Ni/TiO_2 catalyst after LTR and HTR, de Bokx et al.[173] arrived at a similar conclusion. No shift of the peak of linearly bonded CO has been observed after HTR, while there is a small

Table 8.2. Support effects on the heat of adsorption/desorption of hydrogen upon metal catalysts: selected experimental results.

System	T_R[a] (K)	Physical Characterization		Q_{ad} (Q_{des}) (kJ/mol^{-1})	Method	Reference
		D[b] (%)	d[c] (nm)			
2.1% Pt/SiO_2	723		4.0	116 ± 12.5	Calorimetry (integral heats of adsorption)	143
2.1% Pt/Al_2O_3	723		1.5	116 ± 4.6		143, 145
2.0% Pt/TiO_2	473		1.8	113 ± 10		143, 145
1.5% Pt/TiO_2	773		1.1	23.8		145
2.0% Pt/TiO_2	773	Not given		41–81		143
5% Pt/TiO_2	473[d]	34	2.0	93	Calorimetry (initial heats of adsorption)	146
5% Pt/TiO_2	773[e]	Not given[f]		80		146
Pt/TiO_x model	—	≤1 monolayer PT coverage		Not given	Shift of TPD peaks to lower temperatures	147

10% Ni/SiO_2	LTR	Not given		82	From TPD peaks	157
14% Ni/Al_2O_3	LTR	17[g]	5.7[g]	70, 125	($\theta_H \approx 0.5$)	157
10% Ni/TiO_2	673	Not given		78, 105	downward	157
10% Ni/TiO_2	973	Not given		Not given	shift of the high-temperature TPD peak by 45 K	157
TiO_x/Ni model	Ti evaporated, oxidized and reduced at 620 K	Not given		75, 108, 133		61
					From TPD peaks	
Ni polycrystalline	—	—		89		61
Ni black	673		230	138	Calorimetry and TPD	171
10% Ni/SiO_2	673		71	115		

[a] Temperature of reduction. The extent of reduction was not given.
[b] Dispersion.
[c] Mean particle diameter.
[d] And subsequent treatment in vacuo at 473 K.
[e] And subsequent treatment in vacuo at 773 K.
[f] It is not reported whether the values given for LTR catalysts remain valid.
[g] From Ref. 9.

shift by about 20 cm^{-1} downward[173] indicating a slight weakening of the C–O bond of bridge-bonded CO. The broad band around 1870 cm^{-1} was ascribed as before[141] to CO bonded to nickel near TiO_x adspecies.[173] Following Sachtler[174] the authors believe the CO to be tilted, the C interacting with Ni, and the O with Ti^{n+}.[173] This is just the special interfacial site as proposed by Burch and Flambard[28] to explain the enhanced methanation activity of this catalyst. A C–O bond weakening has been predicted also by Meriaudeau et al.[175] for linearly bonded CO on 1.8 wt% Pt/CeO_2 as inferred from a 16 cm^{-1} downward shift of the stretch frequency. This has been attributed to the formation of Pt–Ce intermetallic compounds.[175] A similar shift was not observed, however, for CO on a 4.8 wt% Pt/TiO_2 catalyst after HTR.[175]

A weakening of the C–O bond is consistent with the frequently proposed electron transfer to the metal in the SMSI state. Assuming C–O dissociation to be the rate-determining step, higher methanation activities become conceivable. This explanation has also been given in the case of alkali-promoted metals (for a detailed discussion of contradictory effects, e.g., the inhibiting role of too strongly held CO, see Ref. 32).

Although there is a huge number of papers investigating also metal–C stretching frequencies (see, e.g., Ref. 176 and references cited therein), the influence of support and preparation (reduction) treatment is rarely addressed (see, e.g., Refs. 141 and 177). From the Pt–C stretching force constants Mink et al.[177] (IR emission) concluded in agreement with calorimetric data[144] that CO is more tightly bound to Pt when it is supported on silica compared to alumina. In Ref. 141 (HREELS) a change of the observed Ni–C stretch frequency with increasing time of reduction is not reported; rather the whole peak vanishes perhaps due to a loss of intensity.

An extended IR investigation on the influence of different support materials on the geminal dicarbonyl species $Rh(I)(CO)_2$ has been published recently.[178] The C–O stretching parameter obtained from the twin band increases in the order NaX < Al_2O_3 < NaY < TiO_2, indicating a decreasing back-bonding ability of supported Rh in the same order.[178] This finding parallels an increased CO hydrogenation activity attributed to an easier removal of the dicarbonyl species from the surface "which at high coverage may block the active sites . . . and act as a kind of poison."[178]

Last but not least a warning seems to be necessary. It concerns the assignment of vibrational frequencies to adsorbed molecules on surface sites of special geometric structures. A broad discussion on this topic is given, e.g., by Blackmond and Ko.[179] First, the stretching frequency of a bridge-bonded CO is lower on Ni(100) than on Ni(111) since Ni atoms in the latter plane are more densely packed and "have fewer *d*-electrons available for back-bonding."[179] Second, a weakening of adsorption and, hence, an upward shift of the C–O stretching frequency of linearly bonded CO may occur due to the interaction with the oxide phase withdrawing electrons (when reduction is performed at low temperatures) or with unreduced Ni species.[179] Even a carbonyl formation comes into play.[179,180] Thus, it is extremely complicated to find out which effects are really due to a direct or indirect support influence.

Summary. Keeping in mind the aim of the present review we may conclude from the available experimental material:

1. The physical blockage of adsorption sites seems to play an essential role for the diminution of H_2 and CO chemisorption capacity on SMSI systems. Besides this, there are direct (activation barrier of dissociative chemisorption, change of strengths of adsorptive bonds at selected surface sites) as well as indirect (creation and filling, respectively, of "new" surface sites) support influences on the electronic properties of the active metallic phase for several systems. The electronic support effects can come about both via the underlying oxide and via partly reduced support species on the metal particle.
2. Support effects are also pronounced for common, "nonreducible" oxidic supports. Their importance increases with decreasing metal loading and increasing dispersion.
3. In either case electronic support effects seem to be short-range in character.
4. The metal–support interaction generally increases in the order $SiO_2 < Al_2O_3 < TiO_2$, although there may be exceptions depending on the preparation and pretreatment conditions.
5. The "microscopic" mechanism of the electronic support effect is still unclear. Does an electron transfer between metal and support play a significant role in the SMSI effect, or what other properties of the metal–support system are essential? The importance of the electron transfer has been questioned since, e.g., with increasing electron density on the metal the heat of adsorption of CO should be increased rather than decreased.[3] Hence, this subject is a profitable field of research for theoretical work.

Support Influence on Activity and Selectivity

We confine our brief survey on a few remarks with respect to experimental results on the CO hydrogenation, the understanding of which may profit from theoretical investigations on support effects on adsorbates. Reviews concerning support influences on different reactions over nickel-containing catalysts[181a] and on hydrogenation properties[181b] have been published recently.

As discussed in the Introduction, TiO_2 supports are frequently found to increase the hydrogenation activity while shifting the selectivity to higher hydrocarbons. There are some suggestions on the mechanism of support effect. Vannice et al.[144] reach the conclusion that the most active Pt methanation catalysts are those with the lowest heats of adsorption for both hydrogen and CO, i.e., in their case Pt/TiO_2(HTR). As an explanation for this finding, the creation of a "unique active site" at the Pt–support interface is assumed, which enhances the rate-determining step and *independently* reduces the adsorption bond strength.[144] These authors reject a simple electronic influence on the Pt surface atoms, such that only heats of adsorption are decreased.[144] Raupp et al.,[182] investigating Ni/TiO_2(HTR) catalysts, point out that due to a blocking of sites which strongly bind CO, more weakly binding CO sites are increasingly populated. On the other hand, the hydrogen heat of adsorption is increased (cf. section titled "The Support Influence on Chemisorption").[182] So hydrogen is better able to compete with CO for the same adsorption sites, which results in a relatively higher hydrogen coverage and in higher methanation activities.[182] Bartholomew et al.[183] show that the higher reaction order for H_2 as well as higher CO dissociation activities are of major importance for the occurrence of the unusually high CO hydrogenation activity. The higher reaction order with respect to H_2 is understandable from the decreased heat of adsorption of H_2 after HTR.[183]

Burch and coworkers,[10,28] however, disclose that the SMSI state as usually defined is not responsible for enhanced methanation activity and shifted selectivity but rather the existence of Ti^{3+}-related interfacial sites occurring also in "normally" prepared catalysts. Tamaru et al.[184,185] arrive at the same conclusion showing that the CO hydrogenation activity is even decreased in the SMSI state. They argue[185] that a high dispersity SMSI catalyst may be reoxidized under the reaction conditions by the water formed. Fang et al.[186] found out that, although the Ni–CO bond strength is decreased by a factor of 2, the saturation coverage of CO is not reduced in the SMSI state. Hence, the enhanced methanation rates are due to other properties of the TiO_2 support.[186] Summing up, the main question turns out to be how the Ti^{3+} cations—present also at lower temperatures and under CO–H_2 reaction conditions[27]—act rather than the question of the need of the "conventional" SMSI state in this connection.

Another interesting aspect has been investigated by Chen et al.[187] Using Cu, a metal of low CO hydrogenation activity, the authors observed an increased activity due to the influence of electron-withdrawing Cr_2O_3 and ZrO_2.[187] The binding energy of adsorbed CO is increased but no change of the C–O stretching frequency has been reported.[187] Possibly a larger amount of adsorbed CO compared with bare Cu leads to higher methane yields. Thus, the designation "inverse SMSI effect"[187] becomes clear.

In conclusion, we are left with a broad manifold of effects and possible explanations that may act partly simultaneously and exhibit a pronounced overlap between WMSI and SMSI behavior as conventionally defined. Santos et al.[188] state that "the acronym 'SMSI' is an umbrella under which a number of different phenomena may be covered."

SURFACE MODELING AND COMPUTATIONAL METHODS

After formulating (1) the requirements on realistic structural models of metal–support interface systems and (2) the open questions concerning the mechanism of the electronic support effects on adsorbates (cf. the preceding section), we give a brief survey of computational methods that should in principle be suited to deal with problems of this kind. To get insight into the electronic properties of the system, we have to solve Schrödinger's equation. This cannot be done in full rigor and generality. First, simplified surface (interface) models are to be considered, and, second, approximations in the Hamiltonian are necessary. Although both these methodological aspects are interrelated to a certain degree, we will discuss them separately.

Solid-State versus Molecular-Cluster Approach

Solid-state theoretical approaches rest, in general, on the supposition of an ideal, infinite crystalline bulk solid and utilize its translational periodicity. The wavefunctions then satisfy Bloch's theorem; the energy eigenvalues are functions of the wavevector **k** establishing energy bands rather than discrete eigenvalues. This approach was extended to consider also surface and interface problems (see, e.g., Refs. 189 and 190). In that case, a two-dimensional translational periodicity parallel to the surface is assumed, which is justified for perfect as well as for relaxed and reconstructed surfaces. Mostly thin films ("slabs" consisting of several layers of atoms) are calculated, or the "superlattice" method is applied.[189] In the latter case slabs of the (adsorbate-covered or clean) surface system are repeated in order to establish an artificial periodicity perpendicular to the surface.[190,191] Hence, computational programs developed for band-structure calculations of bulk solids may be used. A third approach to deal with the electronic structure of surfaces is the direct solution of the Hamiltonian of the semiinfinite crystal, e.g., by matching the wavefunctions at the solid–vacuum interface.[189] The fourth type of method treats the creation of a surface as a perturbation of an otherwise perfect infinite bulk solid.[189]

For example, two semiinfinite solids are created by removing one or several layers of atoms from the solid, and the corresponding eigenvalue problem is solved within the framework of the scattering-theoretic method by means of Green's function technique.[189]

The calculation of a metal–support system using one of these methods would describe idealized limiting cases, since we are, as a rule, not concerned with infinite metal films on support surfaces but with crystallites that are frequently small. Furthermore, we are mostly not interested in the properties of full-monolayer adsorbates. Finally, the justification of a two-dimensional periodicity may be questioned due to the presence of defects at the interface.* Nevertheless, results on metal–semiconductor interfaces gained with these methods are of interest also with respect to the metal–support interaction.

Molecular-cluster approaches are widely used in surface physics and chemistry (see, e.g., Refs. 192–194). Conventional quantum chemical methods are applied. At the first glance it seems that this approach is particularly appropriate to deal with our problem: molecules adsorbed on metal particles which are deposited on support surfaces. Indeed, quantum chemistry is well suited to treat molecules. But the metal particles are often too large to be calculated as a whole. The same holds for the supporting material. Hence, only some pieces termed "clusters" are cut out from the metal particle and supporting material, respectively. This is again a severe simplification of our structural model. To correct for this deficiency at least in part the clusters have to be "embedded" to account for the environment. This is usually done for semiconductors and insulators by means of pseudoatoms† which saturate the dangling bonds at the surface occurring due to the cut-out from the bulk.[193,198,199] Thus, artificial surface states are avoided. Frequently, e.g., for silicon or SiO_2 with fourfold coordinated silicon atoms, hydrogen atoms are used as pseudoatoms saturating both oxygen and silicon dangling bonds at the surface.[198,200,201] But if, e.g., the valency of the oxide-forming cations deviates from its coordination number, it is necessary to use modified hydrogen atoms exhibiting a nuclear charge different from one[198,202] in order to provide for charge neutrality, which is a prerequisite for obtaining qualitatively correct charge distributions in the metal–support clusters.[203] Furthermore, by such a procedure the correct chemical stoichiometry of the support clusters is as a rule ensured.[198]

---

* In principle solid-state theoretical methods are able to account for these more localized phenomena as well but this would increase the unit cells to be handled and, hence, the computational effort.

† Alternatively, for ionic solids an embedding in a "point-charge lattice" is frequently applied,[195,196] and in the case of amorphous materials the cluster-Bethe-lattice approach is used.[197]

In principle, metallic clusters must also be embedded; this has been done until now only in using nonempirical quantum-chemical methods.[198,204,205] Here a central group of atoms is considered on a highly sophisticated level, while the remaining surrounding atoms are treated in a more approximate manner. Such calculations have not been performed on metal–support systems.

Even embedding clusters as previously described necessitates an investigation of the dependence of the properties under study on the cluster size. The major obstacles in fulfilling this requirement are, in particular, when applying nonempirical methods, limitations in computer time and memory.* This limitation is encountered also for solid-state theoretical methods: in this case the dependence of the results on the thickness of the slab is of importance. Fernando et al.[207] mention that using the linear augmented-plane-wave (LAPW) method, one of the "best self-consistent first-principles calculational methods available," for the calculation of a CO monolayer on a three-layer transition-metal film (which is probably still too thin) "several tens (for simplified spectroscopic information) or hundreds (for total energies) of hours on a supercomputer such as a Cray X-MP" are required.

Approximations in the Hamiltonian

The reliability of the results obtained by means of solid-state theoretical and quantum-chemical calculations depends on the approximations involved in the solution of the Schrödinger equation. This concerns (1) approximations in the Hamiltonian, (2) the type of the wavefunction ansatz (mainly, inclusion of electron correlation or not), and (3) the one-electron basis set used.

The rapid progress in computational methods and in computer technology in the last decade has also brought about highly sophisticated quantum-chemical[204,205,208–216]† and, recently, solid-state theoretical[217–221] methods that allow for the calculation of *reliable total energies of transition-metal-containing systems* and, hence, make geometry optimizations feasible. These are, as a rule, first-principle (*ab initio*) methods, basically nonempirical in character in that they do not incorporate empirical input data. At least the treatment of transition-metal-containing systems in general necessitates the inclusion of electron correlation effects, i.e., the

---

* There is an interesting way out: In the framework of the semiempirical Hückel theory, orbital energies and wavefunctions for finite large clusters may be obtained analytically (see, e.g., Ref. 206). This approach is termed the "analytic cluster model."

† For simplification, also, methods using the local-density approximation (LDA)[214b,216] or the Slater $X\alpha$ formula[215] to treat exchange and correlation contributions are termed as quantum-chemical methods here.

application of approaches beyond the Hartree–Fock level.[208] The most advanced nonempirical solid-state theoretical approaches are the strict *ab initio* crystal-orbital procedures [mostly restricted to the self-consistent field (SCF) Hartree–Fock level; for a recent review, see Ref. 198] and procedures that apply the local- (spin) density approximation to exchange and correlation contributions using different types of basis sets (gaussian-type orbitals, muffin-tin orbitals, plane waves) and different approximations to the atomic potentials (muffin-tin potentials, pseudopotentials). There are a few investigations employing these methods for total-energy calculations of adsorbate-covered transition metal surfaces,[222–224] but work devoted to transition metal–semiconductor or to transition metal–support systems is rare up to now.[221,225] As far as the latter systems are concerned, this is valid also for the previously mentioned highly sophisticated quantum-chemical methods. Only recently, the geometry and stabilization energy of isolated Cu and Ag atoms on a silicon cluster have been calculated.[226] Besides several studies of chemisorption at transition metal surfaces (e.g., Refs. 204, 205, 208*b*) some work on surface reactions including dissociation (e.g., Refs. 209, 211, 214) and on homogeneous catalytic reactions (e.g., Refs. 208*d*, 210, 212, 213) have been performed. Activation barriers of dissociative chemisorption[209,214a] and diffusion,[214b] enthalpies of reaction,[211,212] total energies and geometries of reactants, products and transition states[208d,210,213] have been obtained in these papers. As methods we mention here explicitly the complete active space (CAS)–SCF–Configuration Interaction (CI) method[213,227] and the Multireference single and double excitations (MRD)–CI method.[208b,228] Pseudopotential approaches are widely applied.

For many purposes it is not necessary to determine geometric structures and quantitative energetic data of surfaces, adsorbates, or interfaces. Then more approximate methods with respect to the Hamiltonian or to the basis set may be used still yielding a valuable *insight into the electronic aspects of surface phenomena*. In this case "a number of plausible models" with an assumed atomic structure is calculated to decide "which set of results agrees most closely with experiment."[229] This is a common practice within the framework of both solid-state theoretical approaches (e.g., Refs. 229 and 230) and conventional quantum-chemical cluster approaches (e.g., Refs. 201, 203, 231–234). Several perturbation-theoretical calculations are run at about the same level of sophistication (for example, the effective medium approach[235] or the use of a special model Hamiltonian like that of Hubbard[236] or Newns and Anderson[237] treated by means of Green's function techniques). It is beyond the scope of the present chapter to review all these methods and approximations involved in detail. Rather, we intend to give a brief review of their ability to get a *qualitative* understanding of electronic aspects of surface phenomena. The main emphasis is put on those methods that are used in calculations of transition metal–support systems (cf. Table 8.3). Some of these methods are nonempirical and in principle identical to those suited to calculate reliable total energies. Contrary to the calculations discussed previously, these methods may also be run, however, with small basis sets or without considering electron correlation. Other methods are semiempirical; mostly minimal basis sets are employed, the Hamiltonian is partly parametrized, electron correlation is either not at all or at best implicitly included. In some methods correlation contributions are treated empirically (dispersion energy, vide infra) or in a very approximate manner ($X\alpha$ method). In any case only single electronic configurations or states are considered. For simplification we distinguish five groups of methods.

Extended Hückellike Approaches. The Hamiltonian is parametrized; electron–electron repulsions are not explicitly considered.[252] In some cases electrostatic repulsion forces are included to account qualitatively for core–core repulsion; examples are (a) the atom superposition and electron delocalization (ASED) method, cf. Chapter 10 of this volume, and Refs. 245, 247, 248, and (b) the modified extended Hückel molecular orbital (EHMO) method after Anders et al.[253–255] The corresponding solid-state theoretical methods are frequently termed as empirical tight-binding (ETB) methods (cf. Ref. 189).

As a rule these methods should be run with rigid geometries, assumed or taken from experiment.[238,243,256] In some cases a considerable effort has been devoted to the choice of parameters for the systems under study[245,247,248,254,255] in order to justify attempts to calculate equilibrium distances. In the framework of ASED one arrives at sets of parameters being different for the same type of atom in different environments.[245,247,248] The calculated distances for experimentally well-known subsystems are quite reasonable,[248] however, for some transition metal systems electronic states, even if emerging from different configurations, are frequently near-degenerate;[208a] sometimes the ground state is not even identified.[257] Then the determination of parameters rests on the observation that the spectroscopic constants for several states are found to be similar in some systems.[257] In any case calculated distances and interaction energies obtained within these approaches should be interpreted with caution. Keeping this in mind, they are of qualitative value yielding information on *relative* stabilizations of adatoms on different surface sites or on *trends* in changes of bond lengths. As far as transition state energies or geometries are concerned,[254,255] at best a rough guide for each elementary reaction step may be expected. The foregoing discussion holds also for the following three groups of methods.

Table 8.3. Quantum-chemical investigations of transition metal–support systems.

Metal(s)	Support	Calculated Quantities[a]	Method[b]	Reference
Ni	C	LDOS with and without C	EHMO: C_{24}, Ni_7	238
Ni	C	LDOS with and without C	GFT, MTP: C_{13}, Ni_n, $n = 6, 12, 19$, 1 to 3 Ni layers on a C monolayer	239
Au, Cu	C	ΔE_M, Δq, valence band structures (histograms)	CNDO: M_n, $n = 1$–5, C_{16}	234
Cu, Ag	Si(111)	ΔE_M, $r_{M\text{-}S}$, potential curves	SCF–LSD: Si_{20}, M_1	226
Cu, Ag, Au	SiO_2	ΔE_M, Δq, valence band structures (histograms)	CNDO: M_n, $n = 1$–4, Si_2O_4	234
Ni	SiO_2	Δq and its localization, ELS	CNDO: Ni_n, $n = 1, 4$, SiO_4H_3 to $Si_3O_{10}H_6$	201
Ni, Rh, Pt	SiO_x/Si, TiO_2	Gap states, Δq	Qualitative discussion based on a comparison of cluster energy level schemes (M_{13}, $X\alpha$, EHMO) and energetic positions of Si, SiO_2, and TiO_2 band edges	114, 115
Ru, Pt	SiO_2	ELS, gap states, mechanism of H_2 dissociation	SCF–$X\alpha$–SW, M_1, OSiO	240
Pt	SiO_2	ELS, comparison with near edge X-ray absorption: Δq	SCF–$X\alpha$–SW: Pt_1, SiO_4^{8-}	241
Ni	SiO_2	Electronic state of Ni atoms, orbital shape	*Ab initio*–SCF–UHF: embedded SiO_2-cluster (size not given), Ni_2	242
Ru, Rh, Pt, Pd	Al_2O_3	ELS, gap states	EHMO: M_1, AlO_5	243
Pd	Al_2O_3	ELS, gap states	SCF–$X\alpha$–SW: Pd_n, $n =$ 1, 6, 8, AlO_6 to Al_8O_{12}	244
Ni	Al_2O_3	LDOS with and without support, magnetic properties	SCF–LSD–SW: Ni_9, Al_5O_8	231
Ni	Al_2O_3	Δq, gap states, $BS_{M\text{-}S}$	CNDO: $OAl(O\bar{H})_3$, $Al(O\bar{H})_3$, $Al(O\bar{H})_5$, Ni_1, Ni_2	75, 76
Ni	Al_2O_3	ΔE_M, $r_{M\text{-}S}$, binding mechanism	ASED: Ni_{10} (2 layers), AlO_6	245

Table 8.3. (Continued).

Metal(s)	Support	Calculated Quantities[a]	Method[b]	Reference
Fe, Ni, Cu, Ag	Al_2O_3	ELS, $BS_{M\text{-}S}$	$X\alpha$–SW: AlO_6, M_1	246
Pt	Al_2O_3	ΔE_M, $r_{M\text{-}S}$, Δq, binding mechanism	ASED: Pt_{31} (2 layers), Al_9O_{24}, Al_4O_{18}	247
Ni	Al_2O_3	Δq, ΔE_M, $r_{M\text{-}S}$	PP–*ab initio*–UHF: embedded Al_4O_3-cluster, Ni_1	195
Cu	MgO	Δq, ΔE_M, $r_{M\text{-}S}$	PP–*ab initio*–UHF: embedded MgO-clusters (Mg_5O, O_5Vac_{Mg}), Cu_1	196
Pt	TiO_2	Δq, ELS	SCF–$X\alpha$–SW: Pt_1, TiO_5, TiO_6	233
Pt	TiO, FeO, ZnO	Δq, binding energy of dopant molecules	ASED: Pt_{22}, TiO, FeO, ZnO on the metal	248
Ni	TiO_2, TiO, SiO_2	Δq, gap states, $BS_{M\text{-}S}$	CNDO: $Ti(O\bar{H})_5$, $Si(O\bar{H})_5$ Ti_4O_3, $Si(OH)_3$, etc., Ni_1	203
Minimum Support Models				
Ni, Cu	H_2O	ΔE_M, $r_{M\text{-}OH_2}$, Δq	*Ab initio*–SCF–UHF + dispersion energy	249*a*
			Ab initio–SCF–RHF + SDCI	249*b*, *c*
Ni	H_2O	ΔE_M, $r_{M\text{-}OH_2}$, Δq	CASSCF + CCI	227
Cu, Ni	H_2O	ΔE_M, $r_{M\text{-}OH_2}$, H_2O bending frequency shift	SCF, CASSCF, empirical estimate of dipole-induced-dipole attraction	250
Ni, Fe, Cu	H_2O	ΔE_M, $r_{M\text{-}OH_2}$, Δq, μ	*Ab initio*–SCF, CASSCF	251

[a] $BS_{M\text{-}S}$: Bond strengths between metal and support (mostly Wiberg bond index)
ELS: Electronic energy level schemes
ΔE_M: Binding energy of the metal atom M on the support [cf. Eq. (8.3)]
LDOS: Local density of states
Δq: Electron transfer between metal and support
$r_{M\text{-}S}$: Distance between metal and support
$r_{M\text{-}OH_2}$: Distance between metal and a water ligand
μ: Dipole moment

[b] ASED: Atom superposition and electron delocalization
CASSCF: Complete active space SCF
CCI: Contracted CI
CNDO: Complete neglect of differential overlap
EHMO: Extended Hückel molecular orbital
GFT: Green's function technique
LSD: Local spin density approximation
MTP: Muffin-tin potential
PP: Pseudopotential
RHF: Restricted Hartree–Fock
SCF–$X\alpha$–SW: Self-consistent field–$X\alpha$–scattered wave
SDCI: Single and double excitations configuration interaction
UHF: Unrestricted Hartree–Fock

Working with rigid geometries, information on the nature of interfacial or chemisorptive bonds (molecular-orbital analysis in terms of atomic orbitals, symmetry analysis, charge distribution by means of population analyses), on their strengths (overlap populations, bond indices), and on valence band structures [energy level schemes, densities of states (DOS)] are obtained. Again classes of models should be considered in order to have a sufficient number of results at hand for a qualitative understanding of electronic effects. With respect to charge distributions (cf. the corresponding discussion in the experimental section), it should be remarked that they may suffer from the non-self-consistency of these methods. The iterative extended Hückel theory (IEHT) is suited to overcome this deficiency at least in part (see, e.g., Ref. 194).

NDO Approaches. The second group of semiempirical approaches consists of "neglect of differential overlap" (NDO) methods.[258,259] They may be viewed as approximate Hartree–Fock (HF) methods; electron interactions are included. Both unrestricted (UHF) and restricted (RHF) approaches with respect to spin and space symmetry are formulated; for open-shell systems the latter is frequently run in the half-electron version.[260] The (approximate) HF equations are solved to self-consistency. For certain elements of the Fock matrix parameters are introduced, which are to be determined carefully. This has been done, e.g., in the CNDO (complete NDO) framework for transition metals by adjusting to *ab initio* SCF calculations of simple diatomics[261] or by fitting special cluster properties to experimental bulk electronic properties[262,263]; in the latter case "some electron correlation" is indirectly included. Attempts have also been made to extend the INDO (intermediate NDO) method (cf. Ref. 232, 264, and references therein) and the more elaborate MINDO (modified INDO) method[265] to transition metals, but these versions were not applied to transition metal–support systems up to now.

We have used throughout our investigations the CNDO/2 method in the UHF version for open-shell and in the RHF version for closed-shell systems.[75,76,201,203] The suitability of the parameters has been checked for relevant subsystems (SiO_2,[200] Ni_3Ti,[263] TiO_2,[263] Al_2O_3[75]). For further details we refer to the original papers.[200,201,203,263] The geometry parameters are obtained from experiment or from related inorganic compounds (cf. the next section) with the exception of the Ni–H distance, which has been optimized in some cases.[201,203] The results are as in the case of methods of extended Hückel type (vide supra) qualitative in nature. Properly chosen models are required. The binding energy* ΔE_A of an atom or molecule

* ΔE_A is given in eV throughout.

adsorbed on the metal–support cluster X is calculated from the total energies $E(\mathrm{A})$, $E(X)$, and $E(\mathrm{A}X)$ of the individual subsystems ("supermolecule approach"):

$$\Delta E_{\mathrm{A}} = [E(X) + E(\mathrm{A})] - E(\mathrm{A}X) \tag{8.3}$$

It should be noted here that, e.g., the energy of atomic H adsorption ΔE_{H} is not equal to the quantity E_{HM} as defined in Eq. (8.2) (cf. the experimental section), since the latter is exclusively connected with the establishment of the M–H bond. The supermolecule approach, however, does not discriminate between contributions of different bonds to the interaction energy (as with the different experimental methods aimed at the determination of Q_{ad}). Furthermore, ΔE_{H} does not contain any vibrational energy contrary to E_{HM}. Thus, the following relation holds:

$$\Delta E_{\mathrm{H}} \approx E_{\mathrm{HM}} - \tfrac{1}{2}\eta E_{\mathrm{MM}} \tag{8.4}$$

For the meaning of the last term we refer to the discussion of Eq. (8.2). The foregoing argumentation is analogously valid also for adsorption of molecules and will be stressed further in the next section. The local strength of an individual bond (e.g., H–M) is expressed by means of the Wiberg bond index.

$X\alpha$ and Density-Functional (DF) Approaches. These methods use approximate formulas to account for exchange and correlation contributions to the electronic energy.[192,214b,215,216] The extent of correlation is not obvious, and an empirical element comes into play via the "exchange parameter" α. The accuracy of the results strongly depends on the basis set chosen (plane waves, muffin-tin orbitals, gaussian-type orbitals), on its dimension, and on the type of potential used. As a rule scattered wave (SW) $X\alpha$ approaches[233,244] are less reliable than, e.g., SW–LSD (local spin density functional)[231] or even LCGTO (linear combination of gaussian-type orbital)–LSD[214b] methods, the latter yielding even reliable total energies (vide supra). Good quality wavefunctions (more exactly, one-electron densities) are as a rule obtained. The methods are run to self-consistency. Spin-polarized methods [e.g., LSD instead of the simple local density approximation (LDA)][231] allow for the calculation of magnetic properties as semiempirical UHF approximations do.[264] Besides the ETB approaches, the $X\alpha$ and density functional approximations are widely used in solid-state theoretical methods both with respect to interfaces[189,191] and to magnetic properties of surfaces.[266]

Model Hamiltonians and Green's Function Techniques. These methods can be run self-consistently on different levels of approximation[235,236] or within the framework of the non-self-consistent tight-binding formalism.[237] In the latter case there is a strong dependence on the parameters; parameters with a special physical significance can be varied in order to obtain insight into their influence on the calculated electronic and energetic properties (e.g., the "metal-support coupling" parameter γ in Ref. 237). Self-consistent methods are able to compensate somewhat for the dependence on the parameters (e.g., in the Coulomb and exchange potential). Generally, the methods rest on the knowledge of the solution of the unperturbed problem (bulk solid, clean surface).

Nonempirical SCF–MO Methods. Using limited basis sets, the reliability of the calculated energies, properties, and wavefunctions is comparable to those obtained with the four groups of methods discussed previously. In many cases (in particular when transition metals are involved, vide supra) the inclusion of electron correlation contributions is indispensable. With respect to weak intermolecular interactions (van der Waals type, no chemical bonds formed), a three-step procedure has been found to be useful (e.g., Ref. 249, and references therein): Calculate first the SCF potential energy curve using a small (e.g., split-valence) basis set, correct for the basis set superposition error, and then estimate the dispersion energy as the essential part of the intermolecular correlation. In any case this procedure is to be preferred to semiempirical methods for calculations of interface systems, e.g., for the interaction of atoms or clusters with groups of atoms (e.g., bridging oxygen atoms) on the perfect support surface. Qualitative information on the mechanism of bonding, its energetics, and rough bond lengths estimates may be expected.[249]

Classical Methods for Geometry Predictions Using Empirical Potentials

These methods make use of basic relations of classical electrodynamics and mechanics. The former provide for the interaction potentials (mainly pair potentials included), which are either employed in "molecular statics modeling"[267] or are used as input to dynamical simulations. In the latter case classical mechanics contributes the corresponding equations of motion. It is beyond the scope of this chapter to give a detailed review (cf. Ref. 267 for a recent overview with respect to internal interfaces). As examples of statical and dynamical approaches we mention here the determination of total interaction potentials between metal atoms and ionic surfaces[268–270] and the molecular-dynamics computer experiments.[271,272] The total interaction potential between a single adatom and the substrate summing over several hundred substrate ions consists of the dispersion potential, the electrostatic potential, and the repulsive potential.[268–270] The numerical values of the parameters involved are expressed in terms of properties of the adatoms and substrate ions, respectively.[268,269] Surface relaxation can be accounted for.[268,269] Based on these empirical potentials, interaction energies and equilibrium distances of a metal atom on an ionic surface are obtained.[268–270] A similar potential ansatz is taken as a basis for molecular-dynamics simulations. The solutions of Newton's laws of motion depend sensitively both on the values of potential parameters and on the initial conditions (initial positions and velocities of atoms) of the system.[271,272] In a study of Pt adatoms on a vitreous silica surface, no less than about 800 atoms are treated in this way.[271,272] Alternatively, the potential ansatz for mutual interactions between metal atoms and between metal atoms and substrate atoms has been used within a Monte Carlo procedure for the search of the minimum energy configuration of the metallic cluster.[273] With a sufficiently large number of Monte Carlo steps (i.e., geometric configurations in this context), the mean geometric parameters approach the true statistical-mechanical result.

These methods yield valuable qualitative insight into structural aspects of interactions with surfaces and at interfaces (cluster growth, wetting phenomena) and their interrelation with interatomic forces and potentials. Quantitative predictions, however, are unlikely due to the uncertainties connected in particular with the values of parameters.

Although not resting on empirical potentials, we mention here approaches employing free-energy estimates for structural predictions. For example, the kinetics of the formation of an SMSI state in Pt/TiO_2 has been considered in this way.[274] The results are qualitative in character.

METAL–SUPPORT INTERFACES AND SUPPORT INFLUENCE ON CHEMISORPTION: RECENT THEORETICAL RESULTS

In what follows we review mainly our work on nickel interacting with special surface sites of silica, alumina, and titania. The main emphasis will be laid on the interrelation between the results on "WMSI models" and "SMSI models." The work of other authors on the questions under discussion and on related subjects will be included. Simplified structural models have the essential advantage of yielding valuable insight into the connection between structure and electronic properties, thus making transparent how a special geometry (caused, e.g., by the geometric support effect) brings about a special electronic and chemisorptive behavior (termed the electronic support effect).

Structural Models from Calculations

The generation of structural models of transition metal–support systems by means of theoretical methods is a difficult task rarely attacked up to now. We mention here the molecular-dynamics simulation of the deposit of Pt atoms on a vitreous silica surface.[271,272] The parameters for the Pt–Si interaction potential "were arbitrarily selected to be weakly attractive," while those of the Pt–Pt and Pt–O interactions were taken to be strongly attractive.[271,272] When solving the equations of motion, the support influence has been introduced by means of the initial distances of the Pt atoms from the support surface.[271] The calculated distances (Pt–Pt, Pt–O) and coordination numbers have been compared with EXAFS results.[271] The best agreement is obtained when starting the calculation with an fcc cluster or a cluster of randomly distributed Pt atoms "well above" (>0.5 nm) the surface.[271] Hence, one arrives at the conclusion that the occurrence of three-dimensional clusters is connected with "little or no interaction with the substrate."[271] Depositions on "mobile" substrates due to their temperature (e.g., 300 K) "indicated a clear ability for thermal (energy) transfer from the heavy Pt adatoms to light substrate atoms" resulting in a relaxation of the silica surface that creates strongly binding sites near the surface.[272] The latter underlines the importance of dynamical calculations and of temperature effects for structure generations.

Taking into account nearest-neighbor interaction between metal atoms, E_{MM}, and between metal atoms and substrate atoms, E_{MS}, it has been found that for $E_{MM} \gg E_{MS}$ the cluster morphology is not influenced by the substrate as expected.[11] With $E_{MM} \approx E_{MS}$ different cluster shapes including the two-dimensional monolayer coexist.[11] Using a Monte Carlo simulation it has been shown[273] that with increasing metal–support interaction the formation of fcc clusters is favored against that of tetrahedrally grown clusters. $E_{MS} \gg E_{MM}$ produces metallic monolayers on the substrate.[11] Only recently it has been demonstrated, however, by minimizing the total potential energy of a 13-atom cluster on a substrate that three-body interactions are also of importance.[275] As a rule the latter are neglected in calculations of this type. While the authors reached the same conclusion with respect to two-body interactions, they found three-body interactions between adatoms to favor monolayer formation.[275] Three-dimensional cluster growth is becoming more important with increasing three-body interactions between adatoms and substrate atoms and with strong adatom–adatom pair interactions.[275] The results should sensitively depend on the numerical values and ratios of different interaction parameters (which have been assumed without specifying the nature of the atoms) and are, hence, qualitative in character.

Using thermodynamic equilibrium arguments TiO_x layers on Pt in Pt/TiO_2 were found to be stable both in the LTR and HTR regimes but their formation is very slow at low temperatures due to kinetic hindrance.[274] The two mechanisms proposed for the TiO_x layer formation are (a) the reduction of TiO_2 producing a Pt/Ti alloy with a subsequent surface segregation and (partial) oxidation, and (b) the formation of TiO_x at the interface with a subsequent diffusion across the Pt surface.[274] Bond strengths arguments have been used.[274]

Semiempirical quantum-chemical calculations have provided qualitative arguments for the stabilization and diffusion of partially reduced metal oxide particles on metallic surfaces (RuO_x on Ru)[257] and for the formation of flat rather than three-dimensional metal particles on oxidic surfaces in the SMSI state.[75] In the latter case Ni_2 vertically on top of partially reduced alumina has been considered.[75] While the interfacial nickel–alumina bonds are strong (concluded from Wiberg bond indices), the Ni–Ni bonds are considerably weaker than in the free Ni_2 molecule[75] indicating an unstable vertical geometry at high reduction temperatures. A stabilization may be expected by a "lying-down" of Ni_2 allowing for the formation of interfacial bonds of both Ni atoms in the horizontal geometry.

A survey on quantum-chemical calculations of transition metal–support systems is given in Table 8.3. Neither calculations of this type nor solid-state theoretical calculations have been performed on a level of sophistication that allows for the reliable determination of total energies in order to predict geometries at interfaces or of metal particles on a support surface.* Only for single metal atoms on support surfaces binding energies and distances have been obtained (see subsequent section). Furthermore, some total-energy calculations have been carried out for *sp* metal–semiconductor interfaces (cf. Ref. 276, and references therein).

Assumed Structures of Relevant Model Systems

Proposals of model structures for nickel on silica, alumina, and titania surfaces[75,76,201,203] are largely based on the arguments from the experimental section (vide supra). As pointed out previously, we consider mainly single-nickel-atom models. Although this is a severe restriction compared with the actual situation, the well-founded models selected here should allow for preliminary conclusions, which of course must be checked in future work on large clusters. These are models of (a) the completely oxidized support surface (type A models, the corresponding Si and Ti cations have formal charges of +4, the Al cations of +3, cf. Figs. 8.2 and

* There is one exception concerning the Ni/Si(111) interface: The total energy of three different assumed interface geometries has been calculated using the LAPW approach and the thin-film method.[221] But within each of these models all distances and angles were kept constant.[221]

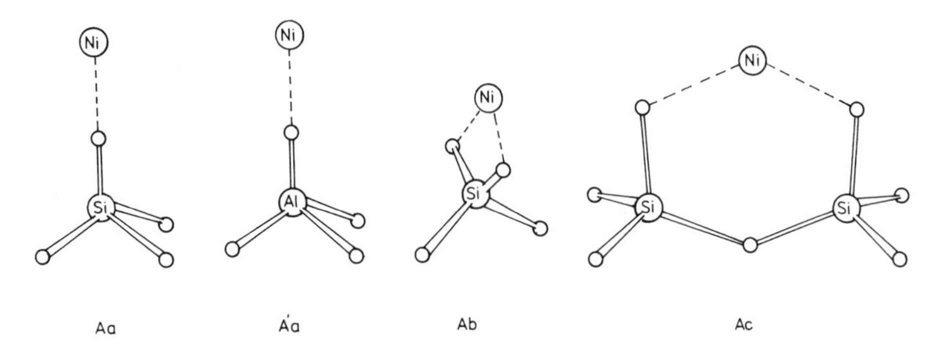

Figure 8.2. Models of nickel bonded via oxygen to the support (type A models). Aa, Ab, and Ac represent Ni on the isolated, geminal, and vicinal Si–O groups. Small circles designate oxygen atoms. Saturating H and H̃ atoms are not drawn. [Adapted in part from Ref. 203. Reprinted with permission from *The Journal of Physical Chemistry* **90:** 4322. Copyright (1986) American Chemical Society.]

8.3), (b) the partly reduced support surface (type C models, the formal charges are reduced by one, cf. Fig. 8.3), and (c) titanium monoxide (type D models, Ti^{2+}, cf. Fig. 8.4).* The geometry parameters, their origin, and further characteristics of surface sites are summarized in Tables 8.4 and 8.5. Nickel is bonded via oxygen (Figs. 8.2 and 8.4) or directly to surface cations (Figs. 8.3 and 8.4). In any case the models may be viewed also as representing oxide particles on top of nickel, thus accounting for the "island" formation of suboxides. Owing to the unreliability of CNDO/2 equilibrium distances, we prefer the use of distances taken from similar inorganic substances or estimated from ionic or covalent radii.† The Ni–O distance is similar to the experimental findings given previously. Some more explanations concerning the individual surface sites with which Ni interacts on SiO_2 and TiO_2 can be found in Ref. 203. These also hold in principle for γ-Al_2O_3,[75] which exhibits Al both in a fourfold and in a sixfold coordination. Altogether we state that the proposed models meet the requirements (with the exception of the nickel cluster size) as derived in the experimental section. There is, however, one omission, namely, a model accounting for the interaction of the metal with bridging oxygen atoms that exists on amorphous (and crystalline) support surfaces. It has been shown that in a zeroth-order approximation, a "minimum support

* Models containing four Ni atoms on SiO_2 (type B) have been calculated[201] using the CNDO/2-CLACK version and are not discussed here.
† The CNDO/2 equilibrium distances are 252 pm, 242 pm, and 244 pm for models Ae, Ce, and Cdpy, respectively. The qualitative conclusions with respect to the electron transfer, the position and character of interfacial states, and the H binding energies are the same as with the estimated distance of 220 pm.

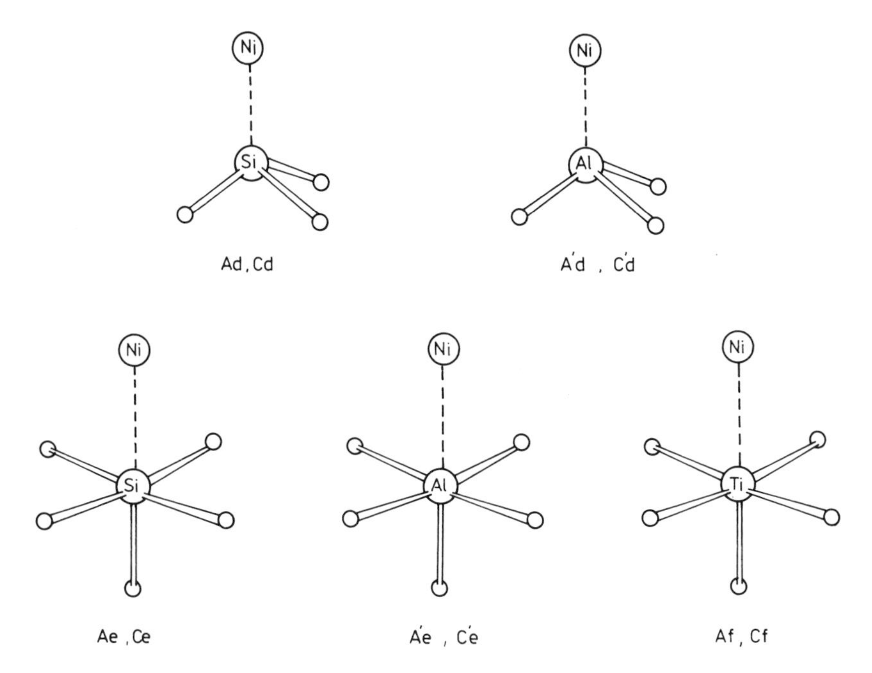

Figure 8.3. Nickel interacting with surface oxygen vacancies. Model structures A′d and A′e are generally assumed to be two different Lewis acid centers on the alumina surface. Ad, Ae, and Af are the corresponding defects on silica and titania surfaces, respectively. The type C models have captured an electron compared with their oxidized isomorphous counterparts. [Adapted in part from Ref. 203. Reprinted with permission from *The Journal of Physical Chemistry* **90:** 4322. Copyright (1986) American Chemical Society.]

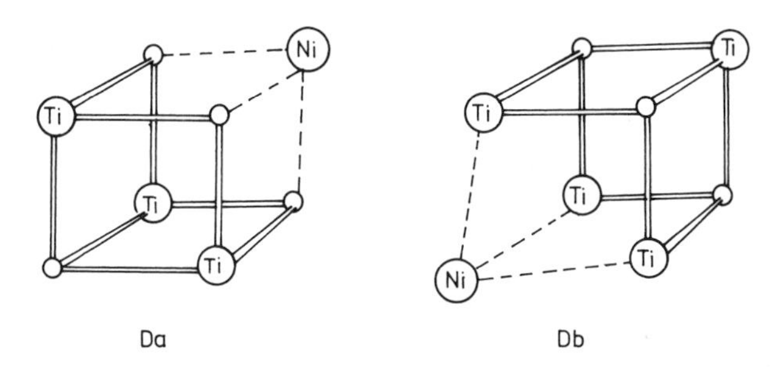

Figure 8.4. Nickel interacting with TiO surface clusters (rock salt structure): O-terminated surface (Da), and Ti-terminated surface (Db). [Adapted from Ref. 203. Reprinted with permission from *The Journal of Physical Chemistry* **90:** 4322. Copyright (1986) American Chemical Society.]

Table 8.4. Properties of clean support clusters.

	Characterization of Models				Calculated Properties			
Model	M^{n+} [a]	Modification[b]	$r_{M\text{-}O}$ [c] (pm)	$q_{\tilde{H}}$ [d] (|e|)	Multiplicity	ε_{OSS} [e] (eV)	ε_{EA} [e] (eV)	q_M [f] (|e|)
Aa	Si^{4+}	β-Cristobalite	155	1	Doublet	—	3.21	1.53
Ab	Si^{4+}	β-Cristobalite	155	1	Triplet	—	2.68	1.49
Ac	Si^{4+}	β-Cristobalite	155	1	Triplet	—	2.84	1.42
Adpl[g]	Si^{4+}	β-Cristobalite	155	2/3	Singlet	—	2.80	1.39
Adpy[h]	Si^{4+}	β-Cristobalite	155	2/3	Singlet	—	0.37	1.35
Ae	Si^{4+}	Stishovite	178	6/5	Singlet	—	2.29	1.64
Af	Ti^{4+}	Rutile	196	6/5	Singlet	—	3.10	1.14
A'a	Al^{3+}	γ-Al_2O_3	171	4/3	Doublet	−13.47	4.66	1.12
A'dpl[g]	Al^{3+}	γ-Al_2O_3	171	1	Singlet	—	2.89	1.12
A'dpy[h]	Al^{3+}	γ-Al_2O_3	171	1	Singlet	—	1.23	1.12
A'e	Al^{3+}	γ-Al_2O_3	198	7/5	Singlet	—	2.85	1.29
Cdpy[h]	Si^{3+}	β-Cristobalite	155	1	Doublet	−7.69	1.58	0.96
Ce	Si^{3+}	Stishovite	178	7/5	Doublet	−5.09	3.26	1.23
Cf	Ti^{3+}	Rutile	196	7/5	Doublet	−3.42	4.26	0.58
C'dpy[h]	Al^{2+}	γ-Al_2O_3	171	4/3	Doublet	−5.89	2.14	0.58
C'e	Al^{2+}	γ-Al_2O_3	198	8/5	Doublet	−3.67	3.83	0.16
Da	Ti^{2+}	TiO	209.5	—	Singlet	−4.39	2.32	0.56
Db	Ti^{2+}	TiO	209.5	—	Triplet	−3.98	2.14	0.12

[a] Surface cation and its formal oxidation state.
[b] Modification of the oxide from which the models are adapted; the bulk coordination numbers of the cations are 4 (β-cristobalite, γ-Al_2O_3) and 6 (stishovite, γ-Al_2O_3, rutile, TiO).
[c] Cation–oxygen distance.
[d] Nuclear charge of the saturating $\tilde{H}$ atoms.
[e] MO level energies of occupied surface states (OSS) and the lowest unoccupied MOs representing the electron affinity (EA) of the surface site.
[f] Net charge of the surface cation.
[g] Planar.
[h] Pyramidal.

model,'' the water molecule, may be used.[249a] Since in a metal atom–water molecule system a van-der-Waals-type bond is established[249] the semiempirical CNDO/2 method is expected to yield misleading results. Hence, the treatment was based on a nonempirical approach yielding also Ni–O distance (vide infra).

To get some more insight into the situation met with defective support surfaces, we consider two examples in more detail. The oxygen vacancy is a characteristic defect at TiO_2 surfaces. XPS investigations indicate a reduced positive charge of Ti surface cations, the chemical shift being similar to that of Ti^{3+} in Ti_2O_3.[277] ESR results suggest a structure[278] similar to that of the bulk E_1' center (O^- vacancy) in SiO_2 (cf. Ref. 200, and references therein). The following procedure for a

Table 8.5. Properties of Ni-support clusters.

	Interface Geometry		Calculated Properties		
Model	$r_{Ni\text{-}S}$[a] (pm)	Taken From[b]	Multiplicity	q_{Ni}[c] (|e|)	$WB_{Ni\text{-}S}$[d]
Aa	209	NiO	Doublet	0.44	0.84
Ab	209	NiO	Triplet	0.32	0.98[e]
Ac	209	NiO	Triplet	0.32	1.09[e]
Adpl	220	Est.	Triplet	0.12	0.44
Adpy	220	Est.	Triplet	0.26	0.59
Ae	220	Est.	Triplet	0.16	0.45
Af	254.5	Ni_3Ti	Triplet	0.17	0.26
A′a	209	NiO	Doublet	0.31	0.97
A′dpl	230	Est.	Triplet	0.15	0.41
A′dpy	230	Est.	Triplet	0.24	0.52
A′e	230	Est.	Triplet	0.18	0.37
Cdpy	220	Est.	Doublet	0.04	0.94
Ce	220	Est.	Doublet	−0.08	0.90
Cf	254.5	Ni_3Ti	Doublet	−0.22	0.95
C′dpy	230	Est.	Doublet	−0.02	0.97
C′e	230	Est.	Doublet	−0.12	0.92
Da	209.5	TiO	Singlet	−0.31	0.93[e]
Db	254.5	Ni_3Ti	Triplet	−0.48	2.29[e]

[a] Distance Ni-support (cf. Figs. 8.1–8.3).
[b] See the corresponding references in Ref. 203; Est.: estimated from ionic and covalent radii.
[c] Net charge of the nickel atom.
[d] Wiberg bond index of the Ni-support bond.
[e] Sum of the indices of all Ni-support bonds.

defect production by means of Ar^+ bombardment has been proposed[278]:

$$Ar^+ + Ti^{4+} \diagup O^{2-} \diagdown Ti^{4+} \rightarrow Ar + O + Ti^{3+} \diagup \square^+ \diagdown Ti^{3+}$$

$$Ti^{3+} \diagup \square^+ \diagdown Ti^{3+} \rightleftarrows Ti^{3+} \diagup \square \diagdown Ti^{4+}$$

Interestingly, this defect has also been created by thermal treatment (1310 K, low oxygen partial pressure).[277] It is obvious that the reducing treatment conditions are analogous to those used in the preparation of catalysts. In the latter case, however, transition metals predissociating the reducing agent H_2 are present leading to lower temperatures of defect production. Evaporating nickel on a surface prepared as described previously may give systems as characterized by means of models Af and Cf.

As a second example we consider Ni/SiO_2. Recalling the picture obtained in the experimental section for "normally" prepared (LTR) catalysts, we obtain the following schematic sketch:

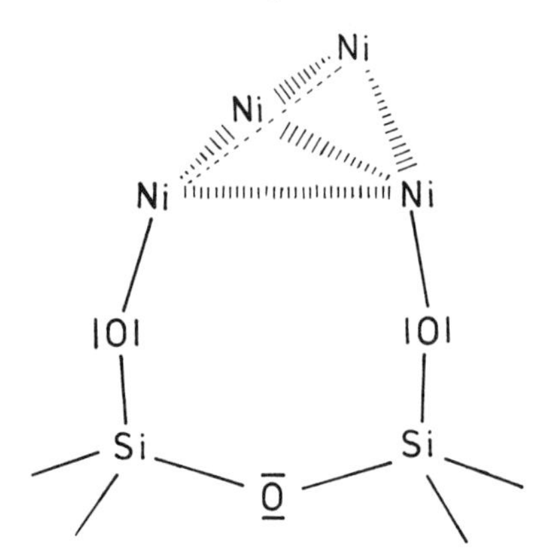

A further reduction in H_2 ("SMSI conditions") may remove the interfacial oxygen atoms resulting in Ni–Si intermetallic bonds and water as a by-product. It is instructive now to have a look at the preparation of an experimental model system as we did previously for TiO_2. In Figure 8.5 we have depicted a possible formation of an oxygen and a silicon-terminating surface center on amorphous SiO_2. In the bulk solid both nonbridging oxygen centers (ESR, luminiscence)[279] and paramagnetic E_1' centers (cf. Ref. 200) are known to exist. There are, however, arguments for their occurrence at the surface [nonbridging O: luminescence,[279] ESR[280]; E_1': ESR after crushing SiO_2 in the ultrahigh vacuum (UHV)[281]]. An inspection of the stoichiometry* further elucidates the use of the terms "fully oxidized" and "partly reduced." The close relationship between the two models is also evident from the fact that the one may be created from the other by oxygen addition/elimination. Hence, we conclude that in evaporating nickel, systems as characterized by models Aa and Cd (cf. Figs. 8.2 and 8.3) should occur.

Models Ad, A′d and Cd, C′d (Fig. 8.3) are the only ones that account for surface reconstruction at least in part. In these clusters an Si and Al dangling bond, respectively, represents the surface. In type A models it is empty (indicating Lewis acidity) but in the type C models it is occupied by a single electron. Thus, reconstruction is expected to be of crucial importance leading to a planar structure in the former cases and to a pyramidal structure in the latter ones.[203] Theoretical investigations have shown that surface relaxation (and reconstruction) may disappear when metals

* The connection between the meaning of the pseudoatoms and the stoichiometry is discussed in Ref. 198.

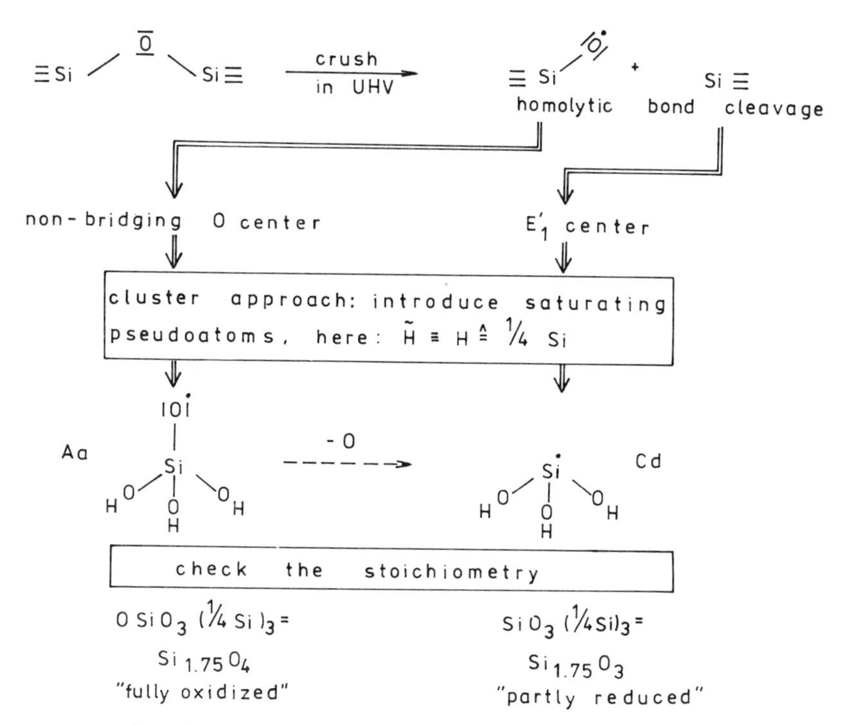

Figure 8.5. Creation and modeling of different surface centers on an amorphous silica surface (see text).

are deposited.[276] Since no definite data are available, we consider both the planar (pl) and pyramidal (py) structures of models Ad and A′d (Table 8.4).

Referring to Figure 8.5 and its discussion we go a step further, keeping in mind our intention to calculate interfacial distances with a reliable quantum-chemical method. As in the case of the water molecule modeling a bridging oxygen atom, the support cluster is made smaller arriving at a "minimum support model." As such we propose to use the silyl radical, $\cdot SiH_3$, i.e., we omit the three support oxygen atoms from model Cd. This molecule has been observed in an Ar matrix by means of ESR[282] rendering it a rough model for the E'_1 surface center. The silicon sp^3 hybridization may be preserved depending on (Ni–)Si–H bonding angle. CNDO/2 calculations yielded qualitatively comparable results from (H . . .)$NiSiH_3$ and model Cdpy.[283] We performed some preliminary calculations with rigid geometries using the MRD–CI method (vide infra).

Finally we compare our assumed model structures with those of other authors (cf. Table 8.3). Both metal bonding via oxygen[195,196,234,242–248] and directly to surface cations[196,233,243,247,248] have been considered. The interaction with reduced oxide particles was investigated only by Horsley[233] (one of his models was used by us as model Cf) and, recently, by Anderson.[248] The bond distances at the interface are in the same range as ours (cf. Table 8.5). Besides metal atoms,[195,196,233,234,238–240,243,244,246] metal clusters on supports were also investigated.[231,234,242,244,245,247,248] The distances between metal atoms are taken from the bulk solids (e.g., 249 pm[231] for Ni–Ni, and 277 pm[247] for Pt–Pt). In some cases a lattice mismatch at the interface is accounted for by slight bond distance variations.[239] The same principles are applied in solid-state theoretical calculations of transition metal–semiconductor and silicide–silicon interfacial systems.[43,284,285]

Electronic Aspects of Metal–Support Interactions and the Bonding Character

Electron Transfer and Clean Surface Sites. A prerequisite for the understanding of bonding at interfaces is a qualitatively reasonable description of the electronic structure of the clean support surface. Surface sites on fully oxidized supports have no occupied surface states in the band gap* (Table 8.4) with the exception of model A′a (vide infra). As previously pointed out in detail,[201,203] our theoretical results for models Aa, Ab, Ac, Ad (SiO_2), and Af (TiO_2) are in agreement with hitherto existing experimental and theoretical findings. Although we are not aware of any experimental or theoretical work concerning the surface of stishovite, the results should be similar to those for the isomorphous rutile. Hence, we believe the characteristics of model Ae (and also of model Ce) to be correct. With respect to alumina surfaces experimental and theoretical information is rare,[286] in particular on the oxygen-terminated surface. Contrary to model Aa some occupied oxygenlike surface states are found slightly above the valence band edge for model A′a (Table 8.4). Using a linear-combination-of-muffin-tin-orbital method, a similar finding was reported for an oxygen-terminated TiO_2(110) surface,[287] but by relaxing the oxygen layer toward the nearest Ti layer, the surface states were shifted down into the valence band.[287] A similar result was not found for model A′a.[75] Although the reason for this unexpected behavior is not yet clear, these states that are not far from the valence band edge will not substantially modify the electronic properties of the surface site compared with model Aa. In the latter case the on-top valence band states are also O_{2p}-derived but delocalized rather than localized at the surface. The Al-terminated Al_2O_3 surface shows an unoccupied sur-

* In spite of the use of small clusters the characteristics of the molecular orbital (MO) energy level schemes are in qualitative agreement with the DOS of the bulk oxides.

face state in the gap,[288] which is ascribed to an empty Al dangling bond state.[288,289] This finding is in accordance with the results for models A′d and A′e and the well-known existence of two Lewis centers of different acidity on the alumina surface.* The latter are ascribed to surface oxygen vacancies at fourfold and sixfold coordinated Al^{3+} cations.[290] This is again in agreement with our results comparing the MO energies of the lowest unoccupied molecular orbitals (LUMO) of models A′dpy and A′e (Table 8.4). Relaxing the A′d model to a planar structure (model A′dpl, the Al–O distances are kept constant) results in an electron affinity comparable to that of model A′e. Hence, it is concluded[75] that a surface relaxation if any does not exclusively produce planar structures. Summarizing, we conclude that the electronic structure of surface sites on fully oxidized surfaces as properly described by our models is governed by the electron-withdrawing properties of the LUMO. Hence, the electron transfer from the nickel atoms to the support (Table 8.5) is understandable. In addition it is in agreement with the experimental findings (cf. the experimental section) on WMSI catalytic metal systems.

Contrary to this, surface sites on the reduced support exhibit singly occupied donor states in the band gap (Table 8.4). The cation dangling bonds of the isomorphous type A models have captured an electron resulting in a reduction of the formal oxidation state by one. Evidence for the proper description of this situation with respect to defective/partly reduced SiO_2 and TiO_2 surfaces has been given recently[203] and will not be repeated here. For the alumina surface a comparable behavior is expected. Consequently, the electron transfer is directed toward the metal in all these cases (Table 8.5) in agreement with the experimental results on metal–support systems in the SMSI state (cf. the experimental section).

There is a tendency to increasing negative nickel charges with decreasing net charges of the cation on *clean* support sites (type C and type D models) and a tendency to increasing positive nickel charges with increasing net charges of the respective cations.[75,203] But a closer inspection of the electron transfer should include the electron affinities and ionization potentials of both subsystems.[75,76] Hence, we arrive at an expression [Eq. (8.5)] equivalent to the well-known Mulliken molecule electronegativity. The electronegativity of the clean support cluster (EC) is defined in terms

$$EC = -\tfrac{1}{2}(\varepsilon_{HOMO} + \varepsilon_{LUMO}) \tag{8.5}$$

of the energies of its frontier orbitals. As shown in Figure 8.6, a nearly linear correlation between this property and the electron transfer is ob-

* The acidity of the alumina surface is considered in more detail by Yoshida[290] (cf. also Chapter 11 of this volume).

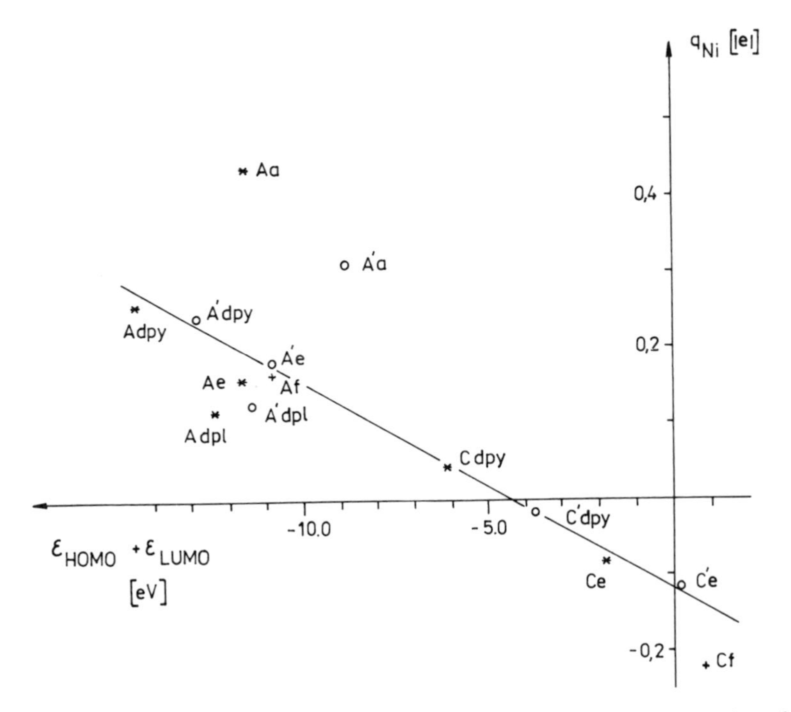

Figure 8.6. The electron transfer between metal and support related to the electronegativity EC of the clean support clusters (see text). Circles, stars, and crosses designate Al_2O_3, SiO_2, and TiO_2 models, respectively.

tained for the models exhibiting single metal–support bonds. The deviation of models Aa and A'a is understandable from differences in the nature and spatial localization of orbitals on oxygen surface atoms compared with surface cations. In any case the electron transfer to the support increases with increasing electronegativity of the surface site as expected.

This simple relationship does not, however, hold for Ni_2 vertically on top of the different surface sites.[75] But, nevertheless, in any case the nickel species are more negative on type C models compared with the corresponding type A models. This applies for both the interfacial nickel atom and the Ni_2 molecule. It should be noted that the huge manifold of nearly degenerate states in Ni_2 makes the calculations difficult. Convergence problems are particularly severe in this case and in the UHF framework a mixing between different states of the metal–support system occurs.

Considering, finally, the role of the chemical nature (SiO_2, Al_2O_3, TiO_2) of the supporting oxide, we have to concede that the CNDO/2 calculations based on *estimated* distances at the interface are too rough to allow for definite conclusions in this respect. However, the effect of the chemical composition seems to be not as marked as that of the degree of support reduction. Hence, isomorphous models of different chemical composition show a similar behavior; but it should be kept in mind that, e.g., the TiO_x model Cf represents a catalyst reduced at considerably lower temperature than catalysts modeled by the corresponding SiO_x and Al_xO_y models Ce and C′e.

The results on the electron transfer presented here are in agreement with other theoretical work cited in Table 8.3. For example, electrons are transferred to the support both for a Pt particle in contact with Al^{3+} cations and with an oxygen terminated surface,[247] whereas the direction of electron transfer is reversed for Pt in contact with TiO_x particles.[233,248] In the case of metal atoms on bridging oxygen sites (Table 8.3, minimum support models) a very small electron transfer *to* the metal has been predicted,[227,249,251] although the "support" is fully oxidized. Hence, with respect to low-temperature-reduced metal catalysts, the question arises whether the XPS chemical shift indicating a slightly positive metal phase (vide supra) is due to incompletely reduced metal clusters, at least in part.

The Structure of Interfacial Bonding. For models exhibiting a single metal–support bond, three groups can be differentiated with respect to the bond strength (Table 8.5, $WB_{Ni\text{-}S}$): (i) The type C models reveal a strong intermetallic chemical bond. This is connected with a doubly occupied nickel-support σ-bonding MO in the gap,[75,203] which can be interpreted as an occupied interfacial state. It is caused by the interaction of the nickel atom with the half-filled surface state (OSS, Table 8.4) in the band gap of the clean support clusters. (ii) The previously mentioned doubly occupied interfacial state is also present in models having radical-like oxygen at the interface (Aa, A′a); a strong Ni–O bond is formed. (iii) The interfacial bond is distinctly weaker for the other type A models revealing fully oxidized cations at the interface. The nickel-support bonding MO in the gap is singly occupied[75,203] leading to lower Wiberg bond indices. The same conclusions can be drawn from the binding energies of the nickel atom (not given here since they do not refer to CNDO equilibrium geometries, vide supra).

It is instructive here to consider calculations of single metal atoms interacting with support surfaces that have been performed using a minimization of the total interaction potential,[268–270] the pseudopotential (PP) *ab initio*,[195,196] and the SCF–LSD[226] cluster approaches. With these methods more reliable binding energies and even equilibrium distances may be expected. The surface has been kept rigid in some of these calculations,[195,196,270] whereas surface relaxation is accounted for in the others.[226,268,269] For completeness also those papers simulating the support by a "minimum support model" (OH,[291] H_2O[227,249,251]) are included. Selected results are reviewed in Table 8.6. Obviously, strongly and weakly binding surface sites are present. The former are point defects (vacancies) on the surface, and the latter are surface sites on the perfect surface [bridging oxygen atoms, closed-shell ions (O^{2-}, Cl^-, Mg^{2+}) on strongly ionic surfaces] establishing van-der-Waals-type bonding to the metal species. Intermolecular correlation (dispersion) has been shown to be of essential importance for stabilization on bridging oxygen sites.[249] Strong chemical bonds are formed also between metal atoms and radical (open-shell) surface oxygen atoms[195,291] or silicon dangling bonds.[226] The radical oxygen atoms are more reactive than their closed-shell counterparts on ionic surfaces. At step sites the bond strength is found to be intermediate.[269] Hence, we have to add to the three groups mentioned previously a fourth one: (iv) Metal atoms form weak van-der-Waals-type bonds on bridging (closed-shell) oxygen atoms. In this case the bond distance is appreciably longer. The same conclusion has been drawn by Anderson et

Table 8.6. Binding energies and distances of single metal atoms M interacting with support models; selected results.

Metal	Support	Surface Site	ΔE_M (eV)	$r_{M\text{-support}}$ (pm)	Reference
Pd	MgO	Mg^{2+}	0.35	235	270
		Mg^{2+} vacancy	2.68	193	270
		O^{2-} vacancy	4.60	124	270
Cu	MgO	O^{2-}	0.04	423	196
	MgO	Mg^{2+} vacancy	9.4	135	196
Ni[a]	Al_2O_3	On hollow over an O_3 mesh	4.7	212	195
Ag	NaCl (flat surface)	Na^+	0.62	251	269
		Cl^-	0.30	346	269
	NaCl (stepped surface)	Near Na^+ on the lower terrace	1.10	260	269
Cu	Si(111)	On-hollow (three-fold)	3.99	234	226
Ag	Si(111)	On-hollow (three-fold)	3.12	267	226
Ni[a]	OH[b]	O	1.76	196	291
Ni[a]	H_2O[c]	Obr[d]	0.47	225	249[b,c]
Cu[a]	H_2O[c]	Obr[d]	0.35	230	249[b,c]
Cu[a]	H_2O[c]	Obr[d]	0.20	226	251
Ni[a]	H_2O[c]	Obr[d]	0.37	217	251

[a] The quantitative results depend on the electronic state considered.
[b] Ni. . .OH linear.
[c] Ni. . .OH_2, Cu. . .OH_2, C_{2v}.
[d] Obr: bridging oxygen atom.

al.[245,247] recently: Pt and Ni interact strongly with Al^{3+} cations and radical surface oxygen atoms, while their interaction with closed-shell oxygen atoms terminating the surface is weak and results in longer bond distances.

Summarizing, we conclude that the four groups of bonding situations are likely to represent the findings reviewed in the experimental section: Group (i) corresponds to the SMSI state of metal–support systems; strong intermetallic bonds are formed as in intermetallic compounds. Group (ii) resembles a system exhibiting incompletely reduced metal species ("redox metal–support interaction"). Extending these models by more metal atoms would produce structures typical for metal catalysts in the WMSI state. Group (iii) should roughly describe the behavior of WMSI systems as well. Group (iv) characterizes the situation met with metals on ceramics, inside zeolite cavities, and on defect-free surfaces exhibiting "inert" surface sites like the bridging oxygen atoms.

Miscellaneous. The support influence on the magnetic properties has been carefully investigated in only one paper (Table 8.3).[231] In particular, the magnetic moments of nickel atoms directly in contact with oxygen atoms on the Al_2O_3 support are drastically reduced.[231] A strong mixing between Ni_{3d} and O_{2p} brings about a destabilization of "a level with significant *d* character hence provoking a transfer from up to down-spin."[231] Corresponding investigations on SMSI systems are missing. For Ni_3Ti spin-polarized SCF–$X\alpha$–SW calculations[122] showed also a reduction of the Ni magnetic moment from their bulk value of $0.5\mu_B$ to $0.31\mu_B$. A "strong loss of magnetic moment in the surface layer" has been predicted using the full-potential LAPW method also for a carbidic carbon phase on Ni(100).[292]

The short-range character of the electronic support effect has been demonstrated[239] investigating one, two, and three nickel layers on graphite (Table 8.3). An appreciable support effect on the local density of states is found in the case of a nickel monolayer, but the effect is "vanishingly small" for two or more layers.[238,239] This result is in accordance with the majority of experimental findings (cf. the experimental section) suggesting a rather localized influence of partly reduced oxides below or upon the active metallic phase on the electronic (and chemisorptive) properties. Hence, in theoretical calculations on the influence of the support it should be sufficient to consider flat metal clusters.

Another valuable tool of geometry prediction should be mentioned. It is the calculation of densities of states of different model structures and their comparison with photoelectron spectra of the respective real system. This approach is mostly applied for transition metal–semiconductor interfaces.[43]

The Support Influence on Chemisorption

The gas-phase species are supposed to approach the nickel–support system vertically from above. A bond is established due to the interaction of the frontier orbitals of both subsystems. For example, in the case of H adsorption, these are the occupied (spin-up) and empty H_{1s} orbitals, occupied $Ni_{3d_{z^2}}$ and Ni_{4s} orbitals, and empty Ni_{4p_z} orbitals. The Ni orbitals are more or less "contaminated" with contributions from "support orbitals." The main contribution to the Ni–H bond order comes from the Ni_{4s}–H_{1s} interaction. These bonding features will not be discussed in detail for each model. Rather we will focus our attention, referring to the questions raised in the summary of the section titled "The Support Influence on Chemisorption," to the origin of the support effect on binding energies, bond strengths, and activation barriers of dissociation. The main emphasis is laid again on a comparison of different types of models. The binding energies have been obtained using Eq. (8.3) (vide supra). An overview on the theoretical results concerning the support influence is given in Table 8.7.

Atomic Hydrogen Adsorption. The dependence of H binding energy ΔE_H on the electron transfer between support and metal is shown in Figure 8.7. We restrict the presentation to models exhibiting a single metal–support bond. Obviously, the correlation is not uniform for all models, but the models fall into three categories, with one set (i) identical to type C models, the second set (ii) identical to models Aa and A′a, and the third set (iii) identical to the other type A models. Interestingly, these three categories are the same as the first three groups introduced in the foregoing section in connection with the structure of interfacial bonding. In all three groups the binding energy is roughly proportional to the electron transfer to nickel. This increase of ΔE_H is further designated as change I. However, comparing H binding energies of models of the oxidized support surface with those of the reduced surface a distinct decrease of this quantity is observed (change II). This decrease is accompanied by a noticeable transfer of electrons to nickel (cf. the preceding section).

It is obvious from the Wiberg bond indices of the Ni–H bonds (Table 8.8) that the local Ni–H bond strength is only slightly changed (with the exception of models Cdpy and C′dpy, vide infra). Recalling the discussion on Eqs. (8.3) and (8.4) in the theory section, ΔE_H contains not only information on making and breaking metal–hydrogen bonds, but also on strengthening and weakening metal–metal and, obviously, metal–support bonds. It should be added that in the models considered here the same Ni orbitals are appropriate to interact both with the H and with the "sup-

Table 8.7. Support influence on chemisorption in transition metal–support systems: theoretical results.

Metal	Support	Adsorbed Species (on top)	Results[a]	Method[b]	Reference
Ni	SiO_2	H	Mostly lowering of ΔE_H, support–orbital involvement in the Ni–H bonding MOs	cf. Table 8.3	201
Ni	TiO_2, SiO_2, Al_2O_3	H	Lowering of ΔE_H, see text	cf. Table 8.3	75, 76, 203
Ni	ZnO	H	Lowering of ΔE_H with increasing metal–support interaction, simultaneously increasing Δq to the support	GFT, TB approach, Newns–Anderson model: one-dimensional model, Ni layers on semiinfinite ZnO	237
Ni	SiO_2	H	Lowering of ΔE_H compared with NiH ($X^2\Delta$) from ≈1.9 eV to ≈1.75 eV	*ab initio*–UHF: H. . .NiO (MSM)	291
Ni	SiO_2	H	Lowering of ΔE_H compared with NiH ($X^2\Delta$) from 2.63 eV to 1.46 eV	PP MRD–CI, SR CI: H. . .$NiSiH_3$ (X^3E, MSM)	293
Ni	Al_2O_3	CO	Decrease of $BS_{Ni\text{-}CO}$ as inferred from the position of the Ni–CO 5 $\tilde{\sigma}$ antibonding level	cf. Table 8.3	231

Table 8.7. (Continued).

Metal	Support	Adsorbed Species (on top)	Results[a]	Method[b]	Reference
Pt	TiO, FeO dopants	CO	Tilting of CO: lowering of CO vibrational frequency and increase of ΔE_{CO}. Untilted CO: increase of BS_{C-O} and decrease of BS_{C-Pt} (TiO) but increase of ΔE_{CO} (TiO)	cf. Table 8.3	248
Pt	ZnO dopant	CO	No tilting of CO, no change of BS_{C-O}, slight increase of BS_{C-Pt}, decrease of ΔE_{CO}	cf. Table 8.3	248
Rh	Al_2O_3, TiO_2	2 × CO, dicarbonyl	Increase of BS_{C-O} for TiO_2 as a support compared to Al_2O_3, BS_{Rh-C} not changed	EHMO: Al_2O_{10}, Ti_2O_{10}, 2 × Rh $(CO)_2$	256
Ni	SiO_2, TiO_2 (fully oxidized)	CO	Slight decrease of ΔE_{CO} compared with Ni–CO, see also Table 8.9	CNDO: Figures 8.2 and 8.3, Ni_1	294
Ni	SiO_2, TiO_2 (partly reduced)	CO	Drastic decrease of ΔE_{CO} compared with Ni–CO, see also Table 8.9	CNDO: Fig. 8.3, Ni_1	294

Pd	Ligands: $2 \times H_2O$	H_2	The ligand influence stimulates H–H bond breaking	*ab initio*–CASSCF–SDCI: complexes PdH_2, $(H_2O)_2 PdH_2$ (MSM)	295
Ni, Rh	SiO_2, Al_2O_3, TiO_2	CO (bridge position)	The energy of the transition state of CO dissociation increases in the order Ni < Rh and $TiO_2 < Al_2O_3 < SiO_2$	EHMO: $(Si,Al)_2O_7$, Ti_2O_{10}, Rh_2, Ni_2	254
Ru, Pt	SiO_2	CO, coadsorbed H, CH_2, CH_3, C_2H_5 (several positions)	Ru yields a larger weakening of BS_{C-O}, coadsorbed H weakens BS_{C-O}	EHMO: Si_2O_5, Ru_2, Pt_2, Ru–Si, Pt–Si bonds	255
Ni	Electronegative support	CH_4	CH radical occurs on an on-top site, 3 H atoms bound to the metal	EHMO: Ni_5, 5 pseudoatoms with varying IP represent the support	296
	Electropositive support	CH_4	CH_2 radical formed, 2 H atoms bound to the metal		

[a] BS: Bond strength. See also footnote a of Table 8.3
[b] IP: Ionization potential.
MRD–CI: Multireference single and double excitations–configuration interaction.
MSM: Minimum support model.
SR: Single reference.
TB: Tight binding.
See also footnote b of Table 8.3.

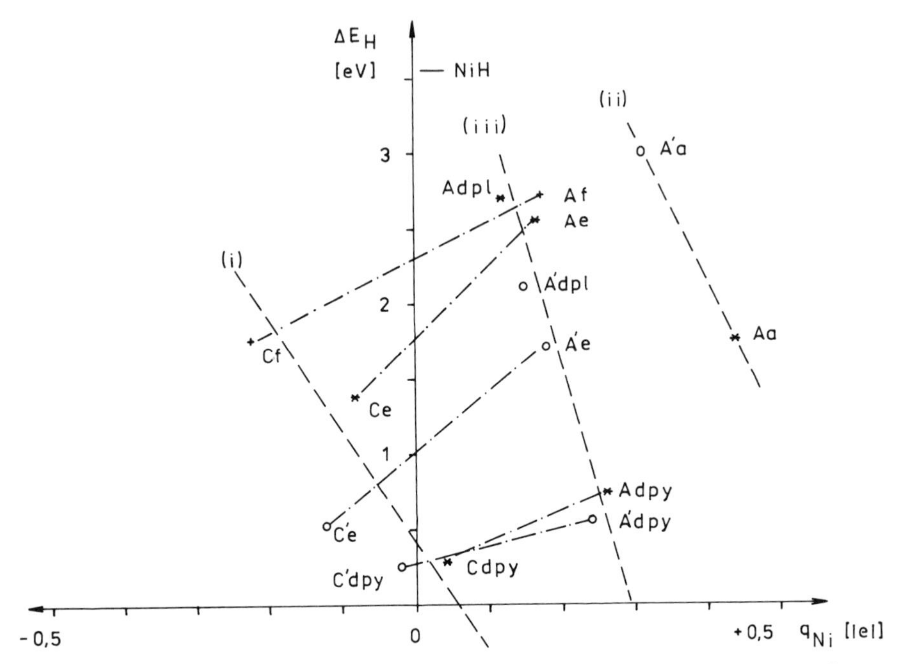

Figure 8.7. The H binding energy ΔE_H related to the electron transfer. Dashed lines indicate change I and dash-dotted lines change II of ΔE_H. Circles, stars, and crosses designate Al_2O_3, SiO_2, and TiO_2 models, respectively. For further explanation see text.

port'' orbitals of proper symmetry. Hence, we compare the Wiberg bond indices of the nickel–support bonds both without (Table 8.5) and with (Table 8.8) H on top of the nickel. This decrease (last column of Table 8.8) is much more pronounced for models Aa, A'a, Ce, Cf, and C'e, which exhibit a doubly occupied interfacial state and a strong bond in the adsorbate-free systems in contrast to the WMSI models that already have a relatively weak metal–support bond before chemisorption. The diminution of ΔE_H in the SMSI models Ce, Cf, and C'e compared with their isomorphous WMSI counterparts (change II) may thus be explained by assuming that the support makes the nickel less reactive against hydrogen owing to a strong nickel–support bond in the former cases. Approaching hydrogen atoms not only establish the Ni–H bond, but also weaken the metal–support bond to a larger extent than in the WMSI models. Consequently, the corresponding loss of energy is not as large in the latter models. Effects of strongly bound chemisorbates on the bonding in substrates are experimentally well known (cf. the experimental section).

Table 8.8. Support influence on atomic H adsorption (adapted in part from Refs. 75, 76, 203).

Model	Multiplicity	r_{Ni-H}[a] (pm)	ΔE_H (eV)	WB_{Ni-H}[b]	WB_{Ni-S}[b]	ΔWB_{Ni-S}[c]
Ni atom	Doublet	159	3.59	0.89	—	—
Aa	Triplet	160[d]	1.77	0.89	0.33	−0.51
A′a	Triplet	160	3.06	0.82	0.38	−0.59
Adpl	Doublet	160[d]	2.70	0.90	0.17	−0.27
Adpy	Doublet	160[d]	0.74	0.92	0.22	−0.37
Ae	Doublet	160[d]	2.54	0.89	0.24	−0.21
Af	Doublet	160[d]	2.73	0.84	0.07	−0.19
A′dpl	Doublet	159	2.12	0.87	0.14	−0.27
A′dpy	Doublet	160	0.56	0.85	0.16	−0.36
A′e	Doublet	160	1.72	0.86	0.11	−0.26
Cdpy	Triplet	203	0.30	0.06	0.93	−0.01
Ce	Triplet	160[d]	1.38	0.86	0.40	−0.50
Cf	Triplet	160[d]	1.76	0.84	0.31	−0.64
C′dpy	Triplet	203	0.25	0.06	0.95	−0.02
C′e	Triplet	163	0.53	0.72	0.43	−0.49

[a] Distance Ni–H.
[b] Wiberg bond indices of the nickel–H and nickel–support bonds, respectively. The latter refer to Ni–O bonds for models Aa, A′a.
[c] Change of WB_{Ni-S} as induced by H adsorption (see text).
[d] Not optimized (cf. Ref. 203).

The situation is different for models Cdpy and C′dpy in that the CNDO optimization yields a considerably increased Ni–H bond length.[75,76,203] The drastic decrease of ΔE_H is due to a weakening of the local Ni–H bond strength, while the Ni–support bond strength is practically the same (Table 8.8).

Change I, outlined above, occurs in all three groups of models. An increasing electron transfer to the support is accompanied by increasing contributions of "support orbitals" to the nickel–hydrogen bonding orbitals. This contribution is characterized by an index $(1 - S)$, S being zero for zero contributions of the support.[75] A correlation of this index with ΔE_H (Fig. 8.8) results in a nonlinear decrease of the latter quantity with increasing "support contributions." Type A models and type C models are clearly distinguished in Figure 8.8 owing to change II as discussed previously. However, they do not differ with respect to the amount of "support orbitals" contributing to the Ni–H bonding MOs. The behavior of model A′a is not well understood in that it exhibits a smaller electron transfer to the support but slightly larger contributions of support orbitals compared to model Aa. But, of course, all these interpretations are quali-

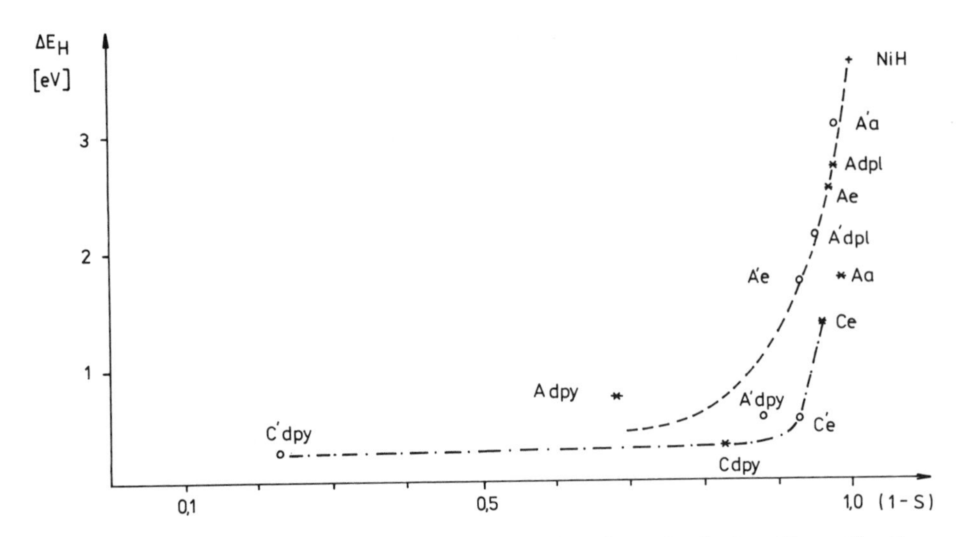

Figure 8.8. The H binding energy ΔE_H related to the index $(1 - S)$. S represents the amount of support orbitals involved in the Ni–H bonding orbitals. Circles and stars designate Al_2O_3 and SiO_2 models, respectively. Dashed and dash-dotted lines connect type A and type C models, respectively.

tative and phenomenological in nature and may serve as a guide to get a feeling of the mechanisms of electronic support effects.

Both changes of ΔE_H observed here are of different character. Change I is continuous and equivalent to the formation of oxidic nickel as in NiO, which is well known not to show any hydrogen adsorption (see, e.g., Ref. 61). This finding is in line with the theoretical results of Kunz,[242] who concluded the presence of Ni_{4sp} hybrid orbitals in nickel on SiO_2 to be responsible for the reactivity of the system against hydrogen. These orbitals have not been found in NiO.[242] The support effects in WMSI and "redox–MSI" systems may originate mainly from change I. Change II, yet, is discontinuous like the change in chemisorption properties of the SMSI system when switching from "normally" reduced to the high-temperature reduced state. However, the suppression of hydrogen adsorption is not necessarily caused by the diminution of binding energy of atomic hydrogen. There are some indications in the literature that hydrogen adsorption energies on SMSI catalysts can even be increased (cf. the experimental section). Hence, we will briefly discuss the support effects on activation barriers of H_2 dissociation below.

Calculations by means of the highly sophisticated PP MRD-CI method have been performed recently[293] using the silyl radical as a "minimum support model'' (vide supra). The result given in Table 8.7 is in qualitative accordance with the CNDO/2 result on model Cdpy in that a drastic decrease of ΔE_H compared with the NiH molecule is obtained. In this study a fixed Ni–Si distance was used (228 pm, optimized for $NiSiH_3$ *). The optimized Ni–H bond lengths in NiH and $HNiSiH_3$ are 144 pm and 156 pm, respectively.†

Comparing models of different chemical composition of the support (SiO_2, Al_2O_3, TiO_2) it is found that, as with the electron transfer, the results concerning chemisorption of atomic hydrogen are qualitatively identical. This is caused by the use of structural isomorphous models and should not be confused with the occurrence of the respective chemisorption properties in the real catalysts under the same conditions of preparation and reduction. Rather, identical chemisorption properties occur under different conditions of preparation and reduction (vide supra). Quantitative differences due to the chemical nature of the support (Fig. 8.7) should be taken with caution, since *estimated* bond distances at the interface are used throughout.

Considering Ni–Ni(–H) vertically on top of the various support surface sites,[75] we recall first the discussion on bare Ni_2 on top of the SMSI models (section titled ''Structural Models from Calculations''). Considerably weakened Ni–Ni bonds due to the formation of strong Ni–support bonds render the vertical Ni_2 configuration unlikely. Similar arguments may hold for the redox–MSI models Aa and A′a. Hence, we restrict the discussion here to the WMSI models. In these cases the lowering of the Ni–Ni bond strength is not as drastic as for the other models. An approaching H atom further diminishes this bond strength (demetallization), but, practically, it does not affect the Ni–support bond. Simultaneously, a Ni–H bond is formed with about the same local bond strength as in unsupported Ni–Ni–H. The energetic effects connected with forming and weakening of the different bonds largely compensate each other resulting in a nearly unchanged ΔE_H as against Ni_2H.

Finally, we mention a work[237] on H adsorption at a Ni/ZnO system (Table 8.7). Increase of the metal–support interaction ''draws charge from the Ni film into the ZnO support at the expense of that transferred to the hydrogen adatom, thus resulting in a lowering of ΔE.''[237] This observation is in accordance with our change I outlined previously; the model with Ni in contact to Zn^{2+} resembles our WMSI models. Another finding of this work is the short-range character of the support effect: it is practically absent for three and more Ni layers.[237]

---

* See Note Added in Proof and Ref. 351.

† The values in this paragraph have not been corrected for the basis set superposition error (BSSE).

Summarizing, we conclude that the H binding energy is diminished by the support influence, but this diminution cannot be simply attributed to electron transfer between the support and the metal. Two effects have been observed, which may serve as a qualitative explanation: First, turning from WMSI to SMSI models we found a discontinuous decrease of ΔE_H related to the formation of a strong nickel–support bond diminishing the reactivity of nickel against hydrogen. Second, a continuous decrease of ΔE_H is connected with increasing contributions of "support orbitals" to the Ni–H bonding orbitals in SMSI, redox–MSI, and WMSI models. In both cases the lowering of ΔE_H is closely related to the weakening of Ni–H bonds or Ni–support bonds or both.

The results are in qualitative agreement with the common picture of the support influence based on experimental findings (cf. the experimental section). It should be realized, however, that we treated mainly single-metal-atom models and exclusively H in on-top geometry. A more complete elucidation also requires consideration of bridge and on-hollow positions of the adsorbate. On the other hand, from a comparison of the theoretical results with experimental observations on the H_2 desorption from modified nickel surfaces[31,159] (cf. also the experimental section) the conclusion appears to be contradictory. Electropositive (electronegative) atomic modifiers donating (withdrawing) charge to (from) the nickel increase (decrease) the hydrogen desorption energy, whereas our electron-donating SMSI models do the opposite. However, as previously outlined there is *no* straightforward correlation between electron transfer and H binding energy in our nickel–support models. Comparing, e.g., the experimental result on electropositive and electronegative modifiers with change I (Fig. 8.7) discussed previously, we arrive at the same trend. Imagine, however, a fictitious open-shell modifier D and its singly ionized closed-shell cation D^+. The former reveals a reactive half-filled dangling bond that produces a strong partly ionic covalent bond when interacting with the "open-shell" metal. The electron density of the shared electron pair is shifted to the metal owing to the electron-donating character of D inducing, say, a positive net charge of $(1 - \delta)$ on the modifier. D^+, however, may have electron-withdrawing character but, owing to its closed-shell nature, a weaker bond to the metal is established and a single electron is shared (cf. vide supra, the WMSI models). The electron density is shifted to the modifier also leaving a positive net charge $(1 - \delta)$ on it. This is approximately the situation we have described by the type C models and their oxidized isomorphous counterparts.

Two related theoretical findings concerning the effect of modifiers should be mentioned here in addition. PP MRD–CI calculations have been performed aimed at the influence of Li on the ScH, CuH, and PdH molecules.[297] Considering the lithide molecules without H atoms, the Li atom makes Pd slightly positive but Sc and Cu negative.[297] Weak bonds are found in PdLi and ScLi and a strong bond in CuLi.[297] The H binding energy is drastically decreased compared with the monohydride molecule in H–Cu–Li, but it is only slightly decreased in H–Pd–Li and it is even increased in H–Sc–Li.[297] Thus, as in the case of our different types of models the change of ΔE_H is not simply related to the electron transfer. Rather, the strength of the bond between metal and modifier and the availability of orbitals exhibiting the appropriate symmetry for interacting with the adsorbate play a significant role.[297]

The second example is delivered by the work of Shustorovich[298] (cf. also Chapter 9 of this volume). Using bond order conservation arguments and Morse-type potentials between adatoms and substrate atoms, he points out that modifiers D, present either as atoms or as parts of monolayers, should always decrease the binding energy of atomic adsorbates A. This is valid for bridge, on-top, and on-hollow sites.[298] It is argued that the observed increase in heats of adsorption "is a clear indication of the prevailing direct A–D attraction."[298] The latter is nicely reflected by the results on the Pd–Li system obtained in the work mentioned above[297]: Contrary to the H–Pd–Li structure both H in a bridge position and on the Li side are more tightly bound than in the Pd–H molecule.[297]

Molecular CO Adsorption. The binding energy of a CO molecule* vertically on top of various Ni–support models is given in Table 8.9. The C end is directed toward the metal. The binding energy is slightly decreased for type A models, but it is drastically diminished for type C models. This result is qualitatively in agreement with the experimental findings for most catalytic systems (cf. the experimental section). Comparing the models Ae with Ce and Af with Cf it is obvious that as in the case of H adsorption the lowering of nickel–support bond strength is involved, being more pronounced in type C models (compare the last columns in Tables 8.5 and 8.9). For type A models ΔE_{CO} is only slightly decreased. While the loss in binding energy in models Ae and Af seems to be mainly due to a further diminution of the nickel–support bond strength accompanying the adsorption, the situation is different in the redox–MSI models Aa, Ab, and Ac. Here the nickel–support bond strength is practically unchanged, whereas a distinct weakening of the Ni–C bond is energetically compensated in part by a strengthening of the C–O bond. A closer inspection of the Ni–C and C–O bond indices shows that the former increases with increasing negative charge at the nickel caused by the support influence,

* The numerical values of ΔE_{CO} are too large throughout compared with the experimental results as expected with the CNDO method. Hence, we discuss relative magnitudes and trends.

Table 8.9. Support influence on molecular CO adsorption.[a]

Model	Multiplicity	ΔE_{CO} in eV	$WB_{Ni\text{-}C}$[b]	$WB_{C\text{-}O}$[b,c]	$WB_{Ni\text{-}S}$[b]
Ni atom	Singlet	6.17	1.12	2.33	—
Aa	Doublet	5.49	0.85	2.51	0.85
Ab	Triplet	5.17	0.82	2.50	1.02[d]
Ac	Triplet	6.04	0.86	2.49	1.08[d]
Ae	Singlet	5.56	1.10	2.34	0.17
Af	Singlet	5.44	1.10	2.34	0.05
Cdpy	Doublet	3.77	0.98	2.42	0.76
Ce	Doublet	3.87	1.09	2.33	0.50
Cf	Doublet	3.81	1.15	2.28	0.29

[a] Fixed Ni–C (176 pm) and C–O (113 pm) distances.
[b] Wiberg bond indices of the Ni–C, C–O, and Ni–support bonds, respectively. The latter refer to Ni–O bonds for models Aa, Ab, and Ac.
[c] The Wiberg bond index of the free CO molecule equals 2.61.
[d] Sum of the indices of all Ni–O bonds.

while the latter simultaneously decreases (Fig. 8.9). Taken together with the ΔE_{CO} results, it discloses that for on-top adsorption a lowering of the binding energy is not necessarily connected with an increase of the C–O bond strength and vice versa. Rather, we find a simultaneous drastic decrease of the CO binding energy on the metal and a slight lowering of

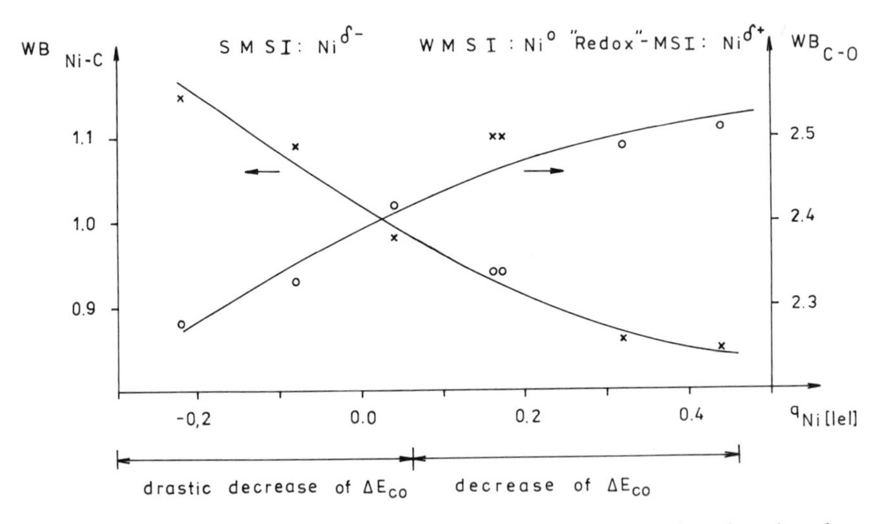

Figure 8.9. Wiberg bond indices of Ni–C (x) and C–O (o) bonds related to the Ni net charges of adsorbate-free nickel–support models.

the C–O bond strength with increasing negative charge at the nickel species (type C models). However, the latter decrease is too small to make enhanced CO dissociation understandable.

Keeping in mind that we are dealing with linearly bonded CO, we look at the experimental results. After HTR a downward shift of the C–O stretch frequency has been reported[175] in only one case, while it is not changed in other cases (vide supra). Experimental results concerning the metal–C stretch frequency are not available for the SMSI state (vide supra), neither are measurements aimed at vibrational frequencies of metal species on the support.

Comparing the C–O bond strengths for all models (Fig. 8.9), we find the known dependence on the oxidation state of nickel (e.g., Ref. 168) well reflected: The C–O bond is strengthened for redox–MSI models as against unsupported Ni–CO, but is still lower than in free CO (cf. Table 8.9). For SMSI models it is in the range of that of Ni–CO. The WMSI models Ae and Af are not in line with the other types of models as far as the Ni–C and C–O bond indices are concerned. The nickel–support bond is almost broken due to the CO chemisorption (Table 8.9) indicating a carbonyl formation. Both bond breaking and carbonyl formation are well known experimentally (vide supra). Interestingly, a diminution of carbonyl formation has been reported to occur in the SMSI state of Ni/TiO_2.[51] We have to concede, however, that our results on the slight changes of bond strengths should not be taken too literally, since they are obtained using rather simplified models and a semiempirical SCF method.* The main point to emphasize is (as with the H adsorption) that the weakening of different bonds including the metal–support bond must be taken into account for an understanding of support effects.

At this point it is useful to consider first the respective theoretical results on the influence of modifiers before continuing the review of support effects. Electropositive additives increase the CO binding energy and electronegative additives decrease it in agreement with experimental findings.[232,235] This has been explained in terms of a decrease of the "local" work function owing to an electron transfer from the co-adsorbed electropositive modifiers to the metal.[232,235] As pointed out by Rodriguez et al.[232]

---

* Using a nonempirical SCF method for CuCO two major effects cancelling each other in part have been detected[299]: A large repulsion "arising from the overlap of the CO 5σ and Cu $4s$ electrons" increases the C–O stretch frequency, while the metal to CO $2\pi^*$ back donation decreases it to a larger extent. The increase due to the CO σ donation to Cu is minor compared to the former effects. The π back donation is quantitatively correctly described only if correlation effects are taken into account. It is worthwhile to note in this context a recent statement of Bauschlicher et al.[300] who pointed out that even the systems Ni_5CO and NiCO "might well be used in model calculations" aimed at surface problems; "however, they must be used cautiously, since they clearly do not represent all aspects of the metal surface."[300]

for CO on Cu clusters, consequently, the metal–C bond strength increases slightly and the C–O bond strength decreases caused by an enhanced π back donation. While the conclusion concerning the bond strengths is consistent with our results on the SMSI models (cf. Table 8.9), we find a lowering of ΔE_{CO}. Interestingly, IR and accompanying evacuation experiments[187] disclosed a more strongly adsorbed CO at Cu on electron-withdrawing ZrO_2 and Cr_2O_3 supports (cf. the experimental section) in agreement with our results (compare WMSI and redox–MSI models with SMSI models). On the other hand, Tanaka et al.[301] using UPS found no CO adsorption on nickel on electron-donating SiO_x/n-Si(111). In our opinion the different amount of lowering of the binding energies of adsorbates is closely connected with a modification of bonding in the substrate (metal + support) rather than with the direction and amount of charge transfer (cf. also the discussion on H adsorption in the foregoing paragraph).

To be somewhat more specific we briefly discuss, based on the jellium model as given by Lang et al.,[302] the electrostatic explanation of the effects of modifiers with respect to our results. In the critical range of distance (i.e., at about the equilibrium distance between metal and molecule) the electrostatic potentials brought about by an electropositive or an electronegative modifier have opposite signs resulting in a stabilization of the electron-accepting molecule in the former case and in a destabilization in the latter.[302] Both in our WMSI and SMSI models, however, the Ti cation when already brought into contact with nickel has about the same positive charge or is even slightly more positive (recall that in Table 8.4 only cation charges at nickel-free surfaces are given) in the electron-withdrawing WMSI models. Hence, a dipole between the support and the metal is established with its negative end toward the metal in either case. So the difference in the contribution of the electrostatic energy as caused by the WMSI and SMSI models may be small compared with the energy loss resulting from a weakening of the metal–support bond. Interestingly, Lang et al.[302] note that the jellium model is less appropriate for treating electronegative modifiers since they as a rule interact more strongly with the metal (as our "electropositive" SMSI models do). Thus, we state that conclusions well established for the modifying effect of electropositive and electronegative additives on metals must not be simply extended to the electronic support effect as one might expect at first glance.

The involvement of both through-metal and through-space influences of modifiers on the electronic state of nickel surface sites[303] and on the properties of adsorbed CO[232] has been pointed out recently. Through-metal influences are more short range in character due to screening.[232,303] Another interesting aspect has been elucidated by Feibelman and Hamann.[304] The valence charge density of a Rh film covered with a sulfur adlayer is influenced only at Rh atoms adjacent to a sulfur atom indicating the extreme short-range character of this effect.[304] The charge transfer is found to be negligible; unexpectedly the work function even decreases (contrary to S-covered Ni), which has been ascribed to a rehybridization effect.[304] On the other hand, however, there is a drastic decrease of the local density of states (LDOS) at the Fermi level above second-neighbor Rh atoms.[304] "This decrease reflects the formation of S–Rh bonding and antibonding states and its long range is possible because a LDOS is not a screened quantity."[304] This finding is in line with our results in that it emphasizes the role of bond formation and frontier orbitals (cf. the interfacial states in different support models) rather than the role of an electron transfer. Similar conclusions have been drawn for sulfur on nickel, although in this case the charge depletion from the CO binding site is slightly more pronounced.[305] The direct result of a rehybridization of the nickel on-top site is "a reduction of the CO–metal interaction" (as elucidated by the contour plot of the 5σ orbital).[305] Li modifiers, however, influence mainly the $2\pi^*$ orbital; the latter is pulled down to the Fermi level and its occupancy is increased.[305] This is in agreement with the Ni–C and C–O bond order indices obtained in our investigations (Fig. 8.9).

Returning to the support influence we consider next the work of Anderson[248] (cf. Tables 8.3 and 8.7). Here TiO, FeO, and ZnO molecules are preadsorbed on a Pt cluster;[248] subsequently CO is adsorbed at an adjacent Pt site. Contrary to our models a direct interaction between the "support molecule" and the adsorbate is possible leading to an increase of ΔE_{CO} for an attractive interaction (TiO, FeO) and a decrease of binding energy for a repulsive interaction (ZnO).[248] As pointed out by Shustorovich[298] only direct attractive interactions with modifiers are likely to increase the binding energies of adsorbed molecules. In this case the attraction results in a tilting of the CO molecule with its O end toward the Ti cation accompanied by a lowering of C–O vibrational frequencies.[248] The latter fact is in line with explanations of different experimental results (cf. the experimental section), although an increase in ΔE_{CO} has not been observed in the SMSI state. These theoretical results should stimulate further investigations of different adsorption sites including various metal–support geometries. So, for example, the calculated C–O bond strengths of rhodium dicarbonyl species (Table 8.7) have been successfully correlated[256] with the respective IR frequencies[178] (cf. the experimental section). In this case Rh(I) is bonded to the support oxygen atoms as in our redox–MSI models. Finally, a paper concerning the magnetic aspects should be mentioned.[231] While after bringing a nickel cluster in contact with an alumina cluster a drastic decrease of the magnetic moments of interfacial nickel atoms has been found (vide supra), further changes due to CO adsorption are minor.[231] In agreement with our results on redox–MSI models the authors[231] predict a decrease of the Ni–CO bond strength that is likely to elongate the Ni–CO distance. Probably, alumina as support will lower this bond strength even to a larger extent than silica as support (compare the results on the Pt–C stretch frequencies given in the experimental section), indicating the stronger metal–support interaction in this case. Experimental results with respect to nickel are, unfortunately, not available.

The Support Influence on Elementary Steps of Surface Reactions

Theoretical investigations on this topic are generally restricted to the dissociation of molecules at supported or modified metals.

H_2 Dissociation. Usually the existence of a molecular precursor state is assumed in which the molecule is weakly adsorbed (cf. Fig. 8.1). There is experimental evidence for such a state in a lot of systems.[306] The respective desorption energy E_p has been determined to be only about 11 kJ mol^{-1} for deuterium (D_2) on Ni(111).[306] The dissociative adsorption proceeds with simultaneously approaching the surface and elongating the molecular bond distance thereby passing a transition state (destabilization of the system, activation barrier), which is characterized by progressive occupation of antibonding H_2 orbital. Finally, two H atoms are chemisorbed at adjacent surface sites. Height, shape, and location of the activation barrier have, analogously to gas-phase processes, decisive significance for the dynamics of the atomic rearrangement. If the barrier occurs in the early stage of the H_2 approach to the surface ("entrance channel"),[306] it can be overcome most easily by increasing the translational energy of the H_2 molecule relative to the surface while in the case of a late barrier (located in the "exit channel")[307] vibrational energy of the molecule is most effective. The detailed dynamics of these processes, even for clean metal surfaces, is still under discussion both from the experimental and the theoretical point of view[308] and no attempt has been made to investigate the support effect. Hence, we will focus our reflections here on static arguments based on electronic hypersurfaces or curves.

Necessary conditions for low or negligible dissociation barriers are (1) the availability of metal orbitals having an appropriate symmetry with respect to the symmetry of the orbitals of the molecule, and (2) not too large energetic separation of the pertinent energy levels.[162] Remember that both occupied and unoccupied frontier orbitals of the substrate and the molecule are involved in breaking the H–H and forming the metal–H bonds. Investigating H_2 dissociation on a Ni(100) surface, e.g., by means of cluster calculations using single and double excitation CI methods,

Siegbahn et al.[209] pointed out the importance of occupied Ni_{3d} orbitals suited by symmetry to interact with antibonding H_2 orbitals. An omission of these orbitals leads to large barriers[209] as in the case of *sp*–metal surfaces. A barrier of about 18 kJ mol^{-1} was found for an on-top approach to a Ni site, which is nearly identical to that for an isolated Ni atom.[209] However, H_2 dissociation at an isolated Cu atom, exhibiting a large *sd*-hybridization energy, requires about 20 times more energy.[309] This large activation barrier was found to be lowered drastically when taking into account "the Herzberg–Teller coupling of the $^2A_1$ state (which arises from the interaction of the Cu atom in its ground state and the H_2 molecule) with the $^2B_2$ state" emerging from an interaction of H_2 and the excited Cu (^2D) state.[210]

Turning now to the influence of ligands it has been shown[295] (cf. Table 8.7) that two water ligands at a Pd atom stabilize the d^9s^1 state as against the d^{10} ground state thus facilitating *sd* hybridization; consequently, covalent Pd–H bonds are formed and the H–H bond is completely broken in this oxidative addition reaction.[295] In case of single metal atoms bonded via oxygen atoms to an SiO_2 support, Johnson et al.[240] could discriminate two cases: Pt reveals singly occupied *d*-derived orbitals of proper symmetry in the energy range of the H_2 antibonding orbital thus facilitating dissociation. This resembles the situation met with our WMSI models. In the case of Ru, however, the corresponding *d* orbital is empty indicating a strong binding to the support; consequently, no dissociation should occur.[240] A similar conclusion was drawn by Horsley[233] with respect to his SMSI model. He pointed out that a singly occupied Pt_s–O_p-derived orbital in the gap does not show the appropriate symmetry to interact with the H_2 antibonding orbital.[233] In the case of our similar model Cf, the situation is slightly different in that the HOMO is doubly occupied (vide supra) but as in Horsley's model it does not possess the necessary symmetry for exerting a bond-breaking effect on the hydrogen molecule that approaches the surface in a parallel arrangement. However, as previously pointed out, the symmetry allows for binding H atoms though weakly due to the strong metal–support bond. At this point it must be emphasized that since in general the semiempirical methods used in several papers as cited previously are not suitable for the estimation of activation barriers, even qualitative predictions given here have to be treated with caution. A nice demonstration of the possibilities of the EHMO method in this context has been given by Saillard and Hoffmann.[230]

Summing up, the general picture derived from the available theoretical studies is in agreement with experimental findings (vide supra). It is interesting to note that activation barriers for H_2 dissociative adsorption have been found not only for catalytic systems (Table 8.1) and for clean metal surfaces of different types (cf. the experimental section), but recently molecular beam experiments gave explicit evidence that small quantities of oxygen adatoms on Ni(100) increase the barrier.[310]

As discussed by Shustorovich,[152,298] there is an interesting relation between the barrier height and the binding energy ΔE_H. Decreasing values of ΔE_H which may result from poisons[298] or from the support will increase the barrier height,* as one may intuitively conclude from Figure 8.1. Direct attractive interactions between molecules and promotors, however, are likely to lower it in special cases.[298]

It has been pointed out that even two barriers may exist, which H_2 has to cross to become dissociatively chemisorbed.[311] Besides a weakly adsorbed molecular precursor, there may be a more strongly bound molecular state (sorption energy E_{mol}) closer to the surface. Kinetic arguments show that the dissociation probability increases with a decrease of the activation barrier for reaching the strongly bound molecular state and/or with an increase of the well depth E_{mol}, which governs the trapping probability in this state.[311] Both effects may originate from a decrease of the "local" work function of the metal caused by promoters[311] or by the support. Hence, there are at least two different effects one should not mix up: First, the support (or some kind of modifier) "suppresses" metal states of proper symmetry to interact with the molecular frontier orbitals thus rendering dissociation unlikely. This effect does not depend on the direction of electron transfer. Second, from electrostatic arguments one reaches the conclusion that an electron transfer to the metal favors the adsorption and subsequent dissociation of acceptor molecules. Interestingly, both these effects have much in common with the explanations of changes II and I of ΔE_H in our various support models (vide supra). Since qualitative arguments are not suitable to decide about one against the other, quantitative calculations on a highly advanced level are needed.

CO Dissociation. Many of the arguments given in the preceding paragraph apply also to CO adsorption. The main differences are that CO is a heteronuclear molecule and that it is, as a rule, strongly adsorbed as a molecule (cf. the experimental section). On the clean Ni(111) surface, a molecular precursor state weakly adsorbed by about 0.4 kJ mol^{-1} has been found recently at a temperature of 6 K.[312] Heating a system of chemisorbed CO on Ni above 475 K yields dissociative chemisorption.[313] Arguments from a "static" quantum-chemical calculation for a given geometry are not able to explain the process of CO dissociation. From our investigations of the support influence on (molecular) CO chemisorption (vide supra), we concluded that in the SMSI models the C–O bond

* Strictly speaking the barrier height as defined in Figure 8.1 is not identical with the experimentally determined activation energy (cf. footnote of Table 8.1).

strength is only slightly decreased with increasing negative charge on the nickel. But this decrease is rather small and may not be sufficient to explain enhanced CO dissociation.

As pointed out in the context of H_2 dissociation predictions of the transition state geometry and the barrier height are not reliable on this level of description. It has been shown that the influence of modifiers on the activation energy of dissociation is rather complex for heteronuclear molecules.[298] Although direct attractive interactions may lower the barrier, a decrease of the chemisorption energies of the atomic constituents results in the opposite trend.[298] The work of Anderson et al.[248] (vide supra) has nicely demonstrated that a direct attractive interaction with a "support molecule" brings about a remarkable weakening of the C–O bond in a tilted geometry. At this point it is valuable to recall the experimental results on the intermetallic compounds.[40] CO reveals a very low sticking probability on $NiSi_2(111)$, "once bound, however, dissociation to surface bound carbon and oxygen occurs in a facile process."[40] This is a hint of an increase of the activation barrier prior to a strong molecular chemisorption state and to a simultaneous decrease of the activation barrier for dissociation compared with the nickel surface. In the case of a $Pt_3Ti(111)$ surface[39] the amount of CO adsorbed was found to decrease due to a dilution of Pt chemisorption sites with Ti and, simultaneously, the CO binding energy is lowered. CO dissociation is facilitated on parts of the surface. A calculation on the latter system has been performed by Mehandru et al.[314] by means of the ASED method. A decrease of binding energy of CO adsorbed perpendicularly to an on-top Pt site has been found for the $Pt_3Ti(111)$ surface compared with the Pt(111) surface.[314] If bound perpendicularly to an on-top Ti site, however, the binding energy is even increased.[314] It is argued that the latter site may be blocked due to preadsorbed CO, which is able to dissociate here leaving the oxygen on the site.[314] More interestingly, most stable is a "lying-down" configuration of CO on a Ti site with the C end toward the threefold hollow Pt site.[314] The stabilization originates both from 5σ and π donation to the surface.[314] This configuration is considered as a precursor state for the CO dissociation.[314] Elongating the C–O bond distance shifts the π^* levels down "thus increasing the interaction with the filled Pt + Ti d bands." At an elongation of 60 pm the transition state is reached exhibiting an activation barrier of about 120 kJ mol^{-1}.[314] Although the quantitative predictions made here should be taken with caution, the picture emerging from the theory is consistent with the experimental results. From an extrapolation to the SMSI state, an explanation is provided for both the decreased chemisorption capacity and the increased CO dissociation ability, the former by means of a combined geometric and ligand effect and the latter by a direct attractive interaction with parts of the support.

Attempts have been made to study also the support influence on the CO dissociation pathway for Ni and Rh on clusters of TiO_2, Al_2O_3, and SiO_2[254] (Table 8.7). Here CO has been located at the bridge site. Assuming the carbide mechanism for CO hydrogenation to be correct, the results on the energies of the transition states (Table 8.7) are in agreement with the experimental activity pattern. Hence, both fully oxidized[254] * and partly reduced[314] Ti cations are likely to facilitate CO dissociation. But a comparison of both types is not possible on this level of calculation.

Another work of the same authors[255] considers CO in different positions adsorbed on Ru/SiO_2 and Pt/SiO_2 (cf. Table 8.7). In these cases the metal atoms are directly bound to the cations of the support. Their formal oxidation state is not reported, but, since the models should be neutral and no change of the nuclear charge is indicated, we guess that the oxidation state is +3 as in our model Cdpy (cf. the footnote). Hence, they may be considered to be SMSI models though the authors[255] do not comment on them. CO is adsorbed in on-top configuration. A drastic lowering of the C–O bond strength as against the free molecule value has been reported, which is more pronounced for Ru than for Pt.[255] The C–O overlap population in unsupported Ru(Pt)–CO has not been given. Nevertheless, Ru favors CO dissociation compared with Pt due to the different characteristics of the respective HOMOs.[255] This is in line with the higher CO hydrogenation activity of Ru/SiO_2.[255]

CO Hydrogenation. Using small clusters and the generalized valence bond CI method, Goddard et al.[315] studied the thermochemistry of different mechanisms of CO hydrogenation on unsupported nickel. All hydrogenation steps in the CO insertion mechanism were found to be exothermic, whereas the first step of the carbide mechanism, namely, the CO dissociation, was obtained to be endothermic by 67 kJ mol^{-1}.[6,315] The reaction enthalpy is reduced, however, to +8 kJ mol^{-1} assuming a hypothetical C_2 dimer bound to four nickel atoms.[315] Neither the barrier heights of the different hydrogenation steps nor the support influence were calculated.

Frequently, it is assumed that CO dissociation is facilitated by means of hydrogen coadsorption.[6] This has been investigated for the previously mentioned Pt/SiO_2 model applying the EHMO method and results in a further drastic decrease of the C–O bond strength.[255] The comparison of the total energies of CH_2, CH_3, and C_2H_5 species adsorbed on the Pt/SiO_2

---

* It is not clear, however, from Ref. 254 whether the metal atoms are bound via oxygen or directly[255] to the support cations. In the latter case the formal oxidation state of the cation would not be +4 in SiO_2 and TiO_2 if the saturating pseudoatoms A had nuclear charges of +1.

and Ru/SiO_2 models leads the authors to the conclusion that the production of long-chain hydrocarbons is less favorable on Pt/SiO_2 compared with Ru/SiO_2, which agrees with experimental findings.[255] However, also in these cases neither support-free systems nor supports other than SiO_2 have been investigated.

Baetzold,[316] referring to the carbide mechanism,[6] has estimated the rate constants for the individual steps of hydrocarbon formation on unsupported group VIII metals from computed enthalpies of reaction of intermediate species. An empirical tight-binding method has been used.[316] Although hydrogenation of alkyl groups is found to be the rate-determining step, the heats of adsorption of H_2 and CO on metal surfaces are important parameters in this calculation, while the activation energy of dissociation critically influences hydrocarbon chain growth.[316] Baetzold's theoretical model throws some light also on possible support influences.

CONCLUSIONS

The contributions of theoretical calculations to the understanding of support effects will be restricted in the near future mainly to the elucidation of chemisorptive phenomena at least as far as definite quantum-chemical calculations are concerned (cf. section titled "Surface Modeling and Computational Methods"). The considerations on mechanistic aspects in connection with activity and selectivity (cf. the experimental section) should stimulate theoreticians to study different classes of appropriate models covering both WMSI and SMSI systems. Furthermore, empirical kinetic estimates such as those of Baetzold[316] should be extended to include also support effects. With respect to homogeneous catalytic reactions theory is in a better position.[208*d*,212,213] Optimizations of the geometry of transition states and calculations of activation barriers have become possible by means of highly sophisticated methods.[208*d*] Even ligand effects have already been considered.[208*d*] Using "minimum support models" there is some hope that progress will be made on this line also with respect to support effects.

The main results from hitherto existing theoretical investigations on support effects may be summarized as follows:

(i) The investigation of geometric support effects discloses the importance of the relaxation of support surfaces, temperature effects, and three-body interactions. Thermodynamic and quantum-chemical considerations elucidate the (partial) support reduction and alloy formation as well as the diffusion of suboxides and the flattening of metal particles, respectively, in the SMSI state. Full optimizations of the geometry of metal–support systems by means of methods yielding reliable total energies are still missing.

(ii) Based on experimental structure investigations hypothetical model structures are proposed. Although they are simplified with respect to the real situation, working hypotheses on the electronic support effect are obtained. These hypotheses have to be verified in two directions: First, more realistic structures (e.g., larger metal clusters and larger support fractions) have to be studied. Second, carefully chosen "minimum support models" and calculations on a highly sophisticated level are likely to get insight into the interface geometry and to check the results obtained from the simplified models by means of semiempirical calculations.

(iii) Comparing different classes of models in order to cover different experimental situations the following qualitative statements are offered with respect to the electronic support effect: (a) An electron transfer to the metal is calculated for the interaction of nickel with surface sites on reduced support surfaces. The direction of electron transfer is reversed for nickel interacting with completely oxidized surfaces. Its amount can be roughly correlated with the "electronegativity of clean surface clusters" and with the degree of support reduction. (b) The metal–support bond strength differs depending on the surface site on the support; it is strong for SMSI and redox–MSI models, intermediate for WMSI models, and weak for metals on perfect oxidic surfaces. (c) The support effect is short range with respect to total electron densities, but it is of intermediate range with respect to the characteristics of orbitals localized near the Fermi level and of the corresponding local densities of states. (d) The binding energies of atomic hydrogen and of CO are lowered for SMSI models with respect to their isomorphous WMSI counterparts, but these diminutions cannot be simply attributed to the electron transfer between the support and the metal. Rather, they are connected with a weakening of different bond strengths including the metal–support bond accompanying chemisorption. This weakening is more pronounced for SMSI models resulting in a larger loss of energy. Hence, in the latter models the metal is viewed as poisoned by an electronic support effect. (e) A second explanation for the decrease of H binding energies caused by the support in all types of models has been identified: The mixing of support orbitals to the metal–H bonding orbitals is closely related to a removal of electron density from the metal, which makes the latter similar to its state in oxides. (f) The diminution of CO binding energy is not necessarily connected with an increase of the C–O bond strength. There are indications that this bond strength may be even slightly lowered for linearly bonded CO, but this lowering is more pronounced for CO lying down on the surface. The latter is brought about by direct attractive interactions between support particles and the molecule, and not by an electronic "through-metal" support effect. (g) The effect of the chemical composition of the support (SiO_2, Al_2O_3, TiO_2) seems to be not as marked as that of the degree of support reduction. Isomorphous models of different composition show a similar behavior. Quantitative differences should be considered with caution when using estimated bond distances. (h) There are some indications from calculations on models with metal atoms tightly bound to the support that H_2 dissociation may be an activated process, although definite conclusions cannot be drawn at present. CO dissociation, however, seems to be favored on titania-supported metals compared with SiO_2 and Al_2O_3 supported ones. (i) The influence of modifiers is in some respect similar to support effects, but also reveals distinct differences. This once again sheds light on the unreliability of the electron transfer for explanations of energetic aspects of chemisorption.

Summing up, we conclude that theoretical work on the support influence at present is still in an early stage of development. Nevertheless, some useful results serving as guidelines for future investigations have been obtained. This future work has to check the working hypotheses in the directions previously pointed out. Furthermore, different chemisorption sites on the metal have to be considered, and the investigation of activation barriers of chemisorption by means of highly sophisticated methods is desirable. A permanent refinement of the models based on new experimental results is required. Proceeding in this way, valuable contributions of theory in this area of research may be expected, and we can state with Goddard[212] that "theoretical chemistry comes alive" reaching a position in which it will be a "full partner with experiment."

ACKNOWLEDGMENTS

The author is grateful to Professors M. Bourg (Marseille), S.G. Davison (Waterloo), Ş. Erkoç (Ankara), J. Fišer (Prague), S.G. Gagarin (Moscow), S.H. Garofalini (Piscataway), F. Herman (San Jose), R.V. Kasowski (Wilmington), S.G. Louie (Berkeley), H. Miessner (Berlin), J.K. Nørskov (Copenhagen), D.R. Salahub (Montreal), J. Sauer (Berlin), Xu Yin-Sheng (Dalian), B. Viswanathan (Madras), T. Yanagihara (Koriyama), and A. Zunger (Golden) for sending him preprints and reprints on their works. Thanks are due to Professor S.D. Peyerimhoff (Bonn) for placing at our disposal the MRD–CI program package. Valuable discussions and fruitful cooperation with Professor G. Pacchioni (Milan), Dr. J. Sauer (Berlin), Dr. F. Ritschl (Berlin), and R. Schlesinger (Berlin) are gratefully acknowledged. Furthermore, the author thanks Professor L. Zülicke (Berlin), Dr. D. Gutschick (Berlin), and Dr. H. Miessner (Berlin) for critical reading of the manuscript, A. Matelowski and I. Krüger for technical assistance in its preparation, and Professor G. Öhlmann (Berlin) for continuous interest in this work.

NOTE ADDED IN PROOF

Since the completion of the manuscript in summer 1987 a number of important papers on both experimental and theoretical aspects of metal–support interaction have appeared and it seems necessary to review at least some of them.

Experimental Work

The *effect of catalyst preparation* on structures, catalytic properties, and the occurrence of the SMSI effect has been elucidated for several systems including Ni/Al_2O_3[317,318] and Ni/TiO_2.[319] Whereas in the case of Ni/Al_2O_3 (prepared by wet impregnation) SMSI-type effects with respect to hydrogen adsorption have been observed even at reduction temperatures as low as 773 K,[317] which usually produce SMSI in systems supported on "reducible" transition metal oxides, no SMSI effect has been found in Ni/TiO_2 (ion-exchange).[319] The suppression of hydrogen adsorption on Ni/Al_2O_3 is attributed to a geometric blocking effect.[317] Since the nickel species are only partly covered with Al_xO_y,[317] one may ask, however, whether the diminution of hydrogen uptake could not at least in part be due to an increase of the *activation barrier of H_2 dissociation* originating from an electronic effect of Al_xO_y. The occurrence of such a barrier (67 kJ mol^{-1}) in the case of Ru/SiO_2 and Ru/Al_2O_3 has been ascribed recently to an electronic effect of adsorbed electronegative chlorine species "reducing the electron density at adjacent low coordination sites."[320]

Some new controversial results with respect to the *geometric structure of the interface* have been obtained for Rh/TiO_2.[321,322] The average size of Rh particles was estimated to be 0.65 nm (five-atom cluster).[321] No indications for alloy formation or suboxide migration have been ascertained; interfacial Rh–O and Rh–Ti^{n+} bonds in the HTR state and Rh–O bonds in the LTR state have been deduced from EXAFS measurements.[321] X-ray absorption near edge spectroscopy (XANES), however, revealed the formation of a Rh–Ti alloy.[322]

Meanwhile the first paper on EXAFS investigations of Ni/TiO_2 (and Ni/Nb_2O_5) reduced at 773 K has been published.[323] Short (262 pm) and long (319 pm) Ni–Ti distances have been found,[323] the former being similar to the bond length in the Ni_3Ti alloy (cf. Table 8.5). Ni–Ni bond lengths and Ni–O bond lengths were determined to be 252 pm and 218 pm, respectively.[323]

With respect to the *electronic structure of the metal–support interface*, three papers should be mentioned here. Copper has been deposited on thermally grown Al_2O_3 leading to ionic $Cu^{(+)}$–$O^{(-)}$ bonds at the interface.[324] An electron transfer to the metal is reported both for Rh on $TiO_2(110)$ "containing reduced cations adjacent to O-vacancy defects"[325] and for Rh/TiO_2 catalysts after HTR.[322] On the perfect TiO_2 surface the charge transfer seems to be negligible; a physisorption of Rh on the surface is suggested.[325]

Concerning the *support influence on the heat of adsorption* of hydrogen on supported Pt a correction of some of the results presented in Table 8.2 is necessary. Recently Sen et al.[326] detected systematic errors in their measurements of integral heats of adsorption[143,145] for Pt/SiO_2, Pt/TiO_2, and Pt/Al_2O_3. Corrected values (in kJ mol^{-1}) are given for four of the five values of Q_{ad} (Table 8.2): Replace 116 ± 12.5 by 56.5 ± 1.3, 116 ± 4.6 by 65.2 ± 2.1, 113 ± 10 by 56.1 ± 1.3, and the range 41–81 by 46–65.[326] The authors conclude that there are neither size effects nor support effects on the hydrogen integral heats of adsorption.[326] Owing to the small uptake of hydrogen, the error of the measurements amounts up to 40% in the SMSI state.[326] Further detailed measurements of H_2 heats of adsorption (preferably initial heats) for different systems in their LTR and HTR states are highly desirable.

Evidence for a geometric support effect on CO chemisorption has been found for Rh/TiO_2,[322] LaO_x/Rh/SiO_2,[327] TiO_x/Pt,[328] and AlO_x/Rh,[329] whereas by other authors an electronic support effect on adsorption has been proposed to occur in TiO_x/Pt,[330] TiO_x/Rh,[330] and Rh/TiO_2.[321] The TPD peak of H_2 desorption from lanthana-promoted Rh/SiO_2 is shifted from 405 K down to 385 K with increasing La content, but the H_2 chemisorption capacity is only slightly diminished due to "spillover of H atoms from exposed Rh sites onto the surface of the LaO_x islands."[327] Sadhegi et al.[325] find, in agreement with the conclusions from our theoretical results (section titled "The Support Influence on Chemisorption"), that the Ti → Rh charge transfer interaction "does not lead to a large-scale suppression of CO chemisorption," but they "cannot exclude the possibility that in the SMSI state the modification of the electronic density of states of the metal at E_F may lead to a weakening of the CO–Rh bond in the immediate vicinity of the encapsulating moieties."[325] Both a geometric and an electronic support effect have been considered responsible for the changed CO chemisorption properties in TiO_x/Rh[329] and MnO_x/Ni.[331] A decrease in the CO TPD peak temperature by 35 K due to the support influence has been reported comparing Ni(111) and Ni/Al_2O_3/Al(111) as substrates.[332]

An EELS investigation reveals a *support influence on the CO vibrational characteristics* for the same system.[332] The finding of an upward shift of the stretch frequency of terminal CO by 20 cm^{-1} compared with that of CO on unsupported nickel attributed "to the interaction of the surface oxygen of the $Al_2O_3/Al(111)$ substrate with the deposited Ni particles"[332] is in agreement with our theoretical results presented in Figure 8.9 (redox–MSI). An upward shift of 10 cm^{-1} of this frequency has been ascribed to a strong interaction between the metal species and strong acidic centers in Ni/Y zeolites.[333] Frequently, the observation of low stretch frequencies of CO (1500–1750 cm^{-1}) is attributed to CO adsorbed at interfacial or suboxide perimeter sites, e.g., in MnO_x/Ni[331] and $LaO_x/Rh/SiO_2$,[327] where CO is pointing with its O end toward the cation of the oxide. These geometric configurations are viewed as precursor states for *CO dissociation*.[327,331] Enhanced CO dissociation has been observed recently on Al-promoted Ni and Cu surfaces at 300 K or below[334] and on $Ni/Al_2O_3/Al(111)$ at 375 K attributed in the latter case to an Ni–Al alloy formation.[332]

The *influence of the support on activity and selectivity* of CO hydrogenation has been discussed for Pd/TiO_2[335] and for TiO_x/Rh.[329] In the former system an increased "intrinsic" methanation activity has been found in the SMSI state.[335] The increased hydrogenation activity and the shift of selectivity toward C_{2+} products in TiO_x/Rh has been attributed to the existence of sites of high activity at the oxide–metal interface.[329] Over Ni/Al_2O_3 the rate-determining step in CO hydrogenation is the hydrogenation of CH_x on supported metallic nickel species, but it is the CO dissociation on $NiAl_2O_4$ reaction centers.[318] "Some problems of selectivity in syngas reactions on the group VIII metals" have been reviewed by Lee and Ponec.[336] These authors conclude that a support influence "through the metal" is unlikely, as a rule, rather "the presence of the support material . . . on the metal surface is indispensable" leading to a localized interaction "by forming complexes with reacting molecules or with intermediates."[336] Finally, we refer to a recent short review by Tauster[337] on experimental work concerning strong metal-support interactions.

Theoretical Work

First we mention recent *total-energy calculations* of Ag/Si(111),[338] Ti, Cr, Fe/Si(111),[339] Ni/Si(111),[340] and Ni/Si(100).[340] Further theoretical studies on metal–support systems[341–345] have been conducted on a lower level of sophistication (cf. Table 8.10).

Most of the work devoted to the generation of *structural models from calculations* is concerned with the transition metal–semiconductor interface (cf. Table 8.10). The only paper on a transition metal–oxidic support interface uses a molecular dynamics computer simulation in continuation of earlier work[271,272] in order to study "interactions of model Pt particles and films with the silica surface."[346] Since the results are very sensitive to

Table 8.10. Supplement to Tables 8.3 and 8.7.

Metal(s)	Support	Calculated Quantities[a]	Method[b]	Reference
Pt	C	ΔE_H, LEPS-type potential for H_2 adsorption/dissociation	EHMO: H_1–$Pt_{n_1}C_{n_2}$, $n_1 = 6, 7, 8$; $n_2 = 10, 12, 13$, two-layer system	344
Ag	Si	r_{Ag-Si}, $BS_{Ag-Si,Ag-Ag}$	DV–$X\alpha$ SCC: $Ag_4Si_{13}H_{16}$, $AgSi_7H_{13}$	338
Ti, Cr, Fe	Si	Forces on M, charge density maps, r_{M-S}	SCF–LCAO–$X\alpha$ force calculations: M_1–Si_6H_9	339
Ni	Si	r_{Ni-S}, DOS, potential energy curves	DV–$X\alpha$: Ni_1–Si_n, $n = 4, 6, 7, 8$	340
Ru, Fe, Co, Ni, Pt	Al_2O_3	Chain growth process in Fischer–Tropsch synthesis; CH_x and CO insertion; activity sequence of the metals	EHMO: M_4–Al_4O_7, M_4–$O_8Al_4O_7$	343
Pd, Pt, Ru	Al_2O_3, SiO_2, MgO, TiO_2	ΔE_M	Semiempirical interacting bond method: M_1 on different surface sites	345
Cu	ZnO	Δq, $BS_{Cu-S,Cu-C,C-O}$, q_c, q_0	INDO: Cu_1, Zn_4O_4, $Zn_{13}O_{13}$, CO on top	341
Ni	TiO_x	$BS_{C-O,Ni-C,Ni-Ni,Ni-Ti}$, Δq, AO characterization of MOs, frequency shift due to the presence of TiO	DV–$X\alpha$: Ni_4TiO, $Ni_{11}TiO$; CO, $(CO)_2$, on hollow	342
Cu	AlN	r_{Cu-S}, DOS, band structure	SC PSF: four-layer AlN(110) films; Cu layer on both sides	225

[a] AO: atomic orbital.
DOS: density of states.
LEPS: London–Eyring–Polanyi–Sato potential.
MO: molecular orbital.
q: net charge.
See also footnote a of Tables 8.3 and 8.7.

[b] DV–$X\alpha$: Discrete variational $X\alpha$.
INDO: Intermediate neglect of differential overlap.
SCC: Self-consistent charge.
SC PSF: Self-consistent pseudofunction method.
See also footnote b of Tables 8.3 and 8.7.

the potential parameters,[346] these calculations, although valuable, are to be considered as qualitative "computer experiments" rather than *a priori* calculations of structural data.

With respect to *assumed structures of relevant model systems* we conclude that recent experimental results on structural data reviewed above support the choice of our different models (cf. section titled "Assumed Structures of Relevant Model Systems"). Sadhegi et al.[347] meanwhile have proposed detailed mechanisms for the reaction of the precursor with the support surface for Rh/TiO_2 as well as for its reduction to the WMSI and SMSI state, respectively. In the former state the TiO_2 surface is partly hydroxylated leading to both M . . . O_{br} (model M–OH_2) and M . . . Ti^{4+} units (model Af); in the latter state "half of the Rh atoms have taken up the bridging hydroxyl positions and the other half have moved to valley positions over cations"[347] (M . . . Ti^{3+} units, model Cf).

As to the *electronic structure of clean surface sites and the electron transfer* some new results are available as well. Tsukada et al.[348] using the DV–$X\alpha$ method and a Ti_4O_{15} cluster found a Ti^{3+}-related occupied gap state originating from an oxygen vacancy at the reduced TiO_2 surface (cf. model Cf, Table 8.4). Munnix and Schmeits,[349] however, using a non-self-consistent ETB method ascribed a deep donor state to a subsurface oxygen vacancy at the relaxed $TiO_2(110)$ surface. Since the bridging oxygen above the subsurface vacancy in their model is also missing, the defect should be viewed as a dioxygen vacancy combined with two fourfold-coordinated Ti^{3+} cations.

With respect to the question on a O_{2p}-derived shallow surface state in the gap on oxygen-terminated Al_2O_3 (model A′a, section titled "Electronic Aspects of Metal–Support Interactions and the Bonding Character"), we note that Fleisher et al.[350] using an $OAl(OH)_3$ cluster and the CNDO method find even a deep O_{2p}-derived donor state. As can be expected as a consequence of the CNDO approximation, their Al–O distance of 210 pm obtained by geometry optimization is considerably longer than our one (171 pm) taken from bulk crystal structure data.

Finally, we mention briefly some of our recent PP MRD–CI results on the "minimum support" models $^{\cdot}SiH_3$ and AlH_3.[351] The ionization potentials (IP) are 8.84 eV and 10.46 eV and the electron affinities 0.30 eV and 0.54 eV, respectively. The corresponding (CNDO) values of models Cdpy and A′dpy amount to 7.69 eV and 14.10 eV and to −1.58 eV and −1.23 eV, respectively, showing the same graduation. This is obvious also from the electronegativities of the clean surface clusters [EC, cf. Eq. 8.5), in eV]: 3.06 (Cdpy, CNDO), 6.43 (A′dpy, CNDO), 4.57 ($^{\cdot}SiH_3$, PP MRD–CI), and 5.50 (AlH_3, PP MRD–CI). Furthermore, when going from the reduced support surface (Cdpy, $NiSiH_3$) to the oxidized one (A′dpy,

$NiAlH_3$) both the PP MRD–CI results of "minimum support" models and the CNDO results of models Cdpy and A′dpy show a decrease of the *Ni-support bond strength.* The optimized Ni-support bond lengths and interaction energies are 228 pm and 2.03 eV for $NiSiH_3$ and 248 pm and 0.67 eV for $NiAlH_3$, respectively. The corresponding Wiberg indices of models Cdpy and A′dpy are given in Table 8.5.

Some more examples of evidence for the existence of strongly and weakly bonding surface sites are given in recent papers: Strong metal–support bonds have been found to exist between copper and the surface oxygen of ZnO[341] and between Ni and the Ti cation of contaminating TiO.[342] In the former case the electron transfer is directed to the support and in the latter one to the metal. Using a very large basis set and SDCI calculations earlier results (Table 8.6) on the existence of a weak bond between Ni and the bridging O in H_2O have been confirmed.[352]

Three papers have been dealing with the *support influence on chemisorption* of hydrogen[344] and CO,[341,342] respectively. "EHT calculations predict that the stability of H on different sites on the Pt(111) surface decreases in the sequence on hollow > bridge > on top . . . ."[344] Passing to H interacting with $Pt_{n_1}C_{n_2}$ clusters is associated with "a decrease in the bridge and on-hollow adsorption energies but leads to a slightly increased adsorption energy for the on-top site thus resulting in a reversed order of the stability of H . . . : on top > on hollow > bridge."[344] The authors attribute this change mainly to the "electron transfer between the graphite support and the Pt monolayer." The bond strength between metal and support has not been discussed. Interestingly, the height of the activation barrier to H_2 dissociation of about 10 kJ mol^{-1} is slightly increased due to the support influence.[344]

The INDO results presented for CO on-top adsorption on Cu/ZnO[341] (cf. Table 8.10) are in agreement with our CNDO results (Table 8.9). In either case the adsorption site, i.e., a copper and a nickel atom, respectively, is linked to supporting oxygen atoms withdrawing charge from the metal. As a consequence, the metal-to-CO π-back-donation is diminished increasing the C–O bond order but decreasing the cluster–C bond order compared with the support-free situation. It has been pointed out by the authors[341] that the HOMO of their Cu/ZnO cluster has "only small Cu character" making "this orbital not ideal for bonding overlap with adsorbed species." This fact is a reminder of the redox metal–support interaction occurring in our models Aa, Ab, Ac. Neither CO adsorption energies nor metal–support bond orders have been given.

Considering CO chemisorption on a fourfold hollow site on a nickel cluster, it has been found by DV–$X\alpha$ calculations that a TiO molecule beneath the nickel layer though transferring charge to the metal even "decreases the ability of metal atoms to donate electrons to adsorbed CO molecules.''[342] The Ni–C bond is weakened and the C–O bond is strengthened.[342] This finding contradicts our results obtained for CO on-top adsorption (Table 8.9, model Cf). TiO adsorbed on a surface site neighboring CO adsorption sites, however, is likely to decrease the C–O bond strength owing to direct interaction with CO.[342] A direct (mainly electrostatic) interaction between electropositive modifiers and CO also has been proposed to cause CO activation based on molecular models and nonempirical calculations including electron correlation[353] as well as on MS–$X\alpha$ cluster calculations.[354] On the other hand, ''correlations between the XPS and desorption data suggest that Pt $\rightarrow$ W valence charge transfer is accompanied by an increase in the strength of W–CO binding, presumably by enhanced d $\rightarrow \pi^*$ interaction'' indicating an electronic effect of neighboring Pt atoms on W adsorption sites in a surface alloy.[355] Covalent ''through the metal'' effects of electronegative modifiers (cf. section titled ''The Support Influence on Chemisorption'') on adsorbed CO due to its strong bonding to the metal have been observed recently using MS–$X\alpha$ calculations.[354,356] Sulfur near the adsorption site on a nickel cluster leads to a depletion of charge from E_F both ''weakening CO chemisorption by localizing charge that would have been available for bonding to CO in covalent S–Ni bonds'' and ''inhibiting the incipient response of the surface.''[356] The former effect is similar to the ''poisoning'' of the metal by the support in our SMSI models.

Using EHMO cluster calculations Zhang Xiao-Guang et al.[343] study the ''chain growth process in Fischer–Tropsch synthesis'' on different group VIII metals supported on Al_2O_3. They propose the insertion of the first CH_2 fragment into the metal–CH_x bond to be the rate-determining step. The influence of the supporting oxide on the *mechanisms of the reaction* has not been discussed.

In summarizing we may state that, notwithstanding some minor modifications or corrections which are necessary by the most recent experimental and theoretical results, all main conclusions drawn in this chapter including the several open questions remain valid.

REFERENCES

1. Selectivity in Heterogeneous Catalysis, *Faraday Disc. Chem. Soc.* **72** (1981).
2. Moss, R.L., in *Specialist Periodical Reports: Catalysis*. London: Royal Society of Chemistry, 1981, Vol. 4, p. 31.
3. Bond, G.C., Burch, R., in *Specialist Periodical Reports: Catalysis*. London: Royal Society of Chemistry, 1983, Vol. 6. p. 27.
4. Burch, R., *Specialist Periodical Reports: Catalysis*. London: Royal Society of Chemistry, 1985, Vol. 7, p. 149.
5. Imelik, B., Naccache, C., Coudurier, G., Praliaud, H., Meriaudeau, P., Gallezot, P.,

Martin, G.A., Vedrine, J.C. (eds.), *Metal-Support and Metal-Additive Effects in Catalysis*. Amsterdam: Elsevier, 1982.
6. Henrici-Olivé, G., Olivé, S. *The Chemistry of the Catalyzed Hydrogenation of Carbon Monoxide*. Berlin, Heidelberg, New York, Tokyo: Springer-Verlag, 1984.
7. Vannice, M.A., Twu, C.C., *J. Catal.* **82:** 213 (1983).
8. Vannice, M.A., Garten, R.L., *J. Catal.* **56:** 236 (1979).
9. Bartholomew, C.H., Pannell, R.B., Butler, J.L., *J. Catal.* **65:** 335 (1980).
10. Anderson, J.B.F., Bracey, J.D., Burch, R., Flambard, A.R., in *Proceedings of the 8th International Congress on Catalysis, Berlin(West), 1984,* Frankfurt/Main: DECHEMA, 1984, Vol. V, p. 111.
11. Foger, K., in *Catalysis. Science and Technology* (J.R. Anderson and M. Boudart, eds.). Berlin, Heidelberg, New York: Springer-Verlag, 1985, Vol. 6, Chap. 4, p. 227.
12. Shalvoy, R.B., Davis, B.H., Reucroft, P.J., *Surf. Interf. Anal.* **2:** 11 (1980).
13. Narayanan, S., Uma, K., *J. Chem. Soc., Farad. Trans. I* **81:** 2733 (1985).
14. Arai, M., Ishikawa, T., Nishiyama, Y., *J. Phys. Chem.* **86:** 577 (1982).
15. Blackmond, D.G., Ko, E.I., *Appl. Catal.* **13:** 49 (1984).
16. Montes, M., Penneman de Bosscheyde, Ch., Hodnett, B.K., Delannay, F., Grange, P., Delmon, B., *Appl. Catal.* **12:** 309 (1984).
17. Boudart, M., in *Proceedings of the 6th International Congress on Catalysis*. London: Royal Society of Chemistry, 1977, Vol. 1, p. 1.
18. Bond, G.C., *Surf. Sci.* **156:** 966 (1985).
19. Ponec, V., in Ref. 5, p. 63.
20. Anderson, J.R., *Sci. Progr. Oxf.* **69:** 461 (1985).
21. Boudart, M., Djega-Mariadassou, G., *Kinetics of Heterogeneous Catalytic Reactions*. Princeton, N.J.: Princeton University Press, 1984, p. 207.
22. Bond, G.C., in Ref. 5, p. 1.
23. Tauster, S.J., Fung, S.C., Garten, R.L., *J. Am. Chem. Soc.* **100:** 170 (1978).
24. Solymosi, F., *Catal. Rev.* **1:** 233 (1967).
25. Praliaud, H., Martin, G.A., *J. Catal.* **72:** 394 (1981).
26. Ren-Yuan, T., Rong-An, W., Li-Wu, I., *Appl. Catal.* **10:** 163 (1984).
27. Tauster, S.J., in Ref. 48, p. 1.
28. Burch, R., Flambard, A.R., *J. Catal.* **78:** 389 (1982).
29. Sinfelt, J.H., *Scientific American* **253:** 90 (1985).
30. Resasco, D.E., Haller, G.L., *J. Catal.* **82:** 279 (1983).
31. Goodman, D.W., *Ann. Rev. Phys. Chem.* **37:** 425 (1986); *Acc. Chem. Res.* **17:** 194 (1984).
32. Martin, G.A., in Ref. 5, p. 315.
33. Lee, J., Arias, J., Hanrahan, C.P., Martin, R.M., Metiu, H., *J. Chem. Phys.* **82:** 485 (1985).
34. Sachtler, W.M.H., Ichikawa, M., *J. Phys. Chem.* **90:** 4752 (1986).
35. Nuzzo, R.G., Dubois, L.H., *Appl. Surf. Sci.* **19:** 407 (1984).
36. Houalla, M., Dang, T.A., Kibby, Ch.L., Petrakis, L., Hercules, D.M., *Appl. Surf. Sci.* **19:** 414 (1984).
37. Dubois, L.H., Nuzzo, R.G., *J. Am. Chem. Soc.* **105:** 365 (1983).
38. Bardi, U., Somorjai, G.A., Ross, P.N., *J. Catal.* **85:** 272 (1984).
39. Bardi, U., Dahlgren, D., Ross, P.N., *J. Catal.* **100:** 196 (1986).
40. Nuzzo, R.G., Dubois, L.H., in Ref. 48, p. 136.
41. Flores, F., Tejedor, C., *J. Phys. C: Solid State Phys.* **20:** 145 (1987).
42. Zunger, A., *Thin Solid Films* **104:** 301 (1984).
43. Calandra, C., Bisi, O., Ottaviani, G., *Surf. Sci. Repts.* **4:** 271 (1984).

44. Hauffe, K., Wolkenstein, Th. (eds.)., *Symposium on Electronic Phenomena in Chemisorption and Catalysis on Semiconductors, Moscow, 1968*. Berlin: W. de Gruyter & Co., 1969.
45. Schwab, G.M., *Adv. Catal.* **27:** 1 (1978).
46. Solymosi, F., Tombacz, I., Koszta, J., *J. Catal.* **95:** 578 (1985).
47. Herrmann, J.M., *J. Catal.* **89:** 404 (1984).
48. Baker, R.T.K., Tauster, S.J., Dumesić, J.A. (eds)., *Strong Metal-Support Interactions*. ACS Symposium Series No. 298. Washington DC: American Chemical Society, 1986.
49. Deviney, M.L., Gland, J.L. (eds.)., *Catalyst Characterization Science,* ACS Symposium Series No. 288, Washington DC: American Chemical Society, 1985.
50. Lagarde, P., Dexpert, H., *Adv. Phys.* **33:** 567 (1984).
51. Jiang, X.Z., Hayden, T.F., Dumesić, J.A., *J. Catal.* **83:** 168 (1983).
52. Smith, J.S., Thrower, P.A., Vannice, M.A., *J. Catal.* **68:** 270 (1981).
53. Simoens, A.J., Baker, R.T.K., Dwyer, D.J., Lund, C.R.F., Madon, R.J., *J. Catal.* **86:** 359 (1984).
54. Mustard, D.G., Bartholomew, C.H., *J. Catal.* **67:** 186 (1981).
55. Raupp, G.B., Dumesić, J.A., *J. Catal.* **97:** 85 (1986).
56. Jiang, X.-Z., Song, B.-H., Chen, Y., Wang, Y.-W., *J. Catal.* **102:** 257 (1986).
57. Derouane, E.G., Simoens, A.J., Vedrine, J.C., *Chem. Phys. Lett.* **52:** 549 (1977).
58. Marcelin, G., Lester, J.E., *J. Catal.* **93:** 270 (1985).
59. Chung, Y.W., Xiong, G., Kao, C.C., *J. Catal.* **85:** 237 (1984).
60. Takatani, S., Chung, Y.W., *Appl. Surf. Sci.* **19:** 341 (1984).
61. Raupp, G.B., Dumesić, J.A., *J. Phys. Chem.* **88:** 660 (1984).
62. Baker, R.T.K., Chludzinski, J.J., Dumesić, J.A., *J. Catal.* **93:** 312 (1985).
63. Dumesić, J.A., Stevenson, S.A., Sherwood, R.D., Baker, R.T.K., *J. Catal.* **99:** 79 (1986).
64. Dumesić, J.A., Stevenson, S.A., Chludzinski, J.J., Sherwood, R.D., Baker, R.T.K., in Ref. 48, p. 99.
65. Kao, C.C., Tsai, S.C., Chung, Y.W., *J. Catal.* **73:** 136 (1982).
66. Coenen, J.W.E., in *Preparation of Catalysts II* (B. Delmon et al., eds.). Amsterdam, Oxford, New York: Elsevier, 1979, p. 89.
67. Powell, B.R., Whittington, S.E., *J. Catal.* **81:** 382 (1983).
68. Keck, K.-E., Kasemo, B., *Surf. Sci.* **167:** 313 (1986).
69. Martin, G.A., Dutartre, R., Dalmon, J.A., *React. Kinet. Catal. Lett.* **16:** 329 (1981).
70. Turlier, P., Martin, G.A., *React. Kinet. Catal. Lett.* **19:** 275 (1982).
71. Wu, M., Hercules, D.M., *J. Phys. Chem.* **83:** 2003 (1979).
72. Hoang-Van, C., Villemin, B., Teichner, S.J., *React. Kinet. Catal. Lett.* **26:** 127 (1984).
73. Turlier, P., Praliaud, H., Moral, P., Martin, G.A., Dalmon, J.A., *Appl. Catal.* **19:** 287 (1985).
74. Kester, K.B., Zagli, E., Falconer, J.L., *Appl. Catal.* **22:** 311 (1986).
75. Schlesinger, R., diploma thesis, Berlin: Humboldt-Universität, Sektion Chemie, 1986 (unpublished).
76. Haberlandt, H., Schlesinger, R., Ritschl, F. in *Proceedings of the 6th International Symposium on Heterogeneous Catalysis* (D. Shopov, A. Andreev, A. Palazov, and L. Petrov, eds.). Sofia: Bulgarian Academy of Sciences, 1987, Vol. I, p. 324.
77. Koningsberger, D.C., van Zon, J.B.A.D., van't Blik, H.F.J., Visser, G.J., Prins, R., Mansour, A.N., Sayers, D.E., Short, D.R., Katzer, J.R., *J. Phys. Chem.* **89:** 4075 (1985).
78. Zamaraev, K.I., Kochubej, D.I., *Kinet. Katal.* **27:** 1031 (1986).
79. Sinfelt, J.H., *J. Phys. Chem.* **90:** 4711 (1986).

80. Haller, G.L., *Appl. Surf. Sci.* **20:** 351 (1985).
81. Apai, G., Hamilton, J.F., Stohr, J., Thompson, A. *Phys. Rev. Lett.* **43:** 165 (1979).
82. Ovsjannikova, J.A., Kraisman, V.L., Starzev, A.N., Yermakov, Yu.I., *Kinet. Katal.* **25:** 446 (1984).
83. Tohji, K., Udagawa, Y., Tanabe, S., Ueno, A., *J. Am. Chem. Soc.* **106:** 612 (1984).
84. Tohji, K., Udagawa, Y., Tanabe, S., Ida, T., Ueno, A., *J. Am. Chem. Soc.* **106:** 5172 (1984).
85. Short, D.R., Mansour, A.N., Cook, Jr., J.W., Sayers, D.E., Katzer, J.R., *J. Catal.* **82:** 299 (1983).
86. Lytle, F.W., Greegor, R.B., Marques, E.C., Sandstrom, D.R., Via, G.H., Sinfelt, J.H., *J. Catal.* **95:** 546 (1985).
87. Bonneviot, L., Che, M., Olivier, D., Martin, G.A., Freund, E., *J. Phys. Chem.* **90:** 2112 (1986).
88. Mörke, W., Drevs, H., *Z. f. Chemie* **25:** 453 (1985).
89. Greegor, R.B., Lytle, F.W., Chin, R.L., Hercules, D.M., *J. Phys. Chem.* **85:** 1232 (1981).
90. van Zon, J.B.A.D., Koningsberger, D.C., van't Blik, H.F.J., Sayers, D.E., *J. Chem. Phys.* **82:** 5742 (1985).
91. Koningsberger, D.C., Sayers, D.E., *Solid State Ionics* **16:** 23 (1985).
92. Sakellson, S., McMillan, M., Haller, G.L., *J. Phys. Chem.* **90:** 1733 (1986).
93. Koningsberger, D.C., Martens, J.H.A., Prins, R., Short, D.R., Sayers, D.E., *J. Phys. Chem.* **90:** 3047 (1986).
94. Bommannavor, A.S., Montano, D.A., *Appl. Surf. Sci.* **19:** 250 (1984).
95. Sankar, G., Vasudevan, S., Rao, C.N.R., *J. Phys. Chem.* **90:** 5325 (1986).
96. Surrat, G.T., Kunz, A.B., *Phys. Rev. B* **19:** 2352 (1979).
97. Den Otter, G.J., Dautzenberg, F.M., *J. Catal.* **53:** 116 (1978).
98. Kunimori, K., Ikeda, Y., Soma, M., Uchijima, T., *J. Catal.* **79:** 185 (1983).
99. Cairns, J.A., Baglin, J.E.E., Coark, G.J., Ziegler, J.F., *J. Catal.* **83:** 301 (1983).
100. Ruckenstein, E., Lee, S.H., *J. Catal.* **104:** 259 (1987).
101. Beard, B.C., Ross, P.N., *J. Phys. Chem.* **90:** 6811 (1986).
102. Sushumna, I., Ruckenstein, E., *J. Catal.* **94:** 239 (1985).
103. Geus, J.W., in *Chemisorption and Reactions on Metallic Films* (J.R. Anderson, ed.). London: Academic Press, 1971, Vol. 1, Chap. 3, p. 129.
104. Gillet, M., Robinson, F., Miquel, J.M., *J. Chim. Phys.* **78:** 867 (1981).
105. Chapon, C., Henry, C.R., Chemam, A., *Surf. Sci.* **162:** 747 (1985).
106. Yermakov, Yu.I., Kuznetsov, B.N., Zakharov, V.A., *Catalysis by Supported Complexes*. Amsterdam, Oxford, New York: Elsevier, 1981.
107. Shpiro, E.S., Djusenbina, B.B., Antoshin, G.B., Tkachenko, O.P., Minachev, H.M., *Kinet. Katal.* **25:** 1505 (1984).
108. Anderson, J.B.F., Burch, R., Cairns, J.A., *Appl. Catal.* **21:** 179 (1986).
109. Sexton, B.A., Hughes, A.E., Foger, K., *J. Catal.* **77:** 85 (1982).
110. Chien, S.H., Shelimov, B.N., Resasco, D.E., Lee, E.H., Haller, G.L., *J. Catal.* **77:** 301 (1982).
111. Fung, S.C., *J. Catal.* **76:** 225 (1982).
112. Minachev, H.M., Antoshin, G.V., Shpiro, E.S., *Kinet. Katal.* **23:** 1365 (1982).
113. Greenlief, C.M., White, J.M., Ko, C.S., Gorte, R.J., *J. Phys. Chem.* **89:** 5025 (1985).
114. Viswanathan, B., Tanaka, K., Toyoshima, I., *Chem. Phys. Lett.* **113:** 294 (1985).
115. Viswanathan, B., in *Advances in Catalysis Science and Technology, Proc. 7th Natl. Symp. Catal., Baroda, India:* 1985, p. 63.
116. Lytle, F.W., Greegor, R.B., Marques, E.C., Biebesheimer, V.A., Sandstrom, D.R., Horsley, J.A., Via, G.H., Sinfelt, J.H., in Ref. 49, p. 280.

117. zur Loye, H.C., Stacy, A.M., *J. Am. Chem. Soc.* **107:** 4567 (1985).
118. Huizinga, T., Prins, R., in Ref. 5, p. 11.
119. Castellani, N.J., Leroy, D.B., Lambrecht, W., *Chem. Phys.* **95:** 459 (1985).
120. Wertheim, G.K., *Z. Phys. B* **66:** 53 (1987).
121. Hatwar, T.K., Chopra, D., *Surf. Interf. Anal.* **7:** 93 (1985).
122. Fischer, T.E., Kelemen, S.R., Wang, K.P., Johnson, K.H., *Phys. Rev. B* **20:** 3124 (1979).
123. Maslenko, S.B., Kozlenkov, A.I., Filin, S.A., Shulgin, A.I., *Phys. Status Solidi B* **123:** 605 (1984).
124. Xu Jian-hua, Xu Yong-nian, *Sol. State Commun.* **55:** 891 (1985).
125. Gagarin, S.G., Teterin, Yu.A., Kulakov, V.M., Falkov, I.G., *Kinet. Katal.* **22:** 1265 (1981).
126. Chechlov, A.N., Ionov, S.I., *Koord. Khim.* **7:** 34 (1981).
127. Gregory, I.P., Moody, D.E., *J. Phys. F* **5:** 36 (1975).
128. Huizinga, T., Prins, R., *J. Phys. Chem.* **85:** 2156 (1981).
129. Conesa, J.C., Malet, P., Munuera, G., Sanz, J., Soria, J., *J. Phys. Chem.* **88:** 2986 (1984).
130. Dwyer, D.J., Cameron, S.D., Gland, J., *Surf. Sci.* **159:** 430 (1985).
131. Paul, J., Cameron, S.D., Dwyer, D.J., Hoffmann, F.M., *Surf. Sci.* **177:** 121 (1986).
132. Vannice, A., Chou, P., *J. Chem. Soc., Chem. Commun.* 1590 (1984).
133. Chou, P., Vannice, M.A., *J. Catal.* **104:** 1 (1987).
134. Chou, P., Vannice, M.A., *J. Catal.* **104:** 17 (1987).
135. Baker, R.T.K., Prestridge, E.B., McVicker, G.B., *J. Catal.* **89:** 422 (1984).
136. Singh, A.K., Pande, N.K., Bell, A.T., *J. Catal.* **94:** 422 (1985).
137. zur Loye, H.-C., Faltens, T.A., Stacy, A.M., *J. Am. Chem. Soc.* **108:** 2488 (1986).
138. Ocal, C., Ferrer, S., *Surf. Sci.* **178:** 850 (1986).
139. Meriaudeau, P., Ellestad, O.H., Dufaux, M., Naccache, C., *J. Catal.* **75:** 243 (1982).
140. Chen, B.-H., White, J.M., *J. Phys. Chem.* **86:** 3534 (1982).
141. Takatani, S., Chung, Y.-W., *J. Catal.* **90:** 75 (1984).
142. Raupp, G.B., Dumesić, J.A., *J. Catal.* **95:** 587 (1985).
143. Vannice, M.A., Hasselbring, L.C., Sen, B., *J. Catal.* **95:** 57 (1985).
144. Vannice, M.A., Hasselbring, L.C., Sen, B., *J. Catal.* **97:** 66 (1986).
145. Vannice, M.A., Hasselbring, L.C., Sen, B., *J. Phys. Chem.* **89:** 2972 (1985).
146. Herrmann, J.M., Gravelle-Rumeau-Maillot, M., Gravelle, P.C., *J. Catal.* **104:** 136 (1987).
147. Belton, D.N., Sun, Y.-M., White, J.M., *J. Phys. Chem.* **88:** 1690 (1984).
148. Belton, D.N., Sun, Y.-M., White, J.M., *J. Phys. Chem.* **88:** 5172 (1984).
149. Belton, D.N., Sun, Y.-M., White, J.M., *J. Catal.* **102:** 338 (1986).
150. Conesa, J.C., Malet, P., Muñoz, A., Munuera, G., Sainz, M.T., Sanz, J., Soria, J., in *Proceedings of the 8th International Congress on Catalysis, Berlin (West)*. Frankfurt/Main: DECHEMA, 1984, Vol. V, p. 217.
151. Ertl, G., in *The Nature of the Surface Chemical Bond* (T.N. Rhodin and G. Ertl, eds.). Amsterdam: North-Holland, 1979, p. 315.
152. Shustorovich, E., *Surf. Sci. Repts.* **6:** 1 (1986).
153. Balooch, M., Cardillo, M.J., Miller, D.R., Stickney, R.E., *Surf. Sci.* **46:** 358 (1974).
154. Karner, H., Luger, M., Steinrück, H.P., Winkler, A., Rendulic, K.D., *Surf. Sci.* **163:** L641 (1985).
155. Christmann, K., Ertl, G., in Ref. 49, p. 222.
156. Ko, E.I., Madix, R.J., in *Advances in the Mechanics and Physics of Surfaces* (R.M. Latanision and R.J. Courtel, eds.). Chur: Harwood Ac. Publ. GmbH, 1981, Vol. 1, p. 153.

157. Weatherbee, G.D., Bartholomew, C.H., *J. Catal.* **87:** 55 (1984).
158. Goodman, D.W., *Appl. Surf. Sci.* **19:** 1 (1984).
159. Sun, Y.-M., Luftman, H.S., White, J.M., *Surf. Sci.* **139:** 379 (1984), and references therein.
160. Praliaud, H., Dalmon, J.A., Mirodatos, C., Martin, G.A., *J. Catal.* **97:** 344 (1986).
161. Scholten, J.J.F., Pijpers, A.P., Hustings, A.M.L., *Catal. Rev.-Sci. Engn.* **27:** 151 (1985).
162. Knor, Z., in *Catalysis. Science and Technology* (J.R. Anderson and M. Boudart, eds.). Berlin, Heidelberg, New York: Springer-Verlag, 1982, Vol. 3, Chap. 5, p. 231.
163. Renouprez A., Fouilloux, P., Moraweck, B., in *Growth and Properties of Metal Clusters* (J. Bourdon, ed.). Amsterdam: Elsevier, 1980, p. 421.
164. van't Blik, H.F.J., van Zon, J.B.A.D., Huizinga, T., Vis, J.C., Koningsberger, D.C., Prins, R., *J. Phys. Chem.* **87:** 2264 (1983).
165. Sachtler, W.M.H., *Surf. Sci.* **22:** 468 (1970).
166. DiNardo, N.J., Plummer, E.W., *Surf. Sci.* **150:** 89 (1985).
167. Avouris, P., Demuth, J., *Ann. Rev. Phys. Chem.* **35:** 49 (1984).
168. Peri, J.B., *J. Catal.* **86:** 84 (1984).
169. Zowtiak, J.M., Bartholomew, C.H., *J. Catal.* **83:** 107 (1983).
170. Weatherbee, G.D., Rankin, J.L., Bartholomew, C.H., *Appl. Catal.* **11:** 73 (1984).
171. Omashev, H.G., Zakumbajeva, G.D., Sokolskij, D.V., *Dokl. Akad. Nauk SSSR* **234:** 1132 (1977).
172. Hedge, M.S., Rajumon, M.K., Rao, C.N.R., *J. Chem. Soc., Chem. Commun.* 323 (1986).
173. de Bokx, P.K., Bonne, R.L.C., Geus, J.W., *Appl. Catal.* **30:** 33 (1987).
174. Sachtler, W.M.H., in *Proceedings of the 8th International Congress on Catalysis, Berlin (West)*. Frankfurt/Main: DECHEMA, 1984, Vol. I, p. 151.
175. Meriaudeau, P., Dufaux, M., Naccache, C., in Ref. 48, p. 118.
176. Vannice, M.A., in *Catalysis. Science and Technology* (J.R. Anderson and M. Boudart, eds.). Berlin, Heidelberg, New York: Springer-Verlag, 1982, Vol. 3, Chap. 3, p. 139.
177. Mink, J., Szilagyi, T., Wachholz, S., Kunath, D., *J. Mol. Struct.* **141:** 389 (1986).
178. Miessner, H., Gutschick, D., Ewald, H., Müller, H., *J. Mol. Catal.* **36:** 359 (1986).
179. Blackmond, D.G., Ko, E.I., *J. Catal.* **96:** 210 (1985).
180. Blackmond, D.G., Ko, E.I., *J. Catal.* **94:** 343 (1985).
181. Červeny, L. (ed.), *Catalytic Hydrogenation.* Amsterdam: Elsevier, 1986. (*a*) Marinas, J.M., Campelo, J.M., Luna, D., Chap. 12, p. 411; (*b*) Pajonk, G.M., Teichner, S.J., Chap. 8, p. 277.
182. Raupp, G.B., Dumesić, J.A., *J. Catal.* **96:** 597 (1985).
183. Bartholomew, C.H., Vance, C.K., *J. Catal.* **91:** 78 (1985).
184. Orita, H., Naito, S., Tamaru, K., *J. Chem. Soc., Chem. Commun.* 993 (1983).
185. Miessner, H., Orita, H., Naito, S., Tamaru, K., *React. Kinet. Catal. Lett.* **28:** 295 (1985).
186. Shiu-Min Fang, White, J.M., Campione, T.J., Ekerdt, J.G., *J. Catal.* **96:** 491 (1985).
187. Hsiu-Wei Chen, White, J.M., Ekerdt, J.G., *J. Catal.* **99:** 293 (1986).
188. Santos, J., Phillips, J., Dumesić, J.A., *J. Catal.* **81:** 147 (1983).
189. Pollmann, J., in *Festkörperprobleme. Advances in Solid State Physics* (J. Treusch, ed.). Braunschweig: Vieweg, 1980, Vol. 20, p. 117.
190. Cohen, M.L., Louie, S.G., *Ann. Rev. Phys. Chem.* **35:** 537 (1984).
191. Herman, F., *Int. J. Quant. Chem: Quant. Chem. Symp.* **19:** 547 (1986).
192. Messmer, R.P., in *Semiempirical Methods of Electronic Structure Calculations, Modern Theoretical Chemistry* (G.A. Segal, ed.). New York: Plenum Press, 1977, Vol. 7B, Chap. 6, p. 215.

193. Zhidomirov, G.M., Chuvylkin, N.D., *Uspechi Chimii* **55:** 353 (1986).
194. Simonetta, M., Gavezotti, A., *Adv. Quant. Chem.* **12:** 103 (1980).
195. Zdetsis, A.D., Kunz, A.B., *Phys. Rev. B* **32:** 6358 (1985).
196. Bacalis, N.C., Kunz, A.B., *Phys. Rev. B* **32:** 4857 (1985).
197. Laughlin, R.B., Joannopoulos, J.D., Chadi, D.J., *Phys. Rev. B* **20:** 5228 (1979), and references therein.
198. Sauer, J., *Chem. Rev.* **89:** 199 (1989).
199. Redondo, A., Goddard III, W.A., Swarts, C.A., McGill, T.C., *J. Vac. Sci. Technol.* **19:** 498 (1981).
200. Haberlandt, H., Ritschl, F., *Phys. Status Solidi B* **100:** 503 (1980).
201. Haberlandt, H., Ritschl, F., *J. Phys. Chem.* **87:** 3244 (1983).
202. Fleisher, M.B., Golender, I.O., Shimanskaya, M.V., *React. Kinet. Catal. Lett.* **24:** 25 (1984).
203. Haberlandt, H., Ritschl, F. *J. Phys. Chem.* **90:** 4322 (1986).
204. Madhavan, P.V., Whitten, J.L., *Surf. Sci.* **112:** 38 (1981).
205. Bagus, P.S., Bauschlicher Jr., Ch.W., Nelin, C.J., Laskowski, B.C., Seel, M., *J. Chem. Phys.* **81:** 3594 (1984).
206. Bilek, O., Kadura, P., *Phys. Status Solidi B* **85:** 225 (1978).
207. Fernando, G.W., Cooper, B.R., Ramana, M.V., Krakauer, H., Ma, C.Q., *Phys. Rev. Lett.* **56:** 2299 (1986).
208. Veillard, A. (ed.). *Quantum Chemistry: The Challenge of Transition Metals and Coordination Chemistry.* Dordrecht: D. Reidel Publ. Co., 1986. (*a*) Bauschlicher Jr., Ch.W., Walch, S.P., Langhoff, S.R., p. 15. (*b*) Pacchioni, G., Koutecký, J., p. 465. (*c*) Pélissier, M., Daudey, J.P., Malrieu, J.P., Jeung, G.H., p. 37. (*d*) Koga, N., Morokuma, K., p. 351.
209. Siegbahn, P.E.M., Blomberg, M.R.A., Bauschlicher, Jr., Ch.W., *J. Chem. Phys.* **81:** 2103 (1984).
210. Garcia-Prieto, J., Ruiz, M.E., Novaro, O., *J. Am. Chem. Soc.* **107:** 5635 (1985).
211. Upton, T.H., *J. Chem. Phys.* **83:** 5084 (1985).
212. Goddard III, W.A., *Science* **227:** 917 (1985).
213. Dedieu, A., Sakaki, S., Strich, A., Siegbahn, P.E.M., *Chem. Phys. Lett.* **133:** 317 (1987).
214. Smith Jr., V.H., Schaefer III, H.F., Morokuma Jr., K. (eds.). *Applied Quantum Chemistry.* Dordrecht: D. Reidel Publ. Co., 1986. (*a*) Nakatsuji, H., Hada, M., p. 93. (*b*) Salahub, R., p. 185, and references therein.
215. Ravenek, W., Baerends, E.J., *J. Chem. Phys.* **81:** 865 (1984).
216. Jörg, H., Rösch, N., Sabin, J.R., Dunlap, B.I., *Chem. Phys. Lett.* **114:** 529 (1985).
217. Chan, C.T., Vanderbilt, D., Louie, S.G., *Phys. Rev. B* **33:** 2455 (1986).
218. Chan, C.T., Louie, S.G., *Phys. Rev. B* **33:** 2861 (1986), and references therein.
219. Kasowski, R.V., Tsai, M.H., Rhodin, T.N., Chambliss, D.D., *Phys. Rev. B* **34:** 2656 (1986).
220. Wei, S.-H., Zunger, A., *Phys. Rev. B* **35:** 2340 (1987-I).
221. Hamann, D.R., Mattheiss, L.F., *Phys. Rev. Lett.* **54:** 2517 (1985).
222. Tomanek, D., Louie, S.G., Chan, C.T., *Phys. Rev. Lett.* **57:** 2594 (1986).
223. Kasowski, R.V., Rhodin, T.N., Tsai, M.-H., *Appl. Phys. A* **41:** 61 (1986).
224. Kasowski, R.V., Rhodin, T.N., Tsai, M.-H., reprint (Wilmington, 1986), private communication.
225. Ohuchi, F.S., French, R.H., Kasowski, R.V., *J. Appl. Phys.* **62:** 2286 (1987).
226. Shih-Hung Chou, Freeman, A.J., Grigoras, S., Gentle, T.M., Delley, B., Wimmer, E., *J. Am. Chem. Soc.* **109:** 1880 (1987).

227. Blomberg, M.R.A., Brandemark, U.B., Siegbahn, P.E.M., Mathisen, K.B., Karlström, G., *J. Phys. Chem.* **89:** 2171 (1985).
228. Buenker, R.J., Peyerimhoff, S.D., *Theor. Chim. Acta* **35:** 33 (1974). Buenker, R.J., Peyerimhoff, S.D., Butscher, W., *Mol. Phys.* **35:** 771 (1978).
229. Herman, F., *J. Phys., Colloq. C5, Suppl. No. 4* **45:** C5-375 (1984).
230. Saillard, J.-Y., Hoffmann, R., *J. Am. Chem. Soc.* **106:** 2006 (1984).
231. Raatz, F., Salahub, D.R., *Surf. Sci.* **156:** 982 (1985).
232. Rodriguez, J.A., Campbell, Ch.T., *J. Phys. Chem.* **91:** 2161 (1987).
233. Horsley, J.A., *J. Am. Chem. Soc.* **101:** 2870 (1979).
234. Baetzold, R.C., *J. Phys. Chem.* **80:** 1504 (1976).
235. Nørskov, J.K., *Physica* **127B:** 193 (1984).
236. See, e.g., Morán-Lopez, J.L., Falicov, L.M., in *Electronic Structure and Properties of Hydrogen in Metals* (P. Jena and C.B. Satterthwaite, eds.). New York, London: Plenum Press, 1983, p. 635.
237. Bose, S.M., Davison, S.G., Sulston, K.W., *Surf. Sci.* **200:** 265 (1988).
238. Baetzold, R.C., *J. Chim. Phys. Phys.-Chim. Biol.* **78:** 933 (1981).
239. Joyner, R.W., Pendry, J.B., Saldin, D.K., Tennison, S.R., *Surf. Sci.* **138:** 84 (1984).
240. Johnson, K.H., Balazs, A.C., Kolari, H., *Surf. Sci.* **72:** 733 (1978).
241. Horsley, J.A., Lytle, F.W., in Ref. 48, p. 10.
242. Kunz, A.B., *Phil. Mag. B* **51:** 209 (1985).
243. Gagarin, S.G., Kritshko, A.A., *Zh. Fiz. Khim.* **52:** 1291 (1978).
244. Gagarin, S.G., Gubskij, A.L., Kovtun, A.P., Kritshko, A.A., Satshenko, V.P., *Kinet. Katal.* **24:** 819 (1983).
245. Anderson, A.B., Mehandru, S.P., Smialek, J.L., *J. Electrochem. Soc.* **132:** 1695 (1985).
246. Johnson, K.H., Pepper, S.V., *J. Appl. Phys.* **53:** 6634 (1982).
247. Anderson, A.B., Ravimohan, Ch., Mehandru, S.P., *Surf. Sci.* **183:** 438 (1987).
248. Anderson, A.B., Dowd, D.Q., *J. Phys. Chem.* **91:** 869 (1987).
249. (*a*) Sauer, J., Haberlandt, H., Schirmer, W., in *Structure and Reactivity of Modified Zeolites* (P.A., Jacobs, N.I., Jaeger, P., Jirů, V.B., Kazansky, and G. Schulz-Ekloff, eds.). Amsterdam: Elsevier, 1984, p. 313. (b) Sauer, J., Haberlandt, H., Pacchioni, G., *J. Phys. Chem.* **90:** 3051 (1986). (c) Haberlandt, H., Sauer, J., Pacchioni, G., *J. Mol. Structure* (*THEOCHEM*) **149:** 297 (1987).
250. Blomberg, M.R.A., Brandemark, U.B., Siegbahn, P.E.M., *Chem. Phys. Lett.* **126:** 317 (1986).
251. Bauschlicher Jr., Ch.W., *J. Chem. Phys.* **84:** 260 (1986).
252. Hoffmann, R., *J. Chem. Phys.* **39:** 1397 (1963).
253. Anders, L.W., Hansen, R.S., Bartell, L.S., *J. Chem. Phys.* **59:** 5277 (1973); **62:** 1641 (1975).
254. Xu Yin-Sheng, Hong Xiao-Le, *J. Mol. Catal.* **33:** 179 (1985).
255. Xu Yin-Sheng, Zhu Tian-Wei, *J. Mol. Catal.* **26:** 277 (1984).
256. Miessner, H. (unpublished results, 1986).
257. Anderson, A.B., Awad, M.K., *Surf. Sci.* **183:** 289 (1987).
258. Pople, J.A., Beveridge, D.L., *Approximate Molecular Orbital Theory*. New York: McGraw-Hill, 1970.
259. Klopman, G., Evans, R.C., in *Semiempirical Methods of Electronic Structure Calculations, Modern Theoretical Chemistry* (G.A. Segal, ed.). New York: Plenum Press, 1977, Vol. 7A, Chap. 2, p. 29.
260. Longuet-Higgins, H.G., Pople, J.A., *Proc. Phys. Soc.* **68:** 591 (1955).
261. Clack, D.W., Hush, N.S., Yandle, J.R., *J. Chem. Phys.* **57:** 3503 (1972).

262. Blyholder, G., *Surf. Sci.* **42:** 249 (1974).
263. Haberlandt, H., *Phys. Status Solidi B* **137:** 581 (1986).
264. Pacchioni, G., Fantucci, P., *Chem. Phys. Lett.* **134:** 407 (1987).
265. Blyholder, G., Head, J., Ruette, F., *Theor. Chim. Acta* **60:** 429 (1982).
266. Weinert, M., Freeman, A.J., Ohnishi, S., Davenport, J.W., *J. Appl. Phys.* **57:** 3641 (1985).
267. Baluffi, R.W., Rühle, M., Sutton, A.P., *Mat. Sci. Engineering* **89:** 1 (1987).
268. Chan, E.M., Buckingham, M.J., Robins, J.L., *Surf. Sci.* **67:** 285 (1977).
269. Yanagihara, T., Yamaguchi, H., *Phys. Status Solidi B* **93:** 67 (1986), and references therein.
270. Paryjczak, T., Bartczak, K., Wysocki, S., *React. Kinet. Catal. Lett.* **14:** 265 (1980).
271. Garofalini, S.H., Levine, S.M., *Surf. Sci.* **163:** 59 (1985).
272. Garofalini, S.H., Levine, S.M., *Surf. Sci.* **177:** 157 (1986).
273. Popescu, M., in *Proceedings of the 4th International Symposium on Heterogeneous Catalysis* (D. Shopov, A. Andreev, A. Palazov, and L. Petrov, eds.). Sofia: Bulgarian Academy of Sciences, 1979, Vol. I, p. 79.
274. Spencer, M.S., *J. Catal.* **93:** 216 (1985).
275. Erkoç, Ş., Halicioğlu, T., Pamuk, Ö., *Surf. Sci.* **169:** L273 (1986).
276. Zhang, S.B., Cohen, M.L., Louie, S.G., *Phys. Rev. B* **34:** 768 (1986).
277. Göpel, W., Anderson, J.A., Frankel, D., Jaehnig, M., Phillips, K., Schäfer, J.A., Rocker, G., *Surf. Sci.* **139:** 333 (1984).
278. Göpel, W., Rocker, G., Feierabend, R., *Phys. Rev. B* **28:** 3427 (1983).
279. Skuja, L.N., Silin, A.R., *Phys. Status Solidi A* **70:** 43 (1982).
280. Radtsig, V.A., Bobyshev, A.A., *Phys. Status Solidi B* **133:** 621 (1986).
281. Hochstrasser, G., Antonini, J.F., *Surf. Sci.* **32:** 644 (1972).
282. van Zee, R.J., Ferrante, R.F., Weltner Jr., W., *J. Chem. Phys.* **83:** 6181 (1985).
283. Ritschl, F., Haberlandt, H. (unpublished results, 1987); see also Ref. 351, Table 7.
284. Bisi, O., Chiao, L.W., Tu, K.N., *Surf. Sci.* **152/153:** 1185 (1985).
285. Xu Yongnian, Zhang Kaiming, Xie Xide, *Phys. Rev. B* **33:** 8602 (1986).
286. Henrich, V.E., *Rep. Progr. Phys.* **48:** 1481 (1985).
287. Kasowski, R.V., Tait, R.H., *Phys. Rev. B* **20:** 5168 (1979).
288. Gignac, W.J., Williams, R.S., Kowalczyk, S.P., *Phys. Rev. B* **32:** 1237 (1985).
289. Ciraci, S., Batra, I.P., *Phys. Rev. B* **28:** 982 (1983).
290. Kawakami, H., Yoshida, S., *J. Chem. Soc., Farad. Trans. II* **82:** 1385 (1986), and references therein.
291. Kunz, A.B., Guse, M.P., Blint, R.J., *Int. J. Quant. Chem.: Quant. Chem. Symp.* **10:** 283 (1976).
292. McConville, C.F., Woodruff, D.P., Kevan, S.D., Weinert, M., Davenport, J.W., *Phys. Rev. B* **34:** 2199 (1986).
293. Haberlandt, H., Pacchioni, G., Ritschl, F. (unpublished results, 1989).
294. Haberlandt, H. (unpublished results, 1985).
295. Brandemark, U.B., Blomberg, M.R.A., Petterson, L.G.M., Siegbahn, P.E.M., *J. Phys. Chem.* **88:** 4617 (1984).
296. Beran, S., Slanina, Z., Neshev, N., Proinov, E., in *Proceedings of the 6th International Symposium on Heterogeneous Catalysis* (D. Shopov, A. Andreev, A. Palazov, and L. Petrov, eds.). Sofia: Bulgarian Academy of Sciences, 1987, Vol. I, p. 330.
297. Beckmann, H.-O., Pacchioni, G., Jeung, G.-H., *Chem. Phys. Lett.* **116:** 423 (1985).
298. Shustorovich, E., *Surf. Sci.* **175:** 561 (1986).
299. Bagus, P.S., Müller, W., *Chem. Phys. Lett.* **115:** 540 (1985).
300. Bauschlicher Jr., Ch.W., Nelin, C.J., *Chem. Phys.* **108:** 275 (1986).

301. Tanaka, K., Viswanathan, B., Toyoshima, I., *J. Chem. Soc., Chem. Commun.* 481 (1985).
302. Lang, N.D., Holloway, S., Nørskov, J.K., *Surf. Sci.* **150:** 24 (1985), and references therein.
303. MacLaren, J.M., Pendry, J.B., Joyner, R.W., Meehan, P., *Surf. Sci.* **175:** 263 (1986).
304. Feibelman, P.J., Hamann, D.R., *Phys. Rev. Lett.* **52:** 61 (1984).
305. MacLaren, J.M., Pendry, J.B., Vvedensky, D.D., Joyner, R.W., *Surf. Sci.* **162:** 322 (1985).
306. Russell, Jr., J.N., Gates, S.M., Yates, Jr., J.T., *J. Chem. Phys.* **85:** 6792 (1986), and references therein.
307. Gadzuk, J.W., Holloway, S., *Chem. Phys. Lett.* **114:** 314 (1985).
308. Lee, Chyuan-Yih, DePristo, A.E., *J. Chem. Phys.* **85:** 4161 (1986).
309. Siegbahn, P.E.M., Blomberg, M.R.A., Bauschlicher, Jr., Ch.W., *J. Chem. Phys.* **81:** 1373 (1984).
310. Hamza, A.V., Madix, R.J., *J. Phys. Chem.* **89:** 5381 (1985).
311. Nørskov, J.K., Holloway, S., Lang, N.D., *Surf. Sci.* **137:** 65 (1984).
312. Shayegan, M., Glover, III, R.E., Park, R.L., *J. Vac. Sci. Technol. A* **4:** 1333 (1986).
313. Tracy, J.E., *J. Chem. Phys.* **56:** 2736 (1972).
314. Mehandru, S.P., Anderson, A.B., Ross, P.N., *J. Catal.* **100:** 210 (1986).
315. Goddard, III, W.A., Walch, S.P., Rappé, A.K., Upton, Th.H., Melius, C.F., *J. Vac. Sci. Technol. 14:* 416 (1977).
316. Baetzold, R.C., *J. Phys. Chem.* **88:** 5583 (1984).
317. Huang, Y.-J., Schwarz, J.A., Diehl, J.R., Baltrus, J.P., *Appl. Catal.* **36:** 229 (1988).
318. Huang, Y.-J., Schwarz, J.A., *Appl. Catal.* **36:** 177 (1988).
319. zur Loye, H.-C., Faltens, T.A., Stacy, A.M., *J. Am. Chem. Soc.* **108:** 8104 (1986).
320. Lu, K., Tatarchuk, B.J., *J. Catal.* **106:** 166 (1987); **106:** 176 (1987).
321. Martens, J.H.A., Prins, R., Zandbergen, H., Koningsberger, D.C., *J. Phys. Chem.* **92:** 1903 (1988).
322. Resasco, D.E., Weber, R.S., Sakellson, S., McMillan, M., Haller, G.L., *J. Phys. Chem.* **92:** 189 (1988).
323. Sankar, G., Vasudevan, S., Rao, C.N.R., *J. Phys. Chem.* **92:** 1878 (1988).
324. Ohuchi, F.S., French, R.H., Kasowski, R.V., *J. Vac. Sci. Technol. A* **5:** 1175 (1987).
325. Sadhegi, H.R., Henrich, V.E., *J. Catal.* **109:** 1 (1988).
326. Sen, B., Chou, P., Vannice, M.A., *J. Catal.* **101:** 517 (1986).
327. Underwood, R.P., Bell, A.T., *J. Catal.* **109:** 61 (1988).
328. Gorte, R.J., Altman, E., Corallo, G.R., Davidson, M.R., Asbury, D.A., Hoflund, G.B., *Surf. Sci.* **188:** 327 (1987).
329. Levin, M.E., Williams, K.J., Salmeron, M., Bell, A.T., Somorjai, G.A., *Surf. Sci.* **195:** 341 (1988).
330. White, J.M., in *Physical and Chemical Properties of Thin Metal Overlayers and Alloy Surfaces, Mat. Res. Soc. Symp. Proc.* (D.M. Zehner and D.W. Goodman, eds.). Pittsburgh: Material Research Society, 1987, Vol. 83.
331. Yong-Bo Zhao, Yip-Wah Chung, *J. Catal.* **106:** 369 (1987).
332. Chen, J.G., Crowell, J.E., Yates, Jr., J.T., *Surf. Sci.* **187:** 243 (1987).
333. Reschetilowski, W., Wendlandt, K.-P., Hobert, H., Wichterlova, B., *Z. Chem.* **27:** 307 (1987).
334. Rao, C.N.R., Rajumon, M.K., Prabhakaran, K., Hedge, M.S., Kamath, P.V., *Chem. Phys. Lett.* **129:** 130 (1986).
335. Taniguchi, S., Mori, Y., Hattori, T., Murakami, Y., *J. Catal.* **108:** 501 (1987).
336. Lee, G.v.d., Ponec, V., *Catal. Rev.-Sci. Engn.* **29:** 183 (1987).

337. Tauster, S.J., *Acc. Chem. Res.* **20:** 389 (1987).
338. Zheng Qing-qe, Zeng Zhi, Han Rushan, *Surf. Sci.* **195:** L173 (1988).
339. Tian Zeng-ju, Chikatoshi Satoko, Shukei Ohnishi, *Phys. Rev. B* **36:** 6390 (1987-II).
340. Xie Xide, Zhang Kaiming, Tian Zeng-ju, Ye Ling, in *Proceedings of the 18th International Conference on the Physics of Semiconductors* (O. Engström, ed.). Singapore: World Scientific, 1987, Vol. 1, p. 303.
341. Rodriguez, J.A., Campbell, Ch.T., *J. Phys. Chem.* **91:** 6648 (1987).
342. Li Yan, Chen Rong, Xiong Guoxing, Wang Hongli, *J. Mol. Catal.* **42:** 337 (1987).
343. Zhang Xia-Guang, Xu Yin-Sheng, Guo Xie-Xian, *J. Mol. Catal.* **43:** 381 (1988).
344. Vojtik, J., Češpiva, L., Fišer, J., Acta Phys. Polon. **A74:** 341 (1988).
345. Galeev, T.K., Dosumov, K., Popova, N.M., *React. Kinet. Catal. Lett.* **33:** 149 (1987).
346. Levine, S.M., Garofalini, S.H., *J. Chem. Phys.* **88:** 1242 (1988).
347. Sadhegi, H.R., Resasco, D.E., Henrich, V.E., Haller, G.L., *J. Catal.* **104:** 252 (1987).
348. Tsukada, M., Shima, N., *Phys. Chem. Min.* **15:** 35 (1987).
349. Munnix, S., Schmeits, M., *J. Vac. Sci. Technol. A* **5:** 910 (1987).
350. Fleisher, M.B., Golender, L.O., Shimanskaya, M.V., *React. Kinet. Catal. Lett.* **34:** 137 (1987).
351. Haberlandt, H., Pacchioni, G., *Chem. Phys.*, in press.
352. Bauschlicher, Jr., Ch.W., *Chem. Phys. Lett.* **142:** 71 (1987).
353. Bonačić-Koutecký, V., Koutecký, J., Fantucci, P., Ponec, V., *J. Catal.* **111:** 409 (1988).
354. MacLaren, J.M., Vvedensky, D.D., Pendry, J.B., and Joyner, R.W., *J. Chem. Soc., Faraday Trans. 1* **83:** 1945 (1987).
355. Judd, R.W., Reichelt, M.A., Lambert, R.M., *Surf. Sci.* **198:** 26 (1988).
356. MacLaren, J.M., Vvedensky, D.D., Pendry, J.B., Joyner, R.W., *J. Catal.* **110:** 243 (1988).

9
Mechanisms and Intermediates of Metal Surface Reactions: Bond-Order Conservation Viewpoint

EVGENY SHUSTOROVICH

INTRODUCTION

Heterogeneous catalysis was and remains one of the most empirical chemical endeavors. Catalyst development is still a domain of predominantly Edisonian research, a matter of trial-and-error efforts, intuitive assessments, and a great deal of luck. The reason for this is that the course of a catalytic reaction is determined by many factors, both thermodynamic and kinetic, and the composition and structure of a catalyst may be crucial as well. In this complex chain of events, there is one link that is common and critical for all the variety of catalytic processes. This link is chemisorption. Whatever the mechanism of a heterogeneous catalytic reaction may be, it includes chemisorption, transformations of chemisorbed species (dissociation, recombination, isomerization, etc.) and, finally, their desorption.[1–3] Thus, understanding heterogeneous catalysis is impossible without the knowledge of chemisorption patterns and, ultimately, the reaction energy profiles. What is the state of affairs in these quarters?

Historically, most theoretical modeling of chemisorption bonding was and is made on small metal clusters geometrically resembling a tiny piece of a metal surface, where an adspecies is treated as a ligand.[4–6] More recently, straightforward band-structure calculations of chemisorption on model metal slabs have begun to appear.[5,7–12] However, the premises of quantum chemistry of metal clusters and quantum physics of bulk metals are so different that, at present, there is no generally accepted theory of chemisorption, but a variety of competing approaches developed to describe separate aspects of chemisorption.[4–13]

Given the slow progress of quantum-mechanical models of chemisorption, a practical alternative appears to be phenomenological modeling. Critical here is the choice of basic assumptions and mathematical formalism. As Einstein put it,[14] nature is the realization of the simplest conceivable mathematical ideas, and the creativity of modeling resides in mathematics. We are free to set the rules of the game, but the rules must be rigid: it is their rigidity alone that makes the game possible.

During the last few years, we have been developing such a model of chemisorption on transition metal surfaces. It is based on bond-order conservation (BOC) and makes use of Morse potentials (MP). This BOC–MP model[15–18] explicitly interrelates a variety of chemisorption phenomena such that all the interrelations are expressed in terms of observables only—the heats of chemisorption and various constants (thermodynamic, structural, numerical, etc.). The scope of the model includes the preferred adsorbate sites[17a,17d]; the activation barriers for adsorbate migration,[17a,17d] dissociation,[17b,17f,17j] and recombination[17f]; various coverage and coadsorption effects such as overlayer phase transitions and island formation[17c]; and promotion and poisoning.[17e] And, what is of special value for catalytic applications, the model allows one to calculate reaction energetics on transition metal surfaces.[16,18]

The aim of this chapter is to show how the BOC–MP model can map surface reactions and, thus, help elucidate mechanisms and possible intermediates of heterogeneous catalytic reactions. We will begin with simple cases such as oxidation of CO ($CO + O \rightleftharpoons CO_2$) and nitrogenation of NO ($NO + N \rightleftharpoons [N_2O] \rightarrow N_2 + O$) and turn to complex processes such as hydrogenation of CO ($CO + H_2 \rightarrow CH_4$ vs CH_3OH) and transformations of C_2H_x species. The text will be organized as follows. First, we will give some background of the BOC–MP model and analytic formulas used to calculate the heats of chemisorption and the activation barriers of dissociation and recombination, the quantities that constitute the major elements of the reaction energy profile. We will provide the representative experimental data to illustrate the qualitative validity and the quantitative accuracy of the BOC–MP projections. After this, we will discuss the mechanisms of the cited reactions and compare our conclusions with experiment and some relevant theoretical results. Finally, we will comment on BOC–MP applications to heterogeneous catalysis.

THE BOC–MP MODEL OF CHEMISORPTION

This will be only a brief reminder since the model formalism and various applications have been extensively published[15–18] and summarized in recent reviews.[15,16]

Basic Assumptions

We consider chemisorption of an adsorbate X, atomic A or molecular AB, on transition metal M surfaces. High coordination of the M atoms (up to

12 nearest neighbors in the close-packed lattices such as fcc and hcp) makes the M–M and M–A forces quasispherical when the total energy E depends on the bond distance r only.

Here the main supportive arguments are the following: (i) both transition and simple s^1 (1A and 1B group) metals have the same densely packed (hcp, fcc, bcc) crystal lattices[1]; (ii) the d-orbital anisotropy is averaged out, and for many purposes the d band may be effectively represented by a degenerate s-type band[19]; (iii) for various cases of transition and simple metallic binding, including *chemisorption*, there exists an apparently universal relation between E and r.[19a,20]

Migration and dissociation of an adsorbate X involves changes in its coordination mode M_n–X (where n is the coordination number) and in the M–X distances r accompanied by changes in the M_n–X total energy E. To describe these processes, one should use a model potential relating E to r or some convenient function of r, for example, the two-center M–A *bond order* x,

$$x = \exp[-(r - r_0)/a] \tag{9.1}$$

which is an exponential function of the M–A distance r, and r_0 and a are constants. The possible chemisorption sites should correspond to the total energy E minima so that the model potential must include *both* attractive and repulsive forces. For two-center interactions, the simplest general potential of this kind is the Morse potential,[21] including only linear and quadratic terms in x; namely,

$$E(x) = -Q(x) = -Q_0(2x - x^2) \tag{9.2}$$

where Q_0 is the M–A equilibrium bond energy. The total energy $E(x)$ [Eq. (9.2)] has only one minimum at the equilibrium distance r_0 when the bond order $x = 1$, by definition Eq. (9.1).

In order to describe many-center M_n–A interactions in the Morse-type fashion, the simplest scheme is pairwise additivity of all the two-center M–A contributions to Q and x; namely,

$$Q_n = \sum_{i=1}^{n} Q_i \tag{9.3}$$

and

$$x_n = \sum_{i=1}^{n} x_i \tag{9.4}$$

Here, the simplest way to proceed is to keep the Morse parameters (Q_0, r_0, a) the same for both the isolated M–A bond and each additive M–A contribution within M_n–A so that these contributions would differ by their bond orders $x_i (i = 1, 2, \ldots, n)$ only. Experimentally, Q_n increases less than linearly with n:

$$Q_0 < Q_n < nQ_0, \qquad n > 1 \tag{9.5}$$

which requires imposing some constraints on the allowed values of $x_i < 1$ (if $x_i = 1$, then $Q_n = nQ_0$). The simplest assumption is that the *total* bond order x_n [Eq. (9.4)] does not change with n; namely, x_n is conserved and normalized to unity:

$$x_n = \sum_{i=1}^{n} x_i = x_0 = 1 \tag{9.4$'$}$$

for any $n \geq 1$ [cf. Eq. (9.1)]. The BOC at unity in its pairwise additive form is our major model assumption. For various linear three-center A . . . B . . . C interactions, such a form ($x_{AB} + x_{BC} = 1$) was already assumed[22] and shown to be very accurate, both computationally[23] and experimentally.[24] In a sense, we simply postulate the similar BOC for many-center M_n–A (spherical) interactions. Within the BOC framework, Morse potentials prove to be very efficient to describe the energetics of chemisorption, which ultimately originates from the zero-energy gap between the occupied and vacant parts of the metal band.[17g]

The last question concerns the values of n in M_n–A. The simplest assumption is to limit n to nearest-neighbor metal atoms. For instance, for A/fcc(100), the maximum $n = 4$ can be reached in the hollow site but $n = 2$ and 1 in the bridge and on-top sites, respectively. The last assumption reflects the known efficiency of the nearest-neighbor approximation in many problems of metallic binding.[13,19a]

In summary, our model assumptions are as follows:

1. Each two-center M–A interaction is described by the Morse potential [Eqs. (9.1) and (9.2)].
2. For a given M_n–A, n two-center M–A interactions are additive.
3. Along a migration path up to dissociation, the total M_n–X bond order is conserved and normalized to unity. [The analytic form of BOC depends on X; namely, Eq. (9.4′) for adatoms X = A, or Eq. (9.7) for admolecules X = AB.[25]]
4. For a given M_n–A, n is limited to nearest neighbors.

The assumptions (1)–(4) are the rules of the game. They are the simplest logical possibilities. The rest is straightforward algebra. The most definitive results have been obtained for chemisorption on flat symmetric surfaces with a regular unit mesh M_n, say, an equilateral triangle M_3 for fcc(111) or hcp(001), a square M_4 for fcc(100), etc. Consider the major findings.

Heats of Chemisorption

Atomic Chemisorption. For atomic chemisorption, the M_n–A bond energy Q_n monotonically increases with n as

$$Q_A = Q_n = Q_{0A}(2 - 1/n) \tag{9.6}$$

where Q_{0A} is the maximum M–A two-center bond energy [cf. Eq. (9.2)]. The value of Q_n reaches the absolute maximum in the hollow n-fold site, so that the observed heat of atomic chemisorption Q_A can be identified with Q_n. The immediate conclusion from Eq. (9.6) is that on flat symmetric surfaces adatoms will always prefer the highest coordination sites in the hollow depressions. Indeed, on surfaces in question, adatoms as varied as H, C, N, O, S, Se, Te, P, As, F, Cl, Br, and I, as well as Na and Cd, have been invariably found in the hollow sites of maximum coordination.[1,12,26]

Weak and Strong Molecular Chemisorption. For molecular AB chemisorption, BOC for M_n–AB can be approximated as

$$\sum_{i=1}^{n} (x_{Ai} + x_{Bi}) + x_{AB} = 1 \tag{9.7a}$$

or

$$x_A + x_B + x_{AB} = 1 \tag{9.7b}$$

where x_A and x_B are the effective polycenter M_n–A and M_n–B bond orders, but x_{AB} is the two-center A–B bond order. In order to obtain the value of Q_{AB}, we should maximize the total M_n–AB bond energy under the BOC conditions of Eqs. (9.7a) or (9.7b). The practical difference of using Eq. (9.7a) vs Eq. (9.7b) is that these BOC conditions introduce *different* Morse constants to describe the effective interaction of an atom A (or B) in a molecule AB with a metal surface, namely, the two-center A–M energy Q_{0A} for Eq. (9.7a) and the polycenter A–M_n energy Q_A for Eq. (9.7b). Since Q_A is almost twice as large as Q_{0A} [cf. Eq. (9.6)], it is more appropriate to use Eq. (9.7a) if the M_n–AB interaction is weak but Eq. (9.7b) if the M_n–AB interaction is strong. The criteria of assigning AB molecules to the weak or strong M_n–AB bonding are based on the general quantum-chemical rules. If the A–B bond significantly exhausts the valence capabilities of the A and B atoms, the M_n–AB bonding can be expected to be relatively weak. The best candidates appear to be AB molecules with a closed electronic shell, such as H_2, N_2, CO, NH_3, H_2O, or with unpaired electrons occupying the substantially delocalized molecular orbitals, such as NO or O_2. An important indicator of the weak M_n–AB bonding is relative insensitivity of Q_{AB} to the coordination site (on-top versus twofold bridge versus n-fold hollow). On the other hand, AB molecular radicals where unpaired electrons mainly retain their atomic character, such as CH, CH_2, NH, OH, OCH_3, are deemed to chemisorb much more strongly, resembling the patterns of atomic chemisorption including the distinct preference for the n-fold hollow coordination.

Analytic Formalism and Representative Examples. We begin with the BOC condition of Eq. (9.7a). The simplest case corresponds to AB perpendicular to a surface with the A end down when, to first approximation, the M_n–B bond order can be neglected [$x_{Bi} = 0$ in Eq. (9.7a)]. For such mono(η^1) coordination via A in the on-top site via one metal atom (μ_1) M–A–B, we then have

$$Q_{AB} \leq \frac{Q_{0A}^2}{Q_{0A} + D_{AB}} \tag{9.8}$$

where the inequality sign reflects neglect of a small negative contribution from the M–B interaction. In the same approximation, for the n-fold M_n–AB coordination ($\eta^1\mu_n$), we arrive at

$$Q_{AB,n} \leq \frac{Q_{0A}^2}{(Q_{0A}/n) + D_{AB}} \qquad \text{for } D_{AB} > \frac{n-1}{n} Q_{0A} \tag{9.9}$$

Here, however, the neglected negative M_n–B contribution is larger in absolute value than the one for M–B so that Q_{AB} may be either larger or smaller than $Q_{AB,n}$ and *all* the M_n–AB chemisorption sites tend to be *close* in energy, the approximation $Q_{AB} \simeq Q_{AB,n}$ becoming more accurate the larger the value of D_{AB}.

Since Eq. (9.9) is inadequate for judging how $Q_{AB,n}$ depends on n, the critical question is which value of n should be chosen in Eq. (9.9) to best reproduce the experimental value of Q_{AB}. Since Eq. (9.9) assumes the bond-order conservation for the coordinated atom A, it infers that, other conditions being equal, the stronger (multiply) bound AB molecules will prefer the lower coordination but the weaker (singly or doubly) bound molecules, the higher coordination. Thus, for the triply bound C≡O with $D_{CO} = 257$ kcal mol^{-1} [40] the value of Q_{CO} should be calculated for $n = 1$, reducing Eq. (9.9) to Eq. (9.8). On the other hand, with $D_{NO} = 151$ kcal mol^{-1},[40] NO is basically doubly bonded as N═O, and in Eq. (9.9) it seems reasonable to use $n = 2$.

Consider chemisorption of CO in more detail. Registries of CO have been studied on various flat surfaces of hcp Re, Ru, and Os and fcc Cu, Ag, Au, Ni, Pd, Pt, Ir, and Rh.[27] As *ground* chemisorption states, the on-top sites are typical but the hollow ($n = 3$) sites are rare [and have never been found for $n > 3$, say, for fcc(100), $n = 4$, or bcc(100), $n = 5$], which is in sharp contrast to the preferred hollow sites for adatoms (see previous discussion). The most important pattern, however, is that the energy differences among the on-top, bridge, and hollow sites are so small that at higher coverages and temperatures, some (or all) of these sites may co-exist.[27] For example, for CO/Pt(111), Q_{CO} decreases in the order on-top > bridge > hollow, but within $\Delta Q < 1$ kcal mol^{-1}.[28] Since $Q_{CO} = 32$ kcal mol^{-1}, the differences $\Delta Q/Q$ do not exceed 3%, which makes all the CO sites on Pt(111) practically isoenergetic. The similar difference of $\Delta Q < 1$ kcal mol^{-1} was found for CO/Rh(100),[29a] CO/Cu(111),[29b] CO/Ni(110),[30a] CO/Ni(100),[30b] and CO/Ni(111).[30c] For CO/Ni(111), the Q_{CO} order hollow > bridge > on-top[30c] is reversed compared with that for CO/Pt(111), which demonstrates the lack of correlation between Q_{CO} and n.[31]

From Eq. (9.9), it immediately follows that in the upright M_n–AB geometry AB will be coordinated to M_n through the atom whose heat of chemisorption is larger, namely, M_n–A–B if $Q_A > Q_B$. For example, since Q_A increases in the order O < N < C (reflecting the number of unpaired valence electrons), one can predict that monocoordination (η^1) of CO should always occur via C and NO via N, in full agreement with experiment.[1,27] Some examples of the CO and NO heats of chemisorption are given in Table 9.1. We see that Eq. (9.9) gives the values of Q_{AB} with the typical error 10–15%. The coordinations via oxygen (M–O–C or M–O–N) would have been less favorable by 10–15 kcal mol^{-1}.[17d]

If AB is coordinated parallel to a surface, via both A and B (dicoordination η^2), the bridge model

```
  A—B
 /   \
M-----M
```

appears to be the general prototype with the bonding energy

$$Q_{AB} = \frac{ab(a + b) + D_{AB}(a - b)^2}{ab + D_{AB}(a + b)} \tag{9.10}$$

Table 9.1. Initial heats of chemisorption Q_{AB} for some weakly bound molecules.[a]

Surface	Coordination Type	AB	Experimental Values of Q_A	Q_B	D_{AB}[b]	Q_{AB} Calculated	Q_{AB} Experimental
Ni(111)	η^1	CO	171		257	29[c]	27
Pt(111)	η^1	NO	116		151	26[d]	27
Pd(111)	η^1	NO	130		151	32[d]	31
Pt(111)	η^1	NH_3	116		279	13[c]	12–15
Ni(111)	η^1	NH_3	135		279	18[c]	20
Pt(111)	η^1	OH_2	85		220	11[c]	12
Pt(111)	η^1	$O{=}CH_2$	85		176	11[c]	11
Pt(111)	η^2	O_2	85	85	119	11[e]	9
Ru(001)	η^2	$O{=}C(CH_3)_2$	100	67	179	15[f]	16
Ni(111)	η^2	$H_2C{=}CH_2$	171	171	355	14[e]	13

[a] See text for notations and explanations, and Refs. 15 and 17f for sources of the experimental values of Q_A, Q_B, Q_{AB}. All energies in kcal mol^{-1}.
[b] Reference 40.
[c] Equation (9.8).
[d] Equation (9.9) for $n = 2$.
[e] Equation (9.11).
[f] Equation (9.10).

where

$$a = Q_{0A}^2(Q_{0A} + 2Q_{0B})/(Q_{0A} + Q_{0B})^2$$

and

$$b = Q_{0B}^2(Q_{0B} + 2Q_{0A})/(Q_{0A} + Q_{0B})^2$$

For a homonuclear A_2 ($a = b = \frac{3}{4}Q_{0A}$), Eq. (9.10) reduces to

$$Q_{A_2} = \frac{\frac{9}{2}Q_{0A}^2}{3Q_{0A} + 8D_{A_2}} \tag{9.11}$$

By comparing [Eqs. (9.9) and (9.10)], one can get an idea which coordination, η^1 or η^2, is more favorable. Although one should be careful with numbers (since the equations have been obtained in somewhat different approximations), some periodic trends can be discerned. For example, for CO the estimates show[17g] that on late transition metals, the monocoordination η^1 (via C) is always preferred, in full agreement with experiment.[27] However, the η^1 vs η^2 energy differences are small ($\Delta Q < 5$ kcal mol^{-1} [17f]) and decrease as one traverses from right to left along the transition metal series, so that in the middle of the period the η^1 and η^2 energies seem to converge and may even be reversed. This model conclusion is consistent with the recent findings that the η^2 coordination is slightly more favorable than the η^1 one for CO/Cr(110),[32a] CO/Fe(100),[32b] and CO/Mo(100).[32c] For O_2/Pt(111), where O_2 is known to be chemisorbed parallel to a surface, we used Eq. (9.11) only. The calculated value $Q_{O_2} = 11$ kcal mol^{-1} agrees well with the experimental value 9 kcal mol^{-1}.[33]

The accuracy of the BOC–MP estimates of Q may be compared with that of the available quantum-chemical techniques, usually of the cluster type. For example, one of the most accurate SCF–CI all-electron calculations of the cluster Ni_5–CO to mimic CO/Ni(100) gave $Q_{CO} = 13$ kcal mol^{-1} [34] to be compared with the experimental value of 30 kcal mol^{-1} [37] (and our estimate of 27 kcal mol^{-1} [17d]). A much simpler and rather popular technique is the ASED–MO method,[36] where the extended Hückel binding energies are complemented by some (empirically scaled) repulsive interactions. The ASED–MO calculations of the Ni_4–CO cluster to mimic CO/Ni(111) gave $Q_{CO} \simeq 12$ kcal mol^{-1} [37] versus the experimental value of 27 kcal mol^{-1} [35] (our estimate is 29 kcal mol^{-1} [17d]). At the same time, the ASED–MO calculations of the Pd_{19}–CO cluster to mimic CO/Pd(111) produced $Q_{CO} = 65$ kcal mol^{-1} [36a] compared to the experimental value of 34 kcal mol^{-1}.[38] The ASED–MO calculations of the Cr_{33}–CO[36b] cluster to mimic CO/Cr(110) projected $Q_{CO} = 134$ kcal mol^{-1} for the parallel geometry and $Q_{CO} = 90$ kcal mol^{-1} for the perpendicular geometry. Experimentally, the two geometries coexist at 120 K and differ in energy by *not* more than 1 kcal mol^{-1}.[32a] Also, although the experimental value of Q_{CO} for Cr(110) is not known (owing to easy dissociation of CO), it can be thought to be within the common range of $Q_{CO} = 25$–35 kcal mol^{-1} [15,38] on transition metal surfaces, including, in particular, Fe(110),[39a] Mo(110),[39b] and W(110).[39c] [For CO/Cr(110), we found $Q_{CO} = 38$ kcal mol^{-1} and ΔQ_{CO} (η^2 vs η^1) = 1–3 kcal mol^{-1}.[17d]]

In Eqs. (9.8)–(9.11), A and B may be both atoms and atomic groups (quasiatoms), so that the model can treat not only diatomic but also polyatomic molecules coordinated via one (η^1) or two (η^2) atoms. For example, AH_x molecules such as NH_3 and H_2O can be treated as a quasi-diatomic AB, where B is H_x, $D_{H_2O} = 220$ kcal mol^{-1}, and $D_{NH_3} = 279$ kcal mol^{-1}.[40] To describe the mono(η^1) coordination of these saturated molecules $O{=}H_2$ and $N{\equiv}H_3$, it seems reasonable to use Eq. (9.9) with $n = 1$. Also, Eq. (9.11) can be applied directly to the di(η^2) coordination of symmetric polyatomic molecules of the A_2 type such as H_xC–CH_x ($x = 2$, 1). For CH_xCH_x, in Eq. (9.11), D_{A_2} stands for the total energy of all bonds formed by each atom C. From $D_{CH_2CH_2} = 538$ kcal mol^{-1} and $D_{CH_2} = 183$

kcal mol^{-1},[40] we find $D_{CC} = 538 - 2 \times 183 = 172$ kcal mol^{-1}, which makes $D_{A_2} = 183 + 172 = 355$ kcal mol^{-1}. For the nonsymmetric coordination

```
   A—B
  /   \
 M—————M
```

however, in Eq. (9.10), the parameters a, b, and D_{AB} will now depend on the way of partitioning into the molecular fragments A and B.

Remember that partitioning of the total bond energy in a polyatomic molecule into the two-center contributions is a rather arbitrary procedure that is a matter of convenience for some practical purpose. In order to use the same formalism to calculate Q_{AB} for *both* diatomic and polyatomic molecules, the simplest and most uniform way to proceed is to define the A–B bond energy D_{AB} as the difference between the total gas-phase energies of AB and the dissociated fragments A and B. The only limitation of this definition is that it cannot be applied to exothermic dissociation of polyatomic molecules, where D_{AB} becomes negative. Since the overwhelming majority of dissociation processes are endothermic, they are within the scope of our modeling.

Now let us turn to the BOC condition of Eq. (9.7b) corresponding to the strong M_n–AB bonding. Although Eq. (9.7b) does not explicitly depend on n, it assumes the best possible coordination within the M_n–AB unit mesh, which reflects in the use of the *experimental* heats of atomic chemisorption, Q_A and Q_B, as the Morse constants in energy calculations. For the monocoordination $\eta^1\mu_n$ in M_n–AB, the variational procedure now leads to an expression

$$Q_{AB} = \frac{Q_A^2}{Q_A + D_{AB}} \tag{9.12}$$

which is an analog of Eqs. (9.8) and (9.9) for the weak M_n–AB bonding employing the parameter Q_A instead of Q_{0A}. As said before, Eq. (9.12) appears to be appropriate to treat chemisorption of molecular radicals such as CH, CH_2, NH, OH, or OCH_3. Table 9.2 lists some examples. Although the direct experimental data are practically not available (radicals usually decompose while being desorbed), the accuracy of these calculations can be verified via the calculated activation barriers ΔE^*, which we will discuss subsequently. Here we only mention that for OH and NH on Pt(111), with $Q_O = 85$[38] and $Q_N = 116$ kcal mol^{-1},[50] Eq. (9.12) gives $Q_{OH} = 39$ and $Q_{NH} = 68$ kcal mol^{-1}, respectively, in excellent agreement with the experimental ranges of 36–45 kcal mol^{-1} for OH on Pt(111)[58a,58b] and 63–69 kcal mol^{-1} for NH on Pt wire,[58c,58d] found in laser-induced fluorescence studies.

Table 9.2. Heats of molecular chemisorption Q_{AB}: strongly bound radicals[a]

System	Coordination Type	D_{AB}[b]	Q_A[c]	Q_{AB}
OH/M	η^1	102	80–125	35–69[d]
OCH_3/M	η^1	90[e]	80–125	38–73[d]
CH/M	η^1	81	150–200	97–142[d]
CH_2/M	η^1	183	150–200	68–104[d]
CH_3/M	η^1	293	150–200	38–62[f]
HCO/M	η^1	274	150–200	40–65[f]
NH/M	η^1	81	115–155	68–102[d]
NH_2/M	η^1	169	115–155	36–57[f]

[a] See text for explanations. All energies in kcal mol^{-1}.
[b] Reference 40.
[c] Reference 38.
[d] Equation (9.12).
[e] The C–O bond energy.
[f] Equation (9.13).

Of course, along with the weak and strong M_n–AB bonding, one can imagine the intermediate one, which may be described by interpolating between the two extremes. In particular, for the monocoordination ($\eta^1\mu_n$) M_n–AB, we will simply average Eqs. (9.9) and (9.12) as

$$Q_{AB} = \frac{1}{2}\left[\frac{Q_{0A}^2}{(Q_{0A}/n) + D_{AB}} + \frac{Q_A^2}{Q_A + D_{AB}}\right] \tag{9.13}$$

The intermediate M_n–AB bond strength may be expected for monovalent radicals AB, where A is a tri- or tetravalent atom, say N or C in NH_2, CH_3, or HCO. Several examples are given in Table 9.2.

From Eqs. (9.8)–(9.13) and Tables 9.1 and 9.2, it clearly follows that the molecular heat of chemisorption Q_{AB} rapidly decreases as the gas-phase dissociation (total bond) energy D_{AB} increases, Q_{AB} being typically smaller than Q_A (Q_B) by a factor of 5–10. For this reason, the periodic changes in Q_{AB} for molecules such as CO, CH_3, NH_3, NO, H_2O, C_2H_4, and C_2H_2 are small and irregular compared with the large and systematic variations in Q_A observed for the relevant multiply-bonded adatoms A.[12,15]

Dissociation and Recombination Barriers

Analytic Formalism. If AB approaches a surface from the gas phase, the activation barrier $\Delta E^*_{AB,g}$ for dissociation $AB_g \rightarrow A_s + B_s$ explicitly depends on the chemisorption energies of the adsorbates. The conceptual advantage of the BOC–MP modeling is that it can effectively treat the molecular AB_s and atomic $A_s + B_s$ energy profiles as *multidimensional* hypersurfaces by assuming some residual bonding A–B with the bond order $c > 0$ in the transition state. Then the variational procedure leads to the dissociation barrier[16,17j]

$$\Delta E^*_{AB,g} = (1 - c)^2 D_{AB} - (Q_A + Q_B) + (1 + c)^2 \frac{Q_A Q_B}{Q_A + Q_B} \quad (9.14)$$

or, for homonuclear A–A dissociation,

$$\Delta E^*_{A_2,g} = (1 - c)^2 D_{A_2} - \frac{(1 - c)(3 + c)}{2} Q_A \quad (9.15)$$

Remember that the conventional Lennard-Jones (LJ) representation of diatomic AB dissociation is one dimensional, where the atomic $A_s + B_s$ and molecular AB_s energy profiles are assumed to intersect[15c] so that $c = 0$, and Eqs. (9.14) and (9.15) become, respectively,

$$\Delta E^{*,LJ}_{AB,g} = D_{AB} - (Q_A + Q_B) + \frac{Q_A Q_B}{Q_A + Q_B} \quad (9.16)$$

and

$$\Delta E^{*,LJ}_{A_2} = D_{A_2} - \tfrac{3}{2} Q_A \quad (9.17)$$

In principle, the values of c may vary from one system AB/M to another. Fortunately, one should not know them explicitly, because the barrier can be averaged as[16,17j]

$$\Delta E^*_{AB,g} = \tfrac{1}{2}(\Delta E^{*,LJ}_{AB,g} - Q_{AB}) \quad (9.18)$$

Obviously, the dissociation barrier $\Delta E^*_{AB,s}$ from a chemisorbed state will be larger than $\Delta E^*_{AB,g}$ (one- or multidimensional) just by the amount of the molecular heat of chemisorption Q_{AB}:

$$\Delta E^*_{AB,s} = \Delta E^*_{AB,g} + Q_{AB} \quad (9.19)$$

Combining with Eq. (9.18), we obtain

$$\Delta E^*_{AB,s} = \tfrac{1}{2}(\Delta E^{*,LJ}_{AB,g} + Q_{AB}) \quad (9.20)$$

For the reverse reaction of recombination of chemisorbed A_s and B_s, the one-dimensional LJ activation barrier is particularly simple

$$\Delta E^{*,LJ}_{A-B} = \frac{Q_A Q_B}{Q_A + Q_B} \tag{9.21}$$

Clearly, if $Q_A \gg Q_B$, $\Delta E^{*,LJ}_{A-B}$ will be close to Q_B, the heat of chemisorption of the weaker bound partner. If $Q_A = Q_B$, we simply have $\Delta E^{*,LJ}_{A-A} = \frac{1}{2}Q_A$.

Obviously, the activation recombination barrier $\Delta E^{*,LJ}_{A-B}$ cannot be smaller than $\Delta H = \Delta H_{AB} - \Delta H_{A+B}$, the difference between the enthalpies of the reactant AB_s ($-\Delta H_{AB} = D_{AB} + Q_{AB}$) and the products A_s and B_s ($-\Delta H_{A+B} = Q_A + Q_B$). Thus, the BOC barrier [Eq. (9.21)] is only the *necessary* (minimal energy) condition for recombination, which may be sufficient if

$$\frac{Q_A Q_B}{Q_A + Q_B} \geq \Delta H = Q_A + Q_B - D_{AB} - Q_{AB} \tag{9.22}$$

but not sufficient if $Q_A Q_B/(Q_A + Q_B) < \Delta H$. In such cases, the recombination barrier may be assumed to be the enthalpy difference, $\Delta E^{*,LJ}_{A-B} = \Delta H$. In general, for the multidimensional dissociation described by Eqs. (9.14) and (9.18), the recombination barriers ΔE^*_{A-B} can be calculated only from the relevant thermodynamic relations. Remember that for the recombination of chemisorbed A_s and B_s to chemisorbed AB_s or gas phase AB_g, the activation barriers $\Delta E^*_{A-B,s}$ and $\Delta E^*_{A-B,g}$ may be the same or different depending on the sign of the gas-phase dissociation barrier $\Delta E^*_{AB,g}$, namely,

$$\Delta E^*_{A-B,s} = \Delta E^*_{A-B,g} \qquad \text{if } \Delta E^*_{AB,g} > 0 \tag{9.23a}$$

and

$$\Delta E^*_{A-B,s} < \Delta E^*_{A-B,g} \qquad \text{if } \Delta E^*_{AB,g} < 0 \tag{9.24a}$$

More specifically, Eq. (9.23a) can be rewritten as

$$\Delta E^*_{A-B,s} = \Delta E^*_{A-B,g} = Q_A + Q_B - D_{AB} + \Delta E^*_{AB,g} \tag{9.23b}$$

and Eq. (9.24a) as

$$\Delta E^*_{A-B,g} = \Delta E^*_{A-B,s} - \Delta E^*_{AB,g} = Q_A + Q_B - D_{AB} \tag{9.24b}$$

Representative Examples. How do these interrelations fit the experiment? Equation (9.17) establishes the linear correlation between the dissociation barrier $\Delta E^*_{A_2}$ and the atomic heat of chemisorption Q_A with the slope of $k = \frac{3}{2}$. It should be stressed that, unlike similar linear relations between the activation barriers and the heats of reactions (Bronsted, Polanyi, Frumkin–Temkin–Semyenov, etc.), Eq. (9.17) is not a postulate but a corollary of the general principle (BOC) applied to the one-dimensional dissociation $AB_s \rightarrow A_s + B_s$. As seen from Table 9.3 for diatomic molecules H_2, O_2, and N_2, dissociated on a variety of metal surfaces (Fe, Ni, Cu, W, Pt, etc.), the experimental values of k lie within the range $k =$ 1.4–1.7, i.e., within 10–15% of the theoretical one-dimensional (LJ) value of $k = 1.5$. Quantitatively, however, the error in $\Delta E^{*,LJ}_{A_2,g}$ may reach (0.15–0.25)$Q_A \simeq$ 10–20 kcal mol^{-1}, which might be chemically unacceptable. So, to describe quantitatively the diatomic AB dissociation, the multidimensional treatment appears to be vital. Table 9.4 illustrates this point and shows that the use of Eq. (9.18) instead of Eq. (9.17) leads to significant improvement, the deviations from the experimental values of $\Delta E^*_{A_2,g}$ now being not more than 2–3 kcal/mol^{-1}.

Equation (9.18) is also superior to Eq. (9.16) in describing $\Delta E^*_{CO,g}$ for CO dissociation on active transition metal surfaces. Here chemisorbed CO thermally dissociates while being desorbed, which hints $\Delta E^*_{AB,g} \leq 0$ [cf. Eq. (9.26)], and only Eq. (9.18) produces consistent results. In particular for CO/Ni(100) and W(110), we find (see Table 9.4) $\Delta E^*_{CO,g} = 0$ and -6 kcal mol^{-1} compared to the experimental estimates of -3[52c] and -15[39c] kcal mol^{-1}, respectively. We project $\Delta E^*_{CO,g} = -6$ kcal mol^{-1} also for Fe

Table 9.3. Observed range of k for A_2 dissociation.[a]

		Experimental Values of			
A_2	Surface	Q_A	D_{A_2}	$\Delta E^*_{A_2}$	k[b]
H_2	Fe(111)	62	104	~0	1.7
	Ni(111)	63		2	1.6
	Ni(110)	62		~0	1.7
	Cu(100)	56		5	1.7
N_2	W(110)	155	228	~10	1.4
	Fe(110)	138		8	1.6
	Fe(100)	140		2.5	1.6
	Fe(111)	139		−0.8	1.4
O_2	Pt(111)	85	119	−1	1.4

[a] See text for notations and Table 11 in Ref. 15a for sources of the experimental values of Q_A, D_{A_2}, $\Delta E^*_{A_2}$. All energies in kcal mol^{-1}.
[b] From $\Delta E^*_{A_2} = D_{A_2} - kQ_A$ [to be compared with the theoretical value $k = 1.5$ in Eq. (9.17)].

Table 9.4. Dissociation and recombination barriers ΔE^* for some surface reactions.[a]

Reaction	Surface	Experimental Values of D_{AB}[b]	Q_A	Q_B	Q_{AB}	ΔE^* Calculated	ΔE^* Experimental
$H_{2,g} \rightarrow H_s + H_s$	Fe(111)	104	62	62	7	2[c]	0
	Ni(111)		63	63	7	1[c]	2
	Cu(100)		56	56	5	7[c]	5
$N_{2,g} \rightarrow N_s + N_s$	Fe(110)	228	138	138	8	6[c]	8
	Fe(100)		140	140	8	4[c]	2
	Fe(111)		139	139	8	5[c]	0
$CO_g \rightarrow C_s + O_s$	Ni(111)	257	171	115	27	6[c]	
	Ni(100)		171	130	30	0[c]	−3[d]
	W(110)		200	125	21	−6[c]	−15[e]
$NO_g \rightarrow N_s + O_s$	Rh(100)	151	128	102	26	−23[c]	−15[f]
	Pt(111)		116	85	27	−14[c]	
$CO_{2,g} \rightarrow CO_s + O_s$	Rh(111)	127	32	102		17[g]	17[h]
	Re(001)		29	127		−5[g]	≤0[i]
$CO_s + O_s \rightarrow CO_{2,g}$	Rh(111)		32	102		24[g]	27[j]
	Pd(111)		34	87		24[g]	25[k]
	Pt(111)		32	85		23[g]	25[k]
	Ag(110)		6.5	80		6.0[g]	5.3[l]
$NO_s + N_s \rightarrow N_2O_s$[m]	Rh(111)		26	128		22[g]	21[n]
	Rh(100)		25	131		21[g]	21[o]
	Pt(111)		27	116		22[g]	20[p]
$N_2O_s \rightarrow N_{2,s} + O_s$	Rh(111)	40	9	102		−63[g]	
	Pt(111)		9	85		−46[g]	

[a] See text for notations and Refs. 15a, 17f, and 38 for sources of the experimental values of Q_A, Q_B, Q_{AB}, and $\Delta E^*_{A2,g}$. All energies in kcal mol^{-1}.
[b] Reference 40.
[c] Equation (9.18).
[d] Reference 52c.
[e] Reference 39c.
[f] Reference 48b.
[g] Equation (9.16) or (9.21), respectively.
[h] Reference 45.
[i] Reference 46.
[j] Reference 42.
[k] Reference 41.
[l] Reference 43.
[m] Followed by nonactivated decomposition $N_2O_s \rightarrow N_{2,g} + O_s$.
[n] Reference 47.
[o] Reference 48.
[p] Reference 49.

and Mo surfaces,[16] in agreement with the experimental estimates of −12 kcal mol^{-1} for Fe(111)[77] and −2 kcal mol^{-1} for Mo(100).[78] As Tables 9.4 and 9.6 illustrate, Eq. (9.18) correctly reproduces the periodic trends in dissociation of CO, when the metal activity rapidly *diminishes* along the series W, Fe > Ni, Ru ≫ Pd > Pt.[59]

We turn to polyatomic molecules, which we will treat as quasidiatomic. Now the BOC condition of Eq. (9.7b) introduces the more composite group terms x_A and x_B, and the use of the effective Morse constants Q_A, Q_B, and D_{AB} further blurs the difference between the one- and multidimensional pictures of AB dissociation. For this reason, it is not clear in advance whether Eq. (9.18) or Eq. (9.16) may prove to be more accurate. For the sake of uniformity, the use of Eq. (9.18) would be of great advantage. Indeed, Eq. (9.18) gives remarkably reasonable results for a broad variety of polyatomic molecules,[16] and we will consider numerous examples later (see Tables 9.6, 9.7, and 9.10). At the same time, we have also found few cases where Eq. (9.16) works better, for example, for dissociation of triatomic linear molecules A–B–C such as O–C–O and N–N–O.

The preference of Eq. (9.16) for CO_2 dissociation may be well anticipated. It has been shown [see, for example, the HREELS studies of CO_2 on Re(001),[60a] UPS/XPS studies of CO_2 on Fe(111), (110),[60b] and computer simulations for CO_2 on Pt(111)[60c]], that the molecule is practically undistorted (symmetric and linear) in the ground chemisorbed state but strongly distorted (nonsymmetric and bent) as an intermediate preceding the dissociation $CO_{2,s} \rightarrow CO_s + O_s$. So, there is a good reason to believe that in the transition state the coordinated C–O bond is strongly expanded (by 0.12 Å[60c]) and becomes very weak. But the weaker the C–O bond ($x_{CO} \simeq 0$), the more accurate is Eq. (9.16). We will return to CO_2 dissociation later.

The Relevance to Real Heterogeneous Reactions. These and other BOC–MP interrelations nicely reproduce a broad range of experimental data[15–17] corresponding to the zero-coverage extreme when the most accurate measurements of Q and ΔE^* exist.[15–17] Real heterogeneous reactions are commonly studied for steady-state conditions that correspond to substantial coverages of coadsorbed species. Coverage effects change the heats of chemisorption and the activation barriers in a complex (nonlinear) way.[17c,17e] In particular, as the coverage increases, the heat of chemisorption decreases, especially for molecules coadsorbed with multiple-bonded atoms, say for CO + S or CO + O.[15,17c,17e] The resulting changes in the activation barriers, for both dissociation and recombination from a chemisorbed state [cf. Eqs. (9.23) and (9.24)], will tend to compensate to some extent, although reliable quantitative estimates are not possible at present. As a consequence, we will proceed with zero-coverage estimates in the hope that they qualitatively reflect the relative energetics of elementary steps comprising the reaction mechanism at the steady-state conditions.

The reaction rate constant r of an elementary process is described in the Arrhenius form by a preexponential factor A and an activation energy

ΔE^*, such that

$$r = A \exp(-\Delta E^*/RT) \tag{9.25}$$

Clearly, in order to make a meaningful comparison between two reactions based solely on the values of ΔE^*, the values of A should be close. This condition is most likely to be met for a given reaction on a given surface, say fcc(111), but of two different metals, say Pd(111) versus Ni(111). Related to this is a comparison of rates, r_{des} versus r_{diss}, of two first-order reactions, desorption $AB_s \rightarrow AB_g$ and dissociation $AB_s \rightarrow A_s + B_s$. Because of the fundamental inequality of the preexponentials $A_{des} \gg A_{diss}$ (usually $A_{des}/A_{diss} \simeq 10^3$–$10^4$ [61]) and since $\Delta E^*_{diss} = \Delta E^*_{AB,s} = \Delta E^*_{AB,g} + Q_{AB}$ [Eq. (9.19)] and $\Delta E^*_{des} = Q_{AB}$, the typical condition of prevailing dissociation is the negative dissociation barrier from the gas phase, that is

$$r_{diss} > r_{des} \qquad \text{if } \Delta E^*_{AB,g} < 0. \tag{9.26}$$

The second-best case to consider is two reactions of the same order differing by one reactant only, for example, $C_s + H_s \rightleftharpoons CH_s$ versus $CH_s + H_s \rightleftharpoons CH_{2,s}$. Finally, if a vacant surface site S is treated as a reactant, one can compare dissociation reactions with recombination reactions, e.g., $S + CO_s \rightleftharpoons C_s + O_s$ versus $H_s + CO_s \rightleftharpoons HCO_s$. Here, however, the conclusions will be the least definitive, since the differences in the nature of surface sites (on-top, bridge, hollow) and in the effective number of the sites involved in the reaction may strongly affect the value of A.

One more comment seems necessary. The Arrhenius expression [Eq. (9.25)] is commonly used to describe the rates of nonelementary reactions including several steps. In this case, the measured value of ΔE^* is the apparent (global) activation energy, which is the resultant of sums and differences (with some coefficients) of activation energies of elementary steps whose rates contribute to the global rate.[62] In our model approach, we calculate ΔE^* for elementary steps only. Thus, there is no direct and simple way to compare our calculated barriers with the apparent barriers of nonelementary processes. This is particularly true for energy estimates made from the thermal-stability thresholds of chemisorbed species.

MAPPING OF SURFACE REACTIONS

The BOC–MP method provides remarkably accurate estimates of the heats of chemisorption Q and the dissociation and recombination barriers ΔE^* for various molecules and molecular fragments. Combined with the knowledge of the molecular total bond (gas-phase dissociation) energies, this allows one to probe potential energy profiles of surface reactions.

Simple Processes

Oxidation of CO. The formation of CO_2 through catalytic oxidation of CO is, at the moment, the best understood catalytic reaction.[41] On platinum-group metal catalysts, the reaction mechanism includes the following elementary steps:

$$CO_g \rightleftharpoons CO_s \tag{9.27}$$

$$O_{2,g} \rightleftharpoons O_{2,s} \tag{9.28}$$

$$O_{2,s} \rightleftharpoons 2O_s \tag{9.29}$$

$$CO_s + O_s \rightleftharpoons CO_{2,s} \tag{9.30}$$

$$CO_{2,s} \rightleftharpoons CO_{2,g} \tag{9.31}$$

The rate-limiting step is Eq. (9.30), and the measured activation energies $\Delta E^*_{\text{O-CO}}$ for Rh,[42] Pd,[41] Pt,[41] as well as Ag[43] are given in Table 9.4.

Consider now the BOC–MP projections on the mechanism [Eqs. (9.27)–(9.31)]. Dissociation $O_{2,s} \rightarrow 2O_s$ should be practically a nonactivated process, as follows from Eq. (9.17), where $D_{O_2} = 119$ kcal mol^{-1}[40] and $Q_O \geq 80$ kcal mol^{-1} for all metals.[15,41,43] So, at least at low coverages, we expect O_2 to easily dissociate, and atomic oxygen O_s to be readily available, in agreement with experiment.[41–44] As explained before, the activation barrier $\Delta E^*_{\text{O-CO}}$ can be directly obtained from Eq. (9.21). The calculated values of $\Delta E^*_{\text{O-CO}}$ are given in Table 9.4 and compared with the experimental data at low coverages. The agreement is excellent. In particular, as Eq. (9.21) predicts, the recombination barrier $\Delta E^*_{\text{O-CO}}$ is very sensitive (and rather close) to Q_{CO} but insensitive to Q_O (since $Q_O \gg Q_{CO}$). For the same reason, at high oxygen coverages, this barrier dramatically drops (by a factor of 2[41] or 3[44]), since Q_{CO} drops this way.[17e,41,44]

For the reverse (dissociation) reaction of Eq. (9.30), the activation barrier $\Delta E^*_{\text{(OC)O}}$ can be calculated from Eq. (9.16). The results are given in Table 9.4 and compared with the available experiment for Rh[45] and Re,[46] in good agreement.

Nitrogenation of NO. A similar but more complex example is N_2 desorption resulting from NO chemisorptive dissociation, which has been carefully studied on Rh[47,48] and Pt.[49] The suggested mechanism of the reaction is

$$NO_g \rightleftharpoons NO_s \tag{9.32}$$

$$NO_s \rightleftharpoons N_s + O_s \tag{9.33}$$

$$NO_s + N_s \rightleftharpoons [N_2O_s] \rightarrow N_2 + O_s \tag{9.34}$$

$$N_s + N_s \rightleftharpoons N_2 \tag{9.35}$$

Here the steps [Eqs. (9.34) and (9.35)] are assumed to be responsible for two observed N_2 peaks at 500 K and 650 K, respectively, where the high-temperature peak (650 K) shows normal second-order kinetics, but the low-temperature peak (500 K) shows the first-order dependence on NO coverage, which was explained by formation and facile decomposition of the N_2O transition state. The activation energies of this step [Eq. (9.25)] for Rh(111),[47] Rh(100),[48] and Pt(111)[49] were found to be within the very narrow range 20–21 kcal mol^{-1}, close to the respective values of Q_{NO} = 25–27 kcal mol^{-1},[47–49] as shown in Table 9.4.

What can the BOC–MP model say about the mechanism [Eqs. (9.32)–(9.35)]? First of all, the model projects that at low coverages, decomposition of NO should be nonactivated (see Table 9.4), so that molecular NO can coexist with atomic N and O only at high enough coverages. Indeed, N_2 desorption has been observed only for high exposures above 0.5 L with $\theta_{NO} > 0.3$,[47–49] and under these conditions the above values of Q_{NO} have been measured.[47–49] From Eq. (9.21) it immediately follows that for a given Q_N, the recombination barrier $\Delta E^*_{N\text{-}NO}$ should be smaller than $\Delta E^*_{N\text{-}N}$, which explains the temperature sequence of the peaks. For quantitative estimates of the recombination barriers, we have to know the magnitudes of Q_N for the surfaces in question. The values of Q_N were suggested only for Pt(111), the latest one being Q_N = 116 kcal mol^{-1}.[50] From Eq. (9.21), it gives $\Delta E^*_{N\text{-}NO}$ = 22 kcal mol^{-1}, in excellent agreement with experiment.[49] Even if a large variation of ΔQ_N = 30 kcal mol^{-1} is allowed, covering the reasonable range of Q_N for Rh, Pd, and Pt,[15,17f] the value of $\Delta E^*_{N\text{-}NO}$ will change only by 1 kcal mol^{-1}, which explains the remarkable similarity of the patterns discussed (which is eventually the consequence of the closeness of Q_{NO} for Rh, Pd, and Pt). Finally, as seen from Table 9.4, N_2O should decompose into N_2 and O without activation, which supports the critical assumption [Eq. (9.34)].

Complex Processes

Hydrogenation of CO. As the first example of a complex process including several competing pathways, let us consider hydrogenation of CO over the platinum-group metals. The process shows distinct periodic regularities; namely, methane CH_4 has been produced on Ni, Pd, and Pt but methanol CH_3OH only on Pd and Pt.[51–57] One can wonder what might be possible pathways for CH_4 and CH_3OH synthesis and how they depend on metal composition. Here, there are two questions of special interest: (1) why the C–O bond cannot be retained on nickel catalysts, and (ii) how does C–O bond cleavage occur, directly from CO or from partially hydrogenated species H_xCO (hydrogen-assisted C–O cleavage)?

Table 9.5. Initial heats of chemisorption (Q) and total bond energies in the gas phase (D) and in chemisorbed ($D + Q$) states on Ni(111), Pd(111), and Pt(111).[a]

		Ni		Pd		Pt	
Species	D[b]	Q	$D + Q$	Q	$D + Q$	Q	$D + Q$
C	—	171	171	160	160	150	150
CH	81	116	197	106	187	97	178
CH_2	183	83	266	75	258	68	251
CH_3	293	48	341	42	335	38	331
CH_4	398	6[c]	404	6[c]	404	6[c]	404
H	—	63	63	62	62	61	61
O	—	115	115	87	87	85	85
OH	102	61	163	40	142	39	141
OH_2	220	17	237	10	230	10	230
OCH_3	383	65	448	43	426	41	424
CH_3OH	487	18	505	11	498	11	498
CO	257	27[d]	284	34[d]	291	32[d]	289
HCO	274	50	324	44	318	40	314
H_2CO	361	19	380	12	373	11	372

[a] All energies in kcal mol^{-1}. See text.
[b] Reference 40.
[c] Taken as the experimental value of $Q_{CH_4} \simeq 6$ kcal mol^{-1} on Rh: Brass, S.G., Ehrlich, G., *Surf. Sci.* **187:** 21 (1987).
[d] Experimental values (Ref. 38).

Table 9.5 lists the total bond energies of reactants and products as well as conceivable intermediates in the gas-phase (D) and chemisorbed ($D + Q$) states on Pt(111), Pd(111), Ni(111). Tables 9.6 and 9.7 summarize the activation barriers of the conceivable elementary steps leading to CH_4 and CH_3OH, respectively. For the computational details, the reader is referred to the review.[16]

As seen from Table 9.5, in the gas phase the methanation reaction $CO + 3H_2 \rightarrow CH_4 + H_2O$ with $\Delta H = -49$ kcal mol^{-1} is much more exothermic than the methanol formation $CO + 2H_2 \rightarrow CH_3OH$ with $\Delta H = -22$ kcal mol^{-1}. Thus, the only chance to selectively produce methanol is to have a catalyst where hydrogenation of H_xCO species (x = 0–3) is kinetically preferred over the H_xC–O bond cleavage. Even then, since the thermodynamic and kinetic factors tend to compete, the same metal catalyst may produce both $CH_4 + H_2O$ and CH_3OH depending on the reaction conditions and the catalyst structure. On the other hand, if the barriers for the H_xC–O bond dissociation are smaller than those for the H_xCO hydrogenation, the thermodynamic and kinetic factors are in

Table 9.6. Zero-coverage activation barriers for forward (ΔE_f^*) and reversed (ΔE_r^*) elementary reactions for methanation $CO + H_2 \rightarrow CH_4 + H_2O$ on Ni(111) and Pd(111).[a]

Reaction	ΔE_f^* Ni	ΔE_f^* Pd	ΔE_r^* Ni	ΔE_r^* Pd
$CO_s \rightleftarrows C_s + O_s$	33	50	35	6
$H_s + C_s \rightleftarrows CH_s$	42	40	5	5
$H_s + CH_s \rightleftarrows CH_{2,s}$	17	15	23	24
$H_s + CH_{2,s} \rightleftarrows CH_{3,s}$	12	9	24	24
$H_s + CH_{3,s} \rightleftarrows CH_{4,g}$	14	9	8	10
$\rightleftarrows CH_{4,s}$	14	9	14	15
$H_s + CO_s \rightleftarrows HCO_s$	23	35	0	0
$HCO_s \rightleftarrows CH_s + O_s$	35	46	23	2
$\rightleftarrows C_s + OH_s$	18	24	28	8
$CO_s + H_s \rightleftarrows CH_s + O_s$	58	81	23	2
$\rightleftarrows C_s + OH_s$	41	59	28	8

[a] See Table 9.5 for the values of Q and $D + Q$ used in calculations of ΔE^*. The dissociation barriers from Eqs. (9.18) or (9.20) and recombination barriers from Eqs. (9.23b) or (9.24b). All energies in kcal mol^{-1}.

Table 9.7. Zero-coverage activation barriers for forward (ΔE_f^*) and reversed (ΔE_r^*) elementary reactions for methanol formation $CO + H_2 \rightarrow CH_3OH$ on Ni(111) and Pd(111).[a]

Reaction	ΔE_f^* Ni	ΔE_f^* Pd	ΔE_r^* Ni	ΔE_r^* Pd
$CO_s + H_s \rightleftarrows HCO_s$	23	35	0	0
$HCO_s + H_s \rightleftarrows H_2CO_s$	33	16	26	9
$H_2CO_s \rightleftarrows CH_{2,s} + O_s$	24	34	23	6
$H_2CO_s + H_s \rightleftarrows CH_3O_s$	5	10	10	1
$CH_3O_s \rightleftarrows CH_{3,s} + O_s$	13	16	21	12
$CH_3O_s + H_s \rightleftarrows CH_3OH_g$	24	7	−5	6
$\rightleftarrows CH_3OH_s$	19	7	13	17
$CH_3OH_g \rightleftarrows CH_{3,s} + OH_s$	−4	10	17	0
$CH_3OH_s \rightleftarrows CH_{3,s} + OH_s$	14	22	13	0

[a] See Table 9.5 for the values of Q and $D + Q$ used in calculations of ΔE^*. The dissociation barriers from Eqs. (9.18) or (9.20) and recombination barriers from Eqs. (9.23b) or (9.24b). All energies in kcal mol^{-1}.

accord, and a catalyst may selectively produce methane only. We begin by considering the pathways for the formation of CH_4 and CH_3OH on Ni.

For Ni(111) and Ni(100), we project the dissociation barriers from the gas phase $\Delta E^*_{CO,g} = 6$ and 0 kcal mol^{-1}, respectively, in agreement with easy thermodissociation of CO on Ni surfaces[52–54] (cf. Table 9.4). Thus, for direct dissociation $CO_s \rightarrow C_s + O_s$ on Ni(111) we predict $\Delta E^*_{CO,s} = 33$ kcal mol^{-1}, the process being slightly exothermic ($\Delta H = -2$ kcal mol^{-1}). The hydrogen-assisted C–O cleavage $CO_s + H_s \rightarrow HCO_s \rightarrow C_s + OH_s$ requires the higher activation energy of 41 kcal mol^{-1}. One should remember, however, that during methanation $CO + 3H_2 \rightarrow CH_4 + H_2O$, oxygen is removed as water. But here the first (atomic) recombination $H_s + O_s \rightarrow OH_s$ is much slower than the second recombination $H_s + OH_s \rightarrow H_2O_s$, the activation barriers differing by 18 kcal mol^{-1}.[16] Thus the hydrogen-assisted C–O dissociation $H_s + CO_s \rightarrow C_s + OH_s$ (producing OH_s) may significantly contribute to the overall methanation kinetics along with the direct dissociation $CO_s \rightarrow C_s + O_s$ (producing O_s). The results are consistent with the fact that carbidic carbon C_s is formed on Ni while heated either in clean CO or in H_2 + CO, but with H_2 the CO dissociation proceeds much faster.[52,54] Once C_s is formed, the progressive hydrogenation $C_s + H_s \rightarrow CH_s + H_s \rightarrow CH_{2,s} + H_s \rightarrow CH_{3,s} + H_s \rightarrow CH_4$ takes place where the activation recombination barrier $\Delta E^*_{H_xC\text{–}H}$ decreases with x; namely, $\Delta E^*_{H_xC\text{–}H} = 42$, 17, 12, and 14 kcal mol^{-1} for $x = 0, 1, 2$, and 3, respectively, for both Ni(111) and Ni(100), which suggests the structural insensitivity of hydrogenation of carbon to methane on Ni surfaces.

Thus, our calculations project that for methanation on Ni, the rate-limiting step (if any) is the first (carbidic) carbon hydrogenation, in agreement with experiment.[52–54] In particular, potassium promotion of CO hydrogenation on Ni(100) decreases the activation barrier of CO dissociation (from 23 to 10 kcal mol^{-1}) but does not alter the apparent activation energy of methanation,[52c] which eliminates CO dissociation as the rate-limiting step. Furthermore, it has been shown that for the hydrogenation of $CH_{3,s}$ compared to the hydrogenation of CO_s on Ni(111), the overall CH_4 production rate per Ni site increases by about six (!) orders of magnitude,[53] which places the rate-limiting step well before the final hydrogenation step. Consistently, the methanation kinetics was found to be virtually identical for single-crystal surfaces Ni(111) and Ni(100) and for Ni/Al_2O_3, which is a highly dispersed supported Ni catalyst.[52]

Our calculations easily explain why Ni is not effective in forming CH_3OH. First of all, the dissociation $HCO_s \rightarrow C_s + OH_s$ with $\Delta E^* =$ 18 kcal mol^{-1} is strongly preferred over the competing hydrogenation $HCO_s + H_s \rightarrow H_2CO_s$ with $\Delta E^* = 33$ kcal mol^{-1}. Furthermore, if the C–O bond could somehow survive up to the formation of methoxide CH_3O_s, it will be easily cleaved as $CH_3O_s \rightarrow CH_{3,s} + O_s$ with $\Delta E^* = 13$ kcal mol^{-1}.

The competitive hydrogenation to methanol $CH_3O_s + H_s \rightarrow CH_3OH_s$ would have required the much larger activation barrier $\Delta E^* = 19$ kcal mol^{-1}, not to mention that CH_3OH_s cannot be desorbed without profound decomposition since $\Delta E^*_{(H_3CO)H,g} = -5$ kcal mol^{-1} [cf. Eq. 9.26)]. Thus, on Ni the kinetic preference of C–O (and O–H) dissociation further enhances the thermodynamic preference of the $CH_4 + H_2O$ over CH_3OH formation.

Hydrogenation of CO on Pd looks rather different. To begin with, on Pd(111) the formation of carbidic carbon is highly endothermic and requires much higher activation barriers, namely, 50 kcal mol^{-1} for $CO_s \rightarrow C_s + O_s$ or 59 kcal mol^{-1} for $CO_s + H_s \rightarrow C_s + OH_s$. Thus, the major route appears to be the progressive hydrogenation $CO_s + H_s \rightarrow HCO_s \rightarrow H_2CO_s \rightarrow H_3CO_s$ with the barriers of 35, 16, and 10 kcal mol^{-1}, respectively. Unlike Ni, now the hydrogenation $HCO_s + H_s \rightarrow H_2CO_s$ is preferred over the dissociation $HCO_s \rightarrow C_s + OH_s$ (the barriers are 16 and 24 kcal mol^{-1}, respectively). Also, in contrast to Ni, hydrogenation of methoxide $CH_3O_s + H_s \rightarrow CH_3OH_s$ requires the lower barrier (7 kcal mol^{-1}) than that (16 kcal mol^{-1}) for the C–O bond cleavage to $CH_{3,s} + O_s$, and now methanol can be effectively desorbed ($\Delta E^*_{(H_3CO)H,g} \gg 0$). Thus, on Pd the thermodynamic and kinetic preferences for the formation of CH_3OH and $CH_4 + H_2O$ appear to be in conflict, which makes it possible for Pd to be a selective catalyst for either CH_3OH or CH_4, especially because differences in the activation barriers for the C–O dissociation and relevant hydrogenation are not large being, for example, only 8–9 kcal mol^{-1} for HCO_s and H_3CO_s (see Tables 9.6 and 9.7). Consistently, Pd metal has been found to be a good methanol synthesis or methanation catalyst, the selectivities varying from 100% to 0% depending on the reaction conditions, the nature of the support, promoters, etc.[55–57] The important point, however, is that unsupported Pd, either as Pd powder[55a] or the single-crystal Pd(110) surface,[55b] may be an active methanol synthesis catalyst. Finally, we project that for the CO methanation the relative significance of hydrogen-assisted H_xC–O bond cleavage (the value of x) increases from Ni to Pd, in agreement with experiment,[57] particularly with H_2/D_2 isotopic studies by Mori et al.[57d] The results for Pt are similar to those for Pd.

Transformations of C_2H_x. Surface reactions of hydrocarbons have drawn a great deal of interest[63,64] because of their fundamental and practical importance, and here C_2H_x species are among the best studied. Some of them, such as ethane (CH_3CH_3), ethylene (CH_2CH_2), or acetylene (CHCH), are stable in the gas phase and therefore quite familiar, but others, such as ethylidyne (CH_3C), vinylidene (CH_2C), or acetylide (CHC), have been identified only in their chemisorbed states. To elucidate C_2H_x transformations, some theoretical approaches have been developed[36c,65,66] and many experimental techniques have been used.[67–86] Let us see what insight can be gained from the BOC–MP modeling.

Similar to hydrogenation of CO, we are mostly interested in periodic (relative) changes in catalytic behavior of transition metals. Specifically, we choose the sequence from Pt to Ni and further to Fe or W, the latter two being simulated as the same model metal Fe/W [with the model parameters averaged over the (close) parameters of real Fe and W]. The BOC–MP calculations have been made for the smoothest (most densely packed) surfaces, namely, for fcc Pt(111), Ni(111), and bcc Fe/W(110).

Particular attention will be given to the following questions: (1) the thermochemistry of C_2H_x species in the chemisorbed versus gas-phase state; (2) the effects of metal composition and the structure of C_2H_x species on the activation energy for C–H and C–C bond cleavage; and (3) the

Table 9.8. Total bond energies in the gas-phase (D) and chemisorbed ($D + Q$) states of C_2H_x on Pt(111), Ni(111), and Fe/W(110).[a]

		$Q_{C_2H_x}$[c]			$D_{C_2H_x} + Q_{C_2H_x} + (8 - x)Q_H$[d]		
C_2H_x	$D_{C_2H_x}$[b]	Fe/W	Ni	Pt	Fe/W	Ni	Pt
$H_3C{-}CH_3$	674	6	5	5	812	805	801
$H_3C{-}CH_2$	576	64	49	39	838	814	798
$H_3C{-}CH$	466	107	85	70	837	803	780
$H_2C{=}CH_2$	538	20	15	12	822	805	794
$H_3C{-}C$	376	141	115	97	847	806	778
$H_2C{=}CH$	421	71	55	44	822	791	770
$H_2C{=}C$	348	110	87	71	854	813	785
$HC{\equiv}CH$	392	25	18	14	813	788	772
$HC{\equiv}C$	259	106	84	69	827	784	755
$CH_3 + CH_3$	586	124	96	76	842	808	784
$CH_3 + CH_2$	476	166	131	106	840	796	765
$CH_3 + CH$	374	204	164	135	842	790	753
$CH_3 + C$	293	262	219	188	885	827	786
$CH_2 + CH_2$	366	208	166	136	838	784	746
$CH_2 + CH$	264	246	199	165	840	778	734
$CH_2 + C$	183	304	254	218	883	815	767
$CH + CH$	162	284	232	194	842	772	722
$CH + C$	81	342	287	247	885	809	755
$C + C$	0	400	342	300	928	846	788
$CH_4 + CH_4$	796	12	12	12	808	808	808

[a] The parameters used: Q_C = 150, 171, 200 and Q_H = 61, 63, 66 for Pt, Ni, Fe/W, respectively. See text for notations and explanations. All energies in kcal mol^{-1}.
[b] From Ref. 40 with corrections and additions specified in Ref. 18b.
[c] Equations (9.9), (9.12), or (9.13).
[d] Normalized for stoichiometry C_2H_8 ($C_2H_6 + H_2 \rightarrow 2CH_4$), when for C_2H_x (or $CH_y + CH_{x-y}$) the rest $(8 - x)$ atoms H are assumed to be atomically chemisorbed.

influence of metal composition on the patterns of C_2H_4 and C_2H_2 decomposition.

Table 9.8 lists total bond energies in the gas phase (D) and chemisorbed ($D + Q$) states for all C_2H_x species ($x = 0$–6). Some comparisons of enthalpies in the gas phase vs chemisorbed states are made in Table 9.9. Finally, we have calculated the activation barriers ΔE^* for C–C and C–H bond cleavage and recombination for chemisorbed C_2H_x species. These results are summarized in Table 9.10. All the discussion below will refer to *chemisorbed* species if not stated otherwise.

Of general model conclusions, perhaps the most important is the following: many reorganizations of C_2H_x species, being highly endothermic in the gas phase, become comparable in total energy and often become exothermic on transition metal surfaces. For example, in the gas phase the ground state of C_2H_3 is vinyl $H_2C{=}CH$, which is lower than ethylidyne $H_3C{-}C$ (the excited state) by 45 kcal mol^{-1}. Under chemisorption, this isomerization becomes distinctly exothermic ($\Delta H = -8, -15, -25$ kcal mol^{-1} for Pt, Ni, Fe/W, respectively). Similarly, isomerization of acetylene to vinylidene $HC{\equiv}CH \rightarrow H_2C{=}C$ is highly endothermic in the gas phase ($\Delta H = 44$ kcal mol^{-1}) but becomes highly exothermic by 13–41 kcal mol^{-1} on the metal surfaces studied. This makes it comprehensible why H_3CC and H_2CC are often observed in the chemisorbed states (unlike the gas phase[72]). Moreover, we can project the relative stability of these and other C_2H_x species depending on metal composition.

For ethylene (C_2H_4) on Pt(111) and Ni(111), our estimates for the heats of chemisorption are $Q_{C_2H_4} = 12$ and 15 kcal mol^{-1}, respectively, in excellent agreement with the experimental range $Q_{C_2H_4} = 11$–13 kcal mol^{-1} for Pt(111),[73] Pd(111),[74] Ru(001),[75] and Ni(110).[76] [For comparison, the recent ASED–MO calculations of C_2H_4/Pt(111) project $Q_{C_2H_4} = 47$ kcal mol^{-1}.[36c]] Because we cannot directly estimate the isomerization activation barriers, we will refrain from mechanistic speculations about how ethylene

Table 9.9. Enthalpies of C_2H_x isomerization reactions: gas-phase versus chemisorbed states.

	Reaction	ΔH (kcal mol^{-1})[a]			
		Gas Phase	Fe/W(110)	Ni(111)	Pt(111)
C_2H_4	$H_2C{=}CH_2 \rightarrow H_3C{-}CH$	72	−15	2	14
C_2H_3	$H_2C{=}CH \rightarrow H_3C{-}C$	45	−25	−15	−8
C_2H_2	$HC{\equiv}CH \rightarrow H_2C{=}C$	44	−41	−25	−13

[a] From Table 9.8.

Table 9.10. Activation barriers for forward and reversed reactions of chemisorbed C_2H_x.[a]

			ΔE_f^{*c}			ΔE_r^{*d}		
C_2H_x	Reaction	D_{CX}[b]	Fe/W	Ni	Pt	Fe/W	Ni	Pt
C_2H_6	$CH_3CH_{3,g} \rightleftarrows CH_3CH_2 + H$	98	−3	5	8	38	19	10
	$CH_3CH_3 \rightleftarrows CH_3CH_2 + H$	98	3	10	13	35	19	10
C_2H_5	$CH_3CH_2 \rightleftarrows CH_3 + CH_2$	100	18	24	33	20	6	0
	$\rightleftarrows CH_2CH_2 + H$	38	16	11	4	0	2	0
	$\rightleftarrows CH_3CH + H$	110	21	24	25	20	13	7
C_2H_4	$CH_2CH_{2,g} \rightleftarrows CH_2 + CH_2$	172	−2	17	36	36	11	0
	$\rightleftarrows CH_2CH + H$	117	−3	7	13	20	8	1
	$CH_2CH_2 \rightleftarrows CH_2 + CH_2$	172	18	32	48	34	11	0
	$\rightleftarrows CH_2CH + H$	117	17	22	25	17	8	1
	$CH_3CH \rightleftarrows CH_3 + CH$	92	19	23	27	24	10	0
	$\rightleftarrows CH_2CH + H$	45	25	21	18	10	9	8
	$\rightleftarrows CH_3C + H$	90	17	19	20	27	22	18
C_2H_3	$CH_2CH \rightleftarrows CH_2 + CH$	157	21	31	38	39	18	2
	$\rightleftarrows CHCH + H$	29	14	9	5	5	6	7
	$\rightleftarrows CH_2C + H$	73	5	7	9	37	29	24
	$CH_3C \rightleftarrows CH_3 + C$	83	5	8	11	43	29	19
	$\rightleftarrows CH_2C + H$	28	2	15	13	9	22	20
C_2H_2	$CHCH_g \rightleftarrows CH + CH$	230	−4	19	36	54	21	0
	$\rightleftarrows CHC + H$	133	−12	1	11	39	18	8
	$CHCH \rightleftarrows CH + CH$	230	21	37	50	50	21	0
	$\rightleftarrows CHC + H$	133	13	19	25	27	18	8
	$CH_2C \rightleftarrows CH_2 + C$	165	20	27	32	49	29	14
	$\rightleftarrows CHC + H$	89	34	32	31	4	3	1
C_2H	$CHC \rightleftarrows CH + C$	178	13	22	29	71	47	29

[a] See text for notations and explanations. The barriers for the gas-phase ethane, ethylene, and acetylene are also added. All energies in kcal mol^{-1}.
[b] The difference between the gas-phase total bond energies of the reactant and products (see Table 9.8).
[c] Equations (9.18) or (9.20). The values of Q_A, Q_B, Q_{AB} are from Table 9.8.
[d] Equations (9.23b) or (9.24b).

C_2H_4 transforms into ethylidyne CH_3C, either via ethylidene CH_3CH or vinyl CH_2CH. As seen from Table 9.8, the formation of CH_3C from C_2H_4 is moderately endothermic on Pt, practically thermoneutral on Ni, and highly exothermic on Fe/W ($\Delta H = 16, -1, -25$ kcal mol^{-1}, respectively). Once ethylidyne CH_3C is formed, however, its fate appears to be sensitive to metal composition. On Pt(111), we predict CH_3C to be rather stable, since the calculated C–C bond scission barrier is $\Delta E^*_{CC} = 11$ kcal mol^{-1}. But on Ni(111) and Fe/W(110), we predict this barrier to become smaller—8 and 5 kcal mol^{-1}, respectively. [These calculated values can be compared with the experimental (TPS SIMS) activation energies of CH_3C decomposition found to be 17 kcal mol^{-1} for Pt(111)[79] and 12 kcal mol^{-1} for Ru(001),[80] the latter surface being intermediate in activity between Pt(111) and Ni(111).] Thus, the model conclusion is that the stability of CH_3C decreases in the order Pt > Ni > Fe/W. Indeed, ethylidyne has been readily observed under decomposition of ethylene on the close-packed surfaces[81] and metal particles[71b,82] of Pt, Pd, Rh, Ru, only at high adsorbate coverages on Ni[83] (and, probably on Co[84]) but not on more active metals. One can add that on Pt(111), the C–C bond cleavage $CH_3C \rightarrow CH_3 + C$ ($\Delta E^*_{CC} = 11$ kcal mol^{-1}) seems to be more favorable than the C–H bond cleavage $CH_3C \rightarrow CH_2C + H$ ($\Delta E^*_{CH} = 13$ kcal mol^{-1}), so that the molecule will retain most of its hydrogen up to the point of C–C bond scission, in agreement with the ^{13}C NMR experiment.[71b,71c]

For a given surface, acetylene chemisorbs stronger than ethylene. Our estimates are $Q_{C_2H_2} = 14, 18, 25$ kcal mol^{-1} compared to $Q_{C_2H_4} = 12, 15, 18$ kcal mol^{-1} for Pt(111), Ni(111), Fe/W(110), respectively. Experimental data on $Q_{C_2H_2}$ (usually from TPD spectra) are not available because C_2H_2 begins to decompose before it desorbs. Consistently, we found the gas-phase C–H bond cleavage to be nonactivated on Ni and especially on Fe/W, namely $\Delta E^*_{CH,g} = 1$ and -12 kcal mol^{-1}, respectively. On Pt, where $\Delta E^*_{CH,g} = 11$ kcal mol^{-1}, the first surface reaction may be the distinctly exothermic isomerization $CHCH \rightarrow CH_2C$ with $\Delta H = -13$ kcal mol^{-1} (unfortunately, as for other isomerization processes, we do not know how to calculate the activation isomerization barrier ΔE^*). Indeed, the formation of the vinylidene CH_2C intermediate on Pt(111) was suggested from EELS spectra[70] and confirmed on Pt particles by the ^{13}C NMR analysis.[71a,71c] However, this isomerization is not a favorable route for C–C bond scission since $CH_2C \rightarrow CH_2 + C$ would require $\Delta E^*_{CC} = 32$ kcal mol^{-1}. So, dehydrogenation $CHCH \rightarrow CHC + H$ will occur instead ($\Delta E^*_{CH} = 25$ kcal mol^{-1}), and then the C–C bond scission $CHC \rightarrow CH + C$ ($\Delta E^*_{CC} = 29$ kcal mol^{-1}). Thus, contrary to decomposition of CH_3C, one can expect a substantial loss of hydrogen before the C–C bond rupture, again in agreement with the ^{13}C NMR data.[71c]

On Fe/W(110), the situation looks rather different because for CHCH not only the C–H but also C–C bond cleavage seems to be nonactivated from the gas phase ($\Delta E^*_{CC,g} = -4$ kcal mol^{-1}). Thus, on Fe surfaces, one can expect acetylene to decompose rapidly into CH_x fragments. This model projection is consistent with the fact that under heating of chemisorbed CHCH, only CH_x intermediates have been observed on various Fe surfaces.[84,85] Between Fe/W and Pt, the decomposition products may be a variety of C_2H_x and CH_x species depending on metal composition and reaction conditions. For example, on Ni surfaces, rapid decomposition of CHCH to (partly) HCC and (mainly) CH_x species was reported.[83,86] At the same time, on Ru(001), whose activity is intermediate between Pt(111) and Ni(111), the whole set of H_xCC species, $x = 1, 2, 3$, resulting from isomerization (CHCH → CH_2C), dehydrogenation (CHCH → CHC + H), and rehydrogenation (CHCH + H → [CH_2CH] → CH_3C), was identified by EELS.[67,68]

We predict that the C–C bond dissociation barrier in H_3C–C is smaller than that in HC–C [e.g., by 14 kcal mol^{-1} for Ni(111)]. Indeed, when both H_3CC and HCC are formed on Ru(001), H_3CC begins to decompose at lower temperatures.[67,69] Also, we find that the C–H bond dissociation barrier for CHCH → CHC + H is persistently larger than that for CH_2CH_2 → CH_2CH + H (by 4–25 kcal mol^{-1} in the series Fe/W $<$ Ni $<$ Pt). Consistently, on all metal surfaces, the decomposition of C_2H_4, while heated, begins earlier than that of C_2H_2.[67–86]

Another model projection is the smallness of most activation barriers for hydrogenation of C_2H_x species. In general, for all hydrogenation reactions, the calculated barriers decrease in the order Fe/W $>$ Ni $>$ Pt, which explains an increase in the hydrogenation ability of transition metals in the direction Fe/W $<$ Ni $<$ Pt making Pt overall the best hydrogenation catalyst.[87]

Our model can also shed light on regularities of hydrogenolysis of C_2H_6. In general, by comparing possible C–C bond scission routes on Pt vs Ni vs Fe/W (cf. Table 9.10), one can easily see that the hydrogen content (x) in the hydrocarbon species C_2H_x undergoing this scission decreases in the order Fe/W $>$ Ni $>$ Pt, in agreement with vast experimental observations.[63]

CONCLUDING REMARKS

The analytic BOC–MP model explicitly interrelates many seemingly disparate chemisorption phenomena, including surface reactivity. Let us stress again that the BOC–MP model is based on a few well-defined assumptions, and within these assumptions, the model interrelations are exact for atomic adsorbates and well defined for molecular adsorbates, the same analytic formalism being used to treat both diatomic and polyatomic molecules. Moreover, these interrelations are expressed in terms of observables only (the heats of chemisorption and various constants), which makes comparison with experiment typically direct and unambiguous.

Since any model is a simplification, the fair questions are: (1) Does the model work and (2) is there a better model? We saw that in most cases when the model projections were definitive and the experimental data reliable, the agreement was good, not only qualitatively but often quantitatively. This efficiency is not simply a consequence of the empirical nature of the model parameters Q_A (Q_B) and D_{AB}, as some people may think. These parameters are transformed into the values of Q_{AB}, ΔE^*_{AB}

($\Delta E^*_{\text{A-B}}$), and other observables *not* in a trivial and self-evident way, but according to a set of specific and rigid rules [cf. Eqs. (9.1)–(9.24)]. The latter is critical, since the empirical scaling of parameters does *not* by itself guarantee the accuracy of results. After all, most of the current approaches to chemisorption and surface reactivity are heavily empirical, and we leave the comparisons to the reader. In any event, the BOC–MP model is phenomenological and, therefore, totally complementary to microscopic quantum-mechanical models.

In conclusion, the BOC–MP modeling provides much insight into both regularities and details of metal surface reactions. We have considered several examples, from simple to rather complex. In principle, any metal surface reaction can be treated this way. And because the BOC–MP model is a truly "back-of-the-envelope" model, the practicing chemist can make direct use of it, trying and judging for himself (herself).

REFERENCES

1. Somorjai, G.A., *Chemistry in Two Dimensions: Surfaces*. Ithaca, New York: Cornell University Press, 1981.
2. King, D.A., Woodruff, D.P. (eds.), *The Chemical Physics of Solid Surfaces and Heterogeneous Catalysis*. Amsterdam: Elsevier, Vol. 2, 1983; Vol. 3, 1984; Vol. 4, 1982.
3. Anderson, J.R., Boudart, M. (eds.), *Catalysis: Science and Technology*. West Berlin: Springer-Verlag, Vol. 3, 1982; Vol. 4, 1983, Vol. 5, 1984.
4. Rhodin, T.N., Ertl, G. (eds.), *The Nature of the Surface Chemical Bond*. Amsterdam: North-Holland, 1979.
5. Smith, J.R. (ed.), *Theory of Chemisorption*. West Berlin: Springer-Verlag, 1980.
6. (a) Veillard, A. (ed.), *Quantum Chemistry: The Challenge of Transition Metals and Coordination Chemistry*. NATO ASI Series. Dordrecht: Reidel, 1986. (b) Sauer, J., *Chem. Rev.* **89:** 199 (1989).
7. Arlinghaus, F.J., Gay, J.G., Smith, J.R., in Ref. 2, Chap. 4.
8. (a) Feibelman, P.J., *Phys. Rev. B* **35:** 2626 (1987). (b) Feibelman, P.J., Hamann, D.R., *Surf. Sci.* **182:** 411 (1987).
9. Soukiassian, P., Rivan, R., Lecant, J., Wimmer, E., Chubb, S.R., Freeman, A.J., *Phys. Rev. B* **31:** 4911 (1985). (b) Freeman, A.J., Fu, C.L., Wimmer, E., *J. Vac. Sci. Technol. A* **4:** 7265 (1986). (c) Chubb, S.R., Wimmer, E., Freeman, A.J., Hiskes, J.R., Karo, A.M., *J. Vac. Sci. Technol. B* **36:** 4112 (1987).
10. (a) Baetzold, R.C., *Solid State Commun.* **44:** 781 (1982). (b) Baetzold, R.C., *J. Am. Chem. Soc.* **105:** 4271 (1983). (c) Baetzold, R.C., *Phys. Rev. B* **29:** 4211 (1984). (d) Baetzold, R.C., *Phys. Rev. B* **30:** 6870 (1984). (e) Baetzold, R.C., *Surf. Sci.* **150:** 193 (1985). (f) Baetzold, R.C., *J. Chem. Phys.* **82:** 5729 (1985). (g) Baetzold, R.C., *Langmuir* **3:** 189 (1987).
11. (a) Shustorovich, E., *Solid State Commun.* **44:** 567 (1982). (b) Shustorovich, E., *J. Phys. Chem.* **88:** 1927, 3490 (1984).
12. Shustorovich, E., Baetzold, R., Muetterties, E.L., *J. Phys. Chem.* **87:** 1100 (1983).
13. For an excellent analysis of theoretical issues of chemisorption, see: Einstein, T.L., Hertz, J.A., and Schrieffer, J.R., in Ref. 5, Chap. 7.
14. Einstein, A., *Ideas and Opinions*. New York: Crown Publishers, Inc., 1982, pp. 274, 292.

15. (a) Shustorovich, E., *Surf. Sci. Repts.* **6:** 1 (1986). (b) Shustorovich, E., *Acc. Chem. Res.* **21:** 183 (1988). (c) Shustorovich, E., *J. Mol. Catal.* **54:** 301 (1989).
16. Shustorovich, E. *Adv. Catal.* **37** (1990).
17. (a) Shustorovich, E., *J. Am. Chem. Soc.* **106:** 6479 (1984). (b) Shustorovich, E., *Surf. Sci.* **150:** L115 (1985). (c) Shustorovich, E., *Surf. Sci.* **163:** L645 (1985). (d) Shustorovich, E., *Surf. Sci.* **163:** L730 (1985). (e) Shustorovich, E., *Surf. Sci.* **175:** 561 (1986). (f) Shustorovich, E., *Surf. Sci.* **176:** L863 (1986). (g) Shustorovich, E., *Surf. Sci.* **181:** L205 (1987). (h) Shustorovich, E., *Surf. Sci.* **187:** L627 (1987). (i) Shustorovich, E., *Surf. Sci.* **205:** 336 (1988). (j) Shustorovich, E., *J. Phys. Chem.* (in press).
18. (a) Shustorovich, E., Bell, A.T., *J. Catal.* **113:** 341 (1988). (b) Shustorovich, E., Bell, A.T., *Surf. Sci.* **205:** 492 (1988). (c) Shustorovich, E., Bell, A.T., *Surf. Sci.* **222:** 371 (1989). (d) Bell, A.T., Shustorovich, E., *J. Catal.* **121:** 1 (1990).
19. See, for instance, (a) Abel, G.C., *Phys. Rev. B* **31:** 6184 (1985). (b) Skriver, H.L., *Phys. Rev. B* **31:** 1909 (1985). (c) Haydock, R., *Philos. Mag.* **35:** 845 (1977); **38:** 155 (1978).
20. (a) Rose, J.H., Smith, J.R., Ferrante, J., *Phys. Rev. B* **28:** 1835 (1983). (b) Rose, J.H., Smith, J.R., Guinea, F., Ferrante, J., *Phys. Rev. B* **29:** 2963 (1984). (c) Ferrante, J., Smith, J.R., *Phys. Rev. B* **31:** 3427 (1985).
21. Morse, P.M., *Phys. Rev.* **34:** 57 (1929).
22. (a) Johnston, H.S., Parr, C., *J. Am. Chem. Soc.* **85:** 2544 (1963). (b) Marcus, R.A., *J. Phys. Chem.* **72:** 891 (1968).
23. See, for instance, (a) Wolfe, S., Mitchell, D.J., Schlegel, H.B., *J. Am. Chem. Soc.* **103:** 7692, 7694 (1981). (b) Dunning, T.H., Jr., Harding, L.B., Bair, R.A., Eades, R.A., Shepard, R.L., *J. Phys. Chem.* **90:** 344 (1986).
24. Dunitz, J.D., *X-ray Analysis and the Structure of Organic Molecules*. Ithaca, New York: Cornell University Press, 1979, pp. 341–360.
25. For coverage effects, there will be another form of BOC.[17c,17e]
26. (a) Van Hove, M.A., in Ref. 3, Chap. 4. (b) Somorjai, G.A., Van Hove, M.A., *Struct. Bonding* (West Berlin) **38:** 1 (1979).
27. For a recent review, see, (a) Hoffmann, F.M., *Surf. Sci. Rep.* **3:** 107 (1983). (b) Biberian, J.P., Van Hove, M.A., *Surf. Sci.* **118:** 443 (1982); **138:** 361 (1984).
28. (a) Hayden, B.E., Bradshaw, A.M., *Surf. Sci.* **125:** 787 (1983). (b) Mieher, W.D., Whitman, L.J., Ho, W., *J. Chem. Phys.* **91:** 3228 (1989).
29. (a) Gurney, B.A., Richter, L.J., Villarubia, J.S., Ho, W., *J. Chem. Phys.* **87:** 6710 (1987). (b) Raval, R., Parker, S.F., Pemble, M.E., Hollins, P., Pritchard, J., Chesters, M.A., *Surf. Sci.* **203:** 353 (1988).
30. (a) Bauhofer, J., Hock, M., Küppers, J., *Surf. Sci.* **191:** 395 (1987). (b) Andersson, S., *Solid State Commun.* **21:** 75 (1977). (c) Tang, S.L., Lee, M.B., Yang, Q.Y., Beckerle, J.D., Ceyer, S.T., *J. Chem. Phys.* **84:** 1876 (1986).
31. However, there is a definitive correlation between n and the C–O stretching frequency,[27a] which can also be explained by our model.[15a,17d]
32. (a) Shinn, N.D., Madey, T.E., *J. Chem. Phys.* **83:** 5928 (1985). (b) Benndorf, C.N., Kruger, B., Thieme, F., *Surf. Sci.* **163:** L675 (1985). (c) Fulmer, J.P., Zaera, F., Tysoe, W.T., *J. Chem. Phys.* **87:** 7265 (1987).
33. Campbell, C.T., Ertl, G., Kuipers, H., Segner, J., *Surf. Sci.* **107:** 220 (1981).
34. Bagus, P.S., Bauschlicher, C.W., Nelin, C.J., Laskovski, B.C., Seel, M., *J. Chem. Phys.* **81:** 3594 (1984).
35. Reference 15a, Table 10, p. 25.
36. (a) Anderson, A.B., Awad, M.K., *J. Am. Chem. Soc.* **107:** 7854 (1985). (b) Mehandru, S.P., Anderson, A.B., *Surf. Sci.* **169:** L281 (1986). (c) Kang, D.B., Anderson, A.B., *Surf. Sci.* **155:** 639 (1985).
37. Tomanek, D., Bennemann, K.H., *Surf. Sci.* **127:** L111 (1983).

38. Ertl, G., in Ref. 4, Chap. 5, Tables 5.3 and 5.5.
39. (a) Broden, G., Gafner, G., Bonzel, H.P., *Appl. Phys.* **13,** 333 (1977). (b) Erickson, J.W., Estrup, P.J., *Surf. Sci.* **167:** 519 (1986). (c) Umbach, E., Menzel, D., *Surf. Sci.* **135:** 199 (1983).
40. *CRC Handbook of Chemistry and Physics*. Boca Raton, Florida: CRC Press, 1984–1985, pp. F171–190.
41. Ertl, G., in Ref. 3, Vol. 5, Chap. 3, pp. 238–250.
42. Weinberg, W.H., *Surf. Sci.* **128:** L224 (1983).
43. Bowker, M., Barteau, M.A., Madix, R.J., *Surf. Sci.* **92:** 528 (1980).
44. Engstrom, J.R., Weinberg, W.H., *Phys. Rev. Lett.* **55:** 2017 (1985).
45. Goodman, D.W., Peebles, D.E., White, J.M., *Surf. Sci.* **140:** L239 (1984).
46. Peled, H., Asscher, M., *Surf. Sci.* **183:** 201 (1987).
47. Root, T.W., Schmidt, L.D., Fisher, G.B., *Surf. Sci.* **134:** 30 (1983).
48. (a) Ho, P., White, J.M., *Surf. Sci.* **137:** 103 (1984). (b) Ho, W., *J. Phys. Chem.* **91:** 766 (1987).
49. Comrie, C.M., Weinberg, W.H., Lambert, R.M., *Surf. Sci.* **57:** 619 (1976).
50. Vajo, J.J., Tsai, W., Weinberg, W.H., *J. Phys. Chem.* **89:** 3243 (1985).
51. For an earlier review, see, (a) Ponec, V., *Catal. Rev.* **18:** 151 (1978). (b) Bell, A.T., *Catal. Rev.* **23:** 23 (1981). (c) Biloen, P., Sachtler, W.M.H., *Adv. Catal.* **30:** 165 (1981). (d) Vannice, M.A., in Ref. 3, Vol. 3, Chap. 3.
52. (a) Goodman, D.W., *Acc. Chem. Res.* **17:** 194 (1984). (b) Goodman, D.W., Kelly, R.D., Madley, T.E., White, J.M., *J. Catal.* **64:** 479 (1980). (c) Campbell, C.T., Goodman, D.W., *Surf. Sci.* **123:** 413 (1982).
53. Yates, J.T., Jr., Gates, S.M., Russell, J.M., Jr., *Surf. Sci.* **164:** L839 (1985).
54. Similar to CO hydrogenation on Ru(001): Hoffmann, F.M., Robbins, J.L., *Electr. Spect. Rel. Phen.* **45:** 421 (1987). (b) Hoffmann, F.M., Robbins, J.L., in *Proceedings, 9th International Congress on Catalysis* (M.J. Philips and M. Ternan, eds.). Ottawa, The Chemical Institute of Canada, 1988, Vol. 3, p. 1144; Vol. 5, p. 373.
55. (a) Ryndin, Y.A., Hicks, R.F., Bell, A.T., Yermakov, Y.I., *J. Catal.* **80:** 287 (1981). (b) Berlowitz, P.J., Goodman, D.W., *J. Catal.* **108:** 364 (1987).
56. (a) Poutsma, M.L., Elek, L.F., Ibaria, P.A., Risch, A.P., Rabo, J.A., *J. Catal.* **52:** 168 (1978). (b) Fajula, F., Anthony, R.G., Lansford, J.H., *J. Catal.* **73,** 237 (1982). (c) Kikuzono, Ya., Kagami, S., Naito, S., Onishi, T., Tamaru, K., *Disc. Faraday Soc.* **72:** 735 (1981).
57. (a) Ho, S.V., Harriott, P., *J. Catal.* **64:** 272 (1980). (b) Wang, S.-Y., Moon, S.H., Vannice, M.A., *J. Catal.* **71:** 167 (1981). (c) Rieck, J.S., Bell, A.T., *J. Catal.* **96:** 88 (1985); **99:** 262 (1986). (d) Mori, T., Miyamoto, A., Niizuma, H., Takahashi, N., Hattori, T., Murakami, Y., *J. Phys. Chem.* **90:** 109 (1986), and references therein.
58. (a) Hsu, D.S.Y., Lin, M.C., *J. Chem. Phys.* **88:** 432 (1988). (b) Hsu, D.S.Y., Hoffbauer, M.A., Lin, M.C., *Langmuir* **2:** 302 (1986). (c) Selwyn, G.S., Lin, M.C., *Chem. Phys.* **67:** 213 (1982). (d) Selwyn, G.S., Fujimoto, G.T., Lin, M.C., *J. Phys. Chem.* **86:** 760 (1982).
59. (a) Reference 51c. (b) Rofer-DePoorter, C.K., *Chem. Rev.* **81:** 447 (1981). (c) Broden, G., Rhodin, T.N., Bruker, C., Benbow, R., Hurych, Z., *Surf. Sci.* **59:** 593 (1976).
60. (a) Asscher, M., Kao, C.-T., Somorjai, G.A., *J. Phys. Chem.* **92:** 2711 (1988). (b) Pirner, M., Bauer, R., Borgmann, D., Wedler, G., *Surf. Sci.* **189/190:** 147 (1987). (c) Kwong, D.W.J., DeLeon, N., Haller, G.L., *Chem. Phys. Lett.* **144:** 533 (1988).
61. (a) Hall, R.B., *J. Phys. Chem.* **91:** 1007 (1987). (b) Ho, W., *J. Phys. Chem.* **91:** 766 (1987). (c) Seebauer, E.G., Kong, A.C.F., Schmidt, L.D., *Surf. Sci.* **193:** 417 (1988).
62. See, for example, (a) Vannice, M.A., Twu, C.C., *J. Catal.* **18:** 213 (1983). (b) Stoltze, P., Norskov, J.K., *J. Catal.* **110:** 1 (1988).
63. Sinfelt, J.H., *J. Phys. Chem.* **90:** 4711 (1986).

64. See also previous reviews: (a) Muetterties, E.L., *Chem. Soc. Rev.* **11:** 283 (1982). (b) Koestner, R.J., Van Hove, M.A., Somorjai, G.A., *J. Phys. Chem.* **87:** 203 (1983). (c) Bertolini, J.C., Massardier, J., in Ref. 2, Vol. 3, Chap. 3. (d) Davis, S.M., Somorjai, G.A., in *The Chemical Physics of Solid Surfaces and Heterogeneous Catalysis* (D.A. King and D.P. Woodruff, eds.). New York: Elsevier, 1982, Vol. 4., p. 217.
65. (a) Gavezzotti, A., Simonetta, M., *Surf. Sci.* **99:** 453 (1980). (b) Simonetta, M., Gavezzotti, A., *J. Mol. Struct.* **107:** 75 (1984).
66. (a) Minot, C., Van Hove, M.A., Somorjai, G.A., *Surf. Sci.* **127:** 441 (1983). (b) Silvestre, J., Hoffmann, R., *Langmuir* **1:** 621 (1985).
67. Parmeter, J.E., Hills, M.M., Weinberg, W.H., *J. Am. Chem. Soc.* **108:** 3563 (1986); **109:** 72 (1987).
68. Jacob, P., Cassuto, A., Menzel, D., *Surf. Sci.* **187:** 407 (1987).
69. Hills, M.M., Parmeter, J.E., Weinberg, W.H., *J. Am. Chem. Soc.* **109,** 597 (1987).
70. Ibach, H., Lehwald, S., *J. Vac. Sci. Technol.* **15:** 407 (1978).
71. (a) Wang, P.-K., Slichter, C.P., Sinfelt, J.H., *Phys. Rev. Lett.* **53:** 82 (1984). (b) Wang, P.-K., Slichter, C.P., Sinfelt, J.H., *J. Phys. Chem.* **89:** 3606 (1985). (c) Wang, P.-K., Ansermet, J.-P., Rudaz, S.L., Wang, Zh., Shore, S., Slichter, C.P., Sinfelt, J.H., *Science* **234:** 35 (1986).
72. For example, the gas-phase activation barrier for $CH_2C \rightarrow CHCH$ is close to zero: Carrington, T., Jr., Hubbard, L.M., Schaefer, H.F., III, Miller, W.H., *J. Chem. Phys.* **80:** 4347 (1984), and references therein.
73. Salmeron, M., Somorjai, G.A., *J. Phys. Chem.* **86:** 341 (1982).
74. Tysoe, W.T., Nyberg, G.L., Lambert, R.M., *J. Phys. Chem.* **88:** 1960 (1984).
75. Hills, M.M., Parmeter, J.E., Mullins, C.B., Weinberg, W.H., *J. Am. Chem. Soc.* **108:** 3554 (1986).
76. Zuhr, R.A., Hudson, J.B., *Surf. Sci.* **66:** 405 (1977).
77. Whitman, L.J., Richter, L.J., Gurney, B.A., Villarrubia, J.S., Ho, W., *J. Chem. Phys.* **90:** 2050 (1989).
78. Semancik, S., Estrup, P.J., *Surf. Sci.* **104:** 261 (1981).
79. Ogle, K.M., Creighton, J.R., Akhter, S., White, J.M., *Surf. Sci.* **169:** 246 (1986).
80. Greenlief, C.M., Radloff, P.L., Zhou, X.-L., White, J.M., *Surf. Sci.* **191:** 93 (1987).
81. See a discussion in Stuve, E.M., Madix, R.J., *J. Phys. Chem.* **89:** 105 (1985).
82. Beebe, T.B., Jr., Yates, J.T., Jr., *J. Phys. Chem.* **91:** 254 (1987).
83. (a) Zhu, X.-Y., White, J.M., *Surf. Sci.* **214:** 240 (1989). (b) Lapinski, M.P., Ekerdt, J.G., *J. Phys. Chem.* **92:** 1708 (1988).
84. Anderson, K.G., Ekerdt, J.G., *J. Catal.* **116:** 556 (1989).
85. Seip, U., Tsai, M.-C., Küppers, J., Ertl, G., *Surf. Sci.* **147:** 65 (1984).
86. Stroscio, J.A., Bare, S.R., Ho, W., *Surf. Sci.* **148:** 499 (1984).
87. See, for example, Peterson, R.J., *Hydrogenation Catalysts*. Park Ridge, NJ: Noyes Data Corp., 1977.

10
Structure and Electronic Factors in Heterogeneous Catalysis: C≡C, C≡O, and C–H Activation Processes on Metals and Oxides

ALFRED B. ANDERSON

Progress in surface science and heterogeneous catalysis is by nature interdisciplinary. There are attempts to draw upon the viewpoint of the X-ray and electron crystallographer, hoping to obtain hard structural data. Yet, diffraction data from surfaces is difficult to obtain and analyze. Low energy electron diffraction (LEED) has played a prominent surface science role during the past 20 years,[1] showing the patterns of well-ordered adsorbate overlayers on single-crystal surfaces of metals and a few covalent and ionic solids, but bond lengths and other structure details have been difficult to determine with this technique. Diffractive helium scattering has found limited use for structure determinations. Extended X-ray absorption fine structure spectroscopy (EXAFS) in its various forms is presently evolving as an interesting new tool for determining local coordination numbers and bond lengths at catalyst surfaces.

Some structural information is being deduced from the analytical techniques of vibrational and electronic spectroscopies. Infrared (IR) spectroscopy has been used for many years to spot characteristic frequencies at surfaces of many highly dispersed catalysts. In the single-crystal surface domain, electron scattering in the form of high-resolution electron energy loss spectroscopy (HREELS) has been developed in recent years into an exciting technique for measuring bond vibrational frequencies, as has surface enhanced Raman spectroscopy (SERS). From such measurements it is possible not only to show there are CH or CO and other bonds in adsorbed molecules, but often to determine their bond orders and lengths from the frequency shifts and their orientations from peak intensities and selection rules. Thus, a reasonably suggestive, but not always unique, picture begins to emerge for what happens when, for example, acetylene is adsorbed molecularly on a platinum surface at liquid nitrogen temperature and then caused to react by warming.

Electronic structure probes such as X-ray photoelectron spectroscopy (XPS), Auger electron spectroscopy (AES), ultraviolet photoelectron spectroscopy (UPS), and inverse photoelectron spectroscopy (IPS) provide data that can sometimes be related to structure. These techniques are more used for analytical characterization: XPS and AES provide atomic oxidation state information; UPS yields valence electronic structures that can be used to tell, for example, whether CO adsorbs molecularly or dissociatively; and IPS provides optical adsorption-like information.

Other techniques for learning about surfaces include ion neutralization spectroscopy (INS), electron stimulated desorption ion angular distribution (ESDIAD), and scanning tunneling microscopy (STM). INS provides information about surface ionization spectra like UPS. ESDIAD is an indirect and uncertain technique for determining bond orientations from the directions in which ions emit from a surface when an atom in the bond is ionized and immediately desorbed. STM is a recent breakthrough that gives direct images of surfaces at the atomic resolution level.

Another facet of the struggle to understand catalysis is the application of chemical concepts and principles of the type taught in organic and inorganic chemistry courses. These are a complex mixture of mechanistic and thermodynamic facts out of which a rationalistic understanding is constructed and which, with the aid of the chemist's octet rule, electronegativity differences, polarizability, hardness and softness, and molecular orbitals, allow predictive extrapolation of reaction mechanisms, product stabilities, and so on. However, the catalyst still enters in as a special, essentially unknown, third body, written over the arrow between reactants and products in the textbook equation. How the catalyst works is a continuing challenge, particularly for those reactions that are heterogeneous.

Much more is known about structures and mechanisms in high-vacuum surface science work than in work with actual catalytic processes. In fact, the catalyst surface structures and compositions while the catalytic reactions are taking place are almost always unknown. Spectroscopies that work only in ultrahigh-vacuum environments, LEED, HREELS, XPS, AES, UPS, and IPS, become inoperable under high-pressure reaction conditions. While *in situ* techniques for catalysis studies such as Mössbauer, EXAFS, IR, and FTIR give part of the picture, a current goal of surface science is to learn as much as possible about the fundamental steps of bond activation and bond stability at catalyst surfaces by using well-defined models. Chemical knowledge and intuition, which have played the key role in the discovery of many catalytic processes, can also be helpful in choosing models.

The quantum theorist has an arsenal of methods for studying and explaining chemical bonding and predicting reaction pathways and the structures and stabilities of reaction intermediates and products. None are easy to apply to heterogeneous catalysis. Any quantum chemist faces the serious question the surface scientist faces: is the surface compositional and structural model that is chosen relevant to the overall catalytic reaction? Yet the quantum chemist has a special advantage, for a wide range of surface structures and compositions are instantly available for theoretical consideration. Since on the surface of a working catalyst there are active sites for, say, CH or CO or CC bond scission, it is in principle possible for the theorist to discover what structures and corresponding electronic properties are most likely to help the reacting molecules over the energy barriers in a catalytic process. Such model studies performed in this lab will be discussed in this chapter. The next section describes the method used.

ATOM SUPERPOSITION AND ELECTRON DELOCALIZATION MOLECULAR ORBITAL (ASED–MO) APPROACH

The ASED–MO theory is semiempirical, employing as input data experimentally determined atomic valence orbital ionization potentials and theoretically determined Slater orbital exponents from the literature. It is used to predict molecular structures, stabilities, force constants, and electronic properties and reaction pathways. The use of semiempirical parametrization holds immense practical importance, for the atom parameters are easily adjusted to suit the atom's chemical environment. The ASED–MO theory is parametrized against bond lengths and ionicity for diatomic fragments and the resulting parameters are used in polyatomic studies. Such adjustments emulate charge self-consistency. Another advantage of being able to employ experimental ionization potentials and adjustable orbital exponents in this way is that the effects of electron correlation and relativistic core shrinking are implicitly included. It cannot be expected that the ASED–MO procedure is capable of producing exactly the structural and other chemical properties listed above in this paragraph; it is completely incapable of doing some things, such as distinguishing between orbital angular and spin momentum states in a predictive way. However if, as is often the case, the spin state is known, then the molecular orbitals can be assigned the corresponding electron occupation and then the chemical properties of such a system can be calculated about as well as for systems with zero spin. In practice there is no good way at present to handle changes in spin states that can occur during the course of a chemical reaction and the effect of such changes on the energy. Consequently, it is usually assumed that the spin is constant.

The ASED–MO theory was originally physically motivated from my studies of vibrational force constants in diatomic molecules,[2] polyatomic molecules,[3] and solids.[4] It was found that, provided the equilibrium bond length R_e was known, free atom density functions could be used for predicting quite accurate (<30% error) force constants, k_e, according to the formula for a diatomic molecule ab:

$$\nabla^2_{R_b} E = 4\pi Z_b \rho_a(R_b) \tag{10.1}$$

or

$$k_e = \left(\frac{d^2E}{dR^2}\right)_{R_e} = 4\pi Z_b[\rho_a(R_b)]_e \tag{10.2}$$

where the origin is on nucleus a, Z_b is the charge of nucleus b, located at R_b, and ρ_a is the charge density function of free atom a, as shown in Figure 10.1. Equation (10.1) is exact provided the total molecular charge density function can be written as a sum of rigid atomic components which "follow" (are centered on) the nuclei and a flexible "non-perfectly-following" component, ρ_{npf}, which, for Eq. (10.1) to hold, must behave like point charges fixed in space. This is explained as follows[5]: beginning with the partitioning of ρ_{mol},

$$\rho_{mol} = \rho_a + \rho_b + \rho_{npf} \tag{10.3}$$

the force on nucleus b, F_b, is calculable by electrostatics and is given by

$$F_b(R_b) = F(R_b, \rho_a) + F(R_b, \rho_{npf}) \tag{10.4}$$

The interaction energy is then, from integration of $-F_b$,

$$E(R_b) = E_{\mathbf{R}}(R_b) + E_{npf}(R_b) \tag{10.5}$$

where

$$E_R = Z_b[Z_a|R_a - R_b|^{-1} - \int \rho_a(\mathbf{r})|R_b - \mathbf{r}|^{-1} d\mathbf{r}] \tag{10.6}$$

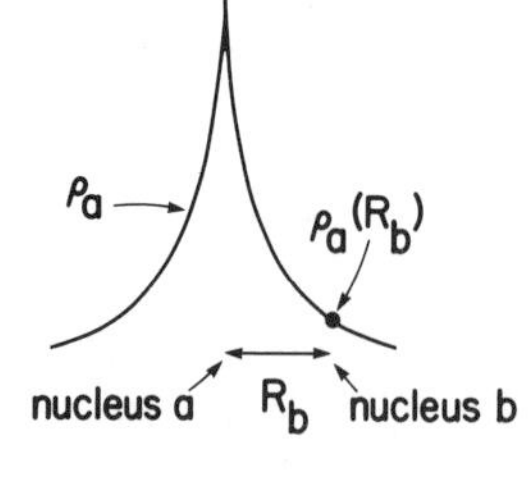

Figure 10.1. Cross-sectional view through atomic charge density distribution function ρ_a as used in Eq. (10.2).

($\mathbf{r}$ is the electronic position) and

$$E_{\mathrm{npf}} = -Z_{\mathrm{b}} \int_{\infty}^{R_{\mathrm{b}}} \int \rho_{\mathrm{npf}}(\mathbf{r}, R'_{\mathrm{b}}) d/dR'_{\mathrm{b}} |R'_{\mathrm{b}} - \mathbf{r}|^{-1} d\mathbf{r}\, dR'_{\mathrm{b}} \qquad (10.7)$$

Equations (10.4) and (10.5) are depicted in Figure 10.2. It is clear from the formula

$$\int f(\mathbf{r}) \nabla_{\mathrm{R}}^{2}(1/R) d\mathbf{r} = -4\pi f(R) \qquad (10.8)$$

that Eq. (10.1) follows from Eqs. (10.5)–(10.7) provided ρ_{npf} is equal to fixed point charges that do not overlap nucleus b.

Equations (10.5)–(10.7) are the basis for the ASED–MO theory. Pairwise atom–atom E_{R} components are easily calculated using atomic orbitals from the literature to generate the densities. E_{npf} cannot be calculated using Eq. (10.7) because ρ_{npf} is unavailable. However, it is an electron delocalization energy that has been found to be generally well approximated by the change in orbital energy which occurs when chemical bonds form, ΔE_{MO}[6]:

$$\Delta E_{\mathrm{MO}} = \sum n_i \varepsilon_i^{\mathrm{ab}} - \sum_{a} \sum_{i} n_i \varepsilon_i^{\mathrm{a}} \qquad (10.9)$$

where n_i are occupation numbers (0, 1, or 2), $\varepsilon_i^{\mathrm{ab}}$ are molecular orbital energies obtained from diagonalizing a hamiltonian whose matrix ele-

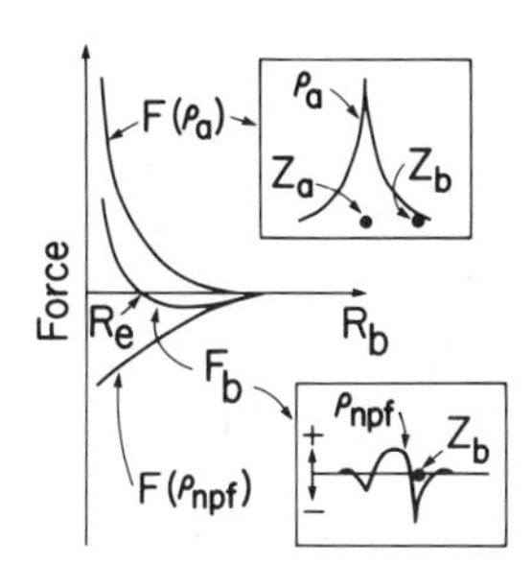

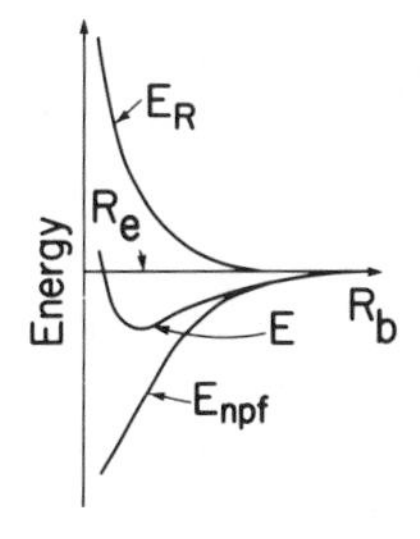

Figure 10.2. Schematic representation of force and energy components in Eqs. (10.4) and (10.5).

ments are

$$H_{ii}^{aa} = -IP_i^a \tag{10.10}$$

$$H_{ij}^{aa} = 0 \tag{10.11}$$

$$H_{ij} = 1.125(H_{ii}^{aa} + H_{jj}^{bb})S_{ij}\exp(-0.13R) \tag{10.12}$$

where IP_i^a is the ith orbital ionization potential of atom a, $S_{ij} = \langle i,\mathrm{a}|j,\mathrm{b}\rangle$, R is the internuclear distance, and

$$\varepsilon_i^a = -IP_i^a \tag{10.13}$$

With the approximation

$$E_{\mathrm{npf}} \simeq \Delta E_{\mathrm{MO}} \tag{10.14}$$

the final working formula becomes

$$E \simeq E_{\mathrm{R}} + \Delta E_{\mathrm{MO}} \tag{10.15}$$

It may be noted that there are two formulas of the form of Eq. (10.6) for E_{R}, and the E_{R} functions are different depending on whether Z_{b} and ρ_{a} or Z_{a} and ρ_{b} are used. When E is calculated by the above procedure, results are usually more accurate when the atom charge density of the more electronegative atom is used in determining E_{R}.

In heteronuclear systems we take charge self-consistency into account in an approximate way, by systematically adjusting the ionization potential and orbital exponent parameters until the charge transfer and bond length of the diatomic molecule are close to measured or estimated values. Presently we employ cluster models of up to 90 atoms with two or more layers to model the surfaces of metals and up to 100 atoms for the oxides. Although we now have operational an ASED band theory program, we find the cluster models offer computational advantages because determining enough $E(k)$ points for integration (k is the wave vector) takes too much computational time for doing the type of surface work discussed here. For ferromagnetic metals we apply our high-spin rule so that all d-band orbitals are occupied with at least one electron. This means, for example for Fe, that some of the more stable d-band orbitals are doubly occupied and the remaining ones are all singly occupied.

The ASED–MO approach offers the ease of calculation and interpretive simplicity of extended Hückel. Unlike extended Hückel, it is not necessary to assume structures; instead they are predicted. The ability of the ASED–MO theory to predict structures is particularly helpful in studies of heterogeneous catalysis because so little is experimentally known about them.

STUDIES IN BOND ACTIVATION

In catalysis strong bonds are rendered weak because of the formation of bonds to the catalyst. It is of primary interest to understand how. A number of systematic studies of C≡C, C≡O, and C–H bond activations have been made in this lab. Most of the work has been on transition metals but methane activation on oxide surfaces has been given special attention because oxides actually catalyze methane dimerization to ethane or oxidation to methanol or formaldehyde and these processes are theoretically explicable due to the small number of steps involved. The work on C≡O activation is generally believed to relate to the initial step in Fischer–Tropsch catalysis but subsequent steps leading to C_n alkanes remain unexplained in detail, although work is underway in my lab. The studies of acetylene adsorption illustrate how structure and electronic concepts develop from systematic studies. Many structure and energy details have been calculated and are tabulated in the papers that are referenced below. This review will focus on the more general conclusions.

ACETYLENE BONDING TO TRANSITION METALS: STRUCTURE AND ELECTRONIC EFFECTS

The primary interest in acetylene adsorption is as a probe of bonding. In the early days of ultrahigh-vacuum surface science, acetylene was a popular candidate for studying chemisorption. For example, it was observed to adsorb weakly and reversibly on Cu in UPS studies,[7] to rearrange near room temperature on Pt according to UPS,[8] and LEED,[9] and HREELS,[10] and to dissociate on Fe according to UPS.[11] It was not until the appearance of HREELS as a standard feature of high-vacuum chambers combined with systematic studies over temperature ranges that structure and reaction details began to become understood.

One of my early surface science studies included acetylene bonding to a Fe_2 model of the bcc (100) surface.[12] C_2H_2 was predicted to dissociate with a nearly zero activation energy into CH fragments binding upright through C to the surface. About that time Rhodin's group was examining acetylene adsorption on this surface and noted a drastic change in the UPS spectrum of acetylene and ethylene in the 98–123 K temperature range.[11] A joint study[13] with the Rhodin group showed how the spectra could be described as resulting from chemisorbed acetylene and ethylene at liquid N_2 temperature decomposing to CH and CH_2 fragments on heating. In this

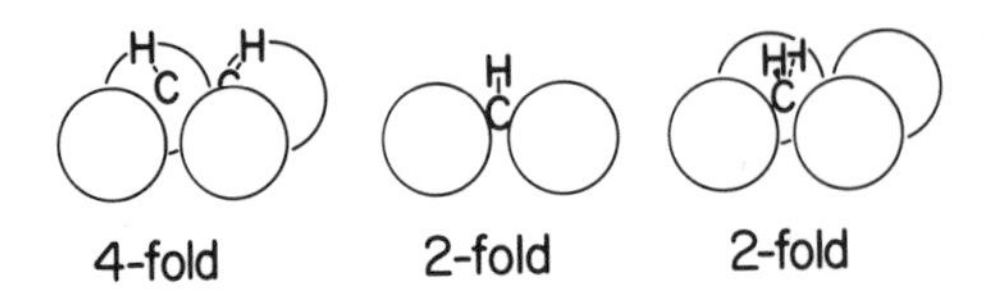

Figure 10.3. Adsorbate structures on Fe(100).

work a larger Fe_5 cluster was used and C_2H_2 was predicted to bind most strongly in the fourfold site. Dissociation barriers were calculated to be close to zero. The CH and CH_2 fragments were predicted to bind in twofold bridging sites. The various sites are defined in Figure 10.3.

The bond weakening leading to easy dissociation of acetylene on Fe(100) is caused in part by substantial metal *d* back-donation to the empty acetylene π^* orbitals. The acetylene π and σ orbitals get involved in the bonding too, and the result is an ethyleniclike π, σ bonding structure for chemisorbed acetylene (Fig. 10.4), with a commensurate lengthening of the CC bond.

Subsequent experimental studies showed that acetylene adsorbed molecularly on the close-packed (110) surface and decomposed above 340 K,[14] and behaved similarly on the open (111) surface.[15,16] To understand why, we made a systematic study of acetylene adsorption on the Fe(100), (110), and (111) surfaces using large clusters of 21 and 22 Fe atoms.[17] By binding pairs of C_2H_2 molecules to the surface models lateral interactions could be established, leading to predictions of overlayer structures, including the C_1 symmetry and p (2×2) coverage for C_2H_2 on Fe(110) that had been determined experimentally—see Figure 10.5. The relative cal-

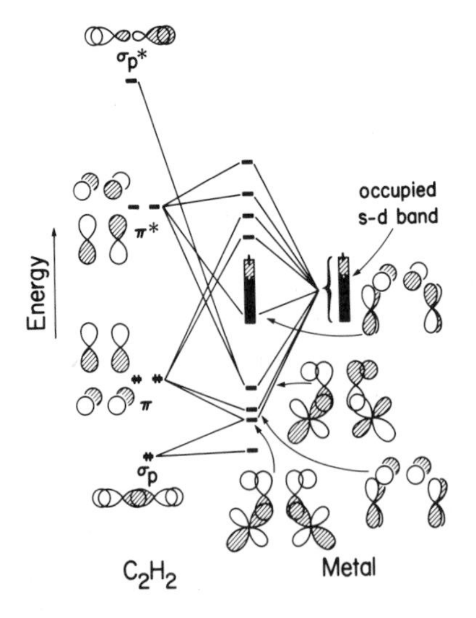

Figure 10.4. Schematic molecular orbital correlation diagram for C_2H_2 adsorbed in a twofold bridging (di-σ) structure on a metal surface. Note that the shaded region of the metal band indicates doubly occupied and the cross-hatched region singly occupied orbitals. The empty metal *s-p* band is omitted for clarity. Empty dangling orbitals on the surface will hybridize with the chemisorption bond orbitals.

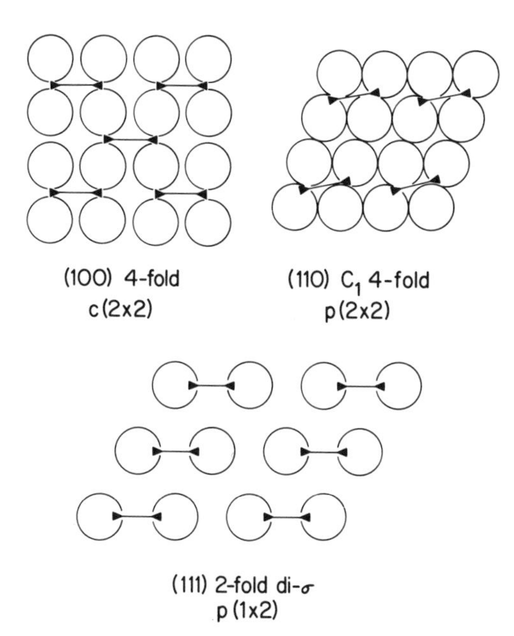

Figure 10.5. Predicted structures for high coverages of acetylene on three Fe surfaces.

culated activation energies for CC bond dissociation to yield adsorbed CH fragments on the surfaces were qualitatively supportive of the experimental determinations. That they were all under 1 eV indicates the high strength of C–Fe bonds because the C≡C bond strength is 10.0 eV. Interestingly, as Figure 10.6 shows, the molecular orbital energy by itself fails to predict the equilibrium and transition state structures. The total energy is needed for such predictions.

On the basis of these results, we proposed in Ref. 17 a general rule based on surface metal atom structure: on surfaces where acetylene

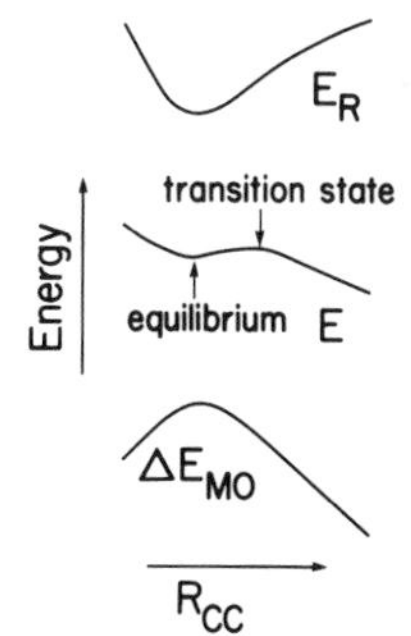

Figure 10.6. Total energy and its components of a function of CC internuclear distance, R_{cc}, on a transition metal surface.

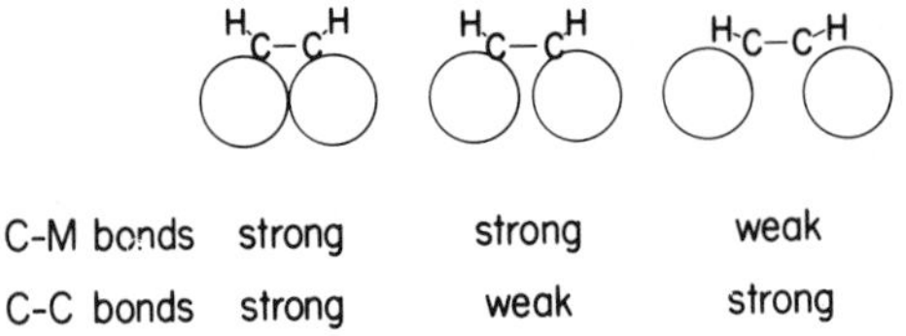

C-M bonds	strong	strong	weak
C-C bonds	strong	weak	strong

Figure 10.7. Relationship between surface metal atom spacing and the C–C bond strength in chemisorbed C_2H_2.

chemisorbs strongly enough to approach an ethylenic structure, CC activation will be lowest on very close packed [e.g., Fe(110)] and very open surfaces [e.g., Fe(111)] and highest on surfaces with intermediate metal atom densities [e.g., Fe(100)]. This was explained as resulting from the competition between maintaining the CC triple bond and making the M–C σ bonds. According to this structure effect, the M–C bond is too weak to cause strong chemisorption or large CC bond stretches and low dissociation activation energies when the surface metal atom spacing is large or small; see Figure 10.7. This principle was found in a subsequent theoretical study to account for acetylene adsorption and activation of acetylene on the Pt close-packed (111) and open (110) and intermediate (100) surfaces.[18] Since Pt is fcc the surface structures are different, as shown in Figure 10.8.

The structure effects for the various Fe and Pt surfaces are similar because the Fe and Pt valence *s* and *d* orbital overlaps with C valence *s*

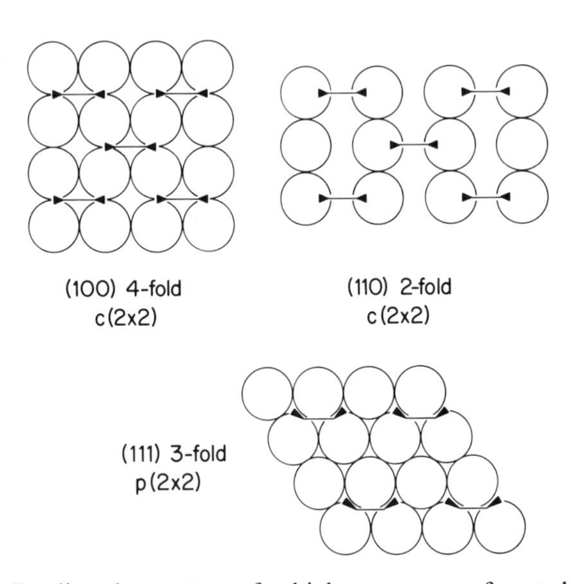

Figure 10.8. Predicted structures for high coverages of acetylene on three Pt surfaces.

and p orbitals at bonding distances are almost the same[19] and because the σ and π donation bonds from C_2H_2 to the metals and the backbonding to the C_2H_2 π^* orbitals are the same on these surfaces. We studied C_2H_2 adsorption on V(100), (110), and (111) surfaces, which are identical in structure to those of Fe, and predicted C_2H_2 should dissociate with zero activation energy on all of them. This difference was traced to the larger overlap between the V and C valence orbitals (Fig. 10.9), which increases the magnitudes of all chemisorption bonding interactions.[19] CH bonds were also predicted to dissociate more easily on the V surfaces because of increased overlap between the H 1s orbital and V valence orbitals.

An electronic effect was examined by modeling acetylene adsorption on Fe(100) and Pt(111) electrode surfaces[20] using our band-shift electrochemical modeling theory.[21] This simple theory is based on the principle that as a potential is applied to an electrode surface in a dielectric medium the Fermi energy, and hence band position, shift on the energy scale by an amount equal to the applied potential. It is clear from perturbation theory and reference to Figure 10.4 that as the band shifts down with anodic charging C_2H_2 π donation to the surface will increase and metal back donation to C_2H_2 π^* orbitals will decrease. The inverse occurs with cathodic charging. For both metals there is a decrease in the CC bond scission activation energy for either anodic charging, where π donation increases, weakening the CC bond or cathodic charging, where back donation to the π^* increases, again weakening the CC bond. The same was found for CH

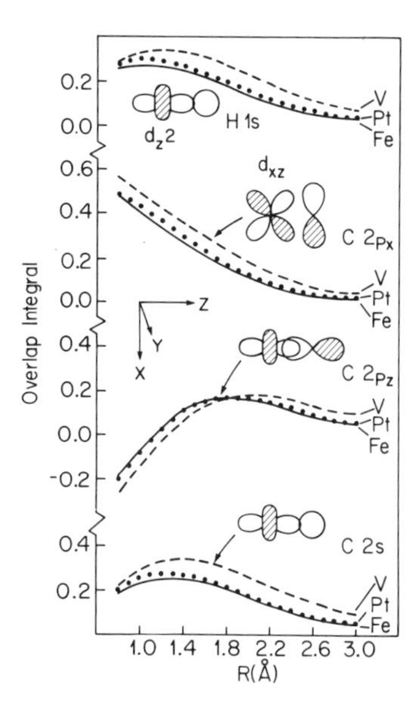

Figure 10.9. Overlap integrals as functions of distance between atom centers.

bond scission in CH fragments formed from C_2H_2 dissociation. The activation of CH bonds will be discussed in greater detail later in this chapter.

CO BINDING TO TRANSITION METAL SURFACES: A NEW INTERACTION

CO, a ligand in innumerable coordination complexes, is an important molecule to the surface scientist for probing metal surfaces. Its heterogeneous catalytic hydrogenation to methane and larger alkanes (Fischer–Tropsch catalysis) and methanol and larger alcohols is of current interest. Many experimental and theoretical surface science papers dealing in CO adsorption on metal surfaces mention the Fischer–Tropsch methanation but very little has been learned about the mechanism. Progress in understanding CO activation has come recently. Less is known about the hydrogenation steps prior to or following dissociation and the mechanism of C_n formation. Neither is the formation of alcohols well understood.

Our work has dealt in surface science type questions of binding site and orientation (whether bonded perpendicular to the surface through the C end or parallel, through C and O) as well as the strong metal support interaction (SMSI)[22] effect where Fischer–Tropsch methanation is promoted on later transition metal surfaces by the addition of early transition metal oxide cations. In our adsorption modeling for single-crystal surfaces we have established the importance of an here-to-fore unappreciated electronic interaction.

Until recently, CO was observed, by a variety of experimental techniques, to bind vertically through C on single-crystal transition metal surfaces. Sometimes it was found at onefold atop positions, as on Pt(111)[23] and other times in twofold bridging sites, as on Pd(100)[24–26] or threefold sites, as on Pd(111)[24] or even 50% on onefold and 50% on twofold sites, as on Pt(100).[24,25] These orientations are defined in Figure 10.10. At high coverage lateral interactions can cause the CO molecules to be shoved by crowding off their most favored lower-coverage positions.[27]

The traditional bonding description for upright CO mentions two interactions, CO 5σ donation to the metal surface band orbitals, and back-donation from these orbitals to the CO π^* as shown in Figure 10.11. This is the Dewar–Chatt–Duncanson[28] description of the coordination of molecules with empty π^* orbital to transition metal atoms, and is known as the

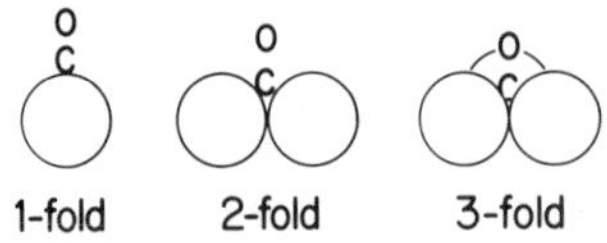

1-fold 2-fold 3-fold

Figure 10.10. Adsorbed CO.

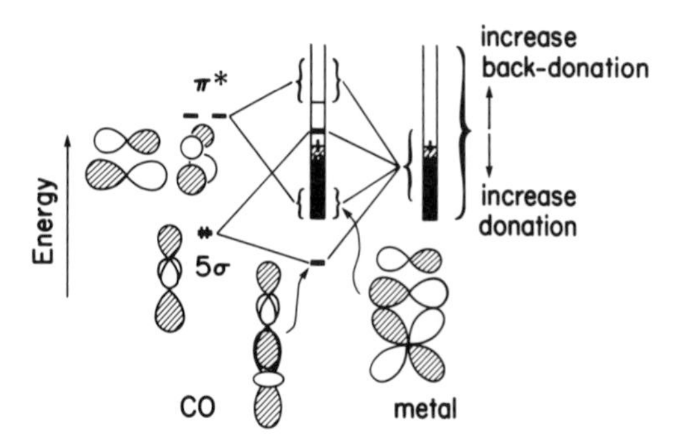

Figure 10.11. 5σ donation and back-donation to π^* for CO adsorbed to a surface metal atom.

Blyholder[29] model in the surface domain. It is generally thought that when the 5σ donation interaction is relatively strong, onefold adsorption is favored, and when back-donation to π^* is strong, higher coordinate sites are more stable. This suggests that metals that are good reducers, i.e., have less stable *d* bands, are more likely to adsorb CO in high coordination.

We demonstrated this idea in a study of CO on (100) and (111) surfaces of Pt and Pd.[30] Calculations showed, in agreement with experiment, that CO adsorbs on the high-coordinate Pd sites and on low-coordinate Pt sites. The Pd valence band lies 1 eV above the Pt band, so Pd is the better donor. Interestingly and significantly, a shift of 1 eV of the Pd valence band to greater binding energy, which is achieved in ASED–MO calculations by increasing the valence state *IP* parameters by 1 eV, makes Pd behave like Pt toward CO. Binding to the onefold site of the (111) surface becomes most stable and binding energies to the one- and twofold sites of the (100) surface become nearly the same. Moving the Pt valence band up 1 eV makes it behave, as expected, like Pd, favoring high coordination for CO.

The above shifts in adsorption site could probably be seen on Pt and Pd electrode surfaces, provided an electrolyte is used that is stable, not undergoing Faradaic reactions, over the necessary potential ranges. In fact, a recent experimental study shows that CO increasingly favors onefold over twofold adsorption sites upon increased anodic charging on a Pt(100) electrode.[31]

The absorption of K at small coverages on Pt(111) has been observed to cause CO to shift from onefold to twofold and possibly threefold bridging sites with increasing coverage.[32] We have modeled this with the Pt valence band shift technique.[33] We postulated that a chemisorbed K atom would donate a substantial fraction of an electron charge to neighboring Pt atoms, making them cathodic, and that with increasing K coverage the cathodic potential would increase. The precise relationship between cathodic shift and K coverage was unknown, but the model led, as discussed above, to a shift to bridging sites with a 1-eV band shift. The predicted rates of decrease in the CO stretching force constant in the various sites with increasing K coverage, slower for onefold and faster for two- and threefold coordination, agreed with the experimentally observed frequency changes.

The band-shift model also accounts qualitatively for experimental observations of the potential dependence of CO vibrations on a Pt electrode.[34] For CO on Pt(111) our calculations yield 58 cm^{-1}/V, fairly close to 50 cm^{-1}/V measured for CO on a Pt electrode in a neutral electrolyte. Interestingly, in acidic electrolytes the potential dependence is much less, 30 cm^{-1}/V. Others have attempted to ascribe the potential dependence to the effects of the field alone in terms of a Stark effect and without considering the band shift.[35] The field was included as a perturbation to the ASED–MO hamiltonian in our recent study of field adsorption of noble gas atoms on tungsten tips.[36] However, it must be remembered that the field is a derived quantity which does not appear in the *ab initio* hamiltonian of the complete physical system. For species adsorbed on an electrode surface the electron donor–acceptor property, which is determined by the applied potential, should provide sufficient understanding just as it does in other areas of chemistry.

Within the band shift model, the increase in CO vibrational frequency with increased anodic charging stems from the decreased back-donation to the CO π^* orbitals as the metal valence band becomes more stable. Because of the reduced Pt–C π bonding, the Pt–C force constant was predicted in Ref. 34 to decrease slightly with potential, but this does not seem to have been measured yet. A similar study of CN^- on an Ag electrode also showed an increasing CN vibrational frequency with increasingly anodic potential, but in this case Ag–C π bonding was weaker and the 5σ donation was stronger.[37] Since the 5σ orbital is antibonding, the increasing donation stabilization strengthened the Ag–C force constant as the potential went anodic. The CN^-/Ag results showed qualitative agreement with SERS measurements.

We discovered the importance of donation from the occupied π bonding orbitals to metal surfaces quite recently.[38] This occurred in an attempt to explain why CO on Ru(001) appeared, according to EELS[39] and ESDIAD[40] measurements, to bind lying-down or parallel to the surface in the presence of 0.1 monolayer of K. More recently, angle-resolved UPS experiments have been used to prove that CO actually binds upright or perpendicular on this surface,[41] but no matter, for our results were actu-

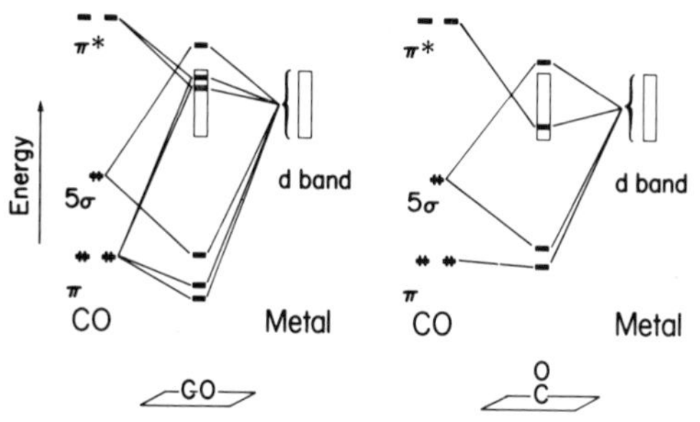

Figure 10.12. Orbital correlation diagram for CO adsorbed lying-down and upright on a transition metal surface.

ally ambiguous in this instance and the principles we discovered are nicely applicable to CO adsorption on other early (*d* electron deficient) transition metal surfaces and atoms.

We have shown that CO can form strong bonds by donation from its occupied π orbitals to the metal *d*-band orbitals, provided there are not too many electrons in the *d* band. See Figure 10.12. The *d* electron deficiency allows the antibonding counterpart orbitals to the π donation interaction to be only partially occupied or even empty. In such cases there is a resulting net bond order between CO and the surface due to the π donation; in *d*-electron-rich metals the antibonding counterpart orbitals to the π donation are doubly occupied and this leads to a closed-shell repulsion which prohibits CO from binding lying down. On the *d*-electron-deficient metals where CO does lie down, so that orbitals on C and O both overlap the surface orbitals, the back donation to π^*, 5σ donation, and even 4σ donation interactions are all increased in strength. As a result, the CO bond stretches ~0.1–0.2 Å from the gas-phase value and its vibrational absorption shifts to lower values by ~300–400 cm^{-1} and its strong 11.2 eV bond dissociates with an activation energy barrier of <1 eV. We have demonstrated this in fine agreement with the experimental studies of CO on Cr(110)[42] (for which we also explain a second upright phase at high coverage), the Pt_3Ti alloy,[43] and in a general way for Pt doped with TiO, FeO, and Fe.[44] Some predicted structures are in Figure 10.13.

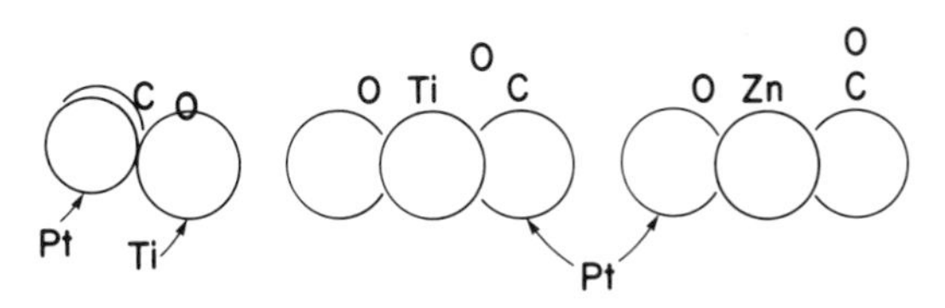

Figure 10.13. Predicted CO orientations on the (111) surfaces of Pt_3Ti and TiO- and ZnO-doped Pt(111).

The doped Pt study was a model for the strong metal support interaction (SMSI) effect in Fischer–Tropsch catalysis. In Fischer–Tropsch catalysis CO is hydrogenated, forming hydrocarbons and higher alcohols and aldehydes. The metal can be Pt and the support can be TiO_2. The SMSI effect is created by exposing the titania supported Pt catalyst system to reduction, whereupon TiO_x species are in intimate contact with reduced Pt. From experimental observations,[45] and our theoretical calculations, CO evidently adsorbs across a Ti^{n+}–Pt site, with C over Pt and O over Ti^{n+} (Fig. 10.13), so that it is activated as discussed above. It is not clear yet whether the CO bond scission is before the hydrogenation commences.

When ZnO, MgO, and CaO supports are used, Fischer–Tropsch alkanes are no longer obtained, but instead methanol is the main product.[45] Thus, these oxides do not activate CO bond scission. For ZnO we calculated a strong closed-shell repulsion between the filled CO π orbitals and d^{10} Zn^{2+}. Perhaps the M^{2+} stabilize the alkoxy intermediates against loss of oxygen. We have shown how Pt itself has too many d electrons to allow CO to bind to it through both the C and the O ends.

Interestingly, the Fe group is in the border line. We predict,[46] in agreement with experiment,[47] that CO lies down on the more open Fe (100) and (111) surfaces but stands up on the close-packed (110) surface. The surface atom spacing plays a key role in this case: when it is large enough, 4σ and 5σ donation and back-donation to CO π^* interactions are strong enough to overcome the π repulsion with the surface. In this case the π repulsion is weak because the antibonding counterparts to the π donation are half-filled. Ru surfaces will probably show a structural effect for CO adsorption too.

CO BINDING TO ZnO AND SURFACE ION RELAXATIONS

ZnO is an interesting catalytic oxide. It can isomerize hydrocarbons and hydrogenate olefins to alkanes[48] and CO to CH_3OH.[49] Our studies have been limited to explaining why H_2 adsorbs heterolytically[50] on ZnO and to explaining the effects of CO adsorption on Zn^{2+} relaxations.[51] The latter work will be discussed here.

ZnO has the wurtzite structure. Each Zn^{2+} cation and O^{2-} anion is tetrahedrally coordinated in the bulk. We studied threefold coordinated Zn^{2+} and O^{2-} on the prism $(10\bar{1}0)$ and the polar (0001) Zn^{2+} and $(000\bar{1})$ O^{2-} covered surfaces using various cluster models of up to 27 atoms. The essential features are the same for Zn^{2+} and O^{2-} on these surfaces. Zn^{2+} relaxes in along the tetrahedral direction 0.5 ± 0.1 Å and ~0.3 Å on the $(10\bar{1}0)$ and (0001) surfaces, respectively, according to LEED work.[52] We calculated 0.4 Å and 0.35 Å, respectively, essentially in agreement with

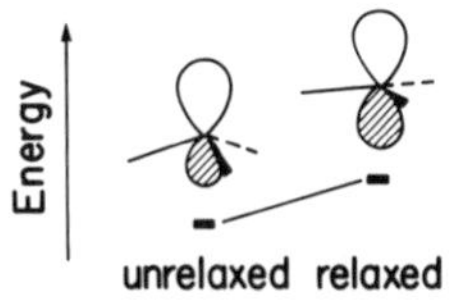

Energy

unrelaxed relaxed

Figure 10.14. Zn^{2+} dangling empty surface orbital.

the LEED determinations. On (10$\bar{1}$0) ZnO the experimental value for O^{2-} relaxation is 0.05 ± 0.1 Å and we obtained 0.15 Å.

The threefold coordinated surface Zn^{2+} have empty dangling $4s$-$4p$ hybridized band gap orbitals on them. The Zn^{2+} relaxation serves to stabilize the occupied bonding Zn $4s$, $4p$–O $2p$ orbitals; the dangling band gap orbital, which is an antibonding counterpart, becomes destabilized, as shown in Figure 10.14. When CO adsorbs, its 5σ interaction with Zn $4s$ and $3d$ bands is closed-shell, but the dangling orbital stabilizes the occupied antibonding counterpart to the 5σ donation (Fig. 10.15), allowing CO to adsorb with a sticking energy of 12 kcal mol^{-1}, in fortuitous exact agreement with experiment. By unrelaxing, the Zn^{2+} dangling s-p orbital drops in energy by over 1 eV, which enhances the mixing with the CO 5σ and strengthens the CO adsorption bond. Our calculated unrelaxation was 0.11 Å above the bulklike position. Unrelaxation had been speculated about on the basis of UPS studies. We also predicted a 0.02 Å decrease in the CO bond length when adsorbed, owing to the stabilization of the weakly antibonding CO 5σ orbital by its bonding overlap with the Zn^{2+} surface dangling orbital. We explained the observed increase in CO vibrational frequency in terms of the calculated bond strengthening, which was dominant, and coupling to the surface, which accounted for one-quarter of the frequency increase.

There is no doubt that surface atom relaxations and even surface restructuring and dangling band gap orbitals will be important features in future characterizations of chemisorption on nonmetals. They must also be expected to be part of the description of catalytic reactions.

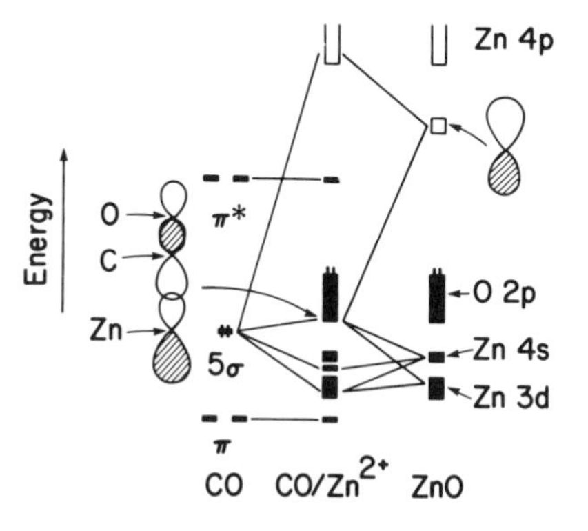

Figure 10.15. Orbital correlation diagram for CO adsorbed to Zn^{2+} on ZnO.

CH ACTIVATION IN ALKANES AND ALKENES

Methane is plentiful in the earth's crust and the equivalent of a million barrels of oil of CH_4 is flared daily at oil fields throughout the world. There is economic incentive to find ways for converting this methane or some of the natural gas that is presently burned for heat into more valuable molecules. The first step in methane conversion is CH bond activation. We have been studying methane activation by metals, metal oxides and, recently, metal sulfides.

The CH bond is a closed-shell σ orbital. It lies beneath metal valence bands and oxide $2p$ bands in energy. The empty σ^* orbital is situated well above these bands. The CH σ bond is strong, with a dissociation energy of about 4.6 eV. In our work we have characterized two types of activation, H· abstraction, and oxidative addition (or metal atom insertion into the CH bond). I shall discuss H^+ abstraction first, which is not activated.

If atom A has a lone pair, such as O in H_2O or N in NH_3, one can write the proton transfer reaction

$$C{:}H + A{:} \rightarrow C{:}^- + A{:}H^+$$

Such reactions are unlikely to occur, at least in the gas phase, not only because of the closed-shell repulsion shown in Figure 10.16, but because anion and cation products are unstable. In the presence of a third body, such as the M^{n+} cation in a metal oxide (A: would then be $:\ddot{\underset{..}{O}}:^{2-}$), heterolytic adsorption products $[M{:}C]^{(n-1)+}$ and $[:\ddot{\underset{..}{O}}{:}H]^-$ might be stable, as we have shown,[53–55] but their formation still has a barrier due to the closed-shell repulsion. Hence, it seems that some sort of activating process will be needed.

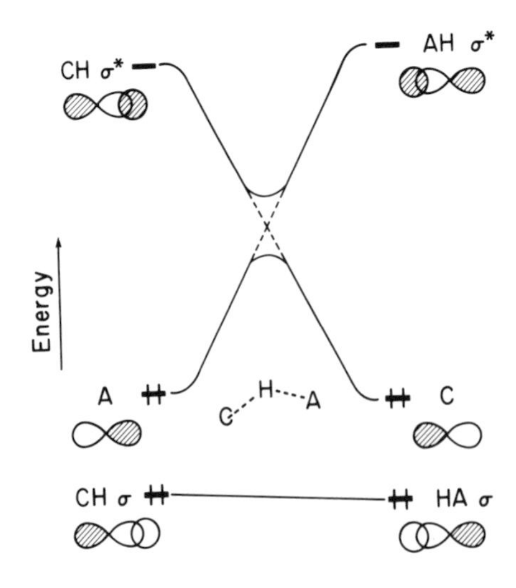

Figure 10.16. Source of the barrier to proton transfer from a CH bond to a lone-pair on A.

To abstract H· one must form a bond that is stronger than the CH bond. In heterogeneous catalysis this excludes everything but oxides (the OH bond strength is approximately <5.2 eV), fluorides (the HF bond strength is 5.9 eV), and possibly nitrides (the NH bond strength is approximately <4.8 eV). Abstraction by O^{2-} produces an electron:

$$H_3C{:}H + {:}\ddot{\underset{\cdot\cdot}{O}}{:}^{2-} \rightarrow H_3C\cdot + {:}\ddot{\underset{\cdot\cdot}{O}}{:}H^- + e^-$$

and this electron will be promoted to the metal cation valence conduction band:

$$e^- + M^{n+} \rightarrow M^{(n-1)+}$$

In metal oxides the valence band typically lies 3 eV (as in MoO_3) or more above the top of the O 2*p* band, so H· abstraction by O^{2-} is a high-energy process that is in addition to the closed-shell repulsion discussed previously: these render O^{2-} inactive for many oxides.

However, O^- is active because no electron is promoted and the initial interaction is no longer closed-shell

$$H_3C{:}H + {:}\dot{\underset{\cdot\cdot}{O}}{:}^{-} \rightarrow H_3C\cdot + {:}\ddot{\underset{\cdot\cdot}{O}}{:}H^-$$

The active O^- surface species can be created chemically[56–59] or by charge transfer photoexcitation.[59] We have carried out ASED–MO studies on three oxides with empty conduction bands, MgO, MoO_3, and $CuMoO_4$ (the last one has interesting properties due to Cu orbitals with energies in the O 2*p*–Mo 4*d* band gap, and will be discussed later).

From the molecular orbital perspective, the initial interaction of the CH σ electron pair with an O^{2-} lone pair is one of closed-shell repulsion. A three-center σ-donation bond begins to form but its antibonding counterpart is doubly occupied and rises rapidly in energy as the OH bond forms and the CH bond breaks, resulting in a high barrier. The CH bond order is 1 and the OH bond order is zero. In the case of O^-, as the antibonding counterpart orbital rises through the O 2*p* band it picks up the hole at the top and becomes singly occupied (Fig. 10.17) so that the CH bond order is

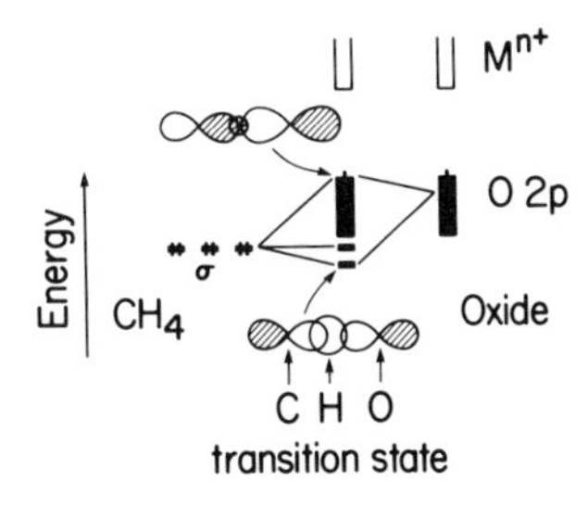

Figure 10.17. Orbital correlation diagram for H· abstraction from CH_4 by O^- on an oxide surface.

three quarters and the OH bond order is one quarter. In the cases of the oxides studied thus far, this event is simultaneous with the occurrence of the transition state. We calculate activation barriers of 0.7 eV for O^- on MoO_3,[53] 0 eV for O^- on MgO,[54] and 1.3 eV for O^- on $CuMoO_4$.[55] Because we slightly overestimate CH and underestimate OH bond strengths, the first and last barrier could be smaller.

The resulting methyl radicals undergo further reactions, with dimerization to ethane being the dominant pathway when the O^- concentration on the surface is low, as on Li-doped MgO[57]; surface methoxy, and CO, CO_2, and CH_3OH form when O^- concentration is high, as when O^- is formed by N_2O decomposition on MoO_3.[60] We find the $CH_3\cdot$ binds weakly to the O^{2-} basal planes of MoO_3 (MoO_3 has a layered structure with O^{2-} on the layer surfaces) with the unpaired radical electron promoted to the Mo^{VI} conduction band (Fig. 10.18). It is strongly trapped as OCH_3 (ads) when it comes upon another O^- and is held for subsequent oxidation or hydrogenation to CH_3OH when H_2O is present. $CH_3\cdot$ does not bind to (100) cleavage plane O^{2-} of the rock-salt-structured MgO crystal except when there is an adjacent corner Mg^{2+} with a low-lying dangling band gap orbital to take the electron. It binds weakly to crystal face Mg^{2+}, so, given the low concentration of surface O^- to trap it as OCH_3^- (ads), it is understood why desorption and dimerization to ethane is the predominant route.

CH_4 can be photoactivated over MoO_3 in the presence of near UV light which creates O^- by charge transfer excitation.[55] This process selectively yields CH_3OH. Doping with CuO, which forms $CuMoO_4$, increases the rate of CH_3OH formation by a factor of three in the presence of broad-band light from an Xe lamp. Part of this increase is due to the absorption of visible light and the remainder is due to prolonged electron hole pair lifetime due to the presence of Cu^{2+}. We have explained these effects respectively in terms of the electronic and geometric structure of $CuMoO_4$.[55] The electronic structure is similar to that of MoO_3 with a gap of ~3 eV between the O 2*p* band and the empty Mo^{VI} 4*d* band, except Cu^{2+} introduces a half-filled Cu 3*d* band near the middle of this gap, as shown in

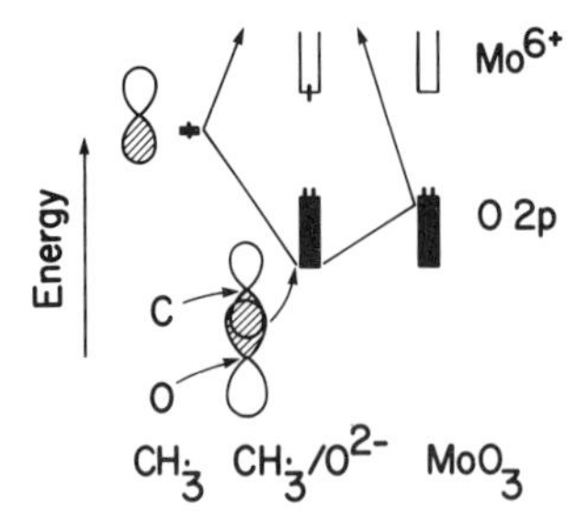

Figure 10.18. Binding of $CH_3\cdot$ to anion basal plane on MoO_3.

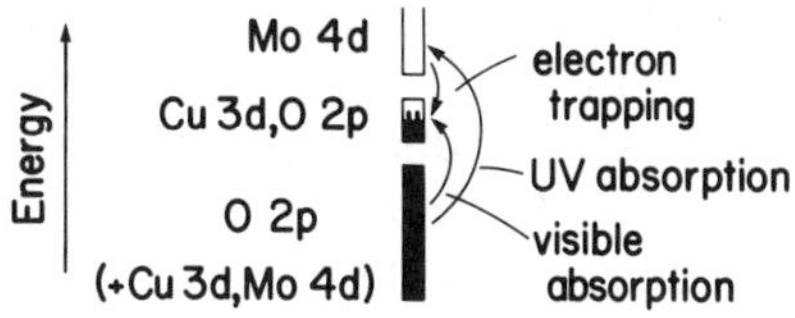

Figure 10.19. Optical absorption and electron trapping mechanism for $CuMoO_4$.

Figure 10.19. This allows O^- formation by visible O 2*p* to Cu 3*d* charge transfer excitations ($CuMoO_4$ is green). Similar excitations occur in CuO but CuO (which is also green) is inactive. This suggests that surface O^- bound to Mo^{VI} forms and is stabilized according to

$$O^{2-}\text{–}Mo^{VI}\text{–}O^{2-}\text{–}Cu^{II} \overset{h\nu}{\leftrightarrows} O^{2-}\text{–}Mo^{VI}\text{–}O^{-}\text{–}Cu^{I}$$

$$\rightleftarrows O^{-}\text{–}Mo^{VI}\text{–}O^{2-}\text{–}Cu^{I}$$

and this O^- activates CH_4. Electron–hole pair recombination will be slow because of the increased distance between the surface O^- and Cu^I. It is reasonable for holes to migrate to the surface because they become more stable as they rise in the O^{2-} band and the surface O^{2-} lone pair dangling orbitals are at the top of the band. A similar scheme would account for the prolonged lifetime of electron–hole pairs created by UV charge transfer excitations:

$$O^{2-}\text{–}Mo^{VI}\text{–}O^{2-}\text{–}Cu^{II} \overset{h\nu}{\leftrightarrows} O^{-}\text{–}Mo^{V}\text{–}O^{2-}\text{–}Cu^{II}$$

$$\rightleftarrows O^{-}\text{–}Mo^{VI}\text{–}O^{2-}\text{–}Cu^{I}$$

In this case the electron promoted to the higher-lying Mo^{VI} band falls into the half-filled Cu^{II} band in the gap where it is trapped to form Cu^I—see Figure 10.19.

Abstraction of the α (methyl) hydrogen from propylene is thought to be the rate-limiting step, with an activation energy of 19–21 kcal mol^{-1} [61] for acrolein formation over bismuth molybdate and for allyl radical dimerization to form hexa-1,5-diene over Bi_2O_3. The presence of O^- at the surface is apparently not necessary for α CH activation in these cases. We have shown that the same features which make the out-of-plane α CH bond weak, namely, the stabilization of the allyl radical product due to π bonding, cause the relatively low activation energy over O^{2-}.[62] As the out-of-plane CH bond is stretched toward the transition state for H transfer to O^{2-}, the antibonding counterpart to the three-centered σ donation bond is stabilized by mixing in of the higher-lying empty π^* orbital at the olefinic end. Ultimately, this orbital becomes the half-filled π_2 radical orbital. On Bi_2O_3 low-lying dangling surface orbitals in the band gap are relatively easily reduced, allowing CH_4 activation by H abstraction.[63]

So far it has been shown how for hydrogen abstraction from alkanes by oxides there is activation when the antibonding counterpart to the C:H–$:\ddot{\underset{\cdot\cdot}{O}}:^{2-}$ closed-shell interaction is either partially emptied, as for $:\dot{\underset{\cdot\cdot}{O}}:^{-}$ and low-lying cation surface states, or orbitally stabilized, as in propylene. Metals cannot abstract H· because M–H bonds are weaker than C–H bonds. However, often M–H + M–C bond strengths are greater in sum than the C–H bond strength. Therefore reactions of the type

$$\underset{\text{metal}}{\overline{\qquad}} + H_3C\text{—}H \rightarrow \underset{\text{metal}}{\overline{\overset{H\quad CH_3}{|\qquad|}}}$$

are thermodynamically possible, and they must proceed as oxidative addition reactions. We have examined a number of reactions of this type. In some, such as the loss of α-H from propylene on Pt(111)[64] and the loss of the first axial H in the cyclohexane ⇆ benzene reaction on Pt(111)[65] the transition state is nearly three-centered (C · · · H · · · Pt) and the antibonding counterpart to the three-centered σ donation bond transfers its two electrons to the metal conduction band early in the reaction as shown in Figure 10.20. This transfer in combination with enhanced metal–carbon bonding results in low barriers. In the case of acetylene, the C that is losing H is already bound to the surface and so part of the barrier comes from tilting the CH bond nearly parallel to the surface, which costs energy because it perturbs the stable ethylenic structure[12,17–21,66] (see Fig. 10.21). A similar distortion for adsorbed ethylene is required and contributes to the barrier by perturbing the stable ethanelike structure.[66] CH fragments also must be tilted in the dehydrogenation process.[12,19,20]. Barriers are low for both dehydrogenations and in these cases it can be said that a surface metal atom is inserting into the CH bond because at the start of stretching to dissociate the bond, following tilting, the CH orientation is bridging across a metal atom. Once the bond is broken oxidative addition is complete, with H· and R· bonded to the surface as $H^-:M^+$ and $R^-:M^+$.

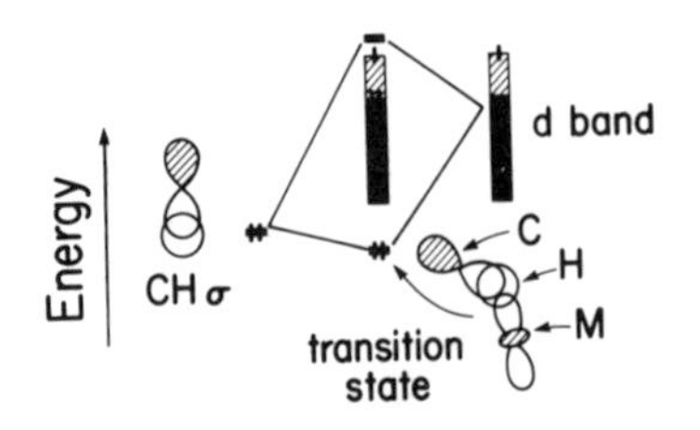

Figure 10.20. Schematic orbital correlation diagram for hydrogen abstraction from C_3H_6 (α) and C_6H_{12}.

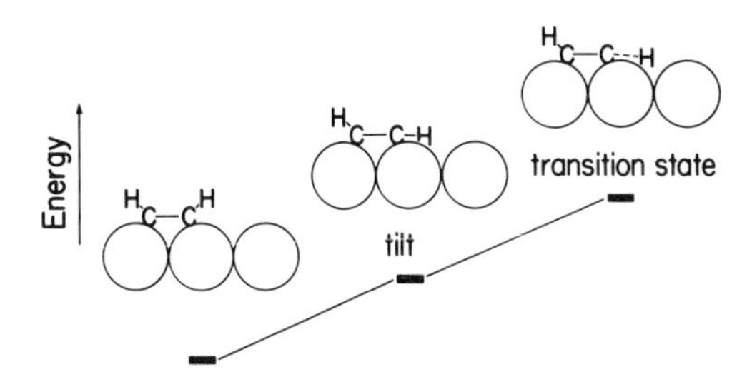

Figure 10.21. Steps in C_2H_2 dehydrogenation on a transition metal surface.

We have examined the oxidative addition of CH_4 to open and close-packed Fe(100) and (110) and Ni(100) and (111) surfaces.[67] In each case the activation can be viewed as having a single surface metal atom insert into the CH bond and transition state CH stretches are 0.4–0.5 Å, which is ~0.1 Å more than for acetylene and ethylene. The binding in the triangular transition states on the Fe surfaces is interesting. There are two well-defined σ donation bonds, one of the

C
H
M

type, and the other involving mostly C and M:

C
H
M

There are two because two orbitals from the CH_4 valence *t* set can mix. The antibonding counterpart is stabilized by having the CH σ^* orbital mixed in so as to remove the antibonding H–M interaction. This half-filled orbital has C · · · M bonding character:

C
H
M

All three orbitals are in Figure 10.22. For CH_4 activation on the Ni surfaces the transition states are better characterized as σ donation with the antibonding counterpart electrons donated to the conduction band.

We find the activating ability of the surfaces is inversely proportional to the metal atom packing density. On Ni(111) we calculate an activation energy of 67 kJ mol^{-1} (51,[68,69] 53 ± 5[69]) and on Ni(100) we calculate 29 kJ mol^{-1} (37,[69,70] 27 ± 5[70]), where experimental results are in parentheses.

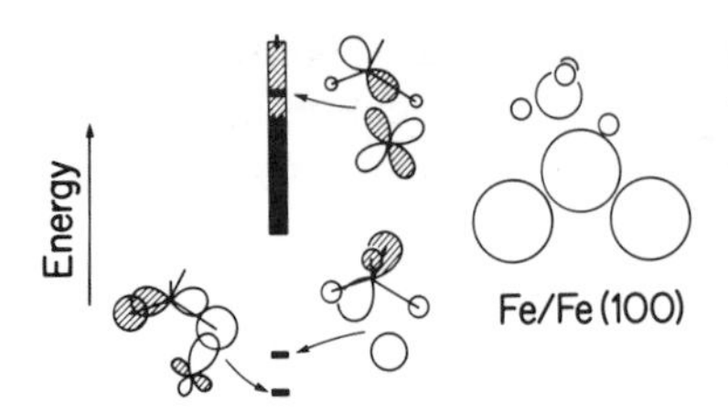

Figure 10.22. Transition state bonding when a surface Fe atom inserts into a CH bond in CH_4.

This can be correlated with metal–metal bond orders in terms of Mulliken overlap populations: when the sum of all overlaps of a surface atom with its neighbors is higher, as for close-packed surfaces, the surface orbitals are less available for activating CH in methane. For Fe(110) we calculate an activation energy of 118 kJ mol^{-1} and 32 kJ mol^{-1} on Fe(100). The barrier drops to 27 kJ mol^{-1} on a rough site consisting of an Fe adatom on Fe(100). There are no experimental data for these surfaces, but the low barriers are expected because Fe particles catalyze the formation of graphite fibers from methane.[71]

Recently we have shown that coordinatively unsaturated Mo^{IV} at edge sites of the layered sulfide MoS_2 insert into the methane CH bond with an activation energy of only 0.2 eV.[72] We are presently studying Fischer–Tropsch mechanisms on this important catalyst.[73]

CONCLUSIONS

I believe I have shown that a lot of surface chemistry can be predicted and explained using the transparently simple ASED–MO theory. I have shown how structure and electronic effects can be identified through systematic studies. The papers that are the basis for the preceding discussion contain much more detail concerning adsorbate structures and adsorption-induced orbital energy level shifts and bond vibrational frequency shifts than is given here. The purpose of this review has been to give the explanations for structures and activation without all of the detailed results. Some of the electronic effects were the effect of *d* electron count on the orientation of CO on a transition metal surface, the effect of the metal ionization potential on the adsorption site preference, the effect of metal *d* orbital size on C–M and C–H bonding, and the effect of O^- hole formation on oxide surfaces toward activating CH bonds. Some of the structure effects were the effect of surface metal atom spacing on the distortions of adsorbed acetylene and the influence of surface atom packing density on the CH activation energy in methane. Of course, structure and electronic effects are not truly distinguishable because all chemical interactions are electronic, but I believe structure effects such as these are conceptually good and useful for the surface scientist, for whom structures are known, and for the catalysis chemist, once structures become known. All of these effects are in accord with chemical intuition, but the final confirmation must ordinarily come from theoretical calculations.

ACKNOWLEDGMENTS

I would first like to thank the Gas Research Institute whose support paid for the excellent typing (Anne Bruening) and drafting work (Leslie Bagdasarian). The efforts of many fine coworkers contributed to the studies. Those involved directly in the calculations were Drs. S.P. Mehandru, R. Kötz, Y. Kim, D.Q. Dowd (now Prof. Dowd), and J.J. Maloney; Profs. N.K. Ray, D.P. Onwood, and J.A. Nichols; D.B. Kang (now Dr. Kang), Md. K. Awad, K. Nath (now Dr. Nath), and D.W. Ewing. Collaborations with Profs. R. Hoffmann, T.N. Rhodin, H.J. Kreuzer, and E. Yeager and Drs. C.F. Brucker, P.N. Ross, J.F. Brazdil, R.K. Grasselli, M.D. Ward, and J.D. Burrington were invaluable to a number of the studies, as were discussions with Profs. W.K. Hall and W.M.H. Sachtler. I am grateful for research support for the work mentioned herein from NSF, PRF, NASA, Standard Oil, and the Gas Research Institute, as well as fellowship support from Tanta University for Md. K. Awad.

REFERENCES

1. There are numerous reviews of surface science techniques. A current one is *The Structure of Surfaces* (M.A. Van Hove and S.Y. Tong, eds.). Berlin: Springer-Verlag, 1985.
2. Anderson, A.B., Parr, R.G., *Theoret. Chim. Acta* **26:** 301 (1972).
3. Anderson, A.B., *J. Chem. Phys.* **57:** 4143 (1972).
4. Anderson, A.B., *Phys. Rev. B* **8:** 3824 (1973).
5. Anderson, A.B., *J. Chem. Phys.* **60:** 2477 (1974).
6. Anderson, A.B., *J. Chem. Phys.* **62:** 1187 (1975); Anderson, A.B., Hoffmann, R., *J. Chem. Phys.* **60:** 4271 (1974).
7. Demuth, J.E. (private communication).
8. Demuth, J.E., *Surf. Sci.* **80:** 367 (1979); Albert, M.R., Sneddon, L.G., Eberhardt, W., Grenter, F., Gustafsson, T., Plummer, E.W., *Surf. Sci.* **120:** 19 (1982).
9. Kesmodel, L.L., Dubois, L.H., Somorjai, G.A., *J. Chem. Phys.* **70:** 2180 (1979).
10. Ibach, H., Lehwald, S., *J. Vac. Sci. Technol.* **15:** 407 (1978).
11. Brucker, C.F., Rhodin, T.N., *J. Catal.* **47:** 214 (1977).
12. Anderson, A.B., *J. Am. Chem. Soc.* **99:** 696 (1977).
13. Rhodin, T.N., Brucker, C.F., Anderson, A.B., *J. Phys. Chem.* **82:** 894 (1978).
14. Erley, W., Baro, A.M., Ibach, H., *Surf. Sci.* **120:** 273 (1982).
15. Mason, R., Textor, M., *Proc. Roy. Soc. (London) A* **356:** 47 (1977).
16. Yoshida, K., Somorjai, G.A., *Surf. Sci.* **75:** 46 (1978).
17. Anderson, A.B., Mehandru, S.P., *Surf. Sci.* **136:** 398 (1984).
18. Mehandru, S.P., Anderson, A.B., *Appl. Surf. Sci.* **19:** 116 (1984).
19. Kang, D.B., Anderson, A.B., *Surf. Sci.* **165:** 221 (1986).

20. Mehandru, S.P., Anderson, A.B., *J. Am. Chem. Soc.* **107:** 844 (1985).
21. Anderson, A.B., *J. Catal.* **67:** 129 (1981).
22. Tauster, S.J., Fung, S.C., Baker, R.T., Horsley, J.A., *Science* **211:** 1221 (1981).
23. Froitzheim, H., Hopster, H., Ibach, H., Lehwald, J., *Appl. Phys.* **13:** 147 (1977); Krebs, H.I., Luth, H., *Appl. Phys.* **14:** 337 (1977); Norton, P.R., Goodale, J.W., Selkirk, E.B., *Surf. Sci.* **83:** 189 (1979); Garfunkel, E.L., Crowell, J.E., Somorjai, G.A., *J. Phys. Chem.* **86:** 310 (1982).
24. Bradshaw, A.M., Hoffmann, F.M., *Surf. Sci.* **72:** 513 (1978); Brown, A., Vickerman, J.C., *Surf. Sci.* **124:** 267 (1983).
25. Biberian, J.P., Van Hove, M.A., *Surf. Sci.* **118:** 443 (1982).
26. Packard, R.L., Madden, H.H., *Surf. Sci.* **11:** 158 (1968); Conrad, H., Ertl, G., Koch, J., Latta, E.E., *Surf. Sci.* **43:** 462 (1974); Behm, R.J., Christmann, K., Ertl, G., Van Hove, M.A., Thiel, P.A., Weinberg, W.H., *Surf. Sci.* **88:** L59 (1979); Ortega, A., Hoffmann, F.M., Bradshaw, A.M., *Surf. Sci.* **119:** 79 (1982); Brown, A., Vickermann, J.C., *Surf. Sci.* **151:** 319 (1985).
27. Tracy, J.C., *J. Chem. Phys.* **56:** 2748 (1972).
28. Dewar, M.J.S., *Bull. Soc. Chim. Fr. C* **79** (1951); Chatt, J., Duncanson, L., *J. Chem. Soc.* 2239 (1955).
29. Blyholder, G., *J. Phys. Chem.* **68:** 2772 (1964).
30. Anderson, A.B., Awad, Md.K., *J. Am. Chem. Soc.* **107:** 7854 (1985).
31. Kitamura, F., Takahashi, M., Ito, M., *J. Phys. Chem.* **92:** 3320 (1988).
32. Garfunkel, E.L., Crowell, J.E., Somorjai, G.A., *J. Phys. Chem.* **86:** 310 (1982).
33. Ray, N.K., Anderson, A.B., *Surf. Sci.* **125,** 803 (1983).
34. Mehandru, S.P., Anderson, A.B., *J. Phys. Chem.* **93:** 2044 (1989).
35. Lambert, D.K., *Solid State Commun.* **51:** 297 (1984); Bagus, P.S., Nelin, C.J., Müller, W., Philpott, M.R., Seki, H., *Phys. Rev. Lett.* **58:** 559 (1987).
36. Nath, K., Kreuzer, H.J., Anderson, A.B., *Surf. Sci.* **176:** 261 (1986).
37. Anderson, A.B., Kötz, R., Yeager, E., *Chem. Phys. Lett.* **82:** 130 (1981).
38. Anderson, A.B., Onwood, D.P., *Surf. Sci.* **154:** L261 (1985).
39. Hoffmann, F.M., de Paola, R.A., *Phys. Rev. Lett.* **52:** 1697 (1984).
40. Madey, T.E., Benndorf, C., *Surf. Sci.* **164:** 602 (1985).
41. Haskell, D., Plummer, E.W., de Paola, R.A., Eberhardt, W., *Phys. Rev. B* **33:** 5171 (1986).
42. Mehandru, S.P., Anderson, A.B., *Surf. Sci.* **169:** L281 (1986).
43. Mehandru, S.P., Anderson, A.B., Ross, P.N., *J. Catal.* **100:** 210 (1986).
44. Anderson, A.B., Dowd, D.Q., *J. Phys. Chem.* **91:** 869 (1987).
45. Ichikawa, M., *Bull. Chem. Soc. Jpn.* **51:** 2268 (1978); Ichikawa, M., Fukushima, T., *J. Phys. Chem.* **89:** 1564 (1985); Ichikawa, M., Lang, A.J., Shriver, D.F., Sachtler, W.M.H., *J. Am. Chem. Soc.* **107:** 7216 (1985).
46. Mehandru, S.P., Anderson, A.B., *Surf. Sci.* **201:** 345 (1988).
47. Moon, D.W., Bernasek, S.L., Dwyer, D.J., Gland, J.L., *J. Am. Chem. Soc.* **107:** 4363 (1985); Benndorf, C., Krüger, B., Thieme, F., *Surf. Sci.* **163:** L675 (1985); Seip, U., C.-Tsai, M., Christmann, K., Küppers, J., Ertl, G., *Surf. Sci.* **139:** 29 (1984); Erley, W.J., *J. Vac. Sci. Technol.* **18:** 472 (1981).
48. Kokes, R.J., *Acc. Chem. Res.* **6:** 226 (1973).
49. Boccuzzi, F., Garonne, E., Zechina, A., Bossi, A., Camia, M., *J. Catal.* **51:** 160 (1978).
50. Anderson, A.B., Nichols, J.A., *J. Am. Chem. Soc.* **108:** 4742 (1986).
51. Anderson, A.B., Nichols, J.A., *J. Am. Chem. Soc.* **108:** 1385 (1986).
52. Gay, R.R., Nodine, M.H., Henrich, V.E., Zieger, H.J., Solomon, E.I., *J. Am. Chem. Soc.* **102:** 6752 (1980); Duke, C.B., Lubinsky, A.R., Chang, S.C., Lee, B.W., Mark, P., *Phys. Rev. B* **15:** 4865 (1977).

53. Mehandru, S.P., Anderson, A.B., Brazdil, J.F., Grasselli, R.K., *J. Phys. Chem.* **91:** 2930 (1987).
54. Mehandru, S.P., Anderson, A.B., Brazdil, J.F., *J. Am. Chem. Soc.* **110:** 1715 (1988).
55. Ward, M.D., Brazdil, J.F., Mehandru, S.P., Anderson, A.B., *J. Phys. Chem.* **91:** 6515 (1987).
56. Shvets, V.A., Kazansky, V.B., *J. Catal.* **25:** 123 (1972).
57. Driscoll, D.J., Matir, W., Wang, J.-X., Lunsford, J.H., *J. Am. Chem. Soc.* **107:** 58 (1985); Ito, T., Lunsford, J.H., *Nature* **314:** 721 (1985).
58. Boldu, J.L., Abraham, M.M., Chen, Y., *Phys. Rev. B* **19:** 442 (1979).
59. Kazansky, V.B., *Kinet. Katal.* **18:** 43 (1977).
60. Liu, H.-F., Liew, R.-S., Johnson, K.Y., Lunsford, J.H., *J. Am. Chem. Soc.* **106:** 4117 (1984).
61. Callahan, J.L., Grasselli, R.K., Milberger, E.C., Strecker, H.A., *Ind. Eng. Prod. Res. Rev.* **9:** 134 (1970); Schuit, G.C.A., *J. Less.-Common Met.* **36:** 329 (1974).
62. Anderson, A.B., Ewing, D.W., Kim, Y., Grasselli, R.K., Burrington, J.D., Brazdil, J.F., *J. Catal.* **96:** 222 (1985).
63. Mehandru, S.P., Anderson, A.B., Brazdil, J.F., *J. Chem. Soc. Faraday Trans.* **83:** 463 (1987).
64. Anderson, A.B., Kang, D.B., Kim, Y. *J. Am. Chem. Soc.* **106:** 6597 (1987).
65. Kang, D.B., Anderson, A.B., *J. Am. Chem. Soc.* **107:** 7858 (1985).
66. Kang, D.B., Anderson, A.B., *Surf. Sci.* **155:** 639 (1985).
67. Anderson, A.B., Maloney, J.J., *J. Phys. Chem.* **92:** 809 (1988).
68. Lee, M.B., Yang, Q.Y., Tang, S.L., Ceyer, S.T., *J. Chem. Phys.* **85:** 1693 (1986).
69. Beebe, T.P., Jr., Goodman, D.W., Kay, B.D., Yates, J.T., Jr., *J. Chem. Phys.* **87:** 2305 (1987).
70. Hamza, A.V., Madix, R.J., *Surf. Sci.* **179:** 25 (1987).
71. Tibbetts, G.G., *J. Cryst. Growth* **73:** 431 (1985).
72. Anderson, A.B., Maloney, J.J., Yu, J., *J. Catal.* **112:** 392 (1988).
73. Anderson, A.B., Yu, J., *J. Catal.,* **119:** 135 (1989).

11
Application of Band-Structure Calculations to Chemisorption

R.C. BAETZOLD

INTRODUCTION

This chapter is aimed at illustrating the point that theory has an important role to play in catalysis. This role is growing, but is not separated from experiment. In a general sense, one may justifiably argue that theory is not just paper and pencil or even computer exercises, but encompasses well thought-out and carefully analyzed experiments. In this definition all scientists practice their own form of theory. Rather than addressing this broader scope, some particular computational approaches will be emphasized in this chapter. A computational approach involving tight-binding band structure calculations will be the focus of attention in this paper. We will not discuss the area of cluster calculations that have been used to model chemisorption. This extended-Hückel version of tight-binding calculations has been applied to chemisorption in other studies.[1–3] We also note that the computational studies discussed here have been applied to a range of problems in a complementary fashion with various analytical models.[4]

Consider the question of how does a solid catalyst perturb the potentially reactant molecules so as to lower their energy barriers for reaction. Certainly this is a question with unique answers for certain classes of reactions and catalysts. For example, there are the well-known oxidation catalysts[5] which not only chemisorb and activate olefins, but also provide the oxygen from lattice sites which lead to oxidation products. Another distinct class of catalysts is the Fischer–Tropsch[6] type which lead to dissociation of CO and H_2 molecules and provide for a pathway of reaction for the dissociated products leading to hydrocarbon products. A simpler kind of reaction[7] involves H/D exchange in saturated hydrocarbons that must involve another distinct activation mechanism. Common to all of these classes of reaction is chemisorption and preparation of chemisorbed species for appropriate reaction by the catalyst. A continuum of interaction strengths may be envisioned. We consider physisorption with weak interactions, chemisorption with stronger interactions, and, finally, reaction through the chemisorbed state along a new reaction direction. One central thesis concerns the idea that one of the chemisorbed state can resemble the transition state, at least for reactive cases, and that an idea of how the catalyst functions can be obtained from the geometry and energetics preceding the transition state.

METHOD OF CALCULATION

Computational Details

The tight-binding method[8] has been used in solid-state physics for a long time to study the properties of solids. We will describe how to adopt the simplest molecular type of calculation, the extended Hückel method,[9] to the tight-binding framework. We will consider a two-dimensional solid with a layer of chemisorbed molecules on one or both sides.

Atomic valence orbitals are taken as Slater orbitals

$$\chi_i = Ar^{n-1}e^{-\alpha r}Y_{l,m}(\theta, \phi) \tag{11.1}$$

where

n = principal quantum number
α = exponent
A = normalizing constant
r = distance from the atomic center
$Y_{l,m}(\theta, \phi)$ = spherical harmonic representing angular part of wavefunction

Typically, exponents from atomic SCF calculations are employed.[10] Single Slater orbitals are used to represent the s,p orbitals while a linear combination of two Slater orbitals represents the d orbitals.

Consider a film of M parallel planes, each having the same two-dimensional structure. The translation vector connecting the origins of adjacent planes is $\bar{\tau}$, $\bar{k}$ is the wavevector, and $\bar{R}$ is the position vector of atoms in a plane. The Bloch function can now be written in terms of the atomic orbitals as

$$\phi_1(\bar{r} - \bar{R}_1 - m\bar{\tau}) = \frac{1}{N^{1/2}} \sum_{m=1}^{M} \sum_{R_1} e^{i\bar{k}\cdot(\bar{R}_1 + m\bar{\tau})}\chi_1(\bar{r} - \bar{R}_1 - m\bar{\tau}) \tag{11.2}$$

Now the usual variation process is applied by developing a wavefunction with coefficients C_{mi}:

$$\psi_1(\bar{r} - \bar{R}_1 - m\bar{\tau}) = \sum_{m=1}^{M} \sum_{i-1}^{m} C_{lmi}\phi_i(\bar{r} - \bar{R}_1 - m\bar{\tau}) \tag{11.3}$$

Variation of coefficients to minimize the energy gives the secular equation

$$\sum_{m=1}^{M} \sum_{i=1}^{m} (H_{m'lmi} - E_l S_{m'lmi})\, C_{lmi} = 0 \qquad (11.4)$$

where m' is the layer number containing l orbital. The matrix elements are given by

$$H_{m'l,mi} = \sum e^{i\bar{k}\cdot(\bar{R}_1 + m'\tau - m\tau)} \int \chi_1(\bar{r} - \bar{R}_l - m'\bar{\tau})\, \hat{H}\chi_i(\bar{r} - \bar{R}_l - m'\tau)\, dV \qquad (11.5)$$

The matrix elements are evaluated within the framework of the extended Hückel method[9]:

$$H_{mi,mi} = -IP_i \qquad (11.6)$$

$$H_{m'l,mi} = \tfrac{1}{2} K S_{li}(IP_l + IP_i) \qquad (11.7)$$

where IP_i is the atomic ionization potential, $K = 1.75$, and S_{li} is an overlap integral,

$$S_{li} = \int \chi_l(\bar{r} - \bar{R}_l - m'\tau)\chi_i(\bar{r} - \bar{R}_l - m\bar{\tau})\, dV \qquad (11.8)$$

In practice, the lattice sums in Eq. (11.5) are evaluated for particular surface mesh arrangements such as square, rectangular, or hexagonal for fcc (100), (110), or (111) surfaces, respectively. We construct a series of unit meshes filling space around a central unit mesh containing all interactions up to at least three atom diameters. Then the matrix element becomes

$$H_{m'l,mj} = \sum_t e^{ik\cdot(\bar{R}_l - \bar{R}_j + (m'-m)\bar{\tau})} H_{lj}^{0,t} \qquad (11.9)$$

where the sum over t includes all the space filling unit meshes. The $H_{lj}^{0,t}$ element refers to orbital l in the origin unit mesh and orbital j in the t unit mesh. A similar expression holds for the overlap matrix element.

We note that the elements of the secular equation are complex numbers which can be written in the form[11]

$$\begin{pmatrix} H_R & -H_I \\ H_I & H_R \end{pmatrix} \begin{pmatrix} C_R \\ C_I \end{pmatrix} = \begin{pmatrix} E \end{pmatrix} \begin{pmatrix} S_R & -S_I \\ S_I & S_R \end{pmatrix} \begin{pmatrix} C_R \\ C_I \end{pmatrix} \qquad (11.10)$$

where the real and imaginary components are denoted by R and I, respectively. The coefficients are used to calculate the electron population of each orbital within the unit mesh.

$$P_{tt} = 2 \sum_{k} \sum_{s=1}^{\text{occ}} \alpha_{s,tt} + \sum_{u \neq t} \alpha_{s,tu} S_{tu}^{0,0} \tag{11.11}$$

where

$$\alpha_{s,tu} = C_{s,t}^{\text{R}} C_{s,u}^{\text{R}} + C_{s,t}^{\text{I}} C_{s,u}^{\text{I}}$$

The index s refers to a sum over occupied Bloch functions and the sum over k involves an averaging over k space. A bond order between orbitals in the origin unit mesh is calculated from

$$\text{O}_{tu} = 4 \sum_{\bar{k}} \sum_{S}^{\text{occ}} \alpha_{s,tu} S_{tu}^{0,0} \tag{11.12}$$

A reduced bond order population between atoms in the unit mesh is obtained by summing over elements of Eq. (11.12) containing the appropriate orbitals. Another quantity of interest for our calculations is the dipole moment perpendicular to the surface obtained from hybridization of p, d functions:

$$D_{p,d} = \sum_{s}^{\text{occ}} \sum_{i,j} \alpha_{s,ij} \langle pi|z|dj \rangle \tag{11.13}$$

The matrix element in Eq. (11.13) is purely an atomic term.

The energy of adsorption is calculated by keeping a constant number of electrons per unit cell in the free and chemisorbed state. The chemisorption energy is calculated as a difference of the sums

$$Q = \left(\sum_{s} n_s \mathscr{E}_s \right)^{\text{free}} - \left(\sum_{s} n_s \mathscr{E}_s \right)^{\text{chemisorbed}} \tag{11.14}$$

where n_s is the occupancy of the s Bloch function. Finally, we note that the development of our method of calculation[12] has followed that for polymers,[11] but a similar version has also been developed in other chemisorption calculations.[1–3]

Extended surfaces are a desirable starting point for chemisorption because their properties are, in general, better determined than those of clusters. In particular, extended surfaces are models of catalyst particles often used in experiment. The particular property of significance is the ionization potential. We note that the ionization potential of a single metal atom is typically 7–8 eV, while the analogous bulk property, the work function, is typically 5–6 eV. This makes extended surfaces a much stronger electron donor than the single metal atom. Clusters, of course, are somewhere in between these extremes and even though, in some cases, the ionization potentials have been measured, they are usually not correlated with a definite structure. Thus, it is not possible to start with a cluster substrate for chemisorption measurements that is well defined experimentally.

There is also the question to be considered of at what size do cluster properties approach the bulk values. Measurements of the cluster density of states using UPS (ultraviolet photoelectron spectroscopy) or XPS (X-ray photoelectron spectroscopy) have been made for a variety of metals including Ag, Au, Pd, Pt, and Ni. It is clear that the size of roughly 100 atoms is required for the spectra to resemble closely the bulk.[13] Difference spectra have also been reported for Cl adsorbed to Ag clusters.[14] In this case, the approach of difference spectra to those measured on extended surfaces requires mean cluster sizes of $\approx$40 atoms. Thus, the photoelectric techniques indicate that very large clusters would be required to represent bulk properties.

Now let us consider the parameters needed to model metal atoms in an extended surface. The Slater orbital exponents and coefficients are available from atomic calculations.[10] Typically, a linear combination of two Slater orbitals is needed to represent d orbitals. The center of d band energy (ε_d) is the other important parameter, as well as the Fermi energy (E_F). The Fermi energy is taken from the experimental work function.[15] The center of the d band relative to the Fermi energy is known from the bulk band structure calculations[16] analyzed by Varma and Wilson.[15] A plot such as that in Figure 11.1 is available for each of the three transition series.

We note that the tight-binding method of calculation is one of the simplest. Experience has shown that the d orbitals of transition metals are well represented by the tight-binding method.[8] The interactions of these functions with adsorbate orbitals are thought to be of key importance for transition metals. Second, the parameters of a metal are available from the background of calculations[16] and experimental data available for these systems. One serious flaw of the method is its deficiency at calculating bond lengths. Accordingly, we treat metal–adsorbate bond lengths as input parameters determined as sums of covalent radii or as experimental bond lengths. Self-consistency in the chemisorbed state is crudely achieved by choosing metal and adsorbate parameters such that adsorbate

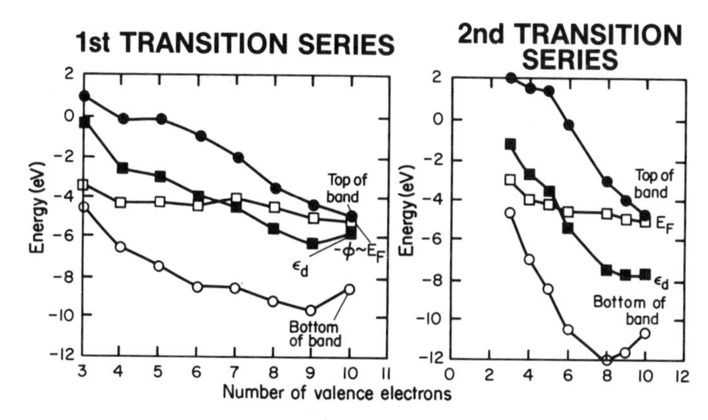

Figure 11.1 Plots of the d band structure versus number of valence electrons for the first- and second-transition metal series. Values taken from Refs. 15 and 16. E_F = Fermi level = work function, ε_d = center of d band.

levels are positioned at energy separations from the Fermi energy consistent with photoelectron spectroscopy data.

The objective in applying the LCAO–MO tight-binding calculations is to obtain a picture of the important causal relationships in chemisorption. Quantitative numerical predictions are outside the scope of this method. Thus, trends in the heats of adsorption[15] and mechanisms of electron transfer are key topics for calculation.

Film thicknesses have been varied between two and five layers in our calculations. The surface unit mesh is a four-atom square for (100) fcc surfaces or a 60° trapezoid for (111) fcc surfaces. The adsorbate is placed in any of the high symmetry sites, including on-top, bridging, or hollow sites, at appropriate bond lengths as described before. Simulation of trends across a transition metal series is accomplished using parameters consistent with Figure 11.1. We note that the center of the d band decreases sharply with movement to the right in the transition series, while the Fermi level decreases only slightly. A typical parameterization of the bare surface is shown in Table 11.1 for *five*-layer films showing how changes in metal s, p, d occupancy are simulated.

Perturbation Theory

Perturbation theory is one of the tools that can be used to gain a qualitative understanding of how molecules or atoms interact with surfaces. The theory up to second order is explained in texts on quantum mechanics. One of the most readable applications of the theory was made for mole-

Table 11.1. Properties of metal films (111) fcc.

Parameters		
4*s*	Exponent	2.75
4*p*	Exponent	2.75
3*d*	Exponent	5.983, 2.613
3*d*	Coefficients	0.5264, 0.6372
4*s*	Energy	−3 eV
4*p*	Energy	−2 eV

$-\varepsilon_d$ (eV)[a]	$-E_F$ (eV)[b]	N_d[c]
5.0	5.0	4.80
6.5	5.5	7.84
8.0	6.0	8.70
9.5	6.5	9.58

[a] Center of *d* band.
[b] Fermi energy.
[c] Average metal *d* occupancy.

cules by Hoffmann.[17] Consider the interaction of two levels ε_{i0}, ε_{j0} which have solutions to the Schrödinger equation ψ_{i0} and ψ_{j0}, respectively. In the presence of an interaction H', the new energy of any ith level interacting with a jth level is

$$E_i = \mathscr{E}_{i0} + \sum_{j \neq i} \frac{|H'_{ij}|^2}{\varepsilon_{i0} - \varepsilon_{j0}} \tag{11.14}$$

and the new wavefunction is

$$\psi_i = \psi_{i0} + \sum_{j \neq i} \frac{H'_{ij}}{\varepsilon_{i0} - \varepsilon_{j0}} \psi_{j0} \tag{11.15}$$

where

$$H'_{ij} = \int \psi^*_{i0} H' \psi_{j0} \, dV \tag{11.16}$$

Band Picture

We may use perturbation theory to understand how a given orbital in an atom or molecular orbital in a molecule interacts with the Bloch functions of a transition metal surface. Consider the band model in Figure 11.2 where we identify the components of a *d* band on the surface of some transition metal. The Fermi level represents the highest occupied molecu-

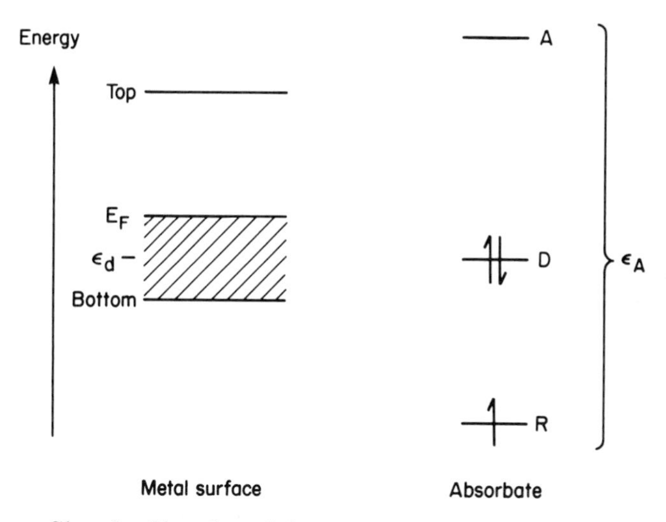

Figure 11.2. Sketch of band model versus energy for hypothetical acceptor (A), donor (D), or radical (R) level interacting with the metal surface.

lar orbital and the center of the *d* band ε_d corresponds to the atomic *d* level. An adsorbed species consisting of an acceptor (A), donor (D), or radical (R) is schematically shown relative in energy to the *d* band. The energy levels are either atomic or molecular orbitals in the case of molecules.

Some interesting patterns of interaction result.[18] Consider isolated adsorbate levels which lie below the Fermi level in Figure 11.3. When they

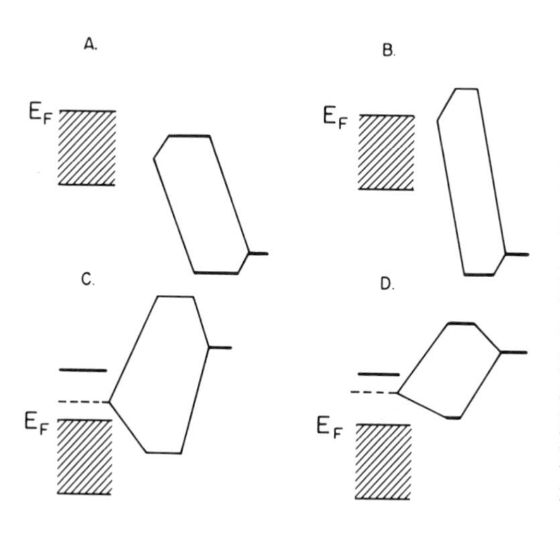

Figure 11.3. Sketch of interaction of adsorbate levels with band states and resulting energy level structure. A,B—adsorbate/filled metal levels leading to occupied metal levels (A) or unoccupied metal levels (B). C,D—adsorbate/filled metal levels leading to occupied metal levels (C) or unoccupied metal levels (D).

interact with filled Bloch states, the Bloch states move toward the vacuum level and the adsorbate level moves deeper in energy. This interaction may be fairly simple as in A or lead to depopulation of the Bloch function as in B. In this circumstance another Bloch function must become occupied to keep the electron count constant. Now these effects must be integrated over the entire band to determine the total interaction with the surface. We did not consider the interaction of adsorbate states with unoccupied Bloch states, since this does not lead to any significant reoccupation pattern. Now consider an acceptor level interacting with unoccupied Bloch states as in case C and D. Here the acceptor may cause a previously unoccupied Bloch function to be occupied if the interaction is sufficiently strong as in C. In cases of weaker interaction (D), no major repopulation of the surface occurs. There are also interactions with the occupied Bloch states which do not lead to major repopulation effects.

Charge transfer can occur through the mixing of levels through Eq. (11.15). Generally, acceptor levels become occupied through mixing with metal levels that lie below the Fermi level as in C. On the other hand, radical and donor levels lose electrons to the metal through mixing with metal levels that lie above the Fermi level as in B. Of course, in the case of radicals there is also the process of electron transfer from the metal to the half-occupied level that lies below the Fermi level. This amount of energy ($E_F - \varepsilon_A$) leads to an expression relating the heat of adsorption of a radical Q^R and donor Q^D species having the same ε_A, adsorbate level:

$$Q^R = Q^D + (E_F - \varepsilon_A) \tag{11.17}$$

ATOMIC ADSORBATES

Heat of Adsorption

Now consider explicit band calculations for five-layer films where we examine the trends in adsorption energy versus position in a transition metal series. Figure 11.4 shows the adsorption energy for radical, acceptor, and donor species, where the calculation is done in A using the parameters of Table 11.1 to represent N_d and in B at constant Fermi energy. In the case of radical and donor species both methods show only minor changes in adsorption energy with N_d. The changes in acceptor heat of adsorption are more sensitive to parameters.

Experimentally, it is difficult to test all of these predictions, especially since the acceptor and donor models are rather simplified cases of molecular analogs. The radical case is more straightforward. The rather insensitive nature of the initial heat of adsorption of hydrogen to position of metal within a transition series has been noted.[19] This behavior is quite

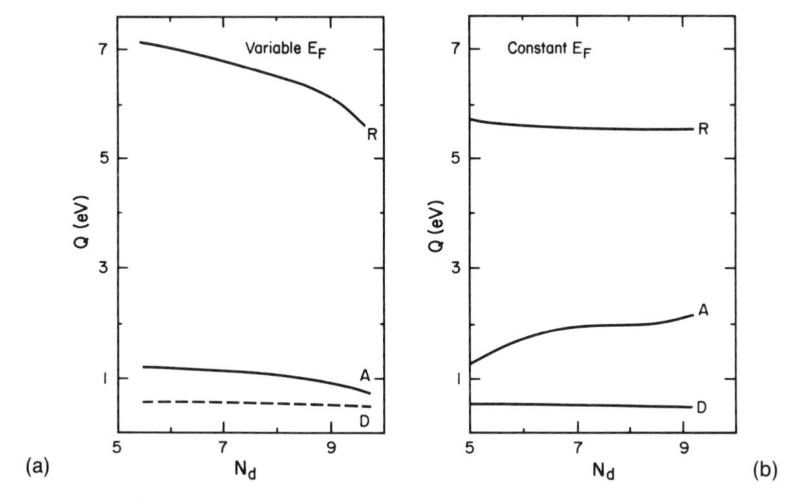

Figure 11.4. Plot of the heat of adsorption for radical (R), acceptor (A), and donor (D) adsorbates. In (a) the metal parameters are from Table 11.1 and $\varepsilon_A = E_F - 4$. In (b) the Fermi level is constant at all N_d values, the average number of d electrons per metal atom.

similar to that shown in Figure 11.4. We will postpone further comparisons with experiment until we consider molecular bonding.

Surface Rehybridization

One of the important properties of metal surfaces that influences chemisorption is the electron occupancy. Metals have a short screening length so that significant charge imbalances between the surface and the next deeper layers do not develop. Instead, there is a rehybridization mechanism that leads to different *d*, *s*, *p* electron occupancy at the surface and the interior. An example of the driving force for this effect is shown for the non-self-consistent five-layer film calculations of an ending transition metal film shown in Table 11.2. Note that the outer layers 1 and 5 have a significantly increased *d* population relative to the interior layers. At the same time, the *s*, *p* population is smaller on the outer layers compared to the interior layers. Of course, the calculations in Table 11.2 exaggerate the effect because they are non-self-consistent, but as we will discuss later the effect has been measured experimentally.

A qualitative model for the surface rehybridization of transition metals has been known for some time.[20,21] It is based on the fact that a reduction in coordination number at the surface leads to narrower bands. Consider

Table 11.2. Electron population of (111) fcc film.

Layer Number	Population	
	sp	*d*
1	0.24	8.64
2	0.36	8.22
3	0.40	8.19
4	0.36	8.22
5	0.24	8.64
	Early Transition Metal	
1	0.25	2.37
2	0.36	2.54
3	0.49	2.48
4	0.36	2.54
5	0.25	2.37

the projection of the *d* density of states on a surface and bulk atom sketched in Figure 11.5. The bulk projection is broader than the surface projection. Now, if the Fermi level (E_F) is near the top of the band, a greater percentage of the surface projection will be occupied and the surface atom will have a larger *d* occupation. The reverse is true if the Fermi level is located near the bottom of the band. If the Fermi level

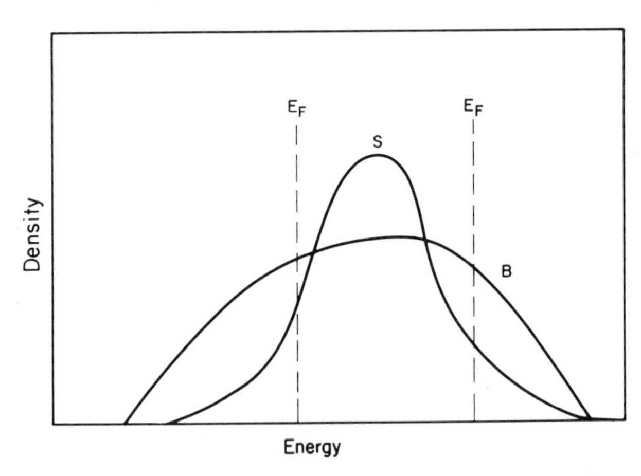

Figure 11.5. Schematic plot of the projection of the *d* band of a transition metal on a typical bulk (B) and surface (S) atom. States to the left of the Fermi energy E_F are occupied showing the greater bulk occupation for less than half-filled bands and reverse for more than half-filled bands.

occurs at the midpoint, there will be roughly equal surface and bulk atom *d* populations. Thus, a crossing over in surface/bulk *d* population occurs with increasing *d* band filling.

These effects are manifested experimentally in surface core-level binding energy shifts. The 5*d* transition metals are particularly suited for this measurement because of their positioning of a localized 4*f* level, which can be accurately probed by photoemission techniques at a storage ring. Studies of clean ordered 5*d* metal surfaces show that the emission from 4*f* levels may be decomposed into a surface and bulk component having different binding energy. The experimental studies in Table 11.3 show a shift of the surface component to higher binding energy for transition metals with less than approximately half-full *d* bands (Ta),[22,23] with the reverse true for metals with more than half-full *d* bands (W, Ir, Pt, Au).[22–29] A shift of the surface component is interpreted as due to a difference in the valence *d* surface population versus the bulk population. Higher *d* populations lead to lower binding energy.

Surface Polarization

Adsorption can result in interesting patterns of polarization of metal surface atoms. This polarization is the result of rehybridization of *p* and *d* orbitals on the surface atoms. This mechanism is shown in Figure 11.6 and has been discussed in analytic models[4b] as well as in computations.[12b] We focus on the orbitals perpendicular to the surface p_z, d_{z^2}, which can cause the largest dipole moment in this direction. Their in-phase or out-of-phase mixing causes formation of a negative dipole or positive dipole

Table 11.3. Surface core binding energy shifts versus bulk.

Surface	Surface Shift (eV)	Reference
Ta(111)	+0.40	22
Ta(111)	+0.39	23
W(111)	−0.43	22
W(111)	−0.43	23
W(110)	−0.30	24
W(100)	−0.35	22
W(100)	−0.36	23
Ir(111)	−0.50	28
Pt(111)	−0.4	26
Pt(111)	−0.37	27
Au(111)	−0.35	25
Au polycrystalline	−0.4	29

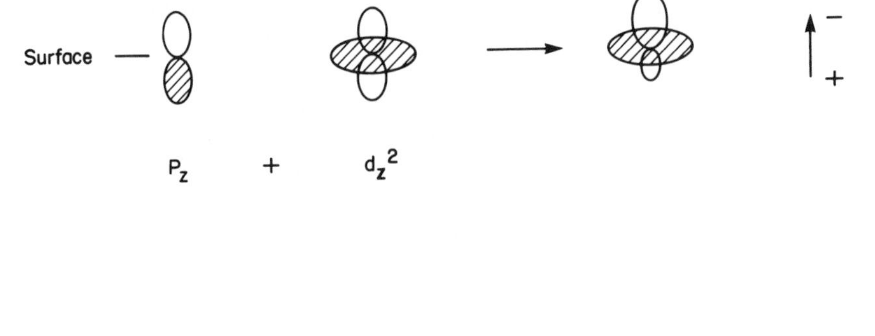

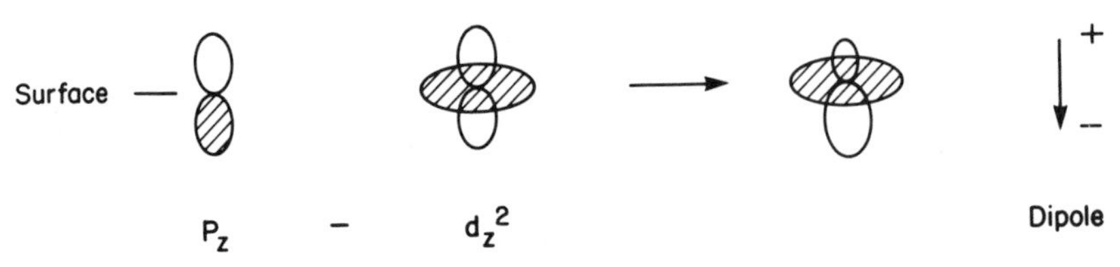

Figure 11.6. Surface polarization mechanism resulting from mixing p_z and d_{z^2} orbitals at a surface.

perpendicular to the surface, respectively. This dipole contributes to a decrease or increase, respectively, in the work function, which can be measured. The work function includes this contribution as well as the component due to charge exchange with an adsorbate.

An example of the change in dipole induced by adsorption of a Cl- and H-type electron withdrawing adatom is shown in Figure 11.7. Here the change in surface dipole caused by the rehybridization mechanism is plotted versus a perturbation parameter analogous to that in Eq. (11.15), which represents the degree of orbital mixing. We see that, for small values of the perturbation parameter up to 1.0, the dipole decreases and leads to a component of the work function that would decrease the work function. This direction of change is opposite to the contribution induced by the electron-withdrawing properties of the adatom. This effect is most pronounced on surfaces of low atomic density as compared to surfaces of high atomic density. At larger values of the perturbation parameter the effect reverses. Estimates[12b] of the magnitude of this rehybridization-induced work function change indicate that in some cases it is dominant and can lead to observed chemisorption-induced work function changes contrary to those expected from electronegativity arguments alone.

We have discussed[2b,12b] several examples where chemisorption-induced work function changes require the internal polarization for understanding.

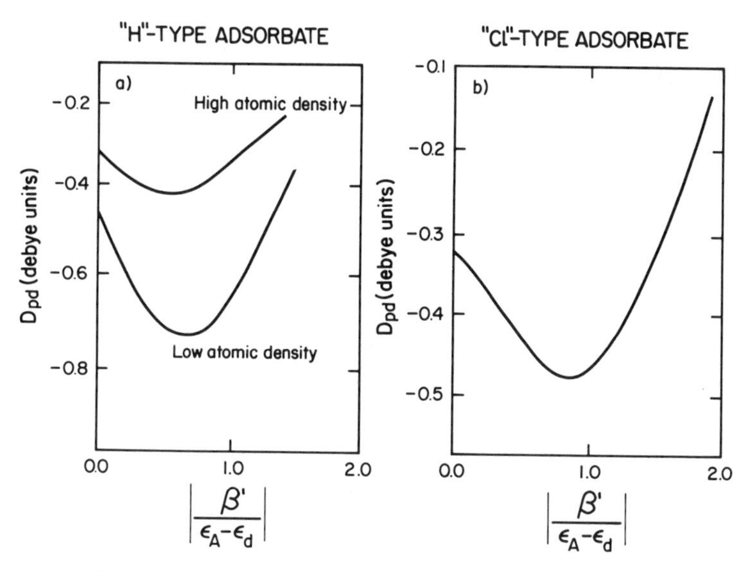

Figure 11.7. Change in p,d dipole moment for adsorption of (a) "H"-type adsorbate or (b) "Cl"-type adsorbate. Dipole moment is plotted versus the perturbation parameter for high-atomic-density fcc surface [e.g., (111)] or low atomic density fcc surface [e.g., (100)].

The normal dipole created by charge separation in the case of halogens adsorbed in W(100) is a good example. Chemisorption of I decreases the work function, but the more electronegative Br and Cl cause an increase in work function.[30] An increase in work function is expected from the charge dipole, but a decrease needs the polarization effect for explanation. Another good example is Cl chemisorbed to Pt(111), where the work function initially decreases.[31] Many more examples could be cited, but we wish only to note that work function changes are not a reliable guide to the direction of electron flow upon chemisorption.

MOLECULAR STEREOCHEMISTRY

Concepts

A relationship between the stereochemistry of chemisorbed closed-shell diatomic molecules and their acceptor donor character has been presented in an analytic model.[4e] We note that there have been several theoretical studies aimed at establishing relationships of the stereochemistry of ligands in complexes[32] and on extended surfaces.[2a] In simplest form, the analytic model considers the antibonding ψ^* and bonding ψ molecular orbitals of the H_2 molecule, which are the acceptor and donor orbitals, respectively:

$$\psi^* = \frac{1}{\sqrt{2(1 - s)}}(Y_1 - Y_2) \tag{11.18}$$

$$\psi^* = \frac{1}{\sqrt{2(1 + s)}}(Y_1 + Y_2) \tag{11.19}$$

where

$$s = \text{overlap integral}$$
$$Y_1 = \text{H atomic orbital}$$

An effective interaction integral β_i is used to represent the interaction between atomic orbital i and the metal surface Bloch functions. Thus, for adsorption with the H_2 molecule perpendicular to the surface with atom 1 closest to the metal, we have the respective donor and acceptor interactions

$$\beta = \frac{1}{\sqrt{2(1 + s)}}(\beta_1 + \beta_2) \approx \frac{\beta_1}{\sqrt{2(1 + s)}} \tag{11.20}$$

$$\beta^* = \frac{1}{\sqrt{2(1 - s)}}(\beta_1 - \beta_2) \approx \frac{\beta_1}{\sqrt{2(1 - s)}} \tag{11.21}$$

Note that because the molecule is normal to the surface with atom 2 further from the surface, $\beta_1 >> \beta_2$ (particularly true for molecules with long bond lengths). The acceptor contribution is dominant because s is quite large for most molecules (0.5–0.7):

$$(\beta^*)^2 >> (\beta)^2 \tag{11.22}$$

Finally, recalling Eq. (14), note that the energy stabilization terms go as β^2, thus enhancing the acceptor contribution. We postpone discussion of the energy denominator until later. We conclude this paragraph by noting that a dominant acceptor function on the part of the diatomic molecule correlates with the perpendicular geometry.

Now consider chemisorption of a diatomic parallel to the surface in a symmetric configuration about the bond midpoint. Under this circumstance, $\beta_1 = \beta_2$ and we have

$$\beta = \frac{2\beta_1}{\sqrt{2(1 + s)}} \tag{11.23}$$

$$\beta^* = 0 \tag{11.24}$$

Clearly, chemisorption parallel to the surface correlates with dominant donor action on the part of the diatomic molecule.

Computations

We wish to investigate the behavior of the stereochemical predictions in our band structure calculations. We examine first a hypothetical H_2-type molecule having bonding $1s$ orbitals of variable ionization energy (ε_H). We have considered all three surface sites, on-top, bridging, and hollow, on the (111) fcc surface of five-layer films. A constant metal–hydrogen distance was maintained for all three sites. We find that adsorption at the three-fold site was favored and Figure 11.8 shows the heat of adsorption (Q) and molecular charge (q) as a function of the average number of metal d electrons (N_d). On any given curve the normal adsorption geometry correlates well with most negative charges on the H_2 molecule. We see that as the molecule becomes more electronegative (ε_H farther below E_F), the perpendicular geometry of adsorption prevails.

We may consider each adsorption site independently. Figure 11.9 is a plot of the difference in heat of adsorption for parallel minus normal geometries for different N_d values of the metal. We consider two ex-

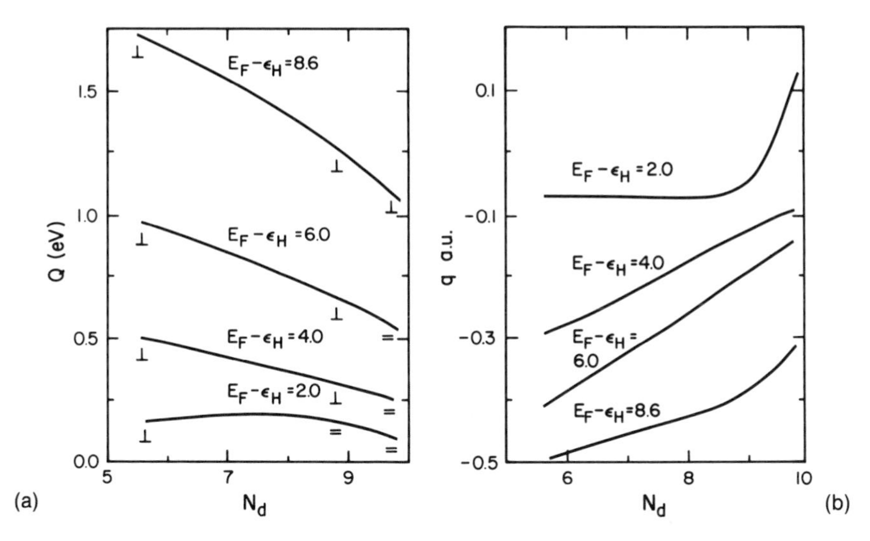

Figure 11.8. Heat of adsorption for a homonuclear diatomic molecule bound by S orbitals ("H_2"-type) is shown versus N_d in (a) with corresponding charge in (b). The difference in energy (eV) of the Fermi energy and adsorbate level is noted. The preferred adsorption geometry whether normal to the surface ($\perp$) or parallel ($=$) is noted.

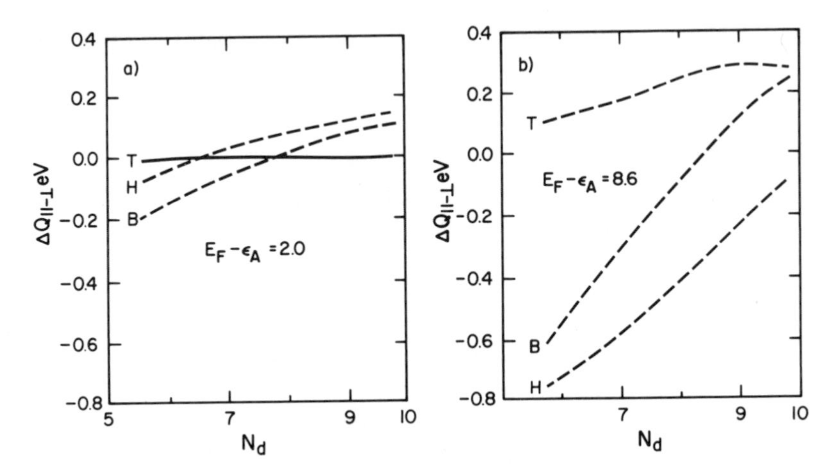

Figure 11.9. Difference in heat of adsorption for parallel and normal geometry in top (T), bridge (B), and hollow (H) fcc (111) is shown versus N_d for H_2-type adsorbates of different electronegativity ($E_F - \varepsilon_A$).

tremes in electronegativity of the adsorbed molecule. We observe generally that the multiple metal atom site promotes the normal geometry with a decreasing tendency as the site coordination decreases. Each site obeys the relationship of greater acceptor character correlating with the normal geometry.

One may question whether diatomic molecules possessing a strong component of *p* bonding character would obey the relationships found for the simpler H_2-type molecules. This point is addressed in Figure 11.10, where a N_2-type molecule of variable electronegativity is considered. The preferred orientation is shown. The parallel geometry is predominant, but with increasing electronegativity of the N_2-type molecule, we see the appearance of some perpendicular bonding geometries. This follows the general trend observed for the H_2 molecule. Heteronuclear molecules show a predominant normal bonding geometry with the most electropositive atom closest to the surface. This behavior is shown in Figure 11.10 for a CO-type molecule involving both *s*- and *p*-type bonding orbitals. When the molecule is inverted so as to bring the electronegative end closest to the surface, a considerable diminution in the adsorption energy takes place.

These results give us a general picture of dominant acceptor bonding associated with the bonding of both *s* and *sp* diatomic species with axis normal to the surface. We may ask about the possibility of dissociation from this geometry. Certainly electron donation to the antibonding molec-

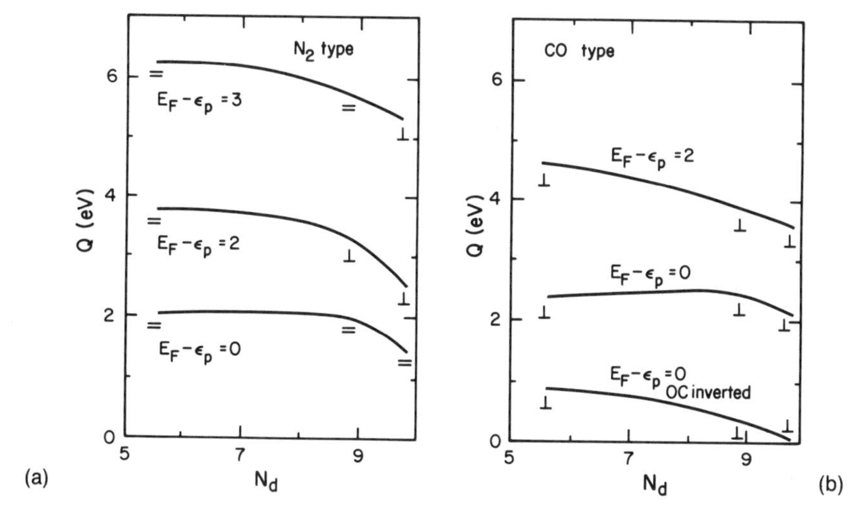

Figure 11.10. Heat of adsorption for a sp^3-bonded diatomic homonuclear molecule in (a) and heteronuclear diatomic molecule in (b) is shown versus N_d. Preferred geometries are noted for various electronegativities of the adsorbed atom ($E_F - \varepsilon_p$), where ε_p is the p valence level.

ular orbitals of H_2, CO, or N_2 decreases the molecular bond strength, but the normal geometry is not productive for dissociation.

Relationship to Experiment

An early suggestion[33] for the mechanism of activation of diatomic molecules involves bending from the normal geometry leading to a way point or intermediate in dissociation.

O–C → C–O → C O (11.25)

This model was based on observations of cluster complexes involving CO ligands. Bending the ligand from the normal geometry to the parallel geometry may thus account for a major part of the activation barrier for reaction. We note that recent experimental observations seem to be consistent with the idea that the parallel bonding species represents an intermediate point in dissociation. Some examples are the precursors CO horizontally bonded to Cr(110)[34a–34c] and Fe(100)[34d] as well as N_2 bonded horizontally to Fe.[35]

The stereochemistry of diatomic molecules adsorbed to metal surfaces is dominated by electronic interactions. Consider CO adsorbed to transition metal surfaces below its dissociation temperature.[36] On the vast majority of transition metal surfaces it is known to bond perpendicular to the surface in either the on-top or bridge site. Exceptions do exist. Two examples are Cr(110)[34a–34c] and Fe(100)[34d] where a "π-bonding" or lying-down adsorption geometry has been reported under some conditions. These are surfaces where dissociation of CO takes place below room temperature and the explanation of the horizontal geometry in terms of "enhanced electron back donation into the antibonding $2\pi^*$ molecular orbital thereby weakening the intramolecular bond"[34c] is certainly consistent with the position of these molecules in the periodic series. Their work function is smaller than metals at the right end, and so enhanced back donation is expected. We will demonstrate this point more clearly in the next section.

The adsorption geometry reported for N_2 on various transition metal surfaces is complex. On Ni(110) a N_2 species bonded with molecular axis normal to the surface is reported,[37] while this terminal and a π-bonded species are reported on Fe(111).[35] A species bonded normal to the surface is reported on Ru(001).[38] Interestingly, alkali metal enhances sticking of N_2 to form the π-bonded N_2 species on Fe(111), but it inhibits adsorption on Ru(001) surfaces when present above a critical level. On the latter surface the chemisorbed N_2 is interpreted to be a net electron donor.[39] The tendency for normal geometry near the end of the transition series with parallel geometry near the middle of the transition series is consistent with Figure 11.10. Certainly the bonding geometry of N_2 is much more delicately balanced between parallel and normal than for CO.

CO ADSORPTION

Acceptor Properties

We consider the adsorption properties of a molecule like CO to transition metal surfaces. This molecule is well known to be an electron acceptor,[40] but we would like to analyze its donor and acceptor contributions to the adsorption energy. We consider that the total adsorption energy can be written as a sum of acceptor and donor parts,

$$Q = Q_A + Q_D \tag{11.26}$$

We determine the Q_A part by projecting the appropriate acceptor molecular orbitals from the occupied Bloch function. This involves determination of the coefficients α_{it}, where

$$\sum_t \alpha_{it} \psi_t = \sum_l C_{il} Y_l \tag{11.27}$$

ψ_t is the free adsorbate molecular orbital, and the sum over l contains the adsorbate component of the l Bloch function. Use of orthonormality of the ψ_t set leads to an expression for the coefficients:

$$\alpha_{it} = \sum_l C_{il} \sum_j b_{tj} S_{lj} \tag{11.28}$$

where S_{lj} is an overlap integral and b_{tj} is a known coefficient in the free adsorbate molecular orbital. Occupation of the Bloch functions up to the Fermi energy gives an acceptor component

$$Q_A = 2 \sum_i \alpha_{it}^2 \mathscr{E}_i \tag{11.29}$$

where t defines the acceptor molecular orbital.

The qualitative mechanism of CO bonding to transition metal surfaces is understood in terms of the Blyholder model[40] considered in organometallic complexes. In this model, CO functions as a donor through electron donation from 5σ, 4σ, and 1π occupied molecular orbitals to the metal *sp* orbitals, and as an electron acceptor through electron donation from metal *d* orbitals to $2\pi^*$ molecular orbitals of CO. We may analyze this behavior on extended surfaces by projecting the $2\pi^*$ molecular orbital component from each occupied Bloch molecular orbital. This number of electrons multiplied by their energy defines the acceptor contribution to CO bonding (Q_A), as shown in Eq. (11.29). The donor contribution may be negative (implying repulsion) as a result of destabilization involving the interaction of closed-shell CO molecular orbitals with occupied parts of the *d* band of the metal.

Chemisorbed Properties

We examine the trends in adsorption energy for CO on various transition metal surfaces in Figure 11.11. Clearly, the acceptor component of chemisorption associated with the $2\pi^*$ level makes the dominant contribution to chemisorption. Absolute energy values depend on parameters and are not discussed. Both the on-top and bridge sites of adsorption, which are predominately observed experimentally, show decreasing acceptor bonding as N_d increases. This is because the Fermi level is dropping as N_d

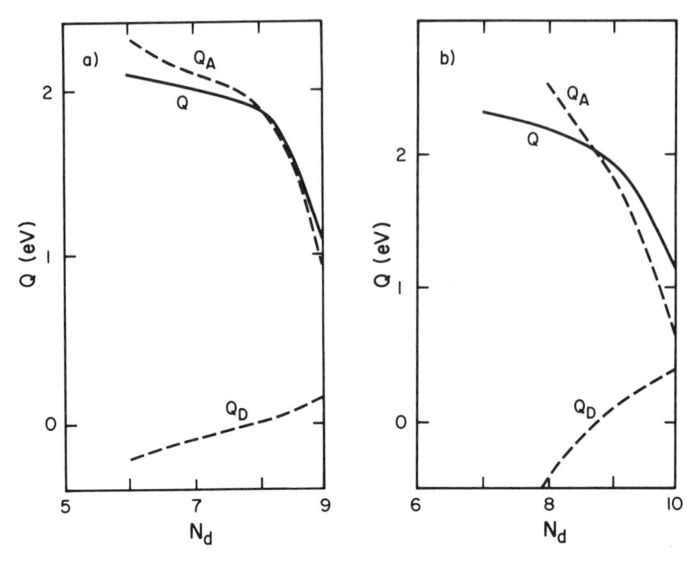

Figure 11.11. The total heat of adsorption (Q), the acceptor component (Q_A), and the donor component (Q_D) is plotted versus N_d for on-top (a) and bridge (b) site adsorption of CO.

increases, and this leads to an increased energy separation between the metal levels and the $2\pi^*$ acceptor level in CO. The reverse effect is true for the donor bonding mechanism.

Several changes in metal properties are observed upon CO chemisorption. Some trends are persistent and therefore parameter independent. Upon CO adsorption, the metal atom has an increase in *sp* population and a decrease in *d* population. This is an orbital-rehybridization effect, which we have discussed previously. The decrease in *d* population and increase in *sp* population are consistent with the Blyholder model and recent measurements of $4f_{7/2}$ surface core shifts reported for Pt(111) + CO.[26]

A change in the surface dipole moment of metal atoms is induced by CO adsorption, as shown in Figure 11.12. The total change in *p*,*d* dipole moment perpendicular to the metal surface is sensitive to the CO adsorption site. Note that a decrease in this internal dipole moment corresponds to decreasing electron density on the vacuum side of the surface and correlates with a decrease in the work function.[12] Of course, the net work-function change is composed of this internal contribution and the external dipole moment change determined by charge transfer. Adsorbed CO has a slight negative charge owing to its acceptor character, which would increase the work function so that the internal and external dipole contributions oppose one another. The more negative change of internal dipole

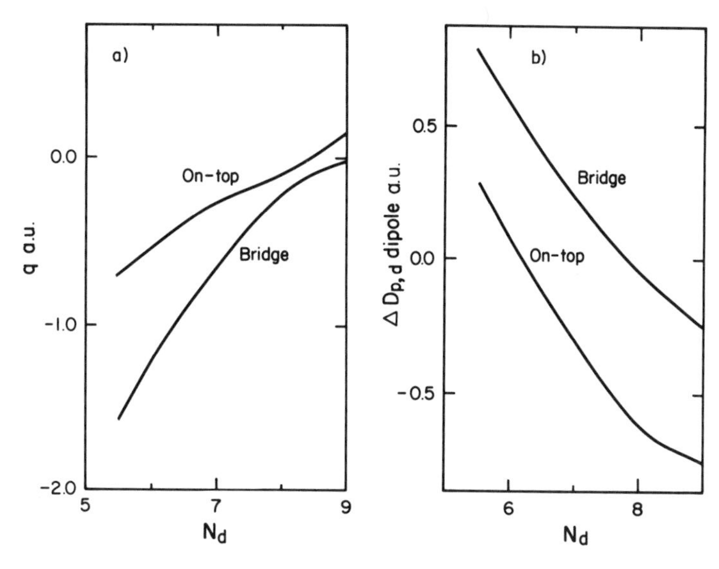

Figure 11.12. The charge (a) and change in *p*,*d* dipole moment (b) is plotted versus N_d for CO adsorption in on-top and bridge sites on fcc (111) metal surfaces.

moment for on-top adsorption sites compared to bridge or hollow sites could lead to a different sign of work function change for these two sites. Measurements of work function change for CO/Pt(111) show first a decrease, then an increase at higher coverage.[41] On-top sites are filled first and then the bridge. Apparently, this internal dipole effect is large enough to explain the different work function changes which have been reported.

The direct interactions of adsorbed species in mixed CO-alkali, -halogen, or -sulfur layers dominate changes in the heat of adsorption.[41] We demonstrate this effect by considering a six-atom trapezoidal unit mesh containing four hollow sites. The CO molecule is placed in one on-top site at the origin and the impurity at one of the hollow sites, with various distances from the CO molecule. Only a short-range effect up to ≈5 Å is found in the calculations. Alkali metal increases the CO heat of adsorption while Cl or S decreases the heat of adsorption.

The nature of the short-range attractive and repulsive forces is not just electron transfer to and from CO as moderated by the metal after charge exchange with the impurity. A mechanism involving direct covalent interactions between the impurity and CO is proposed for the primary factor controlling the heat of adsorption. Figure 11.13 shows the interaction between the 5σ occupied CO lone pair with the valence orbitals of halogen or alkali metal atoms. Perturbation-theory arguments appropriate for

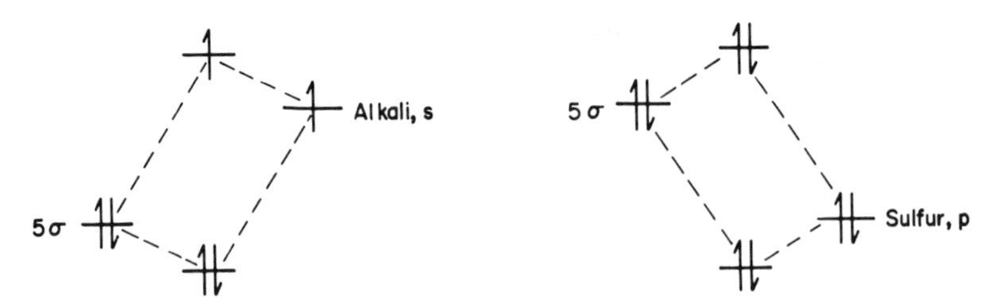

Figure 11.13. Schematic showing orbital interactions involving the 5σ level of CO with the valence *s* level of alkali atom or valence *p* level of sulfur atom.

the interaction of molecular orbitals have been developed to treat the interaction of a molecular orbital with a metal band. It is a strict result of these arguments that the antibonding combination of the molecular orbital metal band is destabilized more than the bonding combination is stabilized. With the alkali metal atom, net stabilization occurs because of the single occupancy (at most) of the alkali metal *s* valence molecular orbital, which is destabilized less than the doubly occupied 5σ level of CO is stabilized. The situation with halogen–CO interactions is different, since the halogen valence *p* level is double occupied. This is a case of two closed shells interacting, which is always destabilizing in molecular orbital terms. Thus the destabilization of the 5σ molecular orbital dominates the stabilization of the halogen *p* orbital. This mechanism for stabilization or destabilization of CO by an adsorbed impurity does not involve significant direct charge transfer. The charge on CO is not significantly changed in this mechanism because the filled CO 5σ level remains well below the Fermi energy of the metal. Of course, the dominant factor in bonding of CO to transition metal surfaces remains the population of the $2\pi^*$ levels.

SATURATED HYDROCARBONS

The mechanism by which saturated hydrocarbons interact with metal surfaces is one of fundamental interest. The question posed by the late Earl Muetterties was simply whether methane was a net donor or acceptor on metal surfaces. The question is not easily answered experimentally or theoretically. Addressing the question has led us[44] to probe several interesting properties of adsorbed hydrocarbons. We note that a very comprehensive analysis and set of calculations[2a] for chemisorbed methane has been presented which is in general accord with the principal points presented here.

CH_4 Geometric Preference

Several of the high-symmetry bonding configurations for methane chemisorbed to the (111) and (100) surfaces of transition metal films were surveyed. Figure 11.14 sketches these sites and defines a notation we will use. The (*x*-*y*) notation corresponds to *x* hydrogen atoms on methane interacting directly with a total of *y* metal atoms (M) on the surface where the gas-phase molecular geometry is retained. To further specify some of the cases, we state that the carbon atom lies along the perpendicular bisector of an M–M bond (1–2 and 2–2), or it is placed perpendicular to the center of a metal–atom triangle (1–3, 2–3, and 3–3). In the special case (S1–1) the C–H axis is parallel to a metal–metal bond with only one hydrogen atom interacting significantly with the metal surface. The metal–H bond lengths are taken as 1.7, 1.75, and 1.85 Å for single, double, and triple coordination, respectively, to metal atoms. These particular values are based upon data observed in organometallic compounds.[45] Nevertheless, our computed trends are not dependent on accuracy of the particular M–H distance chosen. We find the same order of geometric preference upon proportionate increases in the bond length.

The chemisorption energies in Table 11.4 show that methane adsorbs preferentially on the (100) and (111) surfaces with a minimal number of M–H contacts. The (1–1) geometry is preferred for this metal with a nearly full *d* band. We attach greatest significance to the trends in values among different sites rather than the specific absolute values.

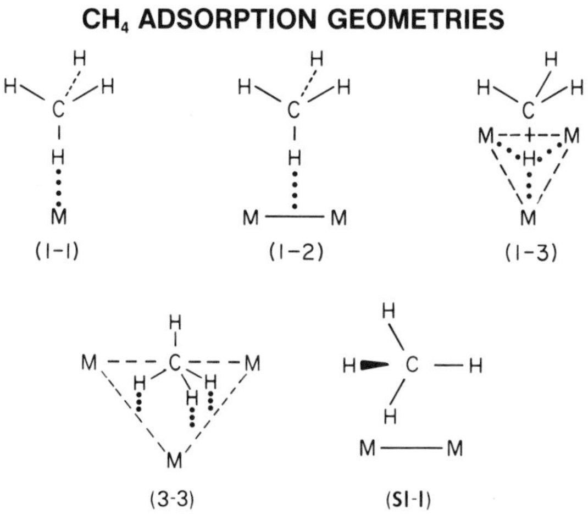

CH_4 ADSORPTION GEOMETRIES

Figure 11.14. Models of CH_4 interacting with metal atoms.

Table 11.4. Results for CH_4 chemisorption of surfaces of metal $s^{0.35}p^{0.37}d^{8.70}$.[a]

			Populations	
			CH_4	M[c]
			Δp	Δd
Geometry	Q^a (eV)	$\Delta\langle p_z\|z\|d_{z^2}\rangle^b$ (a.u.)	(a.u.)	(a.u.)
111 Surface				
1–1	0.05	−0.045	+0.08	−0.15
1–2	−0.46	−0.025	+0.13	−0.12
1–3	−0.71	−0.008	+0.17	−0.07
2–2	−0.32	−0.041	+0.14	−0.13
2–3	−0.69	−0.030	+0.20	−0.12
3–3	−0.96	−0.045	+0.20	−0.18
100 Surface				
1–1	−0.10	−0.043	+0.07	−0.11
1–2	−0.22	−0.009	+0.05	−0.09
1–4	−1.167	−0.24	+0.12	−0.12
2–2	−0.27	−0.007	+0.05	−0.06
3–4	−0.92	−0.017	+0.07	−0.09

[a] Heat of adsorption.
[b] Chemisorption-induced change in dipole moment.
[c] Metal atom(s) in direct contact with hydrocarbon.

Dipole Moment. The dipole moment of a surface metal atom becomes more negative upon CH_4 chemisorption. The data in Table 11.4 show that the change in the dominant p_z, d_{z^2} matrix element is negative as a result of the CH_4 interaction. The methane molecular orbitals mix more of the metal p orbitals into the occupied states, causing rehybridization of the surface metal atom similar to that observed in earlier calculations for H atoms interacting with surface atoms.[12b] Apparently, atomic and molecular chemisorption are equally effective in changing the dipole moment. A decrease in dipole moment perpendicular to the surface can be associated with a decrease in work function. A rough estimate of the work function change is 1 eV/0.06 a.u. change in dipole moment.[46] A decrease in work function of −0.5 eV has been observed for cyclopropane adsorbed on the Ir(110) surface.[47] This direction is consistent with our calculations and follows the lines of a perturbation analysis[4c] presented earlier.

Electron Population. An increase in the electron population of CH_4 is observed at the expense of the d-electron population of metal atoms with which there is direct CH_4 interaction as shown in Table 11.4. This effect occurs with each geometry on both (100) and (111) surfaces. It is the result of mixing antibonding $\sigma^*_{C\text{-}H}$ molecular orbitals with the occupied metal d

orbitals in dominance to the mixing of $\sigma_{C\text{-}H}$ molecular orbitals with unoccupied metal levels. Accompanying the decrease in *d* population of the metal atom(s) directly interacting with the hydrocarbon is a smaller increase in *sp* population. The *d* population of atoms in the second layer, not in direct contact with the hydrocarbon, has only minor sensitivity to the presence of adsorbate. Thicker films show that film thickness is not a major factor in determining this direction of net electron transfer.

Variation of the metal *d* occupation results in different geometric preferences and heats of chemisorption. Figure 11.15 compares the heat of chemisorption for four different sites as a function of N_d. We observe a preference for the (1–2) geometry compared to (1–1) at sufficiently small values of *d* occupation of the metal. There is a considerable increase in stability of the multicenter bonding configurations with decreasing values of N_d. In addition, the heat of chemisorption increases as N_d decreases. This behavior is consistent with the dominant acceptor mechanism leading to electron population of $\sigma^*_{C\text{-}H}$. This behavior was reported earlier for atomic adsorbates and is due to the fact that the energy separation between $\sigma^*_{C\text{-}H}$ and E_F increases as N_d increases. Thus, the partial electron transfer to $\sigma^*_{C\text{-}H}$ becomes smaller and N_d increases.

The computed net charge on the CH_4 molecule is negative for the values of N_d studied in Figure 11.15. The change in slope of the CH_4 charge is a reflection of two competing mechanisms: the number of *d* electrons of the metal available for transfer to CH_4 and the energy separation between E_F and $\sigma^*_{C\text{-}H}$, described previously. The first mechanism becomes more likely with increasing N_d, but the second mechanism becomes less likely, as described previously.

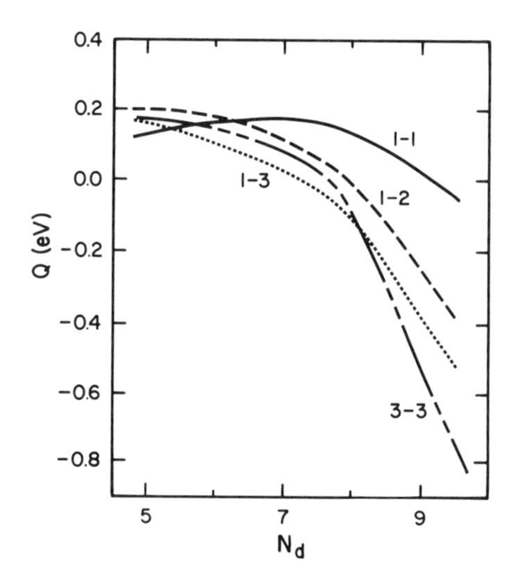

Figure 11.15. The heat of CH_4 adsorption to various sites on the fcc (111) surface is shown versus N_d.

Metal–H Interactions

The overlap population gives a measure of the interaction strength of H atoms in CH_4 with the metal surface. Figure 11.16 shows a positive value for various geometries at different values of N_d. This effect is in accord with the earlier work of Muetterties.[43,48] Along with the positive metal–H overlap population we find a decrease in C–H overlap population for the corresponding H atoms. This effect is due to the population of the σ^*_{C-H} molecular orbitals and should manifest in the vibration frequencies of the chemisorbed molecule. Note the crossover in adsorption energy and metal–hydrogen overlap population for the 1–3 and 3–3 configurations in Figures 11.15 and 11.16 near N_d = 8. This complex behavior shows the metal favoring binding the strongest acceptor geometry (3–3) of methane at N_d values less than 8.

Ethane

Ethane has been examined when chemisorbed to transition-metal surfaces.[47] Ethane was studied in the bridging site with molecular axis parallel to a metal–metal bond on the (111) fcc surface. Both the configurations with one H pointing toward the metal surface or two H pointing toward the surface were examined. The configuration with one H pointing toward the surface is favored on the ending transition metal surfaces while the

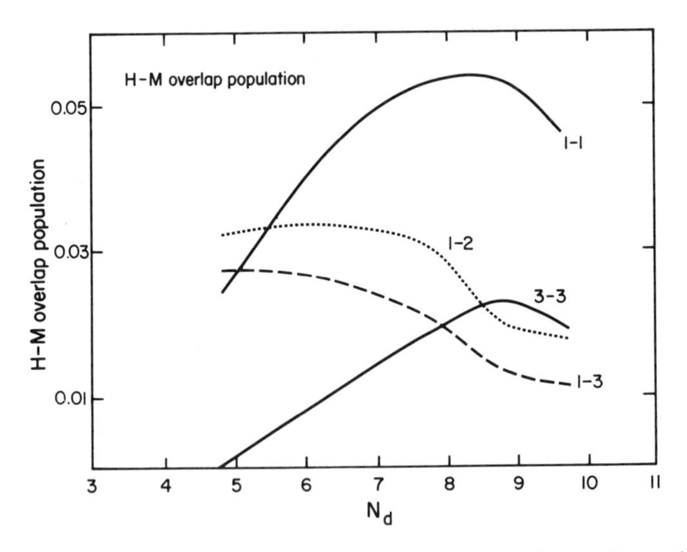

Figure 11.16. The H–M overlap population is shown for various sites on the fcc (111) surface versus N_d.

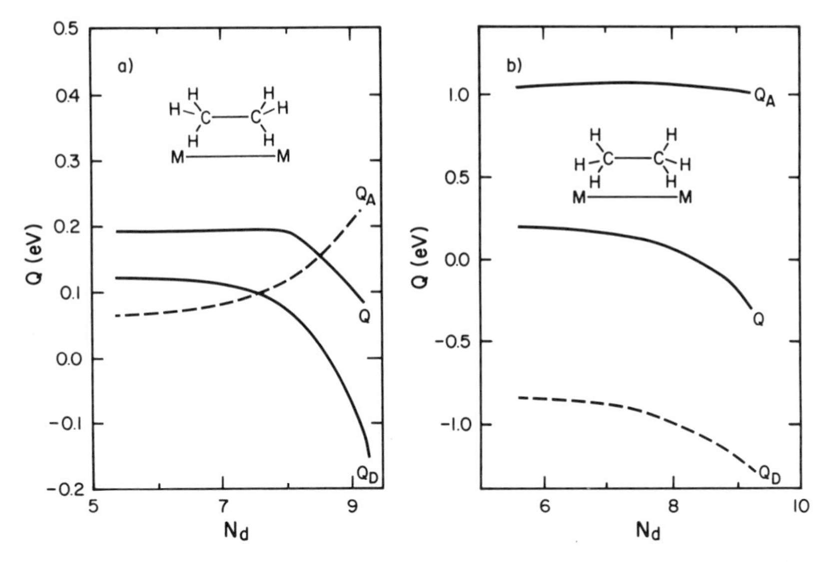

Figure 11.17. Total heat of adsorption (Q), acceptor component (Q_A), and donor component (Q_D) are shown for ethane chemisorbed to three layer metal films versus N_d. The C–C axis of ethane is parallel to the surface and a metal–metal bond. One hydrogen per carbon points to the surface in (a) or two in (b).

two H down species is favored on the middle transition metal surfaces. A typical plot of the heat of adsorption versus N_d is shown in Figure 11.17. Generally, the strongest bond forms to the early rather than the late transition metal surfaces. Interestingly, the acceptor role increases with N_d while the donor role decreases. We see from Figure 11.17 that the acceptor contribution to the heat of adsorption is dominant on ending transition metal surfaces.

The acceptor contribution to the heat of adsorption comes about through population of the antibonding molecular orbitals of ethane. Figure 11.18 is a projection of the density of these molecular orbitals for ethane. Interaction of ethane with the metal surface causes formation of broad bands, which mix normally unoccupied ethane molecular orbitals into states below the Fermi level schematically as in C of Figure 11.3. Because the antibonding orbital interacts more strongly with the surface than with the bonding level, the acceptor function predominates. This mechanism can be traced to the normalizing coefficient of the various molecular orbitals as was done leading up to the discussion of Eqs. (11.20) and (11.21). The bonding levels corresponding to the example of Figure 11.18 do not show any significant dispersion. Thus, the mechanism of

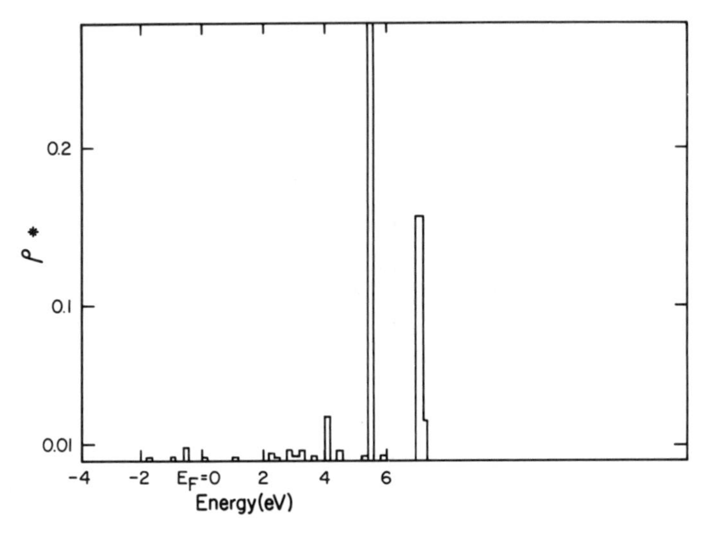

Figure 11.18. Projection of density of first antibonding $\sigma^*_{\text{C-H}}$ level from total density of states for ethane bound with one H down to fcc (111) metal surface with $N_d = 8.7$.

donation by mixing the ethane bonding levels into unoccupied Bloch functions does not seem to be significant.

Qualitative Model

A simplified qualitative argument can be made for the comparison of relative donor or acceptor roles of an adsorbate. We consider CH_4 and the relative mixing of σ_{CH} or σ^*_{CH} molecular orbitals with the d_{z^2} atomic orbital of a single atom:

$$\sigma_{CH} \sim p_z + s$$

$$\sigma^*_{CH} \sim p_z - s \tag{11.30}$$

We use perturbation theory [Eq. (11.15)] to consider the relative mixing coefficients of σ_{CH} and σ^*_{CH} with d_{z^2}:

$$d_{z^2} + \lambda^*\sigma^*_{\text{C-H}} + \lambda\sigma_{\text{C-H}} \tag{11.31}$$

$$\lambda^* = \frac{(\sigma_{\text{C-H}}|H'|d_{z^2})}{\varepsilon_{\sigma^*_{\text{C-H}}} - \varepsilon_d} \tag{11.32}$$

$$\lambda = \frac{(\sigma_{\text{C-H}}|H'|d_{z^2})}{\varepsilon_d - \varepsilon_{\sigma_{\text{C-H}}}} \tag{11.33}$$

Thus the ratio λ^*/λ determines whether C–H is an acceptor (>1) or a donor (<1)

$$\frac{\lambda^*}{\lambda} = \frac{(\sigma^*_{\mathrm{C-H}}|H'|d_{z^2})}{(\sigma_{\mathrm{C-H}}|H'|d_{z^2})} \frac{\varepsilon_d - \varepsilon_{\sigma_{\mathrm{C-H}}}}{\varepsilon_{\sigma^*_{\mathrm{C-H}}} - \varepsilon_d} \tag{11.34}$$

The first ratio in Eq. (11.34) is given approximately by the ratio of the coefficients of the 1s orbital of H in the σ^* and σ levels of CH_4. This has the value 2, as derived from extended Hückel calculations. Note that this value is much greater than one because of the different normalization constants in the σ^* and σ molecular orbitals. Thus the σ level computed at -15.4 eV and the σ^* level computed at 3.6 eV define a point -9.1 eV, above which ε_d values produce an acceptor molecule and below which ε_d values give a donor molecule. This simple example shows that for most transition metals CH_4 will be an acceptor because $\varepsilon_d > -9.1$ eV. Of course, this is a highly simplified model of the metal film designed only to show the physics of the interaction.

Relationship to Experiment

There is support for the dominant role of antibonding hydrocarbon molecular orbitals with metal centers in reactions of metal clusters. A large number of cases in which alkane C–H bonds react with metal clusters involve oxidative addition[49,50] with some degree of electron transfer from metal to the hydrocarbon fragments. An early example of this type of reaction was provided by Janowicz and Bergman[51] who photolyzed Cp^*IrLH_2 ($Cp^* = C_5Me_5$, $L = PMe_3$) to give alkane addition with H_2 elimination. Several other mechanisms can operate in cases of alkane activation[49,50] involving primarily interaction with the σ_{CH} electrons of the alkane. Theoretical analysis of these systems has appeared.[52,53] Despite the possibility of widely different mechanisms of alkane activation by metal complexes, the oxidative addition route seems to be firmly established providing support for the potential role of the acceptor levels of alkane in the activation process.

Chemisorption of alkanes to various ordered metal surfaces of Pt and Ir has shown an interesting reactivity on reconstructed (110) surfaces.[54] When the corresponding close packed (111) surfaces are studied with the hydrocarbons ethane, propane, n-butane, and n-pentane, no dissociative adsorption was observed. Estimates of the activation barrier for C–H cleavage of n-alkanes on these surfaces are greater than 16–17 kcal mol^{-1} on Pt(111) and Ir(111) compared to 7–8 kcal mol^{-1} on Ir(110)–(1 $\times$ 2) and 11–12 kcal mol^{-1} on Pt(110)–(1 $\times$ 2). The enhanced reactivity on the (110) reconstructed surfaces is consistent with the $\sigma^*_{\mathrm{C-H}}$ acceptor mechanism.

The reconstructed surfaces consist of ridges and valleys providing edge metal atoms of low coordination number. These edge atoms should possess an enhanced *d* electron density compared to bulk atoms using the arguments leading up to Figure 11.5. Thus, this enhanced *d* electron density would be available for donation to the empty antibonding orbitals of the alkane and thereby account for the enhanced reactivity of alkanes on the (110) surfaces.

Isotopic exchange reactions of alkanes with deuterium have been studied extensively.[55] Examples have included methane and ethane exchange over metal films and single-crystal surfaces. Slightly larger activation energy values are reported for complete exchange to CD_4 than single exchange to CDH_3,[7,57] but typical values for the ending transition metal elements lie in the range of 10–20 kcal mol^{-1} as shown in Table 11.5. The older studies seemed to indicate a structure insensitive reaction, but newer work under carefully controlled ultrahigh vacuum conditions has given evidence for differences in H/D exchange rate for *n*-hexane on various Pt single crystal surfaces.[59] The activity was slightly higher on (111) surfaces than on some stepped and kinked surfaces.

Various mechanisms have been proposed for the H/D exchange reaction involving methane. One early proposal involved formation of a CH_5 intermediate[57] in which H on the surface reacts with methane prior to

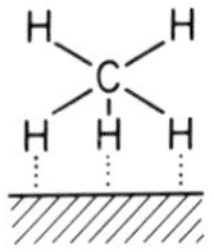

(I)

dissociation. Our calculations indicate a severely strained molecule, and while precise energy comparisons are difficult, this does not seem to be a

Table 11.5. Activation energy for isotopic D_2/alkane exchange reactions.

Alkane	Metal	Reference	ΔE (kcal mol^{-1})
CH_4	Rh film	57	13
	Mo film	57	10
	Re film	57	9
C_2H_6	Pt(111)	58	19
CH_4	Ni film	7	23.8
	Pd film	7	22.0
	Pt film	7	20.8
C_2H_6	Pd film	7	21.4
	Pt film	7	12.5

likely intermediate. Rather, the proposal by Lebrilla and Maier[60] seems more likely. They consider a "bond-slippage" mechanism in which methane from a 1–1 geometry bends its H toward the hollow site as the remaining CH_3 species interacts with one surface atom.

(2) → (3)

Our calculations of bending of the H toward the hollow site while gradually elongating the C–H bond indicate modest activation barriers of the order of 1 eV for this process. It is difficult to describe this reaction coordinate simply so an alternative simpler type of process was modelled. We start with the Sl geometry in Figure 11.14 and elongate the C–H bond and compute the energy at regular intervals along the sequence. Figure 11.19 shows this sequence where we consider two cases of different energy separations of the C $2p$ level and Fermi level $E_F - \varepsilon_p$. We find an

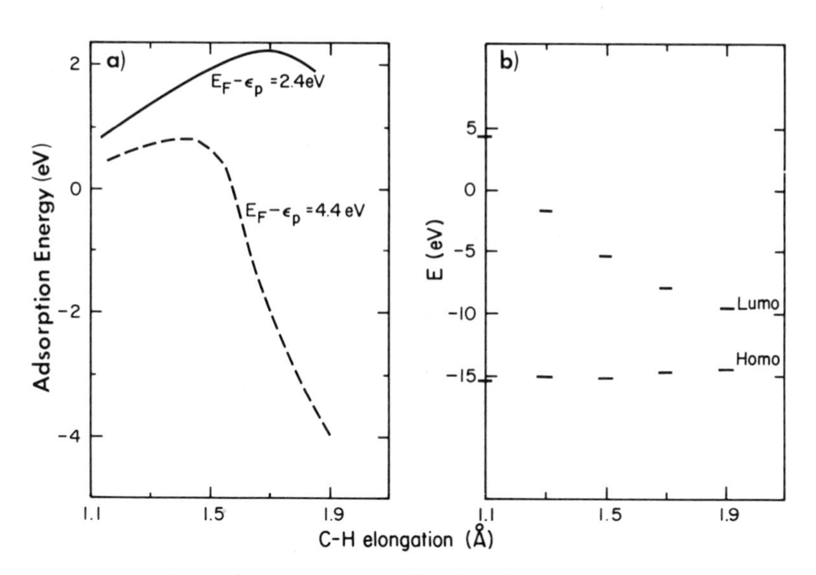

Figure 11.19. Adsorption energy profile for C–H elongation in the S1–1 configuration are shown in (a) and the positions of C–H highest occupied (HOMO) and lowest unoccupied (LUMO) molecular orbital versus distance of C–H is shown in (b). We consider the profiles for various values of $E_F - \varepsilon_p$, where ε_p is the $2p$ valence level in atomic carbon.

activation barrier whose magnitude depends upon the energy separation above. As the Fermi level gets closer to the carbon $2p$ level, the barrier increases. This behavior seems qualitatively correct in that early transition metals with large $E_F - \varepsilon_p$ values most easily dissociate CH_4. We also note the dramatic effect that stretching the CH bond has on the LUMO level of methane. Clearly, as this level drops in energy approaching the Fermi level, it becomes more highly populated, weakening the C–H bond. The factors reinforce one another.

ETHYLENE ADSORPTION

The chemistry of simple olefins chemisorbed to metal surfaces has been extensively studied. Spectroscopic studies of the simple olefin C_2H_4 have been carried out under ultrahigh vacuum (UHV). A low-temperature intact molecule species is observed on transition metal surfaces by various electron energy loss (EELS) and ultraviolet photoemission spectroscopic (UPS) studies. For example, on Pt(111) evidence[61–65] has been given to support a di-σ-bound species with sp^3 hybridization of carbon atoms. The evidence is consistent with the C–C bond parallel to the surface plane. This situation contrasts somewhat with Pd surfaces where the ethylene is relatively undistorted from gas-phase geometry on Pd(111)[66] and Pd(110)[67] but more distorted toward sp^3 for terminal carbons on Pd(100).[68]

The experimental studies have been supplemented by a variety of theoretical analyses[63,65,69] supporting the rehybridized molecule. This rehybridization contrasts to chemisorption on the noble metal surfaces such as Ag(110) or Cu(100) where the C_2H_4 molecule retains carbon sp^2 hybridization, probably indicative of a weaker interaction.[70,71] In addition to the di-σ-bound species on transition metals, the presence of purely π-bound (minor carbon rehybridization) species has also been reported. When the metal surface is varied, a change in the C hybridization has been noted through the vibration frequencies as a basis of the σ/π classification developed by Stuve and Madix.[68] An analysis[72] of the vibrational spectra from chemisorbed ethylene at low temperature has concluded that spectra of basically two patterns exist. These have been assigned to di-σ-adsorbed ethylene **(4)** and π-adsorbed ethylene **(5).**

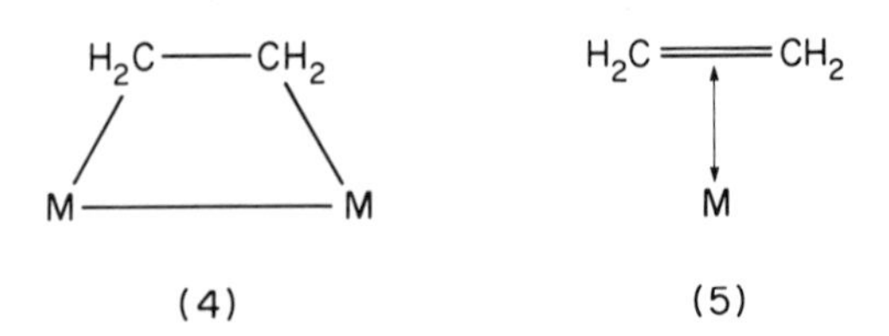

(4) (5)

Computed Properties

We consider the bonding of ethylene to fcc (111) surfaces. The on-top **(5)** and bridging sites **(4)** determined from the vibrational studies are considered in Figure 11.20 with the metal–carbon distance 1.9 Å. We observe that the heat of adsorption is largest early in the transition metal series and decreases to the right for both sites. We also observe that the C–C overlap population is weakened early in the transition series and that a considerable C–M overlap population exists between the corresponding atoms with the shortest bond length.

The mechanism of interaction of ethylene with the surface is shown for the bridge site in Figure 11.21. We project the π and π^* molecular orbital components from the density of states. The plot shows that both molecular orbitals of ethylene are considerably dispersed as a result of the interaction. The π component remains completely below the Fermi level and thus retains an occupancy of 2.0. The π^* level is dispersed sufficiently to give components below the Fermi level so that it becomes partially occupied. This occupancy weakens the C–C bond and is responsible for forming the M–C bond.

A perturbation on the ethylene molecule is caused by the chemisorption to the metal surface. The perturbation is measured in the extent of rehybridization of the molecule from sp^2 toward sp^3 at the C atoms. The two sites of chemisorption considered on the (111) surface were sketched

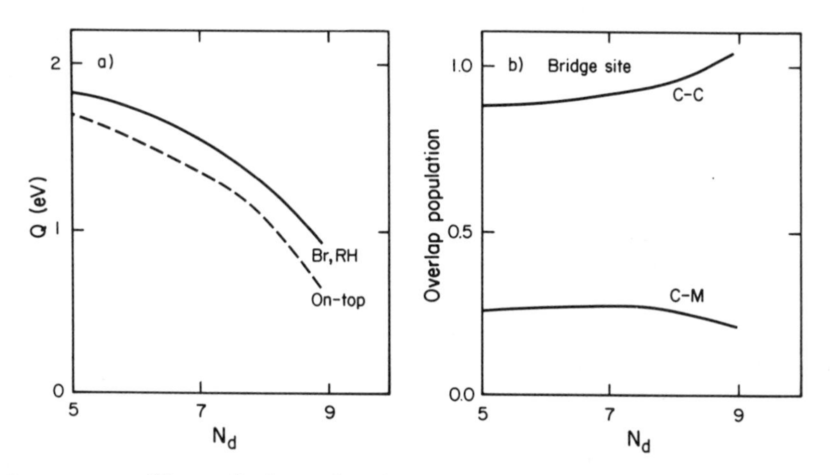

Figure 11.20. Heat of adsorption for ethylene chemisorbed to on-top and bridge sites of fcc (111) metal surfaces versus N_d. In (b) the overlap population for C–C and metal–C bonds of ethylene adsorbed in the bridge site is plotted versus N_d.

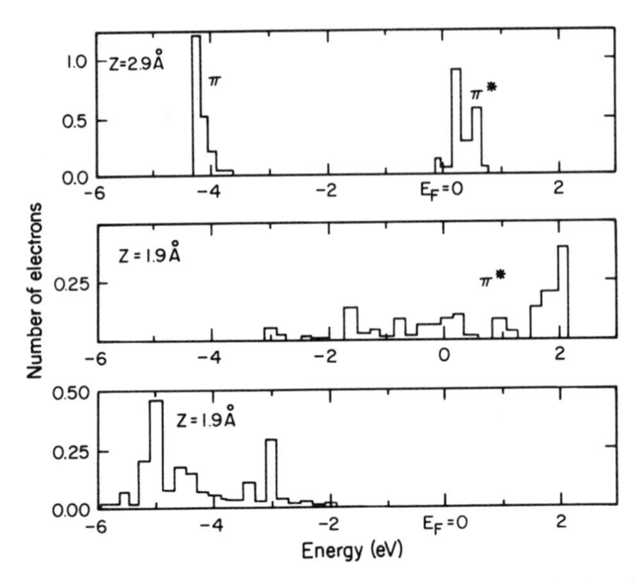

Figure 11.21. The projection of π and π^* molecular orbitals of ethylene from the total density of states is shown at two distances from the surface for the bridge site.

before in Schemes **4** and **5,** where the C atoms and closest metal atoms are shown. The C atoms were considered in a plane parallel to the metal surface plane at a height 1.9 Å in all calculations. The H atoms are bent away from the surface to achieve hybridization changes from sp^2 to sp^3. We consider three equal increments for this change. Figure 11.22 shows how the heat of adsorption depends on the degree of rehybridization of the ethylene molecule. There are dramatic differences at the on-top and bridging sites. Significantly more rehybridization of the molecule is observed at the bridging site approaching sp^3. The origin of this effect lies in the population of the π^* molecular orbitals of ethylene, which has long been known for cluster complexes.[73] This same effect occurs on the extended surfaces that we show in Figure 11.23.

The trends in the properties of chemisorbed ethylene with N_d present a consistent picture. The planar molecule has an increasing π^* population as N_d decreases. This tracks well with a reduction in the C–C overlap population and an increase in both the heat of adsorption and the metal–carbon overlap population. Clearly, the early transition metal surfaces destabilize the C–C bond most and provide the greatest adsorption energy. We note that the rehybridization of ethylene from sp^2 toward sp^3 is accompanied by a weakening of the C–C bond (reduced overlap popula-

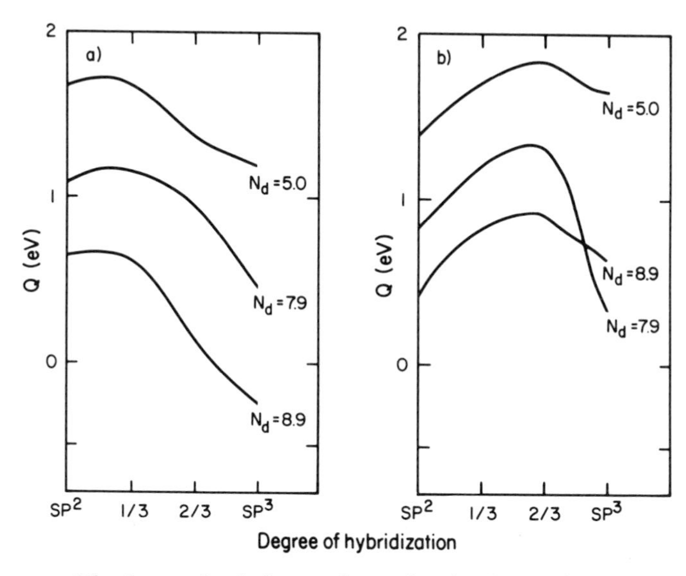

Figure 11.22. The heat of ethylene adsorption is shown for adsorption to the on-top site (a) and the bridge site (b) versus the degree of carbon rehybridization.

tion) and a strengthening of the metal–C bond (increased overlap population).

The strength of C–metal interactions and C–C bond weakening in rehybridized ethylene depends strongly on the metal Fermi level position. Figure 11.23 shows the relevant overlap population densities as a function of energy. The metal–carbon overlap has reached a maximum with the

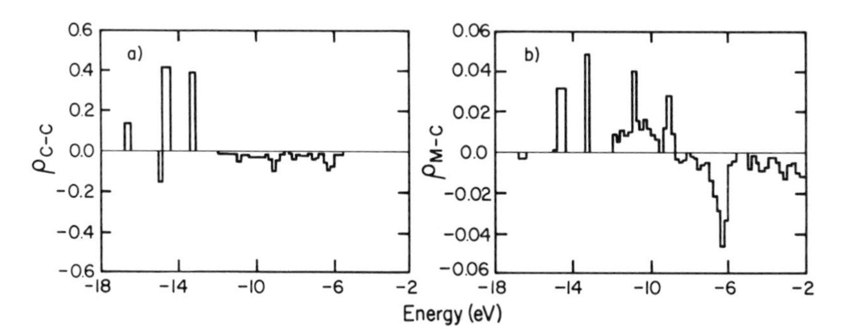

Figure 11.23. The density of C–C and C–metal overlap population is shown versus energy for the partially rehybridized ethylene species on the fcc (111) bridge site.

approximate $d^{7.9}$ filling. Increasing or decreasing this value would cause a reduction in the strength of this interaction. Clearly, the major contributions below the Fermi level in Figure 11.23 must arise from interactions with the bonding levels of ethylene. The C–C overlap population shows major decreases as the *d* band fills, and this trend continues with degree of filling, unlike the trend for the metal–carbon bond strength.

Related Experimental Aspects

The complete literature of ethylene adsorption patterns to transition metal surfaces is vast with different behavior patterns on different metal surfaces. We note that associative adsorption at low temperature is firmly established. Rehybridization of ethylene to configurations approaching sp^3 also seems to be firmly established with a variety of measurement techniques, including ultraviolet photoemission,[65] high-resolution electron energy loss spectroscopy,[64] and X-ray absorption fine structure[74] techniques. Particularly relevant to this discussion are the vibrational and thermal studies on Pd(100).[68] Both a π^- and a di-σ-bonded species are observed simultaneously at low temperature. The π species is reported to desorb with elevated temperature, but the di-σ species dehydrogenates into coadsorbed atomic hydrogen and a stable intermediate species suggested to be $CHCH_2$. Of course, the dehydrogenation of ethylene has been extensively studied on other surfaces, particularly Pt(111) where the ethylidyne species (CCH_3) has been detected in a number of independent spectroscopic studies.[75] An interesting species resulting from further heating of the intermediate to slightly above room temperature on Pd(100) surfaces is CH coadsorbed with H. Stability of this species in the presence of H may be particularly significant for other hydrocarbon reactions.

CHEMISORBED BUTADIENE

Let us consider the electronic structure of free butadiene before treating its interaction with metal surfaces.[76] Figure 11.24 shows the frontier orbitals of the *trans* isomer. The two π and two π^* orbitals composed of p_z atomic orbitals perpendicular to the planar ground state are sketched with regard to their nodal character. The nodal character is important in considering the symmetry relations involved in interactions with metal surfaces. For example, the symmetrical in-phase combinations of metal d_{z^2} orbitals located at the bottom of the *d* band will interact strongly with the lowest π_1 state of either isomer. The other interactions are less clear as they depend on the atomic spacings and position of the adsorbed molecule, but the out-of-phase combinations in π_3^* and π_4^* will generally inter-

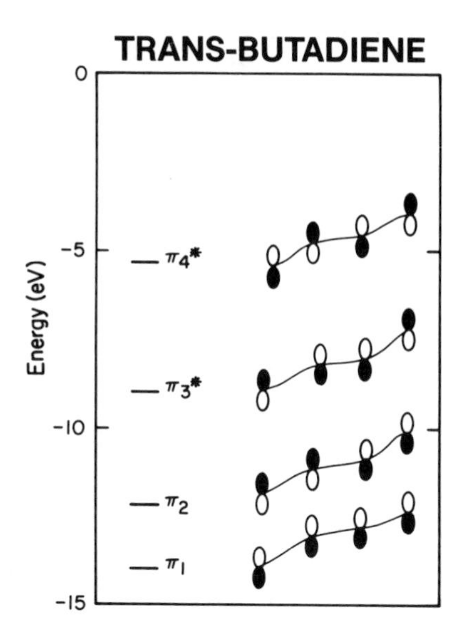

Figure 11.24. Sketch of frontier π and π^* molecular orbitals of *s-trans* butadiene.

act most strongly with the out-of-phase d_{z^2} orbitals at the top of the d band.

Chemisorption Properties

The butadiene is chemisorbed to a surface unit mesh consisting of six metal atoms as shown in Figure 11.25 for the *s-trans* configuration. A film consisting of two metal layers is considered. One might ask about the

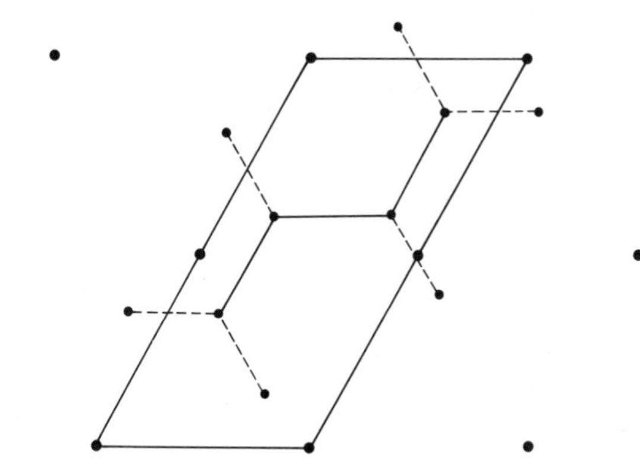

Figure 11.25. Sketch of *s-trans* butadiene chemisorbed to six atoms surface unit mesh.

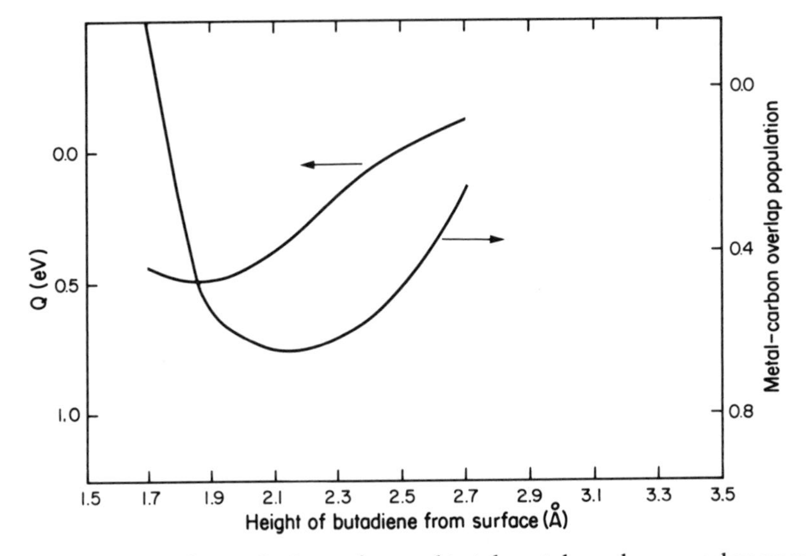

Figure 11.26. The heat of adsorption and total metal–carbon overlap population of *s-trans* butadiene chemisorbed as in Figure 11.25 to the fcc (111) surface with $N_d = 7.9$ as a function of distance above the metal plane.

ability of the method to produce reliable potential energy curves for chemisorption, which we have already indicated was doubtful. Figure 11.26 shows the energy of adsorption and metal–carbon overlap population versus distance of carbon to the nearest surface atoms for the center two atoms of butadiene placed in the bridging site as in Figure 11.25 where the molecule is parallel to the surface. We observe a distinct minimum in the overlap population curve at a reasonable internuclear M–C distance. A weaker minimum is found in the energy curve and it is a somewhat shorter distance than for the overlap population curve.

We may consider the mixing of butadiene molecular orbitals with the metal surface through projecting these molecular orbitals from the total density of states. Figure 11.27 shows these projections for the free ligand, the ligand at a large distance from the surface (2.9 Å), and the ligand strongly interacting with the surface 1.9 Å above. At the large distance from the surface (2.9 Å) the ligand molecular orbitals begin to spread in energy, although their center of gravity remains at the position of the free ligand. When the interaction is strong, the ligand molecular orbitals are spread considerably in energy. The π_1 and π_4^* centers of gravity are shifted deeper and shallower in energy, respectively, while π_2 and π_3^* have completely lost their original shape. The stronger interaction with π_2 and π_3^* compared to π_1 and π_4^* is a consequence of their difference in energy with the metal d orbitals (ε_d) and follows directly from second-order perturbation theory.[9] In this particular calculation, there is roughly equal interaction of π_2 and π_3^* with the metal d orbitals. This type of interaction leads to pushing the π_2 orbitals above the Fermi level, providing a donor function for the ligand, while the π_3^* orbitals are pushed below the Fermi level and provide an acceptor function for the ligand. The latter function predominates for transition metal elements.

The interaction of π and π^* ligand molecular orbitals with the surface is strongly dependent on the position of the center of the metal d band (ε_d). For middle transition metal elements, the ε_d value is closer to the Fermi level than for late transition metal elements. When the middle transition metal surface is considered, the π_1 and π_2 orbitals are perturbed little by the metal surface, but the π_3^* and π_4^* levels show major perturbation. This situation leads to significant net electron acceptor character by butadiene.

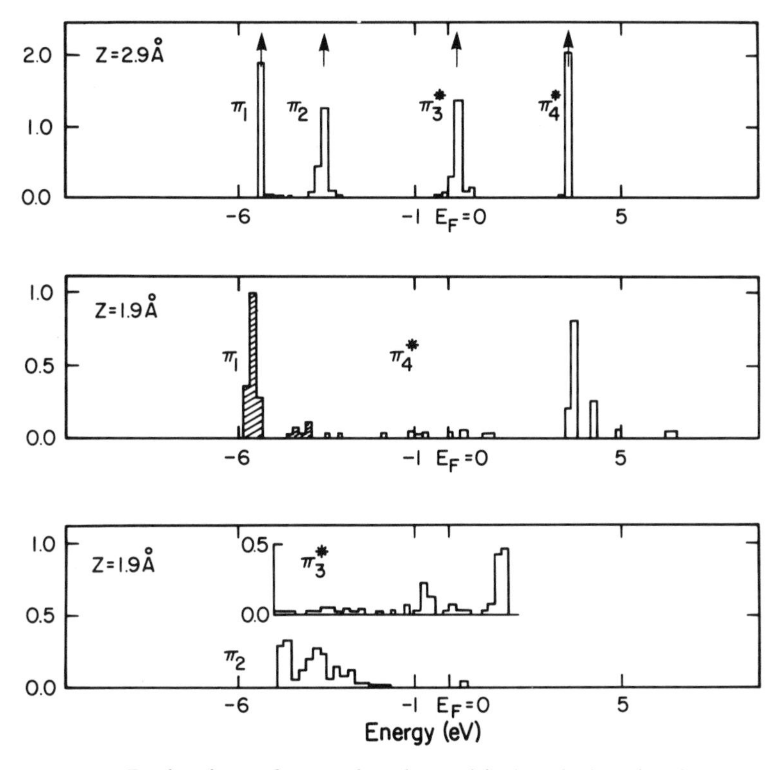

Figure 11.27. Projection of π molecular orbitals of chemisorbed *s-trans* butadiene from the total density of states for the $N_d = 7.9$ fcc (111) metal surface.

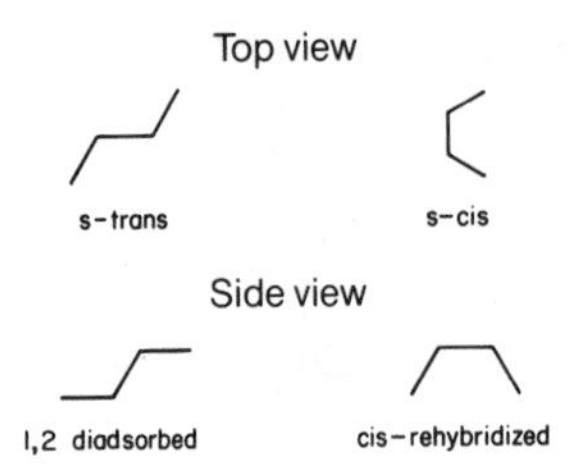

Figure 11.28. Sketch of geometric structures of butadiene considered.

Let us consider four basic configurations of butadiene bound to the metal surface. We have *s-cis* and *s-trans* isomers bound flat to the surface with no rehybridization. These species interact with the metal surface through both π systems. We also consider a species interacting through only one π system bound vertically to the surface. Also, a completely rehybridized (sp^3 at C) *cis* species bound vertically to the surface is considered. These species are sketched in Figure 11.28.

The heats of adsorption computed for the four geometries of chemisorbed butadiene versus N_d are shown in Figure 11.29. The 1,2 form and the flat geometries are almost equal in stability and greater than the rehybridized structure. The adsorption energy decreases sharply with N_d, reflecting the decreased net acceptor character of butadiene. The total

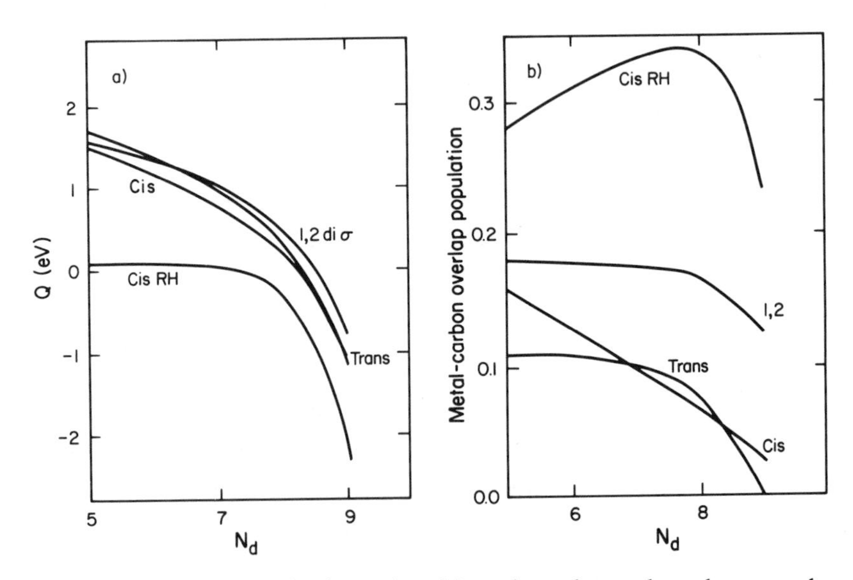

Figure 11.29. The heat of adsorption (a) and total metal–carbon overlap population (b) is shown versus N_d for various chemisorbed structures of butadiene.

metal–carbon overlap population also decreases with N_d, but it indicates a significantly stronger bond with the rehybridized structure. This bond is formed at the expense of the weakened C–C double bonds of butadiene. Finally, we note a significant strength for the 1,2 diadsorbed species.

The C–C overlap population in butadiene is considerably altered upon chemisorption. Figure 11.30 shows that the overlap population relative to the gas-phase molecule increases for the interior C–C bond and decreases for the exterior C–C bond upon chemisorption. The extent of this effect becomes particularly significant as the terminal C hybridization changes toward sp^3, but it is also apparent even in the absence of rehybridization. This effect is caused by chemisorption leading to greater double-bond character for the interior C–C bond and the reverse for the outer C–C bonds. The cause of this effect is electron population of the π_3^* level as shown in Figure 11.27. This level is characterized as bonding between middle carbons and antibonding between terminal and adjacent carbon atoms.

C–H Scission

We should consider the consequences of C–H bond scission in butadiene chemisorbed to metal surfaces. As we discussed earlier, this reaction is

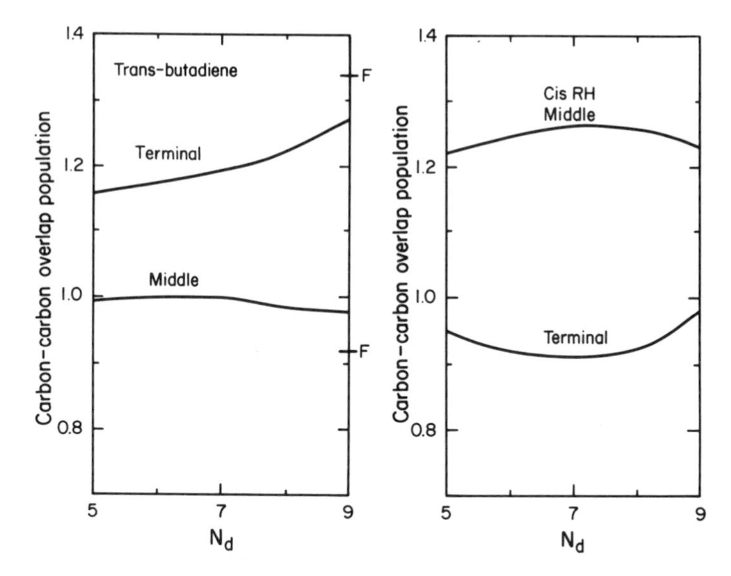

Figure 11.30. The C–C overlap population is shown for middle and terminal C–C bonds in *s-trans* (a) and *s-cis* (b) butadiene versus N_d. The overlap population calculated for the free molecule is denoted as F.

well known in ethylene. This reaction may be anticipated based on the position of the σ^*_{C-H} antibonding levels that lie just above the vacuum level and at this position are more favorably suited to accept electrons from a metal surface than the comparable levels in saturated hydrocarbons that lie at $\approx +5$ eV. The flat adsorption geometry creates intimate contact with the metal surface leading to large interaction integrals in the language of perturbation theory. Thus, population of σ^*_{CH} leading to weakening of C–H bonds and eventual dissociation is expected for butadiene at temperatures probably near room temperature. Our calculations show a population of this antibonding level ranging from 0.02–0.05 electrons for the horizontal geometry in support of this general mechanism of bond activation.

Terminal C–H bonds should be cleaved more easily than interior C–H bonds since this carbon is more electrophilic. We thus consider formation of a C_4H_4 species resulting from dual C–H bond cleavage. Metallacycles of the type bound vertically are energetically the most favorable species

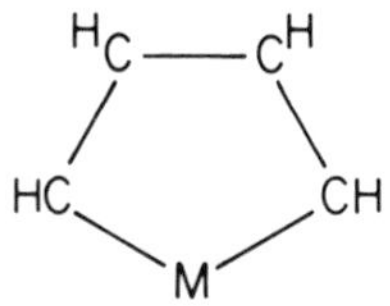

HC —— CH

HC CH

M

(6)

found in the calculations for the C_4H_4 species. There is a definite energetic preference for the ring metallacycle structure. This upright structure could well be an intermediate prior to further dehydrogenation. It is strongly adsorbed to the surface as deduced by comparing its heat of adsorption to the values for the intact butadiene. We note that a ring metallacycle PtC_4H_4 has been postulated from NEXAFS (near edge X-ray adsorption fine structure) studies of thiophene desulfurization on Pt(111).[77] The five-member ring structure has also been postulated in thermal desorption studies.[78]

Related Experiments

There is little structural information to compare with these calculations but infrared spectra of 1,3-butadiene adsorbed to Pd and Ni catalysts at low temperature (−50°C to 0°C) gives evidence for two species.[79] These have been assigned as π-bonded species involving only one or both π systems.

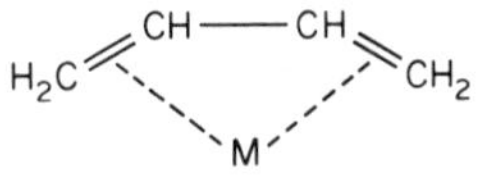

H_2C CH—CH CH_2 M

π_d

(7)

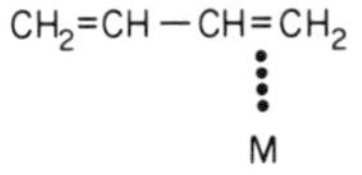

$CH_2=CH-CH=CH_2$

M

π_s

(8)

$CH_2=CH-CH=CH_2$

M M

σ

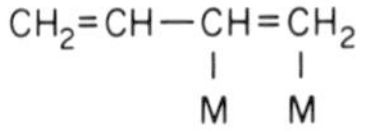

(9)

In addition to these, a σ species was identified along with the π species at room temperature on Co catalyst particles. There are a variety of adsorption patterns on different metals.[79] The hydrogenation of the π species is thought to occur in a 1,2 form on the singly bonded species (π_s) leading to 1-butene and in a 1,4 form on (π_d) leading to 2-butene.

The presence of two forms of π-bound butadiene is also observed in SERS studies of adsorption to silver.[80] This result as well as the infrared results is certainly consistent with the close chemisorption energies of the different forms of butadiene considered here, making possible a variety of simultaneous chemisorbed species. The infrared or kinetic studies have not given evidence for the metallacycle structure proposed here.

Molecular complexes of transition-metal atoms with butadiene offer some means of comparison with the chemisorbed molecule. A review of several of these complexes has recently appeared.[81] An interesting geometry (*s-cis*) reported for *s-cis* ligands involves π bonds from inner C atoms and σ bonds from outer C atoms to the Zr, Hf, or Ta atom as shown below. It is interesting to compare this Zr-Cp_2-C_4H_6 [Cp = $C_5(CH_3)_5$] metallacyclopentene structure with our computed results for chemisorbed butadiene. The computed overlap populations show more double bond character in the center C–C bond than in the outer C–C bonds. This same feature is found for the *cis* rehybridized (sp^3) structure

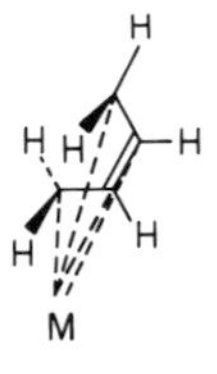

H H H H H H M

(10)

considered here. Also, the complexes undergo a fluxional behavior in which the butadiene passes through a transition state **11** with activation barrier of 6.5–12.4 kcal mol^{-1}. This transition state strongly resembles the sp^3 rehybridized geometry. Despite the fact that distortion of the butadiene to the sp^3 form requires ≈4 eV, the extra stabilization gained by bonding to the metal brings the transition state close in energy to the *s-cis* form. Such an effect is also occurring on the metal surfaces and it is quite possible that the rehybridized geometry may be stable on some surfaces.

(11)

CONCLUDING REMARKS

The band structure calculations discussed here, while simplified in their treatment of electronic interactions, have presented many qualitative pictures of how molecules and atoms bond to metal surfaces. One dominant theme is the important role of the antibonding and typically unoccupied molecular orbitals that interact strongly with the extended surface. These levels are particularly diffuse because of relatively large normalization constants. Even though they lie above the vacuum level in the case of saturated hydrocarbons, they can dominate the interaction with the surface. This effect is possible on metal extended surfaces, unlike clusters, because of the relatively small work function of the electron rich extended surface. This picture leads to a mechanism of C–H bond cleavage involving electron donation to the molecule with weakening of the C–H bond.

Several unexpected changes in properties of the metal surface are presented in these calculations. One is the enhanced *d* electron density of surface atoms relative to bulk atoms of the greater than half-full *d* band atoms. The reverse is true for the elements with less than half-full *d* bands. Also, there is a chemisorption-induced change in polarization of metal surface atoms caused by mixing *p* and *d* states. This effect often produces a dipole moment counter to the charge separation dipole moment expected for electronegative elements. When both contributions are summed, a reversal in expected dipole moment can result, making it very difficult to assign the direction of charge transfer based upon work function measurements.

Low-temperature associatively chemisorbed olefins and diolefins are electron acceptors. In the case of ethylene, the C atom undergoes significant rehybridization toward sp^3 when the molecule is adsorbed in a bridge site. The rehybridization is much less when it is adsorbed in an on-top site. Several sites of adsorption are possible for butadiene chemisorbed to transition metal surfaces. There is a significant reorganization of the π electrons in this chemisorbed molecule, leading to weakened terminal and strengthened center C–C bonds. A model for one high-temperature form involving a metallacycle bonded perpendicular to the surface is proposed.

One of the serious deficiencies of the method is that the computed minimum energy configurations cannot be relied upon. Thus, bond length values must come from various sources of experimental information. This significantly limits the application of the method for reaction profiles. Often simple trajectories are considered, but one is never sure that the correct transition state has been modeled. One possible route of obtaining more reliable bond length information has been through the use of constant bond-order calculations.[82] These calculations preserve a bond order value throughout the reaction coordinate treating the interactions with surface atoms. Perhaps a combination of this approach with the band-structure calculation would lead to an improved method.

ACKNOWLEDGMENTS

My colleagues have played an important role in this work and its development both through active collaborations and discussion. I wish to recognize G. Apai, H. Gysling, J. Monnier and E. Shustorovich for all of their contributions. In particular, J.F. Hamilton has played a dominant role in this work through his many interactions and his example of persistent dedication to following the scientific method to its ultimate ends.

REFERENCES

1. Fassaert, D.J.M., van der Avoird, A., *Surf. Sci.* **55:** 291 (1976).
2. (a) Saillard, J.-Y., Hoffmann, R., *J. Am. Chem. Soc.* **106:** 2006 (1984). (b) Sung, S.S., Hoffmann, R., *J. Am. Chem. Soc.* **107:** 578 (1985). (c) Silvestre, J., Hoffmann, R., *Langmuir* **1:** 621 (1986). (d) Zheng, C., Apeloig, Y., Hoffmann, R. (unpublished). (e) Sung, S.-S., Hoffmann, R., Thiel, P.A., *J. Phys. Chem.* **90:** 1380 (1986).
3. Minot, C., VanHove, M.A., Somorjai, G.A., *Surf. Sci.* **127:** 441 (1982).
4. (a) Shustorovich, E., Baetzold, R.C., *J. Am. Chem. Soc.* **102:** 5989 (1980). (b) Shustorovich, E., *Solid State Commun.* **44:** 567 (1982). (c) Shustorovich, E., *J. Phys. Chem.* **86:** 3114 (1982). (d) Shustorovich, E., Baetzold, R.C., *Appl. Surf. Sci.* **11:** 693 (1982). (e) Shustorovich, E., *J. Phys. Chem.* **87:** 14 (1983). (f) Shustorovich, E., *J. Phys. Chem.* **88:** 1927, 3490 (1984). (g) Shustorovich, E., Baetzold, R.C., Muetterties, E., *J. Phys. Chem.* **87:** 1100 (1983). (h) Shustorovich, E., *J. Am. Chem. Soc.* **106:** 6479 (1984). (i) Shustorovich, E., *Surf. Sci.* **150:** L115 (1985). (j) Shustorovich, E., *Solid State Commun.* **38:** 493 (1981). (k) Shustorovich, E., *Surf. Sci. Rep.* **6:** 1 (1986).
5. Keulks, G.W., *J. Catal.* **19:** 232 (1970).
6. Biloen, P., Sachtler, W.M.H., *Adv. Catal.* **30:** 165 (1981).
7. Anderson, J.R., Kemball, C., *Proc. Royal Soc. (London), Sec. A* **223:** 361 (1954).
8. (a) *Theory of Chemisorption* (J.R. Smith, ed.). West Berlin: Springer-Verlag, 1980. (b) *Solid State Physics* (F. Seitz and D. Turnbull, eds.). New York: Academic Press, 1980, Vol. 35.
9. Hoffmann, R., *J. Chem. Phys.* **39:** 1397 (1963).
10. (a) Clementi, E., Raimondi, D.L., *J. Chem. Phys.* **38:** 2686 (1963). (b) Basch, H., Gray, H.B., *Theoret. Chim. Acta (Berlin)* **4:** 367 (1966).

11. Imamura, I., *J. Chem. Phys.* **52:** 3168 (1970).
12. (a) Baetzold, R.C., *Solid State Commun.* **44:** 781 (1982). (b) Baetzold, R.C., *J. Phys. Chem.* **87:** 3858 (1983).
13. Baetzold, R.C., Hamilton, J.F., *Prog. Solid State Chem.* **15:** 1 (1983).
14. Baetzold, R.C., *J. Am. Chem. Soc.* **103:** 6116 (1981).
15. Varma, C.M., Wilson, A.J., *Phys. Rev. B* **22:** 3795 (1980).
16. Hodge, L., Watson, R.E., Ehrenreich, H., *Phys. Rev. B* **5:** 3953 (1972).
17. Hoffmann, R., *Acc. Chem. Res.* **4:** 1 (1971).
18. An interesting discussion of interaction mechanisms is found in Ref. 2a.
19. Shustorovich, E., Baetzold, R.C., Muetterties, E.L., *J. Phys. Chem.* **87:** 1100 (1983).
20. (a) Citrin, P.H., Wertheim, G.K., *Phys. Rev. B* **27:** 3176 (1983). (b) Feibelman, P.J., Hamann, D.R., *Solid State Commun.* **31:** 413 (1979). (c) Citrin, Ph.H., Wertheim, G.K., Baer, Y., *Phys. Rev. Lett.* **41:** 1425 (1978).
21. Desjonqueres, M.C., Spanjaard, D., Lassailly, Y., Guillot, C., *Solid State Commun.* **34:** 807 (1980).
22. Van der Veen, J.J., Himpsel, F.J., Eastman, D.E., *Phys. Rev. B* **25:** 7388 (1982).
23. Wertheim, G.K., Citrin, P.H., van der Veen, J.F., *Phys. Rev. B* **30:** 4343 (1984).
24. Duc, T.M., Guillott, C., Lasailly, Y., Jugnet, Y., Vedrine, J.C., *Phys. Rev. Lett.* **43:** 789 (1970).
25. Heimann, P., van der Veen, J.F., Eastman, D.E., *Solid State Commun.* **38:** 595 (1981).
26. Apai, G., Baetzold, R.C., Jupiter, P.J., Viescas, A.J., Lindau, I., *Surf. Sci.* **134:** 122 (1983).
27. Duckers, K., Bonzel, H.P., Wesner, D.A., *Surf. Sci.* **166:** 141 (1986).
28. van der Veen, J.F., Himpsel, V.J., Eastman, D.E., *Phys. Rev. Lett.* **44:** 189 (1980); **44:** 553 (1980).
29. Citrin, P.H., Wertheim, G.K., Baer, Y., *Phys. Rev. Lett.* **41:** 1425 (1978).
30. Holzl, J., Schulte, F.K., *Solid Surface Physics*. West Berlin: Springer-Verlag, 1979, pp. 1–150.
31. Erley, W., *Surf. Sci.* **94:** 281 (1980).
32. (a) Dedieu, A., Strich, A., *Inorg. Chem.* **18:** 2940 (1979). (b) Sevin, A., *Nouv. J. Chim.* **5:** 233 (1981). (c) Blomberg, M.R.A., Siegbahn, P.E.M., *J. Chem. Phys.* **78:** 986 (1983). (d) Goldberg, K.J., Hoffman, D.M., Hoffmann, R., *Inorg. Chem.* **21:** 3863 (1982).
33. Muetterties, E.L., Stein, J., *Chem. Rev.* **79:** 479 (1979).
34. (a) Shinn, N.D., Madey, T.E., *Phys. Rev. Lett.* **53:** 2481 (1984). (b) Shinn, N.D., Madey, T.E., *J. Chem. Phys.* **83:** 5928 (1985). (c) Shinn, N.D., Madey, T.E., *Phys. Rev. B* **33:** 1464 (1986). (d) Benndorf, C., Kruger, B., Thieme, F., *Surf. Sci.* **163:** L675 (1985).
35. Whitman, L.J., Bartosch, C.E., Ho, W., Strasser, G., Grunze, M., *Phys. Rev. Lett.* **56:** 1984 (1986).
36. (a) Broden, G., Rhodin, T.N., Brucker, C., Benbow, R., Hurych, Z., *Surf. Sci.* **59:** 593 (1976). (b) Andreoni, W., Varmn, C.M., *Phys. Rev. B* **23:** 437 (1981).
37. Horn, K., DiNardo, J., Eberhardt, W., Freund, H.-J., Plummer, E.W., *Surf. Sci.* **118:** 465 (1982).
38. Anton, A.B., Avery, N.R., Toby, B.H., Weinberg, W.H., *J. Electron Spectrosc. Rel. Phenom.* **29:** 181 (1983).
39. Depaola, R.A., Hoffmann, F.M., *Chem. Phys. Lett.* **128:** 343 (1986).
40. Blyholder, G., *J. Phys. Chem.* **68:** 2772 (1964).
41. (a) Steininger, H., Lehwald, S., Ibach, H., *Surf. Sci.* **123:** 264 (1982). (b) Poelsema, B., Palmer, R.L., Comsa, G., *Surf. Sci.* **123:** 152 (1982).
42. Baetzold, R.C., *Phys. Rev. B* **30:** 6870 (1984).
43. Gavin, R.M., Reutt, J., Muetterties, E.L., *Proc. Nat. Acad. Sci. U.S.A.* **78:** 3981 (1981).

44. Baetzold, R.C., *J. Am. Chem. Soc.* **105:** 4271 (1983).
45. Bau, R., Teller, R.G., Kirtley, S.W., Koetzle, T., *Acc. Chem. Res.* **12:** 176 (1979).
46. $\Delta\phi \simeq e\Delta\langle P_z|z|d_{z^2}\rangle$.
47. Wittrig, T.S., Szuromi, P.D., Weinberg, W.H., *J. Chem. Phys.* **76:** 3305 (1982).
48. (a) Muetterties, E.L., *Chem. Soc. Rev. (London)* **11:** 283 (1982). (b) Friend, C.M., Muetterties, E.L., *J. Am. Chem. Soc.* **103:** 773 (1981). (c) Beno, M.A., Williams, J.M., Tachikawa, M., Muetterties, E.L., *J. Am. Chem. Soc.* **103:** 1503 (1981).
49. Buchanan, J.M., Stryker, J.M., Bergman, R.G., *J. Am. Chem. Soc.* **108:** 1537 (1986).
50. Crabtree, R.H., *Chem. Rev.* **85:** 245 (1985).
51. Janowicz, A.H., Bergman, R.G., *J. Am. Chem. Soc.* **104:** 352 (1982); **105:** 3929 (1983).
52. Rabaâ, H., Saillard, J.-Y., Hoffmann, R., *J. Am. Chem. Soc.* **108:** 4327 (1986).
53. Lichtenberger, D.L., Kellogg, G.E., *J. Am. Chem. Soc.* **108:** 2560 (1986).
54. Szuromi, P.D., Engstrom, J.R., Weinberg, W.H., *J. Phys. Chem.* **80:** 508 (1984); 2497 (1985).
55. Burwell, Jr., R.L., *Catal. Rev.* **7:** 25 (1972).
56. Dalmon, J.A., Mirodatos, C., *J. Mol. Catal.* **25:** 161 (1984).
57. Frennet, A., *Catal. Rev.* **10:** 37 (1974).
58. Zaera, F., Somorjai, G.A., *J. Phys. Chem.* **89:** 3211 (1985).
59. Davis, S.M., Somorjai, G.A., *J. Phys. Chem.* **87:** 1545 (1983).
60. Lebrilla, C.B., Maier, W.F., *J. Am. Chem. Soc.* **108:** 1606 (1986).
61. Albert, M.A., Sneddon, L.G., Eberhardt, W., Greuter, F., Gustafosson, T., Plummer, E.W., *Surf. Sci.* **120:** 19 (1982).
62. Baro, W.M., Ibach, H., *J. Chem. Phys.* **74:** 4194 (1981).
63. Demuth, J.E., *Surf. Sci.* **84:** 315 (1979).
64. Kesmodel, L.L., DuBois, L.H., Somorjai, G.A., *J. Chem. Phys.* **70:** 2180 (1979).
65. Felter, T.E., Weinberg, W.H., *Surf. Sci.* **103:** 265 (1981).
66. Gates, J.A., Kesmodel, L.L., *Surf. Sci.* **120:** L461 (1982).
67. Chesters, M.A., McDougall, G.S., Pemble, M.E., Sheppard, N., *Appl. Surf. Sci.* **22:** 369 (1985).
68. Stuve, E.M., Madix, R.J., *J. Phys. Chem.* **89:** 105 (1985).
69. Hiett, P.J., Flores, F., Grout, P.J., March, N.H., Martin-Rodero, A., Senatore, G., *Surf. Sci.* **140:** 400 (1984).
70. Backx, C., DeGroot, C.P.M., Biloen, P., *Appl. Surf. Sci.* **6:** 256 (1980).
71. Nyberg, C., Tengstal, C.G., Andersson, S., Holmes, M.W., *Chem. Phys. Lett.* **87:** 87 (1984).
72. Sheppard, N., *J. Electron Spectrosc. Rel. Phenom.* **38:** 175 (1986).
73. Blizzard, A.C., Santry, C.P., *J. Am. Chem. Soc.* **90:** 5749 (1968).
74. Stohr, J., Sette, F., Johnson, A.L., *Phys. Rev. Lett.* **53:** 1684 (1984).
75. Salmeron, M., Somorjai, G.A., *J. Phys. Chem.* **86:** 341 (1982).
76. Baetzold, R.C., *Langmuir* **3:** 189 (1987).
77. Stohr, J., Gland, J.L., Kollin, E.B., Koestner, R.J., Johnson, A.L., Muetterties, E.L., Sette, E.L., *Phys. Rev. Lett.* **53:** 2161 (1984).
78. Wexler, R.M., Ph.D. Thesis, University of California, Berkeley, 1983.
79. Soma, Y., *Bull. Chem. Soc. Jpn.* **50:** 2119 (1977).
80. Itoh, K., Tsukada, M., Koyama, T., Kobayashi, Y., *J. Phys. Chem.* **90:** 5286 (1986).
81. Yahuda, H., Tatsumi, K., Nakamura, A., *Acc. Chem. Res.* **18:** 120 (1985).
82. (a) Baetzold, R.C., *Surf. Sci.* **150:** 193 (1985). (b) Baetzold, R.C., *J. Chem. Phys.* **82:** 5724 (1985). (c) Baetzold, R.C., Shustorovich, E.M., *Surf. Sci.* **165:** L41 (1986). (d) For an analytic model development of this method, see Ref. 83.
83. (a) Shustorovich, E.M., *Surf. Sci.* **163:** L645, L730 (1985). (b) Shustorovich, E.M., *J. Am. Chem. Soc.* **106:** 6479 (1984).

12
Quantum-Chemical Studies of the Acidity and Basicity of Alumina

SATOHIRO YOSHIDA

INTRODUCTION

Alumina is one of the most popular materials for catalyst supports. Typical examples are found in Pt–Re/Al_2O_3 (reforming), Co–Mo/Al_2O_3 (dehydrosulfurization), and Pt–Pd–Rh/Al_2O_3 (control of automotive pollutants). Alumina is thermally stable and a material suitable for dispersion of catalytically active components, but it is far from a passive or inert support as indicated by a well-established dual function in the reforming process. The important function of alumina surface is related to its acidic or basic properties or both, which control the dispersion state of active components as well as the performance of catalysts. Thus, there have been many investigations on the acidity and basicity of alumina and the catalysis by it.

Many reactions are known to be catalyzed by alumina, such as dehydration and dehydrogenation of alcohols, double bond migration, and *cis-trans* isomerization of olefins, H–D exchange in hydrocarbons. The mechanistic studies of these reactions are reviewed by Pines and Manassen[1] and by John and Scurrell.[2]

Although the surface acidity and basicity of alumina have been noticed for many years, the microstructures of acidic and basic sites are not well characterized. Alumina has many polymorphs and generally γ- or η-alumina is widely used as a support (or a catalyst). The fact that α-alumina is an inert material for catalysis indicates the essential importance of the bulk structure of alumina. Alumina is obtained by thermal decomposition of an aluminum hydroxide and the crystal structure of alumina depends on the starting hydroxide and also the conditions of thermal decomposition. The dehydration sequences of hydroxides were investigated in detail by Lippens et al. The sequences in air are presented as follows[3]:

$$\text{gibbsite} \xrightarrow{250\ ^\circ\text{C}} \chi \xrightarrow{900\ ^\circ\text{C}} \kappa \xrightarrow{1200\ ^\circ\text{C}} \alpha\text{-Al}_2\text{O}_3$$

$$\text{gibbsite} \xrightarrow{180\ ^\circ\text{C}} \text{boehmite} \xrightarrow{450\ ^\circ\text{C}} \gamma$$

$$\text{bayerite} \xrightarrow{180\ ^\circ\text{C}} \text{boehmite}$$

$$\text{bayerite, nordstrandite} \xrightarrow{230\ ^\circ\text{C}} \eta \xrightarrow{850\ ^\circ\text{C}} \theta \xrightarrow{1200\ ^\circ\text{C}} \alpha\text{-Al}_2\text{O}_3$$

$$\text{diaspore} \xrightarrow{450\ ^\circ\text{C}} \alpha\text{-Al}_2\text{O}_3$$

Gibbsite, bayerite, and nordstrandite are polymorphs of aluminum trihydroxide [$Al(OH)_3$] and boehmite; diaspore are of aluminium–oxidehydroxide [$AlO(OH)$]. The ultimate compound obtained by heating a hydroxide at high temperatures is α-alumina without respect to the starting hydroxide.

The fundamental structure can be described as a closed packed lattice of oxygen atoms with aluminum atoms distributed in octahedral or tetrahedral interstices.[4,5] Alumina can be classified in three groups by the packing mode of the oxygen atoms.[4]

α-Series: α-alumina (corrundom), stacking mode; ABAB · · ·

β-Series: χ-, κ-alumina, stacking mode; ABACABAC · · ·

γ-Series: γ-, η-alumina, stacking mode; ABCABC · · ·

Among these, acid–base catalysis is exhibited by γ- and η-alumina (generally in a low crystallization state). Both γ- and η-alumina have a "defect spinel structure."[5] In spinel ($MgAlO_4$), oxygen atoms form a crystal lattice of cubic close packing and a unit cell is composed of 32 oxygen atoms. Eight of 64 tetrahedral holes are occupied by Mg(II) ions and 16 of 32 octahedron holes are occupied by Al(III) ions. In alumina, one-ninth of cation sites of spinel are vacant (defected) because of charge difference between Mg(II) and Al(III). The distribution of aluminum ions is roughly described by a formula[4] $(Al_{0.67}V_{0.33})_{tetr}(Al_2)_{oct}O_4$ (V: vacant cation sites) but the distribution changes by heating.

γ- and η-alumina have the same structure basically; dissimilarities arise from the difference in order of oxygen packing: oxygen atoms are fairly well ordered in γ-alumina, while the disorder is relatively frequent in η-alumina.[3] Since the difference is not significant in the local structure, we will not distinguish both aluminas hereafter and refer simply to γ-alumina (or simply alumina). The important points for discussion of the local structure are that (1) there are two kinds of environment of an aluminum atom (in an oxygen octahedron or in a tetrahedron) and (2) there are oxygen octa- or tetrahedrons missing aluminum atoms.

ACID–BASE PROPERTIES OF γ-ALUMINA

The existence of acidic and basic sites on the surface of γ-alumina is well established by experimental results such as titration with indicators[6] or infrared spectra of adsorbed molecules[7] and there is no doubt that these sites function as the active sites in catalysis by γ-alumina. The important point in the catalysis is the control of the dehydration state of alumina surface. The activity and selectivity depend significantly on a heat treatment (activation procedure) before use as a catalyst. Madema showed that the activation procedure was more significant for catalysis than the preparation procedures of alumina from aluminum oxidehydroxides,[8] and noted that "the results must be explained in terms of hydroxylated surfaces rather than in terms of catalysts biography."

As for the acidic nature, only Lewis acidity was observed by infrared spectroscopic studies of adsorbed pyridine.[9,10] But later, weak Brønsted acidity was claimed from the results of adsorption of 2,6-di-t-butyl pyridine or 2,6-dimethyl pyridine.[11,12] In addition to infrared spectroscopic studies, identification of Brønsted acid sites has been attempted by NMR techniques. Pearson showed the existence of Brønsted acid sites by the measurement of broad line NMR spectra of deuterated pyridine.[13] The conclusion was criticized by Knözinger[14] and Ripmeester also reported that there is no NMR evidence for protonated pyridine on initial exposure of γ-alumina evacuated at 300°C to pyridine vapor, but pyridinium ions were detected for the sample exposed to air.[15] He supposed the pyridinium ions to be the counter ions of carbonate or bicarbonate species, which can be formed by adsorption of carbon dioxide in air.

Apart from spectroscopic studies, reactions supposed to be catalyzed by Brønsted acid have been reported.[16–18] However, the skeletal isomerization of 3,3-dimethylbut-1-ene proceeded after elimination of surface hydroxyls by treating with $Al(C_2H_5)_3$, so that the active Brønsted acid sites may be produced from the adsorbed olefin.[18] Thus the existence of intrinsic Brønsted acid sites is still controversial.

Two kinds of basic sites on γ-alumina are proposed, surface hydroxyls and oxide ions. Alumina activated at 400–600°C adsorbs carbon dioxide forming bicarbonate ions as evidenced by infrared spectra.[19] The bicarbonate ions are supposed to result from interaction of a surface hydroxyl and a carbon dioxide molecule adsorbed on a Lewis acid site. On the other hand, carbonate ions are formed on the alumina activated above

800°C by adsorption of carbon dioxide on basic oxide ions.[19] It is interesting to note that alumina activated at 500°C does not exhibit any basic property by the indicator method using bromothymol blue as an indicator (pK_a = 7.2), but does when it is exposed to moisture.[20] The result suggests that basic oxide ions would be associated with oxygen vacancies generated by activation at high temperatures (see next section).

As a summary, acidic sites on a surface of γ-alumina are mainly Lewis acid (tri-coordinated aluminum ions), but surface hydroxyls may exhibit weak Brønsted acidity under an appropriate dehydration state. The basic sites are surface hydroxyls on alumina activated at low temperatures and oxide ions on alumina dehydrated at high temperatures.

STRUCTURE OF ACIDIC AND BASIC SITES

The degree of dehydration of γ-alumina is the most important factor controlling the surface acid–base properties as mentioned above. Hindin et al. gave a simple schematic description for the generation of acidic and basic sites by the dehydration.[21]

```
          OH      OH
          |       |
    —O—Al—O—Al—O—   --Δ-->
         /|\     /|\
        / | \   / | \
       /  |  \ /  |  \

                  O⊖
          ⊕       |
      —O—Al—O—Al—O— + H2O
         /|\     /|\
        / | \   / | \
       /  |  \ /  |  \
```

or

```
             ⊖O⊖
          ⊕ /    \ ⊕
       —O—Al—O—Al—O—
         /|\     /|\
        / | \   / | \
       /  |  \ /  |  \
```

They proposed a strained Al–O–Al linkage as an active site and Cornelius et al. argued that lattice distortion is responsible for the generation of acid sites on alumina dehydrated at 538°C.[22]

Contrary to such a single point concept. Peri proposed a different model for an active site comprising a multiple vacancy[23] and showed the possibility of formation of such a multiple vacancy by a computer simulation of a dehydration process.[24] The vacancy refers to a vacant site obtained by removing a surface oxygen atom from the cubic lattice. The fully hydrated surface of γ-alumina is covered by hydroxyls. With heating, a water molecule is desorbed by a combination of an adjacent OH pair leaving an oxide ion and an oxygen vacancy. Under a condition of regular dehydration, that is, if two (or more) oxide ions are not left on immediately adjacent sites and if two (or more) immediately adjacent vacant sites are not generated, the degree of dehydration is limited to 67%. Further dehydration produces multiple vacancies inevitably. Under a restriction of no migration of hydroxyls, the degree of dehydration is as high as 90.4%, and the remaining hydroxyls are isolated from each other. Peri classified the isolated hydroxyls into five groups depending on the number of oxide ions adjacent to the hydroxyl and correlated them to the observed infrared vibration bands.

The concept of multiple vacancies as active sites is supported by experiments to estimate the concentration of active sites of an activated alumina. For example, Hall et al. reported that the concentration of active sites estimated by a poisoning technique is at least one order less than the expected concentration of exposed aluminum ions formed by the regular dehydration.[25]

Peri's work advanced the understanding of the nature of active sites significantly. However, his model has some weak points as criticized by Knözinger.[26] The first point is that Peri treated only the (100) plane on which each hydroxyl is equivalent. The equivalency of a hydroxyl is an essential requirement for a computer simulation of a random removal of a water molecule from two hydroxyls, and preferential exposure of the (100) plane was suggested by Lippens.[27] However, this is obviously an oversimplified model for the surface as Peri himself was aware.[24] The second point is that Peri related the nature of an isolated hydroxyl to the number of neighboring oxide ions. The theory is based on an assumption that an inductive effect of oxide ions on a hydroxyl through four bonds is the main factor controlling the electronic state of the hydroxyl. However, as pointed out by Knözinger, the effect through such a long distance is unlikely to be significant.[26] A molecular orbital calculation showed a minor inductive effect of a halogen atom on a hydroxyl through four bonds in the case of halogenated silica gels.[28]

To avoid these difficulties, Knözinger and Ratnasamy proposed a new theory to discriminate between the surface hydroxyls.[26] They took into consideration (111) and (110) planes in addition to (100) planes and identified five kinds of hydroxyls by the number of aluminum atoms bound to the hydroxyl and the coordination number of the aluminum atoms:

```
 OH        OH           H                    H
 |        \ | /         O                    O
 Al         Al      \ | /   \          \ | /   \ | /
/ | \      / | \     Al       Al          Al       Al
                    / | \   / | \        / | \   / | \

 Ia         Ib           IIa                  IIb

                  H
                  O
             \ | / | \ | /
              Al   |   Al
             / | \ | / | \
                 -Al-
                 / | \

               III (IIIb)
```

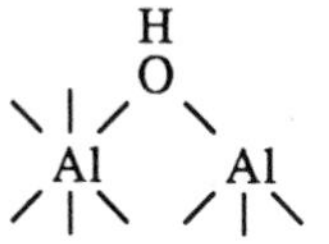

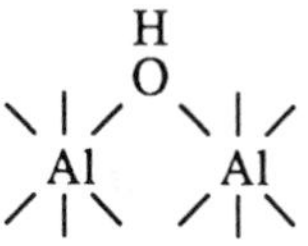

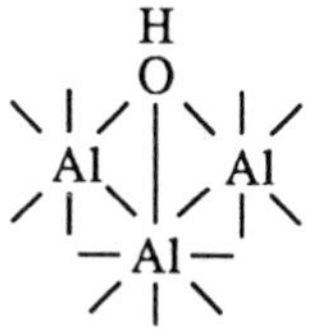

Assuming that the bonds between atoms are purely ionic and the electronic charges of an aluminum atom and an oxygen atom are +3 and −2, respectively, they calculated a formal net charge at a hydroxyl as shown in Table 12.1.

It is naturally assumed that a hydroxyl with a positive charge is acidic and a hydroxyl with a negative charge is basic. In the usual dehydration process, a water molecule will be removed by a combination of a hydrogen atom of an acidic hydroxyl and a basic hydroxyl. As a result, Lewis acid sites of exposed aluminum atoms and Lewis base sites of coordinatively unsaturated oxygen atoms are generated. As noted above, the number of catalytically active sites is considerably smaller than the expected number of such Lewis acid and base pairs generated by the dehydration. Thus, Knözinger and Ratnasamy concluded that the active sites are those of multiple vacancies which are formed by further removal of

Table 12.1. Formal net charge at a hydroxyl of different types.

Type	Coordination Number of Surface OH		Net Charge
	Al(VI)	Al(IV)	
Ia	—	1	−0.25
Ib	1	—	−0.5
IIa	1	1	+0.25
IIb	2	—	0
III	3	—	+0.5

hydroxyls at high temperatures and they supposed that the vacancy includes an isolated hydroxyl, referring to Fink's work.[29]

Knözinger's model is more realistic than Peri's model as a result of the inclusion of a variety of configurations of hydroxyls. However, they discussed the electronic states (net charges) of a hydroxyl on the basis of the ionic bond concept, which is obviously too simple for alumina.

Both Peri's and Knözinger's theories treat the nature of surface hydroxyls but not the nature of oxide ions and the acid–base properties of the multiple vacancy explicitly. It appears to be difficult to clarify the nature of the multiple vacancy experimentally, because of the complexity of the surface of activated γ-alumina. On the other hand, a quantum-chemical approach will allow a detailed discussion on the acidity and basicity of the alumina surface from the view point of structural chemistry since a variety of possible models can be assumed for quantum-chemical calculations and hypotheses not readily verifiable experimentally can be tested. This is the theme of this chapter.

On the basis of work mentioned above, we expect to have a detailed insight into the following subjects through quantum chemical studies.

1. How does the circumstance (environment) of a surface hydroxyl influence the acid–base properties of the hydroxyl?
2. How does the dehydration (activation) procedure influence the acid–base properties?

MODELS FOR MOLECULAR ORBITAL CALCULATIONS OF ALUMINA

There are two fundamental theories for quantum-chemical calculations: molecular orbital (MO) theory and valence bond (VB) theory. The former is more straightforward than the latter for a large molecule and has been widely used not only for organic molecules but also for inorganic molecules containing metal atoms. The application to catalysis is reviewed by Baetzold[30] and Kazansky et al.[31] A variety of calculation methods based on the MO theory have been proposed reflecting various levels of approximation: semiempirical methods without self-consistent field calculation (SCF) (e.g., extended Hückel method), semiempirical methods with SCF (e.g., CNDO, INDO, MINDO), nonempirical (*ab initio*) SCF methods without configuration interaction and *ab initio* multiconfigurational (MC–SCF) methods. For quantitative analysis, a sophisticated method such as MC–SCF is desirable, but this requires a long computational time even for a small molecule. At the present stage, MO calculations for metal oxides are limited to a framework of the *ab initio* SCF method, the semiempirical SCF method being especially popular.

To start MO calculations, we set up models. As a model, an infinite periodic model is often adopted in the case of metals[30] but a cluster model comprising a suitable number of atoms is usually adopted for metal oxides as the structure of oxides is generally complex.[31] The validity of application of a cluster model for a surface phenomenon is based on the experimental observation that the phenomenon causes changes in electronic states at a local region on a surface, hence the environment around the cluster will have little influence on the nature of the active point, if the cluster is of sufficiently large size, e.g., hundreds of atoms. However, the actual size is limited to that of 10–30 atoms because of the capacity of computers. Thus, an appropriate treatment is required to introduce the effect of environment around the cluster as well as possible. The most simple and popular technique is to terminate the dangling bonds, which are generated by quarrying a cluster from a solid surface, by hydrogen atoms or hydrogenlike pseudoatoms. Another technique is to locate some suitable electronic charges around the periphery of a cluster.[32] Besides these, an unique technique was proposed by Litinskii et al.[33] as described subsequently. If one adopts a SCF MO method, the cluster should be electrically neutral and in this sense, γ-alumina presents some challenges. The difficulty arises from the difference between the valence of aluminum and the number of oxygen atoms surrounding an aluminum atom in γ-alumina. As described previously, γ-alumina has a defect spinel structure and an aluminum atom is in an oxygen tetrahedron or in an octahedron. Thus, the valence of aluminum is distributed over four or six bonds. If we attach a proton to every terminal oxygen in an AlO_4 model, for example, the obtained model is an anion. To avoid this problem, the following three techniques have been adopted.

1. Stoichiometric orbital cluster models.
2. Modification of terminal atoms by extra core charge.
3. Assembly of neutral units.

The first and second techniques are for semiempirical SCF calculations. The first one was developed by Litinskii et al.[33] The basic idea is to adjust the atomic orbitals of terminal oxygen atoms to the stoichiometry of alumina, Al_2O_3. For example, let us set up a model containing two aluminum atoms and nine oxygen atoms as shown below.

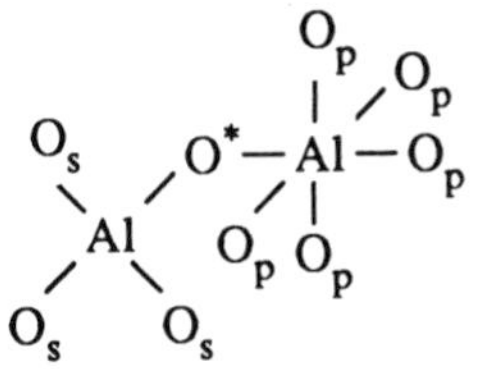

```
                      Op
                      |  / Op
  Os        O*  —  Al — Op
    \      /      / |
     Al        Op  Op
    /  \
  Os    Os
```

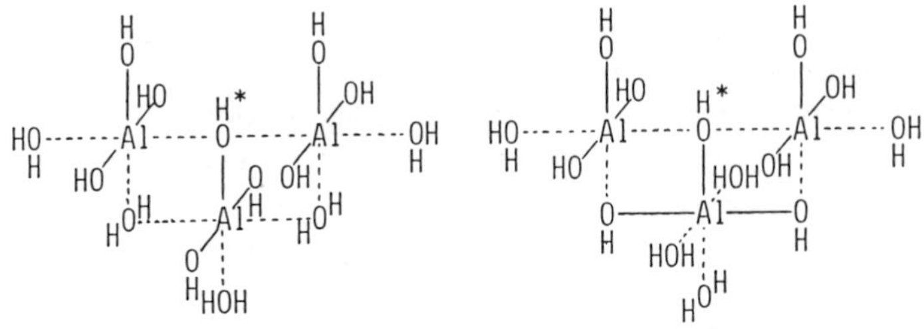

Figure 12.1. Examples of cluster models constructed from an assembly of neutral units. [*J.C.S. Faraday Trans. 2*, **81:** 1119–1127 (1985).]

Not only the number of total electrons is adjusted to the sum of the core charge of atoms but also the terminal oxygen atoms are replaced by virtual atoms having an O_{2s} orbital for O_s and an O_{2p} orbital for O_p in the above model. Thus, the number of atomic orbitals for terminal oxygen atoms is eight, corresponding to the number of valence atomic orbitals of two regular oxygen atoms.

The second technique was adopted by Kazansky's group.[31] The effective electronic charge of a cluster is determined starting with charges −2 for oxygen and +3 for aluminum. Total negative charge is compensated by introduction of the additional positive core charge on the terminal oxygen atoms in order to keep the cluster neutral. In addition, the parameter for ionization potential of the terminal oxygen atoms is adequately adjusted.

The third technique was proposed by Kawakami and Yoshida.[34] A cluster is constructed from an assembly of electrically neutral units such as $Al(OH)_3$, $(OH)_2$–Al–O–Al–$(OH)_2$. For enlarging the oxygen coordination to an aluminum atom, HOH units are located. This technique can be applied to an *ab initio* method and is sufficiently flexible to permit the construction of complex clusters of variable configurations. One disadvantage of this technique is that several clusters with different configurations are possible for one kind of oxygen lattice. For example, two clusters in Figure 12.1 are the models for the same kind of oxygen lattice. However, the calculated values for deprotonation energies at central OH* are 21.1 eV for both models, so that the difference in the models is not a serious problem for qualitative discussion of acidity and basicity of γ-alumina.

TYPES OF SURFACE HYDROXYLS

Peri discussed only the nature of the surface parallel to (100) plane. Knözinger and Ratnasamy included other planes such as (111) and (110) and showed that five kinds of surface hydroxyls should be distinguished according to the configuration as mentioned previously. In an actual alumina, exposure of not only the flat surfaces, but also stepped surfaces are plausible as proposed in magnesium oxide by Tench et al.[35] Such an assumption leads naturally to the existence of six kinds of surface hydroxyls: I_a, I_b, II_a, II_b, III_a, and III_b,[36] where I_a to II_b are the same as those designated by Knözinger and Ratnasamy and III_b is the same hydroxyl as Knözinger's type III. The configuration of III_a is presented schematically below.

H
O
Al Al
Al

IIIa

Figure 12.2 shows a fully dehydrated (100) face of a spinel crystal of alumina (neglecting Al defects) and Figure 12.3 shows an example of a stepped surface developed from the (100) face by cutting a groove on the surface. Generation of such a surface will not be too imaginary, referring to Lippen's work suggesting preferential exposure of (100) plane[27] and

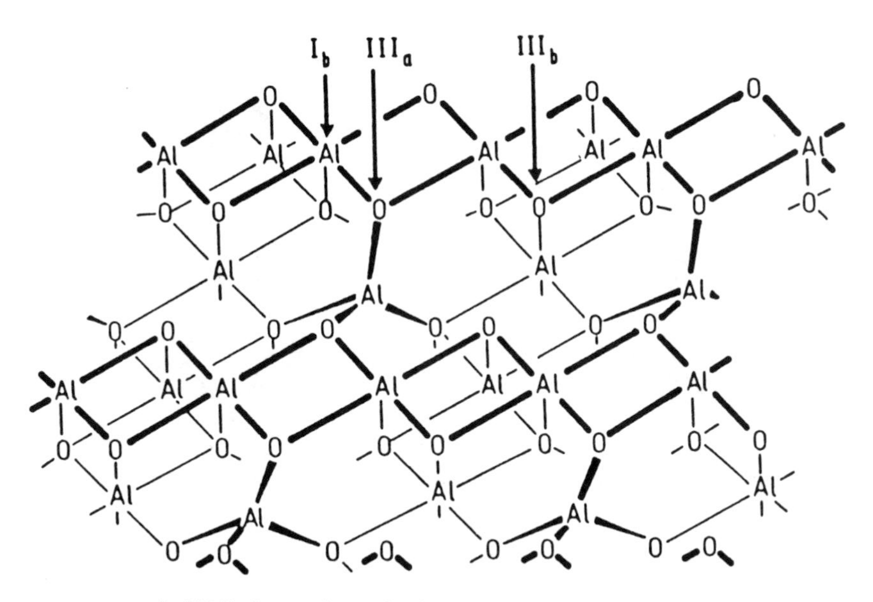

Figure 12.2. A (100) face of a spinel crystal of alumina. [*J.C.S. Faraday Trans. 2*, **81:** 1129–1137 (1985).]

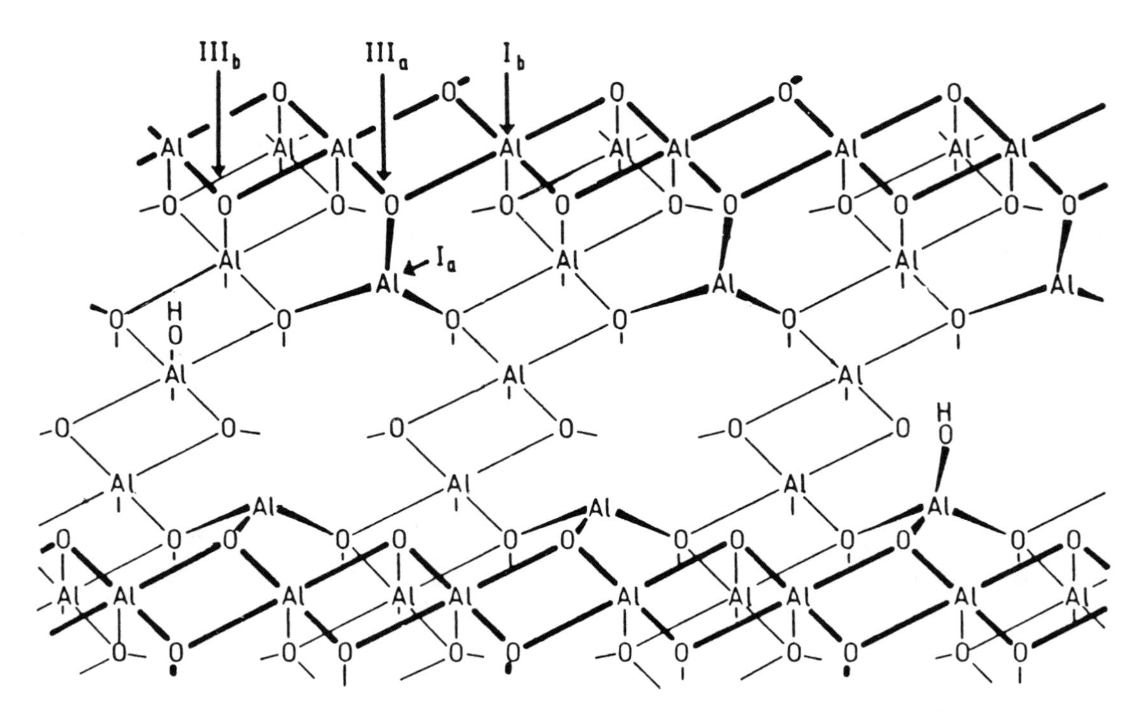

Figure 12.3. A surface obtained by cutting a groove on a (100) face. [*J.C.S. Faraday Trans. 2*, **81:** 1129–1137 (1985).]

noting that cleavage of a spinel crystal parallel to a (100) plane with grooves generates similar surfaces on each side. On the (100) surface only three types of hydroxyls are distinguished as depicted by I_b, III_a, and III_b, while on the grooved surface which is composed of slender faces parallel to (100) and (111) planes, another type of hydroxyl of type I_a appears.

No aluminum defects are depicted in the figures, but as mentioned above, aluminum atoms are absent at one-ninth of the cation sites in γ-alumina. Taking into consideration the defects, two other types of hydroxyls are possible on the surface shown in Figure 12.3; an aluminum vacancy at the octahedral interstice in the type III_a site induces a bridged hydroxyl of type II_a and a vacancy at the tetrahedral interstice of type III_a (or at the octahedral interstice of type III_b) forms a hydroxyl of type II_b. As the occupancy of an octahedral interstice by an aluminum atom is higher than that of a tetrahedral interstice, the number of II_a sites is supposed to be small. The same conclusion is obtained by consideration of a face derived from (111) plane.

BRØNSTED ACIDITY

There is no doubt that the electronic state of an isolated hydroxyl is mainly determined by the local environment, as discussed by Knözinger and Ratnasamy. Thus, clusters modeling type I_a–III_b hydroxyls will be used for discussion of the Brønsted acidity of surface hydroxyls.

Kazansky et al.[31] carried out MINDO/3 calculations for several models in which terminal oxygen atoms are replaced by pseudo atoms such as $(O_x)_3Al{-}OH{-}Al(O_x)_4$ as described previously. The acidity (as estimated by the deprotonation energies as well as the positive charges on hydrogen atoms) increases in the order $I_b < I_a < II_b < II_a < III_a$. (They did not mention III_b.) The order is in agreement with the one predicted by Knözinger and Ratnasamy (Table 12.1). Kazansky et al. also calculated the stretching vibration frequencies of hydroxyls. The frequencies decreased in the order $I_a > I_b \gg II_b > II_a \gg III_a$. These results clearly indicate the primary importance of the number of aluminum atoms bound to a hydroxyl and the secondary effect is observed for the number of oxygen atoms bound to the aluminum atom. The order of frequencies is almost the reverse order of the acidity as expected, but there is one exception (I_a and I_b). Although they observed that the order of frequencies completely agreed with the assignment from experimental data, they did not note the discrepancy between the order of frequencies and that of acidity. The reported frequencies for five kinds of hydroxyls are centered in a small energy region of 3700–3800 cm^{-1}.[10] Thus, the assignment by quantum chemical calculations for small cluster models seems to be doubtful.

Kawakami and Yoshida carried out *ab initio* SCF calculations for models as shown in Figure 12.4.[37] These models are for a fully hydrated state and constructed by a combination of neutral units as described previously. The atom configurations are determined by inspection of a grooved surface shown in Figure 12.3 as well as the surface derived from a (111) face. Models 1–3 are composed of two octahedral aluminum atoms and one tetrahedral aluminum atom. Comparison of these models permitted the influence of the mode of assembly of neutral units on the calculated values to be evaluated. Models 4 and 5 are composed of three octahedral aluminum atoms, and model 6 is of one octahedral and one

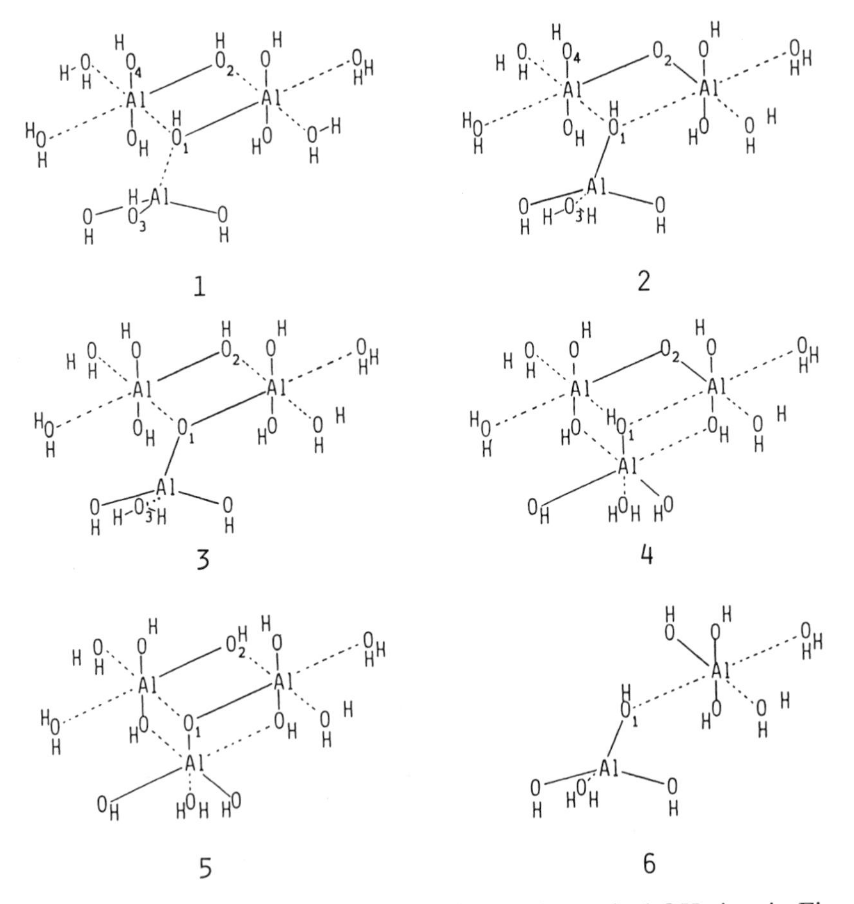

Figure 12.4. Cluster models corresponding to the typical OH sites in Figure 12.3. The units connected by solid lines are fundamental neutral units. [*J.C.S. Faraday Trans. 2*, **82:** 1385–1397 (1986).]

tetrahedral aluminum atom. The most reliable index for Brønsted acidity of a hydroxyl is the energy required to remove a proton (deprotonation energy ΔE_{dp}).[38] The energies were calculated for various OH sites and the representative results are shown in Table 12.2, where the second column shows the site number designated by subscripts of the oxygen atoms in Figure 12.4.

We can deduce an order of acidity depending on the configuration of a hydroxyl, as a general tendency

$$I_b < I_a < II_b < II_a < III_b < III_a$$

This is in good agreement with the order suggested by Knözinger and Kazansky mentioned above. It should be noticed that a deviation of 2–3 eV in ΔE_{dp} is found for a hydroxyl of the same configuration. This would result mainly from the terminal effect of the small cluster. Thus, the results should be referred to as qualitative. For quantitative discussion, we must calculate ΔE_{dp} for much larger clusters but it is impossible at present because of the limitation of computer capacities. We will judge the acid strength of a hydroxyl qualitatively on the basis of the ΔE_{dp} values calculated by the same method for models of silica and silica–alumina. A surface hydroxyl of silica is known to be almost neutral and the value was obtained as ~23 eV while silica–alumina has a very strong acidic site, for which a value of ~15 eV was obtained.[38] Referring to these values, the hydroxyls of type III_a, III_b are concluded to be weak acids and those of type II_a, II_b to be almost neutral or rather basic, while those of type I_a, I_b are basic. The conclusion is somewhat different from Knözinger's prediction, which suggests that hydroxyls of type II_a are acidic and those of type II_b are neutral (Table 12.1). The hydroxyls of type I_a and

Table 12.2. Deprotonation energy for the representative OH site in models 1–6.

Model	Site no.	Type of OH	ΔE_{dp}(eV)
1	1	IIIa	19.3
	2	IIb	23.0
	3	Ia	26.4
2	1	IIIa	20.5
	4	Ib	28.6
3	2	IIb	25.4
4	1	IIIb	21.7
5	2	IIb	24.4
6	1	IIa	22.4

I_b are basic and expected to be removed by dehydration with a combination of protons. The easy removal of type I hydroxyls is supported by calculations for models of more idealized structures.[36]

The preceding discussion is for the alumina of a fully hydrated state. To investigate the influence of dehydration on the acidity, ΔE_{dp} values were calculated for models 7–13 shown in Figure 12.5. These models are derived from models in Figure 12.4 by removing 1–3 pairs of a proton and a hydroxyl (water molecules). Removal of two water molecules from model

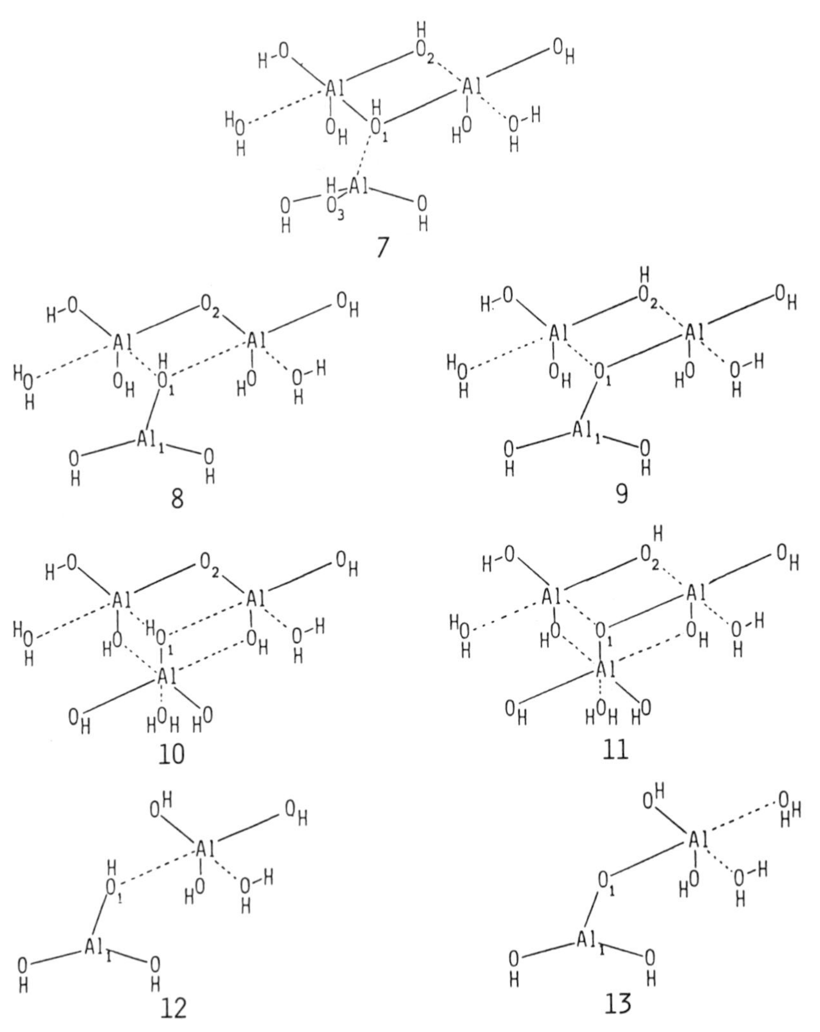

Figure 12.5. Cluster models corresponding to the various and/or O sites in a dehydrated surface. [*J.C.S. Faraday Trans. 2*, **82:** 1385–1397 (1986).] See the captions for Figure 12.4.

Table 12.3. Deprotonation energy for the representative OH site in models 7–12.

Model	Site no.	Type of OH	ΔE_{dp}(eV)
7	1	IIIa	15.2
8	1	IIIa	15.7
9	2	IIb	21.3
10	1	IIIb	19.0
11	2	IIb	22.2
12	1	IIa	18.7

1 results in models 7. Models 8 and 9 are obtained by removing three water molecules from models 1 or 2 and from model 3, respectively. Models 10 and 11 are obtained by removing two water molecules from models 4 and 5, respectively, and models 12 and 13 are derived from model 6.

The results are presented in Table 12.3. Comparing each ΔE_{dp} value with that of the corresponding hydrated one among models 1–6 in Table 12.2, we can see that ΔE_{dp} values are reduced by more than 4 eV by dehydration, i.e., the Brønsted acidity of the sites is strengthened significantly and hydroxyls of type II_a become weakly acidic. The hydroxyl of type III_a is expected to be strongly acidic, if it remained after dehydration, but the situation is hardly expected because the proton will be the most easily removed by dehydration. Thus, hydroxyls of type II_a are possible Brønsted acid sites on a dehydrated alumina. The acid strength is expected to be very weak, judging from the ΔE_{dp} values, which may explain the difficulty of experimental identification of Brønsted acidity of γ-alumina. From the results mentioned previously, a dehydration process is proposed as follows: a water molecule will be formed by the combination of a hydroxyl of type I and a proton in a hydroxyl of type III at low temperatures; especially the combination of hydroxyls of type I_b and type III_a will form a water molecule most easily. By the progress of dehydration, deprotonation will occur from a hydroxyl of type II, which is neutral or rather basic on a hydrated alumina, but becomes a weak acid by hydration. The removal of hydroxyls of type II or III leaving oxygen vacancies will require high temperatures. The removal of hydroxyls leaves behind coordinatively unsaturated aluminum atoms (Lewis acid sites) on the surface. Acid strengths will be discussed later.

BASICITY

As mentioned previously, two kinds of sites are supposed to be basic: an oxide ion bridging two aluminum atoms and a surface hydroxyl. So far, quantum chemical studies on the basicity have been scarce and only one paper has been published by Kawakami and Yoshida.[37]

Let us discuss the basicity of surface hydroxyls at first. Yamadaya et al. reported that basicity was not detectable by an indicator method[20] for an alumina activated at 500°C. The activation temperature evidently is not sufficiently high for generation of basic oxide ions.[19] Parkyns recorded infrared spectra of adsorbed CO_2 on alumina activated at various temperatures starting at 400°C. He observed bicarbonate ions on the alumina activated below 500°C. By activation at temperatures 500–700°C, the absorption bands of the bicarbonate ions decreased and simultaneously, poorly resolved bands assignable to carbonate ions appeared. Above 800°C, no bicarbonate ions were observed and the bands due to carbonate ions became clear. As mentioned in the last section, dehydration by the activation procedure is supposed to remove primarily hydroxyls of type I at low and medium temperatures, and the remaining hydroxyls before rigorous dehydration will be mainly of type II_b (and II_a as a small amount), which will be weakly acidic. Possible basic hydroxyls are of type I. Although these hydroxyls are readily removed, some of them can survive the activation procedure at low temperatures. Thus, we will discuss the basicity of hydroxyls of type I using models 1, 2 (Fig. 12.4), and 7 (Fig. 12.5).

For the index of basicity, the stabilization energy, ΔE_{pr}, for the adsorption of a proton on an oxygen atom of a hydroxyl was employed. The calculated ΔE_{pr} values for models of silica (nonbase) and magnesia (strong base) were ~11 and 20 eV, respectively.[39] Thus, we expect that a hydroxyl should exhibit basicity, if a value greater than 11 eV is obtained for ΔE_{pr}. Table 12.4 shows the calculated ΔE_{pr} values for models 1, 2, and 7. The results indicate that such a hydroxyl has some basicity and also that partial dehydration results in the reduction of basicity as shown from the smaller ΔE_{pr} value for the model 7–(3) site than that of model 1–(3) site. The partial dehydration generates exposed aluminum atoms by removal of a fraction of the hydroxyls of type I. The exposed aluminum atoms are expected to be Lewis acids, so that the proposal by Parkyns for the mechanism of generation of bicarbonate ions is plausible.

Table 12.4. Protonation energy for the OH site in models 1, 2, and 7.

Model	Site no.	Type of OH	ΔE_{pr}(eV)
1	3	Ia	13.4
2	4	Ib	16.2
7	3	Ia	12.6

Table 12.5. Protonation energy for the O site in models 8–13.

Model	Site no.	Type of OH[a]	ΔE_{pr}(eV)
8	2	IIb	15.6
9	1	IIIa	10.2
10	2	IIb	16.4
11	1	IIIb	13.2
13	1	IIa	14.7

[a] Types of hydroxyls before removing protons.

Next, we will discuss the Lewis basicity of surface oxide ions. The models shown in Figure 12.5 can be used to investigate the basicity of an oxide ion bridging aluminum atoms. It is needless to consider non-bridging oxide ions derived by deprotonation from hydroxyls of type I, because these are basic and deprotonation by dehydration cannot be expected. Table 12.5 shows the ΔE_{pr} values for oxide ions derived from hydroxyls of type II_a, II_b, III_a, and III_b. The basicity is in the reverse order of the acidity of the original hydroxyls, as expected. The exposed oxide ions derived from hydroxyls of type III are expected to be generated by dehydration at low temperatures, but the ΔE_{pr} values predict that these have basicity similar to that for silica ($\Delta E_{pr} = \sim 11$ eV). Thus, the quantum-chemical calculations are in agreement with the results of Parkyns mentioned previously. Oxide ions bridging two aluminum atoms are predicted to be basic but not so strong as those in magnesia ($\Delta E_{pr} = 20$ eV). It is noteworthy that an oxide ion derived from a hydroxyl of type II_b is more basic than that from a hydroxyl of type II_a, indicating the effect of the configuration of the bridged aluminum atoms on the basicity.

LEWIS ACIDITY

Lewis acidity has been considered as a surface property of γ-alumina. There is no doubt that exposed aluminum atoms are responsible for the Lewis acidity. Starting from a fully hydrated surface, the aluminum atoms are exposed by removing surface hydroxyls. Various configurations of aluminum atoms are possible depending on the environment but the occurrence will be different. We expect that the removal of hydroxyls will proceed in the order $I_b > I_a > II_b > II_a > III_b > III_a$ as mentioned previously and the corresponding aluminum atom is exposed in this order. Removal of hydroxyls of type I_a and I_b leaves three-coordinated and five-coordinated aluminum atoms, respectively. For the description of the coordination state of an aluminum atom, we will use the notation Al_{3c} and

Al_{5c} for such aluminum atoms. Removal of bridging hydroxyls of type II and III will be accompanied by the generation of multiple vacancies, because the preceding removal of hydroxyls will occur at neighboring sites of type I_a or I_b. The removal of oxide ions bridging three aluminum atoms will result from severe dehydration at very high temperatures.

Gokhberg et al.[33,40,41] investigated the relationship between the Lewis acidity and the local structures using orbital stoichiometric cluster models as mentioned previously. Calculations for the model shown below revealed that acceptor levels localized on the Al_{3c} and Al_{5c} appear and the energy level on Al_{3c} was lower than that on Al_{5c}, indicating that Al_{3c} is a stronger Lewis acid site than Al_{5c}. The same conclusion is obtained by Kawakami and Yoshida by an *ab initio* calculation.[36] Gokhberg et al. examined the effect of neighboring hydroxyls on the Lewis acidity, attaching a hydroxyl on Al_{5c} or Al_{3c} and a proton on the bridging oxygen atom (0*), and found that these enhanced the acidity. They supposed that the reduction of acidity with dehydration at high temperatures results from desorption of the proton attached to O*.

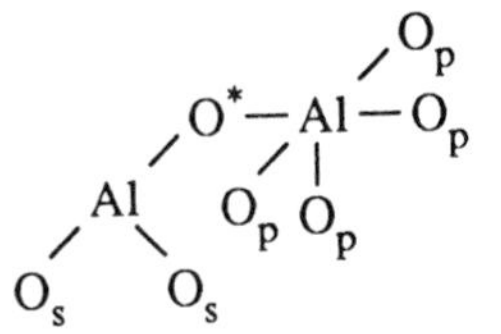

```
                  O
                 / p
           O*— Al — O
          /    / |    p
       Al    O   O
      /  \    p   p
     O    O
      s    s
```

Kawakami and Yoshida carried out detailed calculations by an *ab initio* SCF method for models based on the spinel structure of γ-alumina as shown in Figure 12.6.[37] Clusters 14–19 are models for Lewis acid sites obtained by removing a hydroxyl of type I_a. As mentioned previously, the removal of a hydroxyl of type I_b is believed to precede the removal of the hydroxyl of type I_a, so that the hydroxyls of type I_b are missed in the models. Models 14 and 15 are employed to investigate the effect of coordination of one Al_{5c} unit on the Lewis acidity at Al_{3c} site. In model 14, a hydroxyl of type II_a bridges two aluminum atoms, while the proton of the bridging hydroxyl is removed in model 15. Models 16 and 17 contain another Al_{5c} unit, where the proton of bridging hydroxyl of type III_a is removed. Sites corresponding to these models are expected for alumina activated at medium temperatures. Models 18 and 19 are models for double and triple vacancies, where a bridging lattice oxygen atom is missed. Such a site is expected for alumina activated at high temperatures.

The Lewis acidity is discussed by LUMO levels localized at the Lewis site of the models and also adsorption energy of the OH^- anion (ΔE_{OH}) at the site denoted by an asterisk. The results are shown in Table 12.6 as well as the electronic charge of the aluminum atom (δ_{Al}). It is noteworthy

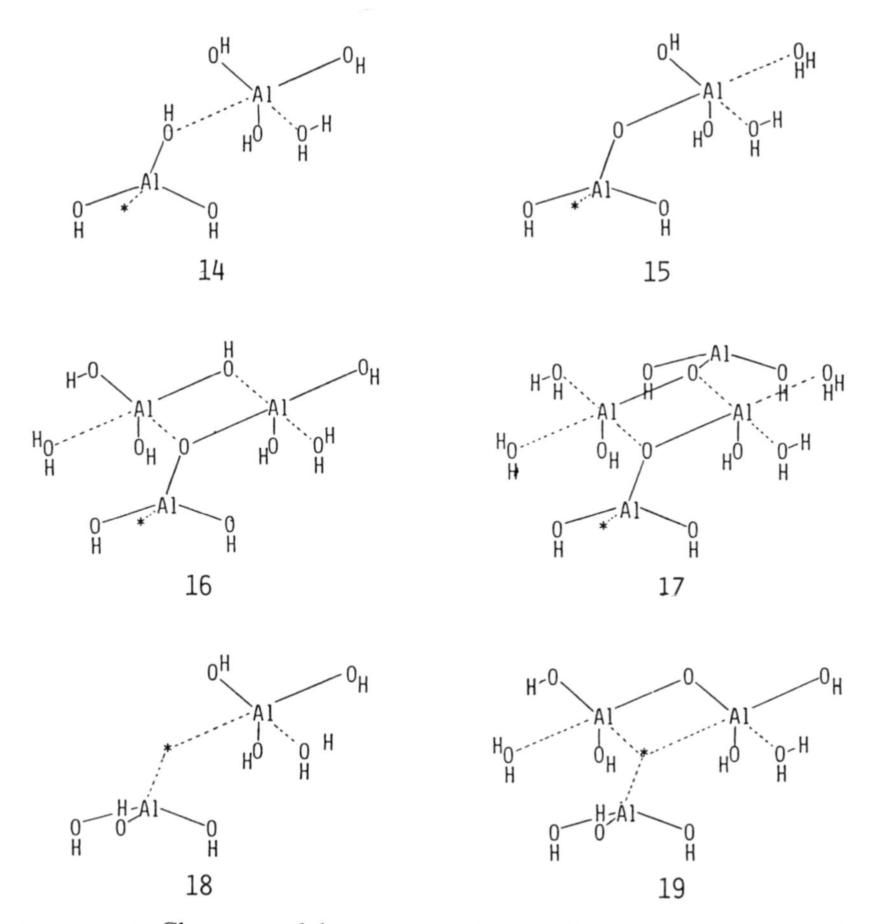

Figure 12.6. Cluster models corresponding to the typical three-coordinated Al sites. [*J.C.S. Faraday Trans. 2,* **82:** 1385–1397 (1986).] See the caption for Figure 12.4.

Table 12.6. Indices of Lewis acidity for models 14–19.

Model	E_{LUMO}	δ_{Al}[a]	ΔE_{OH}(eV)
14	4.7	1.32	10.3
15	7.8	1.12	7.4
16	5.6	1.26	9.6
17	6.0	1.25	9.3
18	3.6	1.18	12.4
19	3.5	1.16	12.2

[a] δ_{Al} is the electronic charge on the aluminum atom.

that a good correlation between the LUMO energy levels and ΔE_{OH} is observed but not for the electronic charge on Al_{3c}, indicating that the charge is not a good index of Lewis acidity.

It is interesting that model 14 shows stronger Lewis acidity than model 15 in agreement with the results by Gokhberg et al. However, the bridging hydroxyl is of type II_a and the concentration is not expected to be high as mentioned previously. Thus, the site corresponding to model 14 is not realistic. Comparison of model 15 and 16 indicates that stronger acidity is generated by attaching two Al_{5c} units to an AlO_3 unit. This suggests that as more aluminum atoms are attached to the AlO_3 unit, a stronger Lewis acidity is generated, although the suggestion could not be confirmed by calculations because too many atoms beyond a computer capacity should be included in the cluster models. Aluminum atoms in the third neighboring sphere from the AlO_3 unit apparently do not influence the acidity as shown from the result for model 17.

Evidently, a site generated by removal of the bridging lattice oxygen atom works as a strong Lewis acid site (models 18 and 19), as suggested previously. The strong acidity is caused by a strong interaction between aluminum atoms through the *p* orbitals. In fact, a calculation showed that only weak Lewis acidity is expected for an assembly of four Al_{3c} atoms, in which the interaction between aluminum atoms is weak. This may be the source of the discrepancy between ΔE_{OH} and the charge on an aluminum atom for models 18 and 19.

In summary, a three-coordinated aluminum atom (Al_{3c}) functions as a Lewis acid as expected generally. The strength depends on the number of coordinatively unsaturated aluminum atoms bound to the Al_{3c} through oxygen atoms. If a bridging lattice oxygen atom is removed, the vacant site will be a strong acid site as a result of the cooperation of two or three aluminum atoms. Such a site will be accompanied by the formation of a multiple vacancy of a new type and will presumably require dehydration at very high temperatures.

So far, we did not consider lattice distortion, i.e., we assume that geometrical parameters of cluster models such as bond distances and bond angles do not change by dehydration. However, lattice distortion may occur and Hindin et al.[21] and Cornelius et al.[22] attributed strong Lewis acidity to generation of lattice distortion by activation at high temperatures. The effect can be estimated by calculations of ΔE_{OH} for $Al(OH)_3$ units with different geometrical parameters. The calculations were carried out as follows, using a model shown in Figure 12.7.

1. The optimum geometry with the minimum total energy for a $Al(OH)_3$ cluster is determined by an *ab initio* MO calculation under a restriction that the distance between point X and the oxygen atom in the basal plane

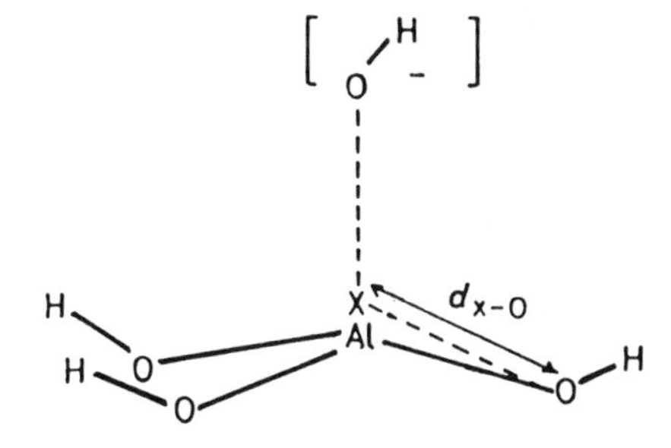

Figure 12.7. The simplest model for the three-coordinated Al site. Point X is the center of the regular tetrahedron including the basal plane of O–O–O triangle. [*J.C.S. Faraday Trans. 2,* **82:** 1385–1397 (1986).]

in the figure, (d_{X-O}), is held fixed, where X is the center of a regular tetrahedron including the basal plane. Thus, the position of the aluminum atom changes with d_{X-O}.

2. The same kind of calculation is carried out for a cluster with an OH^- anion attached to the aluminum atom.

The energy difference is ΔE_{OH} and it was calculated for cluster models with different d_{X-O}. (The distance in γ-alumina is 1.633 Å.) The variation in ΔE_{OH} was found not to be large (within 1.0 eV) among models with d_{X-O} of 1.6–1.7 Å, and is smaller than that induced by a change of environment (Table 12.6). Hence, lattice distortion is not expected to influence the Lewis acidity significantly.

CONCLUSION

The surface of alumina is complex and its characterization is difficult. Recent experimental progress has advanced the understanding of acid–base properties of γ-alumina but more efforts should be devoted to clarification on a molecular level. The quantum-chemical approach is expected to be useful for this purpose and the results presented in this chapter show, at least qualitatively, a clear correlation between the acid–base properties and the local structures of alumina. The present state of quantum-chemical studies on alumina is not sufficiently precise to allow a quantitative description of the nature of the surface because of the limitation of capacity of the currently available computers. The most difficult problem in an application of a molecular orbital method to alumina is in the treatment of dangling bonds at peripheral atoms in a cluster model. For a small cluster model of alumina, this may cause a significant effect on the electronic state of acidic or basic sites or both. If we can handle a sufficiently large cluster model, the effect can be neglected but at present the size is limited to that of 10–30 atoms. The recent advancement in computer technology offers encouragement that calculations for large cluster models may become possible in the near future, thus permitting a quantitative discussion of the nature of the surface of alumina.

It should be noted that the quantum chemical studies are not independent of the experimental studies; the results obtained by experiments are fundamental bases of the quantum chemical studies and on the other hand the results of quantum chemical studies will be reflected in the experimental studies. Cooperative applications of the two approaches have been and will continue to be essential.

ACKNOWLEDGMENTS

The author wishes to express thanks to Dr. H. Kawakami for his cooperation in the quantum-chemical studies of alumina and also to Professor K. Tanabe of Hokkaido University for his valuable suggestions and discussions.

REFERENCES

1. Pines, H., Manassen, J., *Adv. Catalysis,* **16:** 49 (1966).
2. John, C.S., Scurrell, M.S., *Catalysis (Specialist Peri. Rept.)* **1:** 136 (1977).
3. Lippens, B.C., Steggerda, J.J., in *Physical and Chemical Aspects of Adsorbents and Catalysts* (B.G. Linsen, ed.) London: Academic Press, 1970, Chap. 4, p. 171.
4. Haber, J., in *Catalysis-Science and Technology* (J.R. Anderson and M. Boudart, ed.) Berlin: Springer-Verlag, 1981, Vol. 2, Chap. 2, p. 13.
5. Wells, A.F., *Structural Inorganic Chemistry* Oxford: Clarendon Press, 1975, Chap. 12, p. 439.
6. Tanabe, K., *Solid Acids and Bases* Tokyo: Kodansha, 1970, Chap. 4, p. 45.
7. Little, L.H., *Infrared Spectra of Adsorbed Species* London: Academic Press, 1966, Chap. 7. p. 180.
8. Medema, J., *J. Catalysis* **37:** 91 (1975).
9. Parry, E.P., *J. Catalysis* **2:** 371 (1963).
10. Knözinger, H., *Adv. Catalysis* **25:** 184 (1976).
11. Dewing, J., Monks, G.T., Youll, B., *J. Catalysis* **44:** 226 (1976).
12. Corma, A., Rodellas, C., Fornes, V., *J. Catalysis* **88:** 374 (1984).
13. Pearson, R.M., *J. Catalysis* **46:** 279 (1977).
14. Knözinger, H., *J. Catalysis* **53:** 173 (1978).
15. Ripmeester, J.A., *J. Am. Chem. Soc.* **105:** 2925 (1983).
16. John, C.S., Tada, A., Kennedy, L.V.F., *J. Chem. Soc. Faraday Trans. I* **74:** 498 (1978).
17. Irvine, E.A., John, C.S., Kemball, C., Pearman, A.J., Day, M.A., Sampson, R.J., *J. Catalysis* **61:** 326 (1980).
18. Gati, G., Halasz, I., *J. Catalysis* **82:** 223 (1983).
19. Parkyns, N.D., *J. Phys. Chem.* **75:** 526 (1971).
20. Tanabe, K., in *Catalysis-Science and Technology* (J.R. Anderson and M. Boudart, eds.) Berlin: Springer-Verlag: 1981, Vol. 2, Chap. 5, p. 231; Yamadaya, M., Shimomura, K., Kinoshita, T., Uchida, T., *Shokubai*, **7:** 313 (1965).
21. Hindin, S.G., Weller, S.W., *J. Phys. Chem.* **60:** 1501 (1956).
22. Cornelius, E.B., Milliken, T.H., Mills, G.A., Oblad, A.G., *J. Phys. Chem.* **59:** 809 (1955).

23. Peri, J.P., *Actes 2e Consgr. Intern. Catalyse, Paris* **1:** 1333 (1961).
24. Peri, J.B., *J. Phys. Chem.* **69:** 220 (1965).
25. Hightower, J.W., Hall, W.K., *Trans. Faraday Soc.,* **66:** 477 (1970).
26. Knözinger, H., Ratnasamy, P., *Catalysis Rev.* **17:** 31 (1978).
27. Lippens, B.C. Thesis, Delft Univ. Tech., 1961.
28. Yoshida, S., Tai, S., Tarama, K., *J. Catalysis* **45:** 242 (1976).
29. Fink, P., *Rev. Roumaine Chim.* **14:** 811 (1969).
30. Baetzold, R.C., *Adv. Catalysis* **25:** 1 (1976).
31. Zhidomirov, G.M., Kazansky, V.B., *Adv. Catalysis* **34:** 131 (1986).
32. Korsunov, V.A., Chuvylkin, N.D., Zhidomirov, G.M., Kazansky, V.B., *Kinetics and Catalysis* **21:** 311 (1980).
33. Gokhberg, P.Ya., Litinskii, A.O., Khardin, A.P., Lazauskas, V.M., Berzhyunas, A.V., *Kinetics and Catalysis* **21:** 669 (1980).
34. Kawakami, H., Yoshida, S., *J. Chem. Soc., Faraday Trans. 2* **81:** 1117 (1985).
35. Culuccia, S., Barton, A., Tench, A.J., *J. Chem. Soc., Faraday Trans. 1* **77:** 2703 (1981).
36. Kawakami, H., Yoshida, S., *J. Chem. Soc., Faraday Trans. 2* **81:** 1129 (1985).
37. Kawakami, H., Yoshida, S., *J. Chem. Soc., Faraday Trans. 2* **82:** 1385 (1986).
38. Kawakami, H., Yoshida, S., Yonezawa, T., *J. Chem. Soc., Faraday Trans. 2* **80:** 205 (1984).
39. Kawakami, H., Yoshida, S., *J. Chem. Soc., Faraday Trans. 2* **80:** 921 (1984).
40. Gokhberg, P.Ya., Litinskii, A.O., Khardin, A.P., Berzhyunas, A.V., *Kinetics and Catalysis* **22:** 911 (1981).
41. Litinskii, A.O., Gokhberg, P.Ya., Khardin, A.P., Lazauskas, V.M., *Kinetics and Catalysis,* **24:** 193 (1983).

Index

Index